# ENGINEERING MATERIALS

## (Including Construction Materials)

**A Textbook for Engineering Students of B.E., Section-B of AMIE (India), Diploma and Competitive Examinations**

**Er. R.K. RAJPUT**

M.E. (Hons.), Gold Medallist; Grad. (Mech.Engg. & Elect. Engg.); MIE (India);
*MSESI; MISTE; CE (India)*

**Recipient of:**

*"Best Teacher (Academic) Award"*

***"Distinguished Author Award"***

*"Jawahar Lal Nehru Memorial Gold Medal"*
*for an outstanding research paper*
*(The Institution of Engineers–India)*

*Formerly* **Principal**

- *Thapar Polytechnic College;*
- *Punjab College of Information Technology,*

**PATIALA**

# S Chand And Company Limited

**(ISO 9001 Certified Company)**

**S Chand And Company Limited**

**(ISO 9001 Certified Company)**

***Head Office:*** Block B-1, House No. D-1, Ground Floor, Mohan Co-operative Industrial Estate, New Delhi – 110 044 | Phone: 011-66672000

***Registered Office:*** A-27, 2nd Floor, Mohan Co-operative Industrial Estate, New Delhi – 110 044 Phone: 011-49731800

www.**schandpublishing.com**; e-mail: **info@schandpublishing.com**

**Branches**

Chennai : Ph: 23632120; chennai@schandpublishing.com
Guwahati : Ph: 2738811, 2735640; guwahati@schandpublishing.com
Hyderabad : Ph: 40186018; hyderabad@schandpublishing.com
Jalandhar : Ph: 4645630; jalandhar@schandpublishing.com
Kolkata : Ph: 23357458, 23353914; kolkata@schandpublishing.com
Lucknow : Ph: 4003633; lucknow@schandpublishing.com
Mumbai : Ph: 25000297; mumbai@schandpublishing.com
Patna : Ph: 2260011; patna@schandpublishing.com



*First Edition 2000*
*Subsequent Editions and Reprints 2004, 2006, 2007, 2008 (Twice), 2009, 2010, 2011, 2013*
*Revised Fifth Edition 2014*
*Reprints 2015, 2016, 2018 (Twice), 2021*

***Reprint 2022 (Twice)***

**ISBN:** 978-81-219-1960-9 **Product Code:** H3EML46CIVE10ENAE14O

PRINTED IN INDIA

By Vikas Publishing House Private Limited, Plot 20/4, Site-IV, Industrial Area Sahibabad, Ghaziabad – 201 010 and Published by S Chand And Company Limited, A-27, 2nd Floor, Mohan Co-operative Industrial Estate, New Delhi – 110 044.

# PREFACE TO THE FIFTH EDITION

I take the pleasure in presenting the ***"Fifth Edition"*** of this book. The warm reception which the previous edition and reprints have received is a matter of great satisfaction to me.

In this edition, the book has been thoroughly revised and a new ***'Section'*** on **"SHORT ANSWER QUESTIONS"** has been added to make the book still more useful to the students.

I am greatly indebted to the management and the editorial team of S. Chand & Company Pvt. Ltd. for all help and support in publication of this edition.

The constructive suggestions for improvement of the book are most welcome.

**Er. R.K. RAJPUT**
**(Author)**

# PREFACE TO THE FIRST EDITION

This treatise on "Engineering Materials" contains comprehensive treatment of the subject matter in simple and lucid language and envelopes a large number of figures which reinforce the text in the most efficient and effective way.

The book comprises 21 chapters in all, covering various types of engineering materials commonly used in civil works/industry; the last chapter deals with Material Science of Metals. At the end of each Chapter Highlights, Objective Type Questions and Theoretical Questions have been added. All these factors tend to make the book a comprehensive and complete unit in all respects.

The author's thanks are due to his wife Ramesh Rajput for extending all co-operation during preparation of the manuscript and proof-reading of the book. The author also expresses his gratitude to Prof. R.C. Kohli for giving valuable suggestions during preparation of the manuscript.

The author is thankful to the Publisher for bringing out the book in a short span of time and pricing it moderately inspite of heavy cost of paper and printing.

Although every care has been taken to remove all the mistakes and misprints yet some of them might have been remained unnoticed. The author shall feel grateful of these errors are pointed out by the readers. Constructive criticism of the book will be warmly received.

**Er. R.K. RAJPUT**

# CONTENTS

# 1

# Introduction

*1.1 Classification of materials. 1.2 Properties of materials — physical properties, mechanical properties of metals-electrical properties of materials -magnetic properties-chemical properties. 1.3 Properties of building materials. Objective Type Questions – Theoretical Questions.*

The subject of "*Engineering materials*" deals with the study of materials in respect of the following:

- Sources, composition and properties;
- Manufacturing methods and testing;
- Utility in the various fields of engineering and technology;
- Modern techniques being developed for handling and using materials to materialise economical and safer design of structures and machines, etc.

## 1.1 CLASSIFICATION OF ENGINEERING MATERIALS

The engineering materials may be broadly classified as follows:

**A. 1 Civil Engineering Materials**

*Examples*: Building stones, bricks and clay products, lime, cement, concrete, mortar, timber, etc.

**2 Electrical Engineering Materials**

*Examples*: Copper, aluminium, iron and steel, etc. ....... *conductors*;
Silicon, germanium, etc. ....... *semiconductors*;
Asbestos, bakelite, mica, varnishes, air, etc. ....... *insulators;*
Iron, nickel, cobalt, etc. ...... *magnetic materials.*

**3 Mechanical Engineering Materials**

*Examples*: Cast iron, steel, lubricating materials, etc.

**B. 1. Metals**

*Examples*: Iron, aluminium, copper, zinc, etc.

Metals may be further subdivided as:

(*i*) *Ferrous metals* (*e.g.* cast iron, wrought iron and steel) and *alloys,* (*e.g.*, silicon steel, high speed steel, spring steel, etc.)

(*ii*) *Non-ferrous metals* (*e.g.* copper, aluminium, zinc etc.) and *alloys* (brass, bronze, duralumin, etc.)

The iron group which includes all irons and steels is called *ferrous metals* (ferrous-iron), whilst others are specified as *non-ferrous.*

**2. Non-metals**

*Examples*: Building stones, cement, concrete, rubber, plastics, asbestos, etc.

**Difference between Metals and Non-metals:**

Table 1.1 shows the difference between metals and non-metals.

**Table 1.1. Difference between metals and non-metals**

| S. No. | Property | Metals | Non-metals |
|---|---|---|---|
| 1. | *Structure* | All solid metals have crystalline structure. | They exist in amorphic or mesomorphic forms. |
| 2. | *Excitation of valance electron by E.M.F. (electro motive force)* | Easy | Difficult |
| 3. | *State* | Generally solids at room temperature | Gases and solids at ordinary temperature |
| 4. | *Lustre* | Possess metallic lustre. | Do not possess metallic lustre (except iodine and graphite). |
| 5. | *Conductivity* | Good conductor of heat and electricity | Bad conductor of heat and electricity |
| 6. | *Malleability* | Malleable | Not malleable |
| 7. | *Ductility* | Ductile | Not ductile |
| 8. | *Hardness* | Generally hard | Hardness varies |
| 9. | *Electrolysis* | Form anions | Form anions |
| 10. | *Density* | High density | Low density |

**C. 1 Metals and alloys**

**2 Ceramics**

**3 Organic polymers**

**1. Metals and Alloys**

Metals are polycrystalline bodies consisting of a great number of fine crystals ($10^{-1}$ to $10^{-4}$ cm size) differently oriented with respect to one another. Depending upon the mode of crystallization, these crystals may be of various irregular shapes, and in contrast to crystals of regular shapes are called *crystallites* or *grains of the metal.* Metals in the solid state and, to some extent, in the liquid state possess *high thermal* and *electrical conductivity, and a positive temperature coefficient of electrical resistivity.* The electrical resistance of pure metals *increases* with the temperature. Many metals display superconductivity; at temperatures near absolute zero, their electrical resistance drops abruptly to extremely low values. Besides, all metals are capable of *thermionic emission, i.e.* the emission of electrons upon being heated; they are good reflectors of light and lend themselves well to plastic deformation.

*Pure metals* are of low strength and, in many cases, do not possess the required physiochemical and technological properties for some definite purpose. Consequently they are seldom used in engineering. The overwhelming majority of metals used are alloys.

*Alloys* are produced by *melting* or *sintering two or more metals, or metals* and a non-metal, together. Alloys possess typical properties inherent in the metallic state, the substances that make up the alloy are called its *components*. An alloy can consist of two or more components.

Examples: *Steels, copper, aluminium, brass, bronze, invar, superalloys, etc.*

## 2. Ceramic Meterials

These materials are non-metallic solids made of inorganic compounds such as oxides, nitrides, borides, silicides and carbides. They are fabricated by first shaping the powder with or without the application of pressure into a compact, which is subsequently subjected to a high temperature treatment, called *sintering*. Traditional ceramics were made from crude naturally occuring mixtures of materials having inconsistent purity. These have been used essentially in the manufacture of pottery, porcelain, cement and silicate glasses. New ceramics possess exceptional electrical, magnetic, chemical, structural and thermal properties. Such ceramics are now extensively used in the electronic control devices, computers, nuclear engineering and aerospace fields.

Examples: *Silica, sodalime glass, concrete, cement, ferrites, garnets, MgO, CdS, ZnO, SiC, etc.*

## 3. Organic Materials

These materials are derived *directly from carbon.* They usually consist of carbon chemically combined with hydrogen, oxygen or other non-metallic substances. In many instances their structures are fairly complex.

Common organic materials are: *plastics and synthetic rubbers.* These are termed as *polymers* because they are formed by polymerization reaction in which relatively simple molecules are chemically combined into massive long-chain molecules or "three-dimensional" structures.

Examples: *Plastics: PVC, PTFE, polythene; Fibres: terylene, nylon, cotton; Natural and synthetic rubbers, leather, etc.*

**Examples of Composites:**

1. **Metals and alloys and ceramics**
   (*i*) Steel reinforced concrete
   (*ii*) Dispersion hardened alloys.
2. **Metals and alloys and organic polymers**
   (*i*) Vinyl-coated steel
   (*ii*) Whisker-reinforced plastics.
3. **Ceramics and organic polymers**
   (*i*) Fiber-reinforced plastics
   (*ii*) Carbon-reinforced rubber.

## 1.2 PROPERTIES OF MATERIALS

The following properties of materials will be discussed briefly:

1. Physical properties
2. Mechanical properties
3. Electrical properties
4. Magnetic properties, and
5. Chemical properties

### 1.2.1 Physical Properties

1. **The melting or freezing point:** *The melting or freezing point of a pure metal is defined as temperature at which the solid and liquid phases can exist in stable equilibrium.* When a metal is heated to melting point, the liquid phase appears, and if more heat is supplied, the solid melts completely at constant temperature.

The freezing of a pure liquid, on the other hand, may exhibit the phenomenon of super-cooling, the liquid in some cases can be lowered appreciably beyond the melting point without the appearance of crystals. However when crystals do not appear, the mass rapidly assumes the normal temperature of the melting point.

The *use of mercury* in thermometers, manometers and other instruments arises from its low melting point; the use of *tungsten filaments* in incandescent light bulbs is possible because of its extremely high melting point.

2. **Boiling point:** *The boiling point of a liquid is the temperature at which its vapour pressure equals to one atmosphere. The boiling points of the metals except mercury are high.* The boiling points of zinc (907 °C) and cadmium (865 °C) are sufficiently low so that in recovery of these metals from their ores the metals are vapourised and condensed.

3. **Density:** *Mass per unit volume is termed as density.* In metric system it is stated in kg/m$^3$. The low densities of aluminium and magnesium and of their alloys make them particularly valuable in aeronautic and transportation fields.

4. **Linear coefficient of expansion:** *The linear coefficient of expansion of a solid is defined as the increase in length per unit length, for each degree rise in temperature.* These coefficients are important when metals are to be exposed to a considerable range of temperatures as in engine pistons, and other accurately fitting mechanisms.

5. **Thermal conductivity:** *The thermal conductivity of a metal is defined as the number of kilojoules of heat that would flow per second through a specimen one sq. metre in cross-section and 1 metre in length when the temperature gradient is 1 °C.* Silver and copper show the highest thermal conductivities of all metals. Some metals like German silver exhibit very low conductivities and hence find applications where heat losses by metallic conduction should be kept to a minimum.

All metals are conductor of electricity; *silver is the best conductor and copper is next. It should be noted that while volume for volume aluminium has only 61% of the conductivity of copper, nevertheless weight for weight aluminium because of its low density, shows a conductivity nearly twice that of copper.*

6. **Electrical resistivity:** The resistance of a metal is the *reciprocal of its conductivity.* The electrical resistivity of a metal is the resistance of a specimen of 1 cm in length and 1 sq. cm in cross-section. Since these values for metals are very small if expressed in ohms, they are usually given in micro-ohms, where 1 micro-ohms = $10^{-6}$ ohms.

### 1.2.2 Mechanical Properties

**1. Strength:** *The strength of metal is its ability to withstand various forces to which it is subjected during a test or in service.* It is usually defined as tensile strength, compressive strength, proof stress, shear strength, etc. *Strength of materials is a general expression for the measure of capacity of resistance* possessed by solid masses or pieces of various kinds to any cause tending to produce in them a permanent and disabling change of form or positive fracture. Materials of all kinds owe their strength to the action of the forces residing in and about the molecules of the bodies (the molecular forces) but mainly to that ones of these known as *cohesion*; certain modified results of cohesion as toughness or tenacity, hardness, stiffness and elasticity are also important elements and strength is in relation of the toughness and stiffness combined.

**2. Elasticity:** A material is said to be *perfectly elastic* if the whole of the stress produced by a load *disappears completely* on the removal of the load, the modulus of elasticity of young's modulus (*E*) is the proportionality constant between stress and strain for elastic materials. *Young's modulus is the indicative of the property called stiffness; small values of E indicate flexible materials and large values of E reflect stiffness and rigidity.* The property of spring back is a function of modulus of elasticity and refers to the extent to which metal springs back when an elastic deforming load is removed. In metal cutting, modulus of elasticity of the cutting tools and tool holder affect their rigidity. Values of modulus of elasticity for some important metals are given in Table 1.2.

**Table 1.2. Modulus of elasticity of some important metals**

| S.No. | Metals | Young's modulus of elasticity E ($GN/m^2$) |
|---|---|---|
| 1. | Cast Iron | 98 |
| 2. | Wrought iron | 197 |
| 3. | Mild steel | 210 |
| 4. | Aluminium | 72 |
| 5. | Copper | 120 |
| 6. | Zinc | 100 |
| 7. | Tungsten | 430 |
| 8. | Molybdenum | 350 |
| 9. | Tin | 42 |
| 10. | Lead | 18 |

**3. Plasticity:** *Plasticity is the property that enables the formation of permanent deformation in a material.* It is reverse of elasticity; a plastic material will retain exactly the shape it takes under load, even after the load is removed. Gold and lead are highly plastic materials. Plasticity is used in stamping images on coins and ornamental work.

During plastic deformation there is the displacement of atoms within metallic grains and consequently the shapes of the metallic components change.

It is because of this property that certain synthetic materials are given the name "plastics". These materials can be changed into required shape easily.

**4. Ductility:** *It is the ability of a metal to withstand elongation or bending.* Due to this property, wires are made by drawing out though a hole. The material shows a considerable amount of plasticity during the ductile extension. This is a valuable property in chains, ropes, etc., because they do not snap off, while in service, without giving sufficient warning by elongation.

**5. Malleability:** *This is the property by virtue of which a material may be hammered or rolled into thin sheets without rupture.* This property generally increases with the increase of temperature. The metals in order of their ductility and malleability (at room temperature) are given in Table 1.3.

**Table 1.3. Common metals in order of their ductility and malleability**

| Ductility | Malleability |
|---|---|
| Gold | Gold |
| Silver | Silver |
| Platinum | Copper |
| Iron | Aluminium |
| Nickle | Tin |
| Copper | Platinum |
| Aluminium | Lead |
| Zinc | Zinc |
| Lead | Nickel |

6. **Toughness (or tenacity):** *Toughness (or tenacity) is the strength with which the material rupture.* It is due to the attraction which the molecules have for each other; giving them power opposes to resist tearing apart.

*The area under the stress-stain curve indicates the toughness*, (*i.e.* energy which can be absorbed by the material upto the point of rupture). Although the engineering stress-strain curve is often used for this computation, a more realistic result is obtained from a true stress-strain curve. Toughness is expressed as energy absorbed (Nm) per unit volume of material participating in absorption ($m^3$) or $Nm/m^3$. This result is obtained by multiplying the ordinate by the abscissa (in appropriate units) of stress-strain plot.

7. **Brittleness:** *Lack of ductility is brittleness.* When a body breaks easily when subjected to shocks it is said to be brittle.

8. **Hardness:** *Hardness is usually defined as resistance of material to penetration.* Hard materials resist scratches or being worn out by friction with another body.

Hardness is primarily a function of the elastic limit, (*i.e.* yield strength) of the material and to a lesser extent a function of the work hardening coefficient. The modulus of elasticity also exerts a slight effect on hardness.

In the most generally accepted test, an indentor is pressed into the surface of the material by slowly applied known load, and the extent of the resulting impression is measured mechanically or optically. A large impression for a given load and indentor indicates soft material, and the opposite is true for small impression.

The converse of hardness is known as *softness.*

**9. Fatigue:** When subjected to fluctuating or repeating loads (or stresses), materials tend to develop a characteristic behaviour which is different from that (of materials) under *steady loads. Fatigue is the phenomenon that leads to fracture under such conditions.* Fracture takes place under repeated or fluctuating stresses *whose maximum value is less than the tensile strength of the material* (under steady load). Fatigue fracture is progressive, beginning as minute cracks that grow under the action of the fluctuating stress.

*Fatigue failure starts at the point of highest stress.* This point may be determined by the shape of the part; for instance, by stress concentration in a groove. It can also be caused by surface finish, such as tool marks or scratches, and by *internal voids such as shrinkage cracks and cooling in castings and weldments and defects introduced during mechanical working and by defects, stresses introduced by electroplating.* It must be remembered that surface and internal defects are stresses raisers, and the point of highest actual stress may occur at these rather than at the minimum cross-section of highest normal stress. Thus, processing methods are extremely important as they affect fatigue behaviour.

**10. Creep:** *Creep is the slow plastic deformation of metals under constant stress or under prolonged loading usually at high temperature. It can take place and lead to fracture at static stresses much smaller than those which will break the specimen by loading it quickly.* Creep is specially taken care of while designing I.C., engines, boilers and turbines.

The creep at room temperature is known as *low temperature creep* and occurs in load pipes, roofings, glass as well as in white metal bearings. The creep at high temperatures is known as *high temperature creep.* It mainly depends upon *metal, service temperature to be encountered* and the *stress involved.* For studying its effects, the specimens are put under a constant load; the creep is measured during various time intervals and results then plotted to get a creep curve.

### 1.2.3 Electrical Properties

One of the important characteristics of the materials is their ability to permit or resist the flow of electricity. Materials to be used in electrical equipments can be selected on the basis of their electrical properties, such as:

(*i*) Resistivity

(*ii*) Conductivity

(*iii*) Temperature coefficient of resistance,

(*iv*) Dielectric strength,

(*v*) Thermoelectricity, and

(*vi*) Other electrical properties.

**Resistivity**

It is a characteristic of the material of which the conductor is made. It is the electrical property of a material due to which, it impedes or *resists the flow of electricity through it.*

$$\text{Resistivity, } \rho = \frac{R \cdot A}{l}$$

where

$R$ = resistance (ohms) of a conductor

$A$ = the area of the conductor section, and

$l$ = length of the conductor.

**Conductivity**

The conductivity ($\sigma$) is the *reciprocal of electrical resistivity.*

$$\sigma = \frac{l}{\rho} = \frac{l}{R.A}$$

The dimensions of $\sigma$ are $ohm^{-1}$ $cm^{-1}$. Electrical conductivity is that electrical property of a material due to which the *electric current flows easily through the material, i.e.* the material provides an easy path for the flow of electricity through it.

**Temperature coefficient of resistance**

It is usually employed to specify the variation of resistivity ($\rho$) with temperature.

Temperature coefficient of resistance or resistivity

$$\alpha_T = \frac{\rho - \rho_0}{\rho_0} \cdot \frac{1}{T - T_0}$$

where

$\rho$ = resistivity at temperature $T$,

$\rho_0$ = resistivity at temperature $T_0$, and

$T$ and $T_0$ are in degrees kelvin.

**Dielectric strength**

It means the insulating capacity of a material against high voltages. A material having high dielectric strength can withstand sufficiently high voltage field across it before it will break down and conduct. *A dielectric is an insulation.*

**Thermoelectricity**

If two dissimilar metals are joined and this junction is then heated, a small voltage in the millivolt range is produced, and this is known as *thermoelectric effect*. Thermoelectric effect forms the basis of the thermocouple operation.

**Other Electrical Properties of Materials**

Other electrical properties of materials are:

*Electro-chemical phenomena*—as in storage batteries.

*Electro-physical effects*—as in contact potentials.

*Electro-mechanical effects*—as in radars.

**Superconductivity**

Some metals and compounds lose their electrical resistance abruptly before absolute zero is reached and become superconductor. Superconductivity, therefore, refers to the *phenomenon of abrupt drop of resistivity of some metals at a temperature, called superconducting transition temperature,* before absolute zero is reached. This transition temperature is 0.4 K for titanium,

1.17 K for aluminium and 9.2 K for niobium, 14 K for NbH, 1.6 K for $Nb_4$ and 18 K for $Nb_3S_4$.

*Superconductivity state can be abolished by the application of an external magnetic field or produced by a sufficiently large current flowing through the conductor.*

### 1.2.4 Magnetic Properties

*Those materials in which a state of magnetisation can be induced are called "magnetic materials".* Such materials create a magnetic field in the surrounding space.

The magnetic properties of materials arise from the spin of electrons and the *orbital motion of* electrons around the atomic nuclei. In several atoms the opposite spins neutralise one another, but when there is an excess of electrons spinning in one direction, magnetic field is produced. *All substances except ferro-magnetic materials which can form permanent magnets, exhibit magnetic effects only when subjected to an external electromagnetic field.*

Study of the magnetic properties is necessary because the science of magnetism explains many aspects of the structure and behaviour of the matter.

Some of the important properties are:

(*i*) Permeability

(*ii*) Coercive force, and

(*iii*) Magnetic hysteresis.

*Absolute permeability:* It is the *ratio of the flux density in a material to the magnetising force producing that flux density* and is denoted by $\mu$; $\mu = \mu_0\mu_r$, where $\mu_0$ is the permeability of free space having a value of $4\pi \times 10^{-7}$ H/m.

*Coercive force:* It may be defined as the *magnetising force which is necessary to neutralise completely the magnetism in an electromagnet after the value of magnetising force becomes zero.*

Below Curie temperature (it is the rising temperature at which the given material ceases to be ferro-magnetic, or the falling temperature at which it becomes magnetic) all magnetic materials exhibit this phenomenon called *hysteresis* which is defined as the lagging of magnetisation or induction flux density (B) behind the magnetising force (H) or it is that quality of a magnetic substance due to which energy is dissipated in it on reversal of its magnetism.

### 1.2.5 Chemical Properties

A study of chemical properties of matérials is necessary because most of the engineering materials, when they come in contact with other substances with which they can react, tend to suffer from chemical deterioration.

The chemical properties describe the combining tendencies, corrosion characteristics, reactivities, solubilities, etc. of substances. Some of the chemical properties are:

(*i*) Corrosion resistance

(*ii*) Chemical composition, and

(*iii*) Acidity or alkalinity.

**Note:** Corrosion is a gradual chemical or electrochemical attack on a metal by its surroundings so that the metal is converted into an oxide, salt or some other compound. It may be brought about by almost unlimited number of factors of corrosive media such as air, industrial atmosphere soils, acids, bases and salt solutions. It may also occur at elevated temperature in media which are inert when near or below room temperature.

## 1.3 PROPERTIES OF BUILDING MATERIALS

The properties of building materials (governed by the characteristics and their various properties) may be classified into following categories:

1. **Mechanical Properties** (*e.g. elasticity, plasticity, hardness, strength*, etc)
2. **Chemical Properties** (*e.g. chemical composition, acidity, alkalinity, corrosion*, etc.)
3. **Electrical properties** (*e.g. resistivity, conductivity, dielectric strength*, etc)
4. **Magnetic Properties** (*e.g. permeability, hysteresis, coercive force*, etc.)

5. **Optical Properties** (*e.g. light transmission, colour, reflectivity, refractive index*, etc.)
6. **Thermal Properties** (*e.g. specific heat, thermal conductivity, thermal expansion*, etc.).
7. **Physical Properties:**

(*i*) **Bulk density:** The *mass per unit volume of a material in its natural state* is called its "*bulk density*". The bulk densities (kN/$m^3$) of some important materials are:

Granite – 25 to 27; clay brick – 16 to 18; Sand – 14.5 to 16.5; Granel – 14 to 17;
Lime stone (dense) – 18 to 24; Concrete (light) – 5 to 18;
Concrete (heavy) - 18 to 25; steel - 78.5, Pinewood - 5 to 6;
Mineral wool - 2 to 4; Plastic material (porous) - 0.2 to 1.

The mass per unit volume of a homogeneous material is termed as its "*density*".

The *ratio of bulk density of a material to its density* is known as its "*density index*". The *density index* of most of the building materials is *less than unity.*

(*ii*) **Porosity:** The degree by which the volume of a material is occupied by pores is indicated by the term *porosity*. The strength, bulk density, durability, thermal conductivity, etc. of a material depend on is porosity.

(*iii*) **Water absorption:** The water absorption of a material is its ability to absorb and retain water. It mainly depends on the volume, size and shape of pores present in the material.

(*iv*) **Water permeability:** It is defined as *the capacity of a material to permit water to pass through it under pressure.*

(*v*) **Water resistance:** *It is the ability of a material to resist alternating dry and wet conditions without seriously affecting its shape and mechanical strength.*

(*vi*) **Hygroscopicity:** *It is the property of a material to absorb water vapour from air and is governed by the air temperature, number of pores and nature of the substance involved.*

(*vii*) **Coefficient of softening:** *It is the ratio of compressive strength of material saturated with water to that in dry state.* The materials having coefficient of softening equal to 0.8 or more are referred to *water-resisting materials.*

(*viii*) **Fire resistance:** The material should have adequate fire resistance (ability to resist the action of high temperature without losing its load-bearing capacity) to ensure safety and stability under the conditions of a fire.

(*ix*) **Frost resistance:** It depends upon the density of material and its degree of saturation with water. Generally, the dense materials are frost resistant.

(*x*) **Chemical resistance:** *It is the ability of a material to withstand the action of acids, alkalies, gases and salt solutions.*

(*xi*) **Refractorness:** *It is the ability of a material to withstand prolonged action of high temperature without melting or loosing shape.*

(*xii*) **Durability:** *It is the property of a material to resist the combined action of atmospheric and other factors.*

## OBJECTIVE TYPE QUESTIONS

**Fill in the Blanks/Say or 'Yes' or 'No':**

1. The iron group which include all irons and steels are called .............................. metals.
2. ...................................... is used for lagging round steam pipes and steam boilers.
3. Metals posses metallic lustre.
4. Metals are malleable.
5. Non-metals have high density.
6. Non-metals are bad conductor of heat and electricity.
7. All solid metals have crystalline structure.

8. Non-metals exist in amorphic or mesomorphic forms.
9. The excitation of valence electron by e.m.f. in case of non-metals is easy.
10. The electrical resistance of pure metals increases with the temperature.
11. Alloys are produced by melting or sintering two or more metals or metals and a non-metal together.
12. Organic materials are derived directly from ................................
13. Plastics and synthetic rubbers are the common organic materials.
14. ................................... may be defined as the materials which have free electrons.
15. ................................... materials are those materials in which state of magnetisation can be induced.
16. The resistance of a metal is the reciprocal of its conductivity.
17. Plasticity is the property that enables the formation of permanent deformation in a material.
18. .............................. is the ability of metal to withstand elongation or bending.
19. ........................ is the property by virtue of which a material may be hammered or rolled into thin sheets without rupture.
20. Brittleness is the strength with which the material opposes rupture.
21. Lack of ductility is toughness.
22. Hardness is usually defined as resistance of material to penetration.
23. Superconductivity state can be abolished by the application of an external magnetic field or produced by a sufficiently large current flowing through the conductor.
24. A dielectric is an insulation.
25. Fatigue failure starts at the point of .......................... stress.
26. The ratio of bulk density of a material to its density is known as ......................... .
27. The degree by which the volume of a material is occupied by pores is indicated by the term ..............................
28. Water ............................. is defined is the capacity of a material to permit water to pass through it under pressure.
29. .......................... is the property of a material to absorb water vapour from air and is governed by the air temperature, number of pores and nature of the substance involved.
30. ........................... is the property of a material to resist the combined action of atmosphere and other factors.

## ANSWERS

| | | | | |
|---|---|---|---|---|
| **1.** ferrous | **2.** Asbestos | **3.** Yes | **4.** Yes | **5.** No |
| **6.** Yes | **7.** Yes | **8.** Yes | **9.** No | **10.** Yes |
| **11.** Yes | **12.** carbon | **13.** Yes | **14.** conductor | **15.** Magnetic |
| **16.** Yes | **17.** Yes | **18.** Ductility | **19.** Malleability | **20.** No |
| **21.** No | **22.** Yes | **23.** Yes | **24.** Yes | **25.** highest |
| **26.** density index | **27.** porosity | **28.** permeability | **29.** hygroscopicity | **30.** Durability. |

## THEORETICAL QUESTIONS

1. How are engineering materials classified?
2. State the difference between metals and non-metals.
3. Write short notes on:
   Metals and alloys, ceramics and organic polymers.
4. Discuss briefly the properties of materials.
5. Explain briefly the following mechanical properties of metals:
   Strength, elasticity, plasticity, ductility and malleability.
6. Explain briefly the "Physical properties" of the building materials.
7. Explain briefly any three of the following physical properties of the building materials:
   Bulk density, porosity, weather resistance, hygroscopicity and frost resistance.

# 2

# Building Stones

*2.1 Introduction. 2.2 Classification of rocks. 2.3 Common rock -forming minerals. 2.4 Characteristics of good building stones. 2.5 Uses of stones and their selection. 2.6 Deterioration of stones. 2.7 Preservation of stones. 2.8 Testing of stones. 2.9 Common building stones, their composition, properties, uses and occurrence. 2.10 Natural bed of stone. 2.11 Quarrying of building stones – quarrying-location – quarrying terminology – methods of quarrying – explosives – storing and handling of explosives. 2.12 Artificial stones – manufacture of artificial stones – varieties of artificial stones – advantages of artificial stones. 2.13 Dressing of stones. Highlights – Objective Type Question – Theoretical Questions.*

## 2.1 INTRODUCTION

The stones are derived from rocks which form the earth's crust and have no definite shape or chemical composition but are mixtures of two or more minerals.

The **mineral** is a *substance which is formed by the natural inorganic process and possesses a definite chemical composition and molecular structure.*

Following are some important *uses of stones* in *civil engineering works*:

1. Construction of residential and public buildings.
2. Constriction of dams, weirs, harbours, abutment for bridges, etc.
3. Facework of structures where massive appearance, solidity of construction and ornamental features are the primary requirements.
4. Used as road metal and railway ballast.
5. Used as aggregate for concrete.
6. Used in the form of veneers for decorative front and interior of buildings.

## 2.2 CLASSIFICATION OF ROCKS

The rocks from which stones are obtained are classified in the following *four* ways:

1. Geological
2. Physical
3. Chemical
4. Practical.

**1. Geological classification:** Geologically the rocks are classified into *three* different types:

(*i*) Igneous rocks

(*ii*) Sedimentary rocks

(*iii*) Metamorphic rocks.

- **Igneous rocks** are formed by *cooling the molten lava on or inside the earth's surface during the volcanic eruption.* The portion of lava, which comes outside the surface, cools quickly and forms the rock of non-crystaline nature called as *Trap* or *Basalt*. The rest of the portion which remains inside the earth undergoes cooling at a slow rate and results in formation of rock of crystalline variety known as *Granite*.
- **Sedimentary rocks** are formed by gradual deposition of disintegrated rocks (due to the atmospheric action such as rain, wind and temperature), vegetable matter and clay at the bottom of rivers, lakes or sea. These are also called *stratified* because these rocks are formed in layers. *Lime stone* and *sand stone* belong to this category of rocks.
- When sedimentary or even igneous rocks are subjected to great heat and pressure inside the earth, a new variety of rock is formed which is known as **metamorphic rock.** This change of structure is called *metamorphism*. For example, lime stone changes to marble, slate changes to gneiss, etc.

**2. Physical classification:** The basis of the classification are physical properties of rocks, the manner and arrangement of different particles and mass forming a stone. Physically the rocks are classified as:

(*i*) Stratified rocks

(*ii*) Unstratified rocks.

- **Stratified rocks** are those which exhibit distinct layers which can be separated. The plane along which the separation of the layers can be carried out is called *cleavage plane. Lime stone, slate and sandstone* are the examples.
- **Unstratified rocks** do not show any sign of starta and cannot be easily split into slabs. Examples are *granite, trap, marble,* etc.

**3. Chemical classification:** The chemical constituents (in the rocks) are the basis of this classification. Chemically the rocks are classified as:

(*i*) Silicious

(*ii*) Agrillaceous

(*iii*) Calcarious.

- In case of **Silicous rocks** *silica* is the main constituent. The silica in free state is called *sand* and in combined state *silicate*. The rocks containing silica in free form are harder, while those having silica in combined form are likely to be disintegrated. Examples are *sandstone, quartzite,* etc.
- In **Argillaceous rocks** *clay or alumina* is the main constituent. *Slate, laterite, kaolin,* etc., belong to this category.
- **Calcarious rocks** have *calcium carbonate* or *lime* as their leading constituent. These rocks are readily acted upon by even dilute hydrochloric acid. *Limestone* and the *marble* are the important examples.

**4. Practical classification:** Practical classification is based on the usage. Practically stones have been classed as : *Granites, basalts, laterites, marbles, limestones, sandstones, slates.*

## 2.3 COMMON ROCK-FORMING MINERALS

The common rock-forming minerals are described below:

**Silica:** *Quartz is pure or nearly pure silica* and is a hard and glassy mineral. It occurs in granite in grey, white or colourless lumps which are sometimes crystalline It is unaffected by weather. Quartz has a hardness of 7 and specific gravity of 2.66. Due to the presence of small quantities of metallic oxides it has varying transparency and colours.

**Felspar**: Felspars are the *silicates of alumina, with alkaline substances like potassium, sodium and calcium.* In a sound granite, it occurs in the form of grey or reddish brown bright crystals;

in a bad sample of granite the felspar is earthly in appearance. The hardness of felspar is 6 and specific gravity varies from 2.5 to 2.7. A stone readily meets the decay if it contains large proportions of felspar mixed with other softer minerals.

**Mica:** *It contains silicate of aluminium with potassium.* It is of dark grey, black or brown colour. It is very soft and readily affected by atmosphere and chemicals. It has a perfect cleavage and could be split into very thin laminae or flakes. It has a hardness of 2.5 and specific gravity of about 3. The mica occurs in granite as semi-transparent, glistening scales, which are sometimes dark, brown or black.

**Hornblende**: It is very *complex silicate* and its colour varies from dark green to black. It has hardness of 5.5 and specific gravity 3.2.

**Calcite:** It is the leading constituent of the limestone and marbles. Its colour varies from white to grey and lustre varies from vitreous dull to earthy. Hardness is 3 and specific gravity 2.7.

**Dolomite:** It is $CaMg(CO_3)_2$ or magnesium carbonate. It is commonly found in dolomitic limestones and dolomitic marbles. Its stability and weather resisting property is better than pure calcium carbonate.

## 2.4 CHARACTERISTICS OF GOOD BUILDING STONES

Stones find their applications in the construction of heavy structures such as dams, docks and harbours, weirs, bridgepiers, buildings, etc. To find the suitability of stones under different conditions, the following characteristics should be considered:

**1. Appearance and colour:** The stones which are to be employed for decoration work should have uniform and appealing colour and should be free from flaws and clay holes. The use of stones which contain much iron should be discouraged as the formation of iron oxide disfigures them and brings about disintegration. The stones should also have the ability to receive good polish.

**2. Weight:** The specific gravity hence the weight of the good building stone should be high as heavier stones can resist the force of bigger magnitude. A heavy stone possesses more compactness and less porosity.

**3. Porosity and absorption:** Porosity (which is decided by the number of pores) exists in all the stones but if it is present in greater extent it makes the stone unsuitable for building construction because during rain, water seeping into pores contains acids and fumes (absorbed from atmosphere) which destroy the stone. When the climate is cold, water entering the pores may even freeze and split the stone.

**4. Fineness of grain:** The stones which are fine grained are suitable for moulding works. If the stones are non-crystalline they are likely to disintegrate under the action of natural agencies.

**5. Compactness :** Stone's durability is decided by their compactness or density of composition. A compact stone can withstand the effects of external agencies effectively.

**6. Resistance to fire**: For resistance against fire, the stone (*i*) should have homogeneous composition and (*ii*) should be free from calcium carbonate or oxide of iron.

**7. Electrical resistance:** The electrical resistance of a stone decreases when it gets wet. Thus to have steady and high electrical resistance, the stone must be non-absorbent like slate.

**8. Hardness and toughness:** The stones must be adequately hard and tough so that they may resist wear and tear (as in the case of railway ballast, road metal, etc.). Hardness may be tested by scratching with a pen knife which should not be able to produce an impression on hard stone like granite. Toughness of the stone can be tested by subjecting it to a hammer action.

**9. Strength:** Stones used in the structures are usually subjected to compression; so they should have sufficient strength to cope with the requirements. Generally all the stones possess a reasonable degree of strength as far as their use in building work is concerned, but for heavy structures only those stones are suitable which have high strength comparatively.

**10. Durability:** A stone is more durable in case it is (*i*) compact, (*ii*) homogeneous and (*iii*) free from any material affected by dilute hydrochloric and sulphuric acids. It should also have negligible water absorption.

**11. Dressing:** The art of shaping a stone is known as *dressing*. Stones should possess uniform texture and softness so that they may be easily dressed, if it is too hard, finish will be poor and dressing uneconomical.

**12. Cost:** The cost is an important consideration in the selection of building stone. Other things being equal the cost of a stone depends upon the ease with which it can be quarried out, the proximity of the quarry to the place of use, and the transportation facilities available. The subsequent cost of dressing a stone, before it is placed directly in the structure, should also be low.

**13. Seasoning:** Good stone must also be free from *quarry sap*. To ensure this, the stones after quarrying and dressing should be left for a period of 6 to 12 months for proper seasoning, before using in construction work.

## 2.5 USES OF STONES AND THEIR SELECTION

In the table below are given the important uses and the selection of stones for variety of works.

| S. No. | Use | Name of stone | Reason for selection |
|---|---|---|---|
| 1. | *Construction of building exposed to high wind blowing particles.* | Granite and sand stone | Hardness due to presence of silica. |
| 2. | *For general building work, i.e. walls, foundations and super-structure.* | Sand stone | Hard and durable. |
| 3. | *For heavy engineering work such as docks, bridges and light houses.* | Granite and gneiss | Strong, durable, capable of resisting thrust. |
| 4. | *Building exposed to fire.* | Compact sand stone | Fire resisting property. |
| 5. | *For building in industrial town.* | Granite and compact sand stone | Acids and smoke proof. |
| 6. | *For road metal and railway ballast.* | Granite and basalt | Hard, tough and possess abrasion resistance. |
| 7. | *Electrical switch boards.* | Slabs of marble and slate | Possess electric resistance, non-absorbent of water and easily cut into pieces. |
| 8. | *Manufacture of lime.* | Lime stone | Decomposed into quick lime on heating. |
| 9. | *As flux in the manufacture of iron.* | Lime stone | Special characteristics. |

## 2.6 DETERIORATION OF STONES

The various agencies which bring about the deterioration of stones are as follows:

**1. Rain:** A stone gets wet due to rain and becomes dry due to sun; this alternate wetting and drying go a long way in deteriorating the stone. The harmful gases and acids contained in rain water also exercise a damaging effect on the stone.

**2. Temperature:** The stone also gets deteriorated due to frequent changes of temperature of atmosphere (due to these changes expansion and contraction take place frequently, which result in the production of cracks).

**3. Wind:** A strong wind carrying with it grits and dust strikes the stones and brings about the wear of the stones. Also the water particles in the wind penetrate into the pores of the stones and cause dampness which results in deterioration.

**4. Frost:** When the stone is used in the cold climatic conditions, the water in the pores freezes; consequently it expands and thereby splits the stone. Therefore, porous stones should not be used in very cold places.

**5. Atmospheric impurities:** In an industrial town the presence of acids and fumes adversely affect the stones containing carbonate of lime.

**6. Vegetable growth:** Certain acids secreted by the roots of certain plants and trees in the joints of the stone, attack it. The vegetable growth also keep the stone in wet condition; this constant dampness decays the stone.

**7. Living organism**: In the stones, sometimes holes are bored by certain insects due to which they become weak. These insects, are found in sea and thus stones near shores are effected by them.

**8. Movements of chemicals:** If stones of different varieties, such as limestone and sandstone, are used side by side in the same structure, the chemicals formed by the action of atmospheric agencies on one variety may move on the other and cause deterioration of the other

## 2.7 PRESERVATION OF STONES

The durability of stone increases to a considerable extent if it is properly preserved. **Preservation** *means making the stone strong enough to face the atmsopheric agencies which are the root cause of its deterioration and can be effected in the following ways*:

1. Filling up the stone pores.
2. Providing the stone with a coat of preservative to prevent the ingress of moisture into the pores.
3. The use of stones, containing carbonate of lime in industrial town should be discouraged.
4. In a structure, as far as possible, limestones and sandstones, magnesium limestone and granular limestone should not be laid close to each other.
5. Only that binding material, which does not have any adverse effect on a particular stone, should be employed.
6. Growth of plants and trees on the stone should be checked.

Some of the **preservatives** are:

(*i*) Coal tar
(*ii*) Linseed oil
(*iii*) Bartya solution
(*iv*) Solution of alum and soap
(*v*) Paraffin
(*vi*) Paint.

## 2.8 TESTING OF STONES

To determine the suitability of stone for its use in engineering works, the following tests are performed:

1. Hardness test
2. Crushing test
3. Impact test
4. Fire resistance test
5. Attrition test
6. Acid test
7. Electrical resistance/water absorption
8. Smith's test
9. Crystallization test
10. Microscopic Test
11. Freezing and thawing test.

### 1. Hardness test:

Hardness of a stone is tested by a pen knife which will not be able to produce a scratch on a hard stone (like granite).

Hardness is determined with the aid of Moh's scale of hardness. If a pocket knife makes a mark on a fresh surface of a stone say on limestone, the hardness of that limestone may be taken as H = 3, a scratch with the aid of a finger nail indicates a hardness of H = 2. Hard silicious rock which cannot be scratched by a knife represents a hardness of H = 7.

The hardness test, for finding out the **coefficient of hardness,** is carried out as follows:

- A cylinder of diameter 25 mm and height 25 mm is taken out from the sample of stone.
- It is then weighed.
- It is placed in Dorry's testing machine and pressed with a load of 1250 gm.
- The annular steel disc of the machine is then rotated at a speed of 28 r.p.m.
- During rotation, coarse sand of standard specification is sprinkled on top of the disc.
- After 1000 revolutions, the specimen is taken out and weighed.
- Coefficient of hardness is found out from the following equation:

$$\text{Coefficient of hardness} = 20 - \frac{\text{loss in weight in gm}}{3}$$

**2. Crushing test:**

This test, in particular, is performed for the stone which is to be used under compression (such as the stone to be used at the bottom of a heavy structure). It is carried out on blocks in a testing machine as follows:

- The sample of a stone is first cut in cubes of size 40 mm × 40 mm × 40 mm. The sides of the cube are then finely dressed and finished. The minimum number of specimens that can be tested is three.
- The specimens are placed in water for 72 hours prior to the test and thereafter tested in saturated condition.
- The load-bearing surface is covered with a layer of plaster of Paris or a 5 mm thick plywood. Load is then applied axially on the cube with a crushing test machine. The rate of loading is 13.7 N/mm$^2$ per minute.
- The *crushing strength of the stone per unit area is the maximum load at which its sample crushes or fails divided by the area of the bearing face of the specimen.*

For ordinary building works it is not necessary to carrygs out crushing/strength test for stones. For example, the compression test on the stone in buildings does not exceed 1 N/mm$^2$, whereas even the weakest stone possesses much higher compressive strength as illustrated in Table 2.1.

**Table 2.1 Compressive strength of stones**

| S. No. | Name | Compressive strength (N/mm$^2$) |
|---|---|---|
| 1. | Laterite | 2 to 3 |
| 2. | Limestone | 50 |
| 3. | Sandstone | 70 |
| 4. | Granite | 70 to 130 |
| 5. | Slate | 70 to 200 |
| 6. | Diorite, Syenite | 80 to 150 |
| 7. | Basalt | 150 to 200 |
| 8. | Gneiss | 200 to 400 |
| 9. | Trap | 300 to 400 |

**3. Impact test:**

Impact test is carried out on an impact testing machine, to determine the toughness of a stone. This test is carried out as follows:

- A cylinder of diameter 25 mm and height 25 mm is taken out from the sample of stone.
- The sample is then placed on the cast-iron anvil of the machine.
- A steel hammer of 20 N is allowed to fall axially in a vertical direcition over the specimen.

- The height of the first blow is 1 cm; that of the second blow 2 cm; that of third blow is 3 cm; and so on.
- The blow at which the specimen breaks is noted. If it is the *n*th blow *n* represents the *toughness index* of the stone.

**4. Fire resistance test:**

The stone which is free from calcium carbonate can resist fire. The presence of calcium carbonate in the stone can be detected by dropping a few drops of dilute sulphuric acid which will produce bubbles.

**5. Attrition test:**

The attrition test is carried out to determine the rate of wear of stones employed for the construction of road. This test also known as abrasion test, is carried out in Deval's attrition testing machine as follows:

- The sample of stone is broken into pieces of about 60 mm size.
- Such pieces, weighing 50 N, are then put in both the cylinders of Deval's attrition testing machine. The diameter and length of cylinders are 200 mm and 340 mm respectively.
- The cylinders are then closed, their axes making an angle of 30° with the horizontal.
- The cylinders are rotated about the horizontal axis for 5 hours at the rate of 30 r.p.m.
- The contents are then taken out from the cylinders and passed through a size of 1.5 mm mesh.
- The quantity of material which is retained on the sieve is then weighed.
- The percentage wear is worked out at as follows;

$$\text{Percentage wear} = \frac{\text{loss in weight}}{\text{initial weight}} \times 100$$

**6. Acid test:**

In this test, a stone is kept for one week in the solution of sulphuric acid and hydrochloric acid having *1% strength. The corners of stones which have high alkaline content turn roundish and loose particles will get deposited on its surface.* Such types of stones are *unsuitable for smoky atmosphere.*

The stones having high *percentage of lime content exhibit efflorescence when subjected to action of acids.*

**7. Electrical resistance/water absorption test:**

As the electrical resistance of a wet stone is less (and also the porous stone will have less strength), therefore the stone should be non-absorbent. Sandstones should not absorb more than 10% of water, 17% in case of limestones and 1% in granites of their volume of water when dipped for 3-4 hours.

**Water absorption test** is conducted as follows:

- A cube weighing about 0.5 N from a given sample of the stone is prepared. Its actual weight is recorded let it be $W_1$ *N*
- The cube is then immersed in distilled water for 24 hours.
- It is then taken out of water and its surface wiped with a damp piece of cloth. The cube is then weighed again; Let its weight be $W_2$ *N*.
- The cube is suspended freely in water and its weight is recorded; let it be $W_3$ *N*.
- The cube is then kept in boiling water for *five* hours.
- It is then removed and its surface water is again wiped off with a damp piece of cloth. It is then weighed; let its weight be $W_4$ *N*.

From the above data, values of the following properties of the stone are obtained:

$$\%\text{age absorption } \textit{by weight} \text{ after 24 hours} = \frac{W_2 - W_1}{W_1} \times 100 \quad ...(1)$$

$$\%\text{age absorption } \textit{by volume} \text{ after 24 hours} = \frac{W_2 - W_1}{W_2 - W_3} \times 100 \quad ...(2)$$

$$\% \text{ porosity by volume} = \frac{(W_4 - W_1)}{(W_2 - W_3)} \times 100 \quad ...(3)$$

$$\text{Saturation coefficient} = \frac{\text{water absorption}}{\text{total porosity}}$$

$$= \frac{W_2 - W_1}{W_4 - W_1} \quad ...(4)$$

**8. Smith's test:**

This test *indicates the presence of earthly matter.*

- The sample of the stone is broken into small pieces and put into a test tube containing clear water.
- The test tube is then shaken vigorously, the dirty colour will show presence of argillaceous matter.

**9. Crystallization test:**

This test determines the **durability or weathering quality** of a stone.

- A sample of stone is immersed in the solution of sodium sulphate at room temperature and dried in hot air.
- The process of wetting and drying is carried out for two hours; the difference in weight if any is recorded. *Little difference in weight indicates durability and good weathering quality of the stone.*

**10. Microscopic test:**

This is essentially a geologist's test. In this test, the sample of stone is subjected to microscopic examination to study the following properties:

(*i*) Mineral constituents
(*ii*) Texture of stone
(*iii*) Average grain size
(*iii*) Nature of cementing material
(*v*) Existence of fissures, pores, veins and shakes
(*vi*) Presence of any harmful substance etc.

The fractured surface of a durable stone should be bright, clean and sharp with grains well cemented together. Stone likely to decay shows a dull earthy appearance.

**11. Freezing and thawing test:**

This test is carried out as follows:

- The specimen of stone is kept immersed in water for 24 hours.
- It is then placed in a freezing mixture at –12 °C for 24 hours.
- It is then thawed or warmed to atmospheric temperature. This should be done in shade to prevent any effect due to wind, sun rays, rain, etc.
- The above procedure is repeated several times and behaviour of the stone carefully studied.

## 2.9 COMMON BUILDING STONES, THEIR COMPOSITION, PROPERTIES, USES AND OCCURRENCE

### 1. Granite

(*a*) **Composition:** It is *igneous rock.* Its main constituent minerals are *quartz*, (hard and least affected by carbon dioxide and water) *felspar* (crystalline, lustrous, readily attacked by atmospheric agencies) and *mica* (dark grey, black or brown in colour, soft, readily affected by chemicals and atmospheric agencies).

(*b*) **Characteristics:**

(*i*) Its specific gravity is 2.64 and absorption less than 1%.

(*ii*) It has mottled appearance.

(*iii*) Significantly strong and durable.

(*iv*) Finely grained granite takes a fine polish.

(*v*) Crushing strength varies from 110 to 140 MN/m$^2$.

(*vi*) It has least fire resistance as it cracks under a strong fire.

(*vii*) Its colour depends upon that of felspar.

(*c*) **Uses :**

(*i*) Fine grained granite is suitable for ornamental columns, plinth, etc., as it takes a polish.

(*ii*) Owing to the hardness, weight and durability of granite it is considered most suitable for the construction of sea walls, light houses, bridge piers, etc.

(*iii*) Large pieces are used as building blocks, the smaller as road metals or railway ballast and the chippings for the manufacture of concrete or artificial stone.

(*d*) **Occurrence:** Jhansi (U.P), Jabalpur (M.P.), Ajmer (Rajasthan), Dalhousie and Kangra (Himachal). The best specimens are available in Secunderabad (Andhra Pradesh).

**2. Basalt or trap**

(*a*) **Composition:** It is an igneous rock (silicious variety). The main constituent are silica, alumina and felspar.

(*b*) **Characteristics:**

(*i*) Very hard and tough.

(*ii*) Heavier than granite

(*iii*) Has greenish grey to dark grey colour.

(*iv*) Crushing strength varies from 70 to 80 MN/m$^2$.

(*v*) Specific gravity = 2.96.

(*c*) **Uses:**

(*i*) Suitable for paving sets and as a road metal.

(*ii*) Used for the manufacture of artificial stone.

(*iii*) Used as aggregate in concrete.

(*d*) **Occurrence**: Madhya Pradesh and Deccan traps (along the western Coast) and Rajmahal trap (Bihar).

**3. Slate**

(*a*) **Composition:** It is a metamorphic rock (Argillaceous variety). It is composed of *alumina mixed with sand* or *carbonate of lime*.

(*b*) **Characteristics:**

(*i*) Its specific gravity is 2.8.

(*ii*) A good slate is hard, tough and fine grained.

(*iii*) It has grey or dark blue colour.

(*iv*) It can be split into thin sheets.

(*v*) Its crushing strength varies from 60 to 70 MN/m$^2$.

(*vi*) It is non-absorbent.

(*vii*) When struck with a light hammer, it produces a sharp metallic ring.

(*viii*) its durability is good.

(*c*) **Uses:**

(*i*) Used for making electrical switch boards.

(*ii*) Because of its non-absorbent property, it is suitable for use in cisterns, urinals partitions, etc.

(*d*) **Occurrence**: Gurgaon (Haryana), Gurdaspur (Punjab), Cuddadh (Andhra Pradesh), Chanda (Maharashtra), Chindwara (M.P.), Bijapur (Mysore), Alwar (Rajasthan), Chamba, Shimla, Kangra, (Himachal Pradesh), Baroda (Gujarat).

**4. Gneiss**

(a) **Composition:** It is a *metamorphic rock*. It is also sometimes called stratified granite as it occurs in somewhat parallel layers. Its constituents are the same as granite, *i.e. quartz and felspar*.

(*b*) **Characteristics:**

(*i*) It can be readily split into slabs.

(*ii*) It is more easily worked than granite.

(*iii*) As it is available in different colours it is used for ornamental and decorative purposes.

(*iv*) It is reasonably good material for street paving.

(*d*) **Occurrence**: It is largely found in Madras state, Mysore and Bihar.

**5. Sandstone**

(*a*) **Composition:** It is a *sedimentary rock of silicious variety*. It contains *sand* or *quartz cemented by lime, mica, magnesium, aluminium, oxide of iron* or by *a mixture of these materials*. Sometimes fragments of limestone, mica, or felspar are also present.

(*b*) **Characteristics:**

(*i*) Its structure shows sandy grains.

(*ii*) Its usual colours are white, grey, brown, pink, etc.

(*iii*) Its specific gravity is 2.25.

(*iv*) Crushing strength varies from 35 to 40 $MN/m^2$.

(*v*) Fine grained stones are strong and durable.

(*vi*) The sandstone of thin-bedded variety is called flagstone. It splits easily into large slabs along the bedding.

(*vii*) The rock composed of angular sharp edged sand grains is called grit; it is adequately strong and heavy and is used in heavy engineering works.

(*viii*) The sand stone that can be cut easily with mallet and hammer into blocks for building is referred as *free stone*.

(*ix*) Its durability depends upon the nature of cementing material. The quantity of sandstone is poor if it is porous or contains lime.

(*c*) **Uses:**

(*i*) The fine grained and compact variety is suitable for ashlar work, mouldings, carvings, etc.

(*ii*) The rough and coarse grained stone is employed for rubble work.

(*d*) **Occurrence:** The best variety stones which can easily be quarried in large flawless blocks are available in Mirzapur, Chunar, Fatehpur Sikri, Gwalior (M.P), Hemanagar near Ahmedabad (Gujarat), Nagpur (Maharashtra), Kangra and Dharamsala (H.P.), Shahpur and Kasauli Rajpur, and Jabalpore, Dehri, Mandhum and Ranchi (Bihar), Cuttack and Mysore.

**6. Limestone**

(*a*) **Composition:** It is *sedimentary rock of calcarious variety*. When in pure state it contains *$CaCO_3$ but frequently is mixed with $MgCO_3$ and small amount of silica and alumina*. When both $MgCO_3$ and $CaCO_3$ are present nearly in equal proportion it is known as '*Dolomite*'.

(*b*) **Characteristics:**

(*i*) Its specific gravity is 2.6.

(*ii*) It is available in brown, yellow and dark grey colours.

(*iii*) Crushing strength = 52 MN/m$^2$.

(*c*) **Uses:**

(*i*) It is used in large quantities in blast furnaces, bleaching, tanning and other industries.

(*ii*) If compact and fine grained it may be used as stone masonry for walls and paving set in floor.

(*iii*) It is used as road metal (when better materials like trap, basalt or granite are not available.)

(*d*) **Occurrence:** It is found in Alwar (Rajasthan), Hoshiarpur (Punjab), Simla (H.P.), Mirzapur, Ranchi (Bihar) and Mysore, etc.

**7. Marble**

(*a*) **Composition:** It is *metamorphic rock of calcarious variety*. It is changed from limestone. It is crystalline hard compact stone having $CaCO_3$ as the main constituent.

(*b*) **Characteristics:**

(*i*) It is available in white, yellow, grey, green, red, blue and black colours.

(*ii*) It is very hard and takes a fine polish.

(*iii*) Its specific gravity is 2.72.

(*iv*) Crushing strength varies from 50 to 60 MN/m$^2$.

(*v*) It can be easily worked.

(*c*) **Uses:**

(*i*) It is used for carving and decoration work.

(*ii*) It is also used for steps, wall linings, electrical switch boards, table slabs and columns.

(*d*) **Occurrence:** White marbles of high purity are found in Jodhpur and Ajmer (Rajasthan) and Jabalpur (M.P.), Green marbles in Baroda (Gujarat), black marbles in Jaipur (Rajasthan) and yellow marbles in Jaisalmer (Rajasthan).

**8. Kankar**

(*a*) **Composition:** It is an impure limestone containing 30 % of clay and sand.

(*b*) **Characteristics:**

(*i*) It is available in grey or khaki colour.

(*ii*) It has a porous structure.

(*iii*) It occurs in solid layers or as irregular shaped modules a few metres below the surface.

(*iv*) Hard kankar is adequately durable.

(*c*) **Uses:**

(*i*) Nodular kankar when burnt yields excellent hydraulic lime; when clean, hard and tough is used as road metal.

(*ii*) Hard kankar is employed for foundations of buildings.

**9. Laterite**

(*a*) **Composition:** It is basically igneous but not completely metamorphic. It is sandy claystone containing high percentage of iron oxide.

(*b*) **Characteristies:**

(*i*) It has a porous and cellular structure.

(*ii*) It is readily quarried in blocks.

(*iii*) It is of reddish, brown, yellow, dark brown and black colours.

(*iv*) Its specific gravity varies from 2 to 2.2.

(*c*) **Uses:**

(*i*) Laterite blocks are suitable as building stones.

(*ii*) Nodular laterite proves a very good road metal.

(*d*) **Occrrence:** it is available in some districts in Bihar and Orissa, in Madhya Pradesh, in Kangra (Mysore) and in Burma.

**10. Moorum**

(*a*) **Composition:** It is a *decomposed laterite.*

(*b*) **Characteristics:** It has deep brown or red colour.

(*c*) **Uses:**

(*i*) Due to its rich red colour, it is used in surfacing fancy paths and garden walks.

(*ii*) It serves as a fine blindage for metalled rods.

(*d*) **Occurrence:** It is found in some parts of Madhya Pradesh.

**11. Gravel**

It is *mixture of rounded water worn pebbles of any kind of stone with sand.* It is usually available in river beds. The size of the pebble varies from minute grains to 8 cm in diameter. It is suitable for surfacing road and is also used in concrete.

**12. Chalk**

It is *pure, white limestone*. It is *soft and can be easily powdered.*

It is largely used for the *manufacture of Portland cement*. It is also used for marking and as a colouring matter. It is unsuitable for building purposes.

**13. Shingles**

These are large water-worn pebbles found in river beds.

Broken shingles are used in concrete and as road metal when no better stuff is available at reasonable cost.

**14. Quartzite**

It is composed of *silica and is dense, hard and glassy in structure.* Some of the varieties are considered as the hardest and strongest type of building stones.

It is strong and durable and *used as road metal* or *railway ballast* or *in concrete.*

## 2.10 NATURAL BED OF STONE

'*Natural bed of stone' is the original position occupied by the stratified rock from which the stone is obtained.* It is also known as the *plane of cleavage.*

- The strength and durability of the stratified stone depend on its position in a structure, *i.e.* how it is placed. It should always be placed in such a fashion that the *load* or *thrust acts perpendicular to 'natural bed of stone'* or plane of cleavage. If the stone is placed with its natural bed parallel to the direction of load, it will get destroyed by the effects of frost and rain (which scale off the face layer by layer).
- In case of a column or a wall where the load line is vertical, the stones should be placed with the planes of bedding radial so that thrust acts normal to the bedding plane while in cornices with undercut mouldings, the natural bed should be placed *vertically and perpendicular* to the face otherwise the layers of overhanging portions being horizontal will drop of.

## 2.11 QUARRYING OF BUILDING STONES

**Quarrying** *may be defined as an art of extracting from the rock beds, the stones of different varieties as used for general building work and broken stones for roads and concrete work, etc.* The place from which the stone is obtained (by digging or blasting) is known as '*Quarry*'.

*Quarrying* differs from *mining* in which various operations are carried out for exploring minerals, such as *coal, quartzite*, etc., from a *mine* under the ground.

### 2.11.1 QUARRY LOCATION

A good location of a quarry should fulfil the following requirements:

- A large quantity of good quality stones must be available about the earth's surface.
- It should be located nearly roads and railway lines.
- Ample space for installation of crushers, storage of stones and other materials should be available.
- In case the quarrying is to be done by 'blasting' the site of quarry should be away from any permanent structure.
- There should be proper provision for drainage of rain water.

### 2.11.2 Quarrying–Terminology

Important *terms* used in *quarrying* are as follows:

1. **Blasting.** It is quarrying stones (of compact nature) with explosives. The operations in blasting are *boring, charging, tamping* and *firing.*

2. **Blasting or safety fuse.** It is a device consisting of a length of slow and even burning composition, which provides adequate time to the person firing a shot or blast to reach a safe place prior to the occurrence of the explosion, Brickford's fuse, known as the best safety fuse, comprises an inner core of a fine thread of gun powder entrapped in a cotton rope and burns at the rate of about 60 cm per minute.

3. **Detonator.** It is a device whose explosion initiates that of another. The detonators are fired either by fuse or electric spark. It is used when usually dynamite is used as explosive.

4. **Explosive**. An explosive is a mechanical mixture or chemical compound of combustible and oxidising agents which release the energy when fired, almost instantaneously. The blasting powder and dynamite are commonly used as the explosives.

5. **Tamping.** It is an operation of filling the hole containing explosive charge with stiff sandy clay (or other non-inflammable material) with the purpose of enhancing the explosion force.

6. **Grain.** The degree of fineness or coarseness of texture of a stone or rock is termed as grain.

7. **Screening.** It is the process of passing the crushed rock material through one or more screens to separate it into a series of products of varying sizes.

8. **Quarry sap.** It is the moisture (natural) found in the freshly quarried stone.

9. **Line of least resistance.** All the rocks contain lines of bedding fissures or crevices along which these can be split easily. These lines are known as "*lines of least resistance*". Blast holes are normally originated at a little distance from the working face of the rock to the depth of line of least resistance or a few centimetres more. Figure 2.3 shows a line of least resistance.

### 2.11.3 Methods of Quarrying

The different methods of quarrying are:

| | |
|---|---|
| 1. By digging | 2. By heating |
| 3. By wedging | 4. By blasting. |

#### 1. Digging method

Digging or excavation of stones is carried out with the help of tools such as *crowbars* and *pick axes.* Only those stones which occur in the form of *detached nodules* buried in earth can be recovered by this method.

#### 2. Heating method

The use of this method is restored to only in case of those stones which are required in *small pieces* to be employed for *road metal, railway ballast and aggregates*. In this method fuel is collected on the exposed surface of the portion of the rock to be removed and fire is burnt for several hours continuously (ordinary bundle wood is employed for burning). Due to unequal expansion of the rock, it gets separated from the cold mass with dull noise. The detached portion is then removed with the quarrying tools [Fig. 2.1(*a, b, c*)] and then broken into small pieces as per requirements.

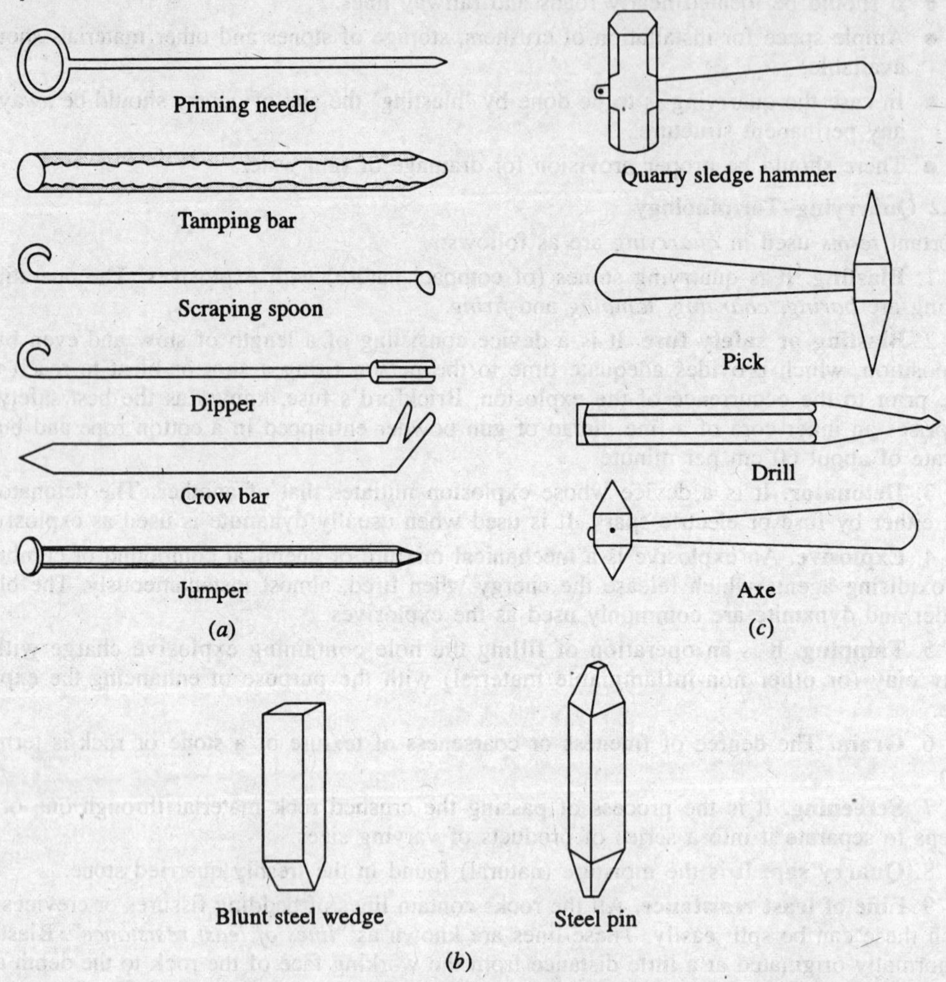


Fig. 2.1 Quarrying tools.

### 3. Wedging

This method of quarrying is employed for the rocks which are in the *form of layers* along which it can be *easily splitted* ( e.g. sedimentary and soft rocks like limestone, marble, slate and laterite, etc.), and the *stone is required in blocks for building purposes*.

Soft stratified rock can be removed with the help of pick axes and crow bars but in case of hard rocks the holes are made and grooves are cut at shorter intervals. Blunt wedges are then inserted in the grooves or conical pins of steel [Fig. 2.1(*b*)] are driven into the holes with a hammer. In place of conical pins sometimes plug and feathers (Fig. 2.2) are used. The plug is a conical wedge and *feathers* are flat wedges with upper ends slightly bent. The plug along with the feathers is applied into the hole and is subjected to hammer blows. If the plugs and feathers are arranged a few centimetres apart and all driven at the same time, the stone will get cracked.

In case of a harder stone the holes are originated by a pneumatic drill.

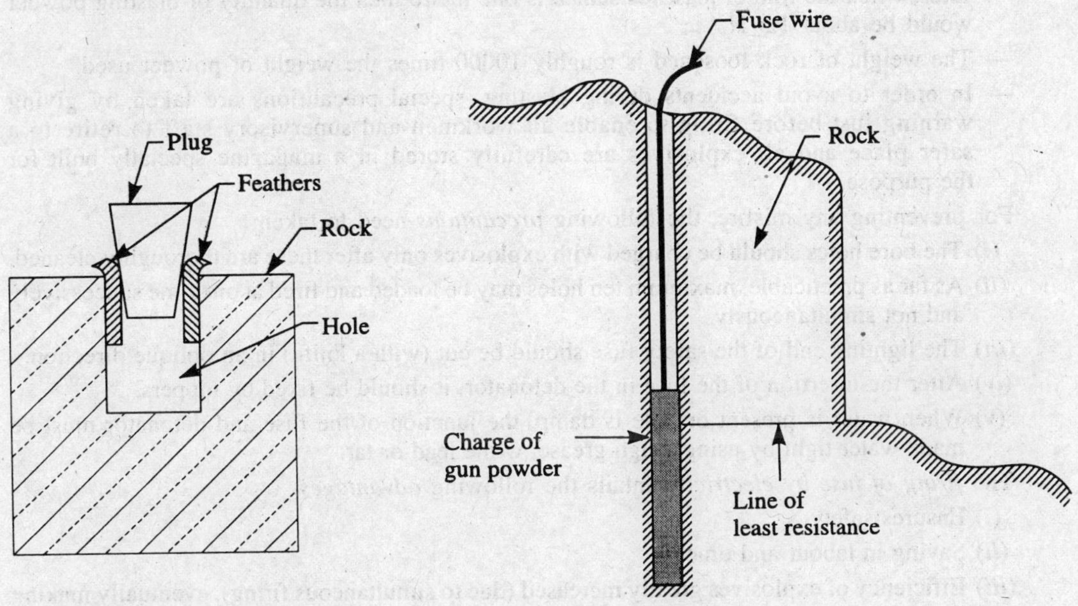


Fig. 2.2 Plug and feathers.

Fig. 2.3 Blast-hole packed with powder.

## 4. Blasting

This method is employed for *quarrying the hard and compact stones.* The various *stages* involved in the method of quarrying by blasting are as follows:

(*i*) *Boring hole in the rock.* The holes are usually made (of desired depth, from 1.25 to 2.5 m deep, and 20 to 40 mm diameter) with a steel bar with knife-edged ends called *jumper* [Fig. 2.1(*a*)]. When large quantity of stones is required, holes may be drilled by a *drilling machine.* During the drilling operation water is used to facilitate the operation. The mud and rock powder produced as a result of drilling are removed by a scraper, or a spoon or by a compressed air blast.

(*ii*) *Charging with explosive.* The drilling of hole is followed by charging it with an explosive. It should be ensured that hole is thoroughly dry before being filled with explosives.

(*iii*) *Tamping.* Tamping is of paramount importance to prevent the reaction of the explosive along the blasting hole itself. While charging the hole with explosive and prior to tamping a fuse of sufficient length is inserted. Tamping consists in filling the hole with stiff sandy clay by a brass rod called the tamping bar (made of brass 15 mm diameter). Occasionally a priming needle of 1.6 mm diameter is inserted in the tamping materials which is removed after tamping. This is followed by the insertion of fuse to help explosive being fired. Oftenly a little quantity of gun powder is put in the hole to link the explosive to the bottom of the fuse. Figure 2.3 shows a blast-hole packed with powder.

(*iv*) *Firing.* The fuse is kept of a sufficient length as to enable the person firing it enough time to retire to a safe place before the explosion of the charge occurs. The use of electrically firing devices is also made to create spark needed for the explosion. When the explosion occurs masses of stones around the hole are removed. A good blast produces a dull sound and the mass of the rock should be *just displaced* without being blown into fragments.

The *quantity of blasting powder* (or dynamite) necessary for each blast *depends mainly on* the *nature of rock*, the *mass of rock to be removed* and the *position of the hole*. As a rough guide, the quantity of blasting powder can be obtained by using the following *thumb-rule*:

*Blasting powder in N* (Newton) = (*Line of least resistance in metre*)$^2$ × *1.50*

— Thus, when the line of least resistance is one metre then the quantity of blasting powder would be about 1.5 N.

— The weight of rock loosened is roughly 10000 times the weight of powder used.

— In order to avoid accidents during blasting, special precautions are taken by giving warning just before firing to enable all workmen and supervisory staff to retire to a safer place and all explosives are carefully stored in a magazine specially built for the purpose.

For preventing any misfire, the following *precautions* need to taken:

(*i*) The bore holes should be charged with explosives only after these are thoroughly cleaned.

(*ii*) As far as practicable, maximum ten holes may be loaded and fired at one time successively and not simultaneously.

(*iii*) The lighting end of the safety fuse should be cut (with a knife) in an oblique direction.

(*iv*) After the insertion of the fuse in the detonator, it should be fixed by nippers.

(*v*) When water is present or hole is damp, the junction of the fuse and detonator must be made water tight by using tough grease, white lead or tar.

The *firing of fuse by electricity* entails the following *advantages*:

(*i*) Ensures safety

(*ii*) Saving in labour and time.

(*iii*) Efficiency of explosives greatly increased (due to simultaneous firing), eventually making operation economical.

(*iv*) Useful for firing fuse under water or in wet places.

(*v*) No danger of misfire.

(*vi*) Proper signalling can be arranged to avoid the occurrence of accidents.

### 2.11.4 Explosives—Composition, Characteristics and Uses

Important *explosives* are:

#### 1. Blasting powder

(*a*) **Composition:** It is a mechanical mixture of 65% potassium nitrate, 20% sulphur and 15% charcoal.

(*b*) **Characteristics:**

(*i*) It is slow in action comparatively.

(*ii*) It possesses high lifting power.

(*iii*) Its shattering power is poor.

(*iv*) It can be ignited easily.

(*v*) It is much cheaper than dynamite.

**Note:** It should be kept dry.

(*c*) **Uses:** It is useful for quarrying large blocks.

#### 2. Dynamite

(*a*) **Compoisition:** It contains 75% nitroglycerine mixed with 25% sandy earth to form a thick paste.

(*b*) **Characteristics:**

(*i*) It is quick acting and *more powerful than blasting powder*. It is *six times as powerful as gun powder.*

(*ii*) Its shattering effect is high.

(*iii*) It can be *used under water and in damp situation.*

(*iv*) It does not necessitate hard tamping.

(*v*) It is fired by means of a detonator which contains 6 to 9 grains of fulminate of mercury in a copper cylinder. The detonator is fired either by electricity or by a slow burning fuse.

(*vi*) It needs protection from direct sun rays.

**Uses:**

(*i*) It is employed for mining and tunneling operations.

(*ii*) It is suitable for use in small bore holes and narrow irregular spaces where a shattering effect is required.

(*iii*) It is used under water and in damp situation.

**3. Gun cotton**

(*a*) **Composition:** It is made by saturating cotton with nitric acid.

(*b*) **Characteristics:**

(*i*) It is the most powerful explosive.

(*ii*) It is highly in flammable but addition of water renders it non-inflammable

(*iii*) Dry gun powder is highly sensitive to sun light.

(*iv*) It can be exploded in a wet condition.

(*v*) It should be stored in a cool place since the rate of decomposition increases with the temperature.

(*vi*) It is soft to handle.

(*c*) **Uses:** It is employed where cutting or shattering effect is required.

**4. Blasting gelatine**

It consists of 93% nitroglycerine and 7% gun cotton. It is prepared by gently heating nitroglycerine in a water bath, then dissolving gun cotton in it.

It is powerful explosive.

It is *used in detonators for initiating high explosion.*

**5. Cordite**

(*a*) **Composition:** It is combination of nitroglycerine and nitrocellulose.

(*b*) **Characteristics:**

(*i*) It is very powerful but smokeless explosive.

(*ii*) It is slow burning explosive.

(*iii*) It can be effectively used under water.

(*c*) **Uses:** It is reasonably good explosive for quarrying.

**6. Rock-a-rock**

(*a*) **Composition:** It contains 79% potassium-chlorate and 21% nitro-benzol.

(*b*) **Characteristics:**

(*i*) It does not explode easily.

(*ii*) Its action under water is more effective.

(*c*) **Uses:** It is employed for blasting under wet condition. It is used in U.S.A.

**7. Gelignite**

(*a*) **Composition.** It contains 65% blasting gelaine and 35% absorbing powder.

(*b*) **Characteristics:**

(*i*) It is very powerful.

(*ii*) It can be used under water.

(*iii*) It is convenient to work with.

(*c*) **Uses:** It is commonly employed for blasting under wet conditions.

### 2.11.5 Storing and Handling of Explosives–Precautions

For storing and handling of explosives, the following *precautions* should be taken:

(*i*) These must be stored in *magazine* (specially built building) and in separate boxes.

(*ii*) Entry of outsiders in the magazine campus should be banned; the authorised persons must wear magazine shoes (or shoes with no nails).

(*iii*) The magazines must be provided with an efficient lightening conductor and be leak proof.

(*iv*) The magazines should be located away from residential and industrial area, public places, petrol and diesel storages, etc.

(*v*) The smoking/firing should be strictly prohibited within a radius of 50 metres from the magazine.

(*vi*) The electric wiring of magazines should be concealed in the walls and oftenly checked.

(*vii*) The magazine under no circumstances should be constructed within 500 m from any working kiln/furnace.

(*viii*) The explosives packages must be handled carefully.

(*ix*) The magazine should be protected by a high barbed wire fencing on its boundary with caution boards and should be guarded round the clock by a watchman.

(*x*) The detonator should be kept entirely away from other explosives.

(*xi*) In case of electric power failure, only torches should be used for lighting.

## 2.12 ARTIFICIAL STONES

These are also known as *cast stones* or *reconstructed stones*. Since it is difficult to obtain durable natural stones at a moderate cost in many localities, many processes have been invented for the manufacture of artificial stone.

Artificial stone cannot be used extensively in ordinary buildings due to its *heavy cost*, but it *can be moulded into most intricate forms easily* and so *it is most economical* when *it is substituted for the carvings and ornamentations in natural stone*. For practical use it is formed into mortar or is used as concrete.

### 2.12.1 Manufacture of Artificial Stone

The procedure of making/manufacturing artificial stone is as follows :

- The natural stone is crushed into size less than 6 mm and the stone dust is removed.
- A mixture of $1\frac{1}{2}$ parts of stones of size 3 mm to 6 mm, $1\frac{1}{2}$ parts of stones of size less than 3 mm and 1 part of cement by volume is prepared.
- To impart colour to the stone, necessary colouring pigments are added to the dry mix. Its proportion should not exceed 15% of cement by weight.
- Water is then added to the dry mix to obtain a mixture of workable consistency.
- The wet plastic mixture is then pressed into moulds (may be of steel or wood, made with utmost care) cured with water and then dried/seasoned in air for suitable length of time.
- The polishing is done, if required.
- In order to produce colour of light shade, white cement may be used in place of ordinary cement.
- It is usual in the manufacture of caststone that the facing or the skin of caststone is 25 to 38 mm in thickness and consists of the above mentioned mix, while the remaining thickness of the stone slab consists of cheaper material (such as lean mix of gravel and cement or lean cement concrete).

### 2.12.2 Varieties of Artificial Stones

The different varieties of artificial stones are:

**1. Cement concrete blocks:**

- These are made from a mixture of cement, fine aggregate, course aggregate and water. They may be *cast-in-situ* as in the construction of piers or *cast-in-moulds* for steps, window sills, etc.
- Artificial paving slabs and stones composed of cement concrete and sometimes treated with sodium silicate solution also come under this class.

**2. Ransome's patent stone:**

- It is made by *mixing dry sand with silica of soda and a small portion of powdered stone or chalk.*
- The mixture while is plastic state, is poured into moulds and the castings on withdrawal are immersed in a solution of chloride of calcium, which cements the particles of sand.
- It can be dressed and carved like natural stone.
- The stone has a compressive strength of 3.2 kN/m$^2$ and weighs 19.2 kN/m$^3$
- Its cost is greater than that of national stone.
- It is used in Nawab's Palace at Murshidabad, and Bombay Post Office.

**3. Artificial marble:**

- It is made by fusing and moulding constituent mixture, which is composed of 80 parts of plaster in powder; 20 parts of pulverised marble, 20 parts of sulphate of potash with a 5 per cent solution of glue mingled together with water.
- It is used for external work. It is mostly used in France.

**4. Terrazo:**

- It is made by *mixing marble chips with white cement and some pigment.*
- It is either pre-cast or laid *in situ.*
- It is used for bathrooms, residential buildings, temples, etc.

**5. Mosaic tiles:**

The pre-cast concrete tiles with marble chips at top surface are known as *mosaic tiles.* They are available in different shades and are widely used.

**6. Reconstructed stone:**

- This stone is made from the debris of limestone quarries by crushing them into grit, mixing it with lime made from dolomite, heating in a closed retort upto 980 °C to drive off $CO_2$, slaking the powdery residue of CaO and MgO, mixing with water, and consolidating under great pressure into blocks.
- It is then dried and $CO_2$ is admitted until the carbonization of hydrate of lime blocks is complete.

**7. Bituminous stone:**

- Diorite and other granite stones are often *impregnated with prepared* or *refined tar* to *form bituminous stones.*
- Such stones are used for durability, noise, wear and dust resistant stone surfaces.

### 2.12.3 Advantages of Artificial Stones

Artificial stones claim the following *advantages over the natural stones:*

1. More durable than the natural stone.
2. Can be easily *cast and seasoned at the site of work* with great promptness and hence avoids the expenses of dressing and transportation.
3. Can be *easily cast into any desired shape* and can be easily and economically moulded to the required ornamental shape.

4. Artificial stone can be made in a single piece and hence the trouble of getting large blocks of stones for lintels, beams, etc., can be avoided.
5. It is comparatively easy to carve artificial stone; it can be worked before becoming hard.
6. The artificial stones which are carefully cast, are free from defects that are likely to be present in natural stones.
7. In the artificial stones cavities may be kept (to convey pipes, electric wires, etc.)
8. Their strength can be regulated by suitably proportionating the ingredients and by using steel reinforcements.
9. Equally good in resisting deterioration and disintegration caused by various atmospheric agencies, (*e.g.* rain, frost, etc.).
10. There is no need to take precaution with regard to the natural bed of stones, since natural bed is absent in artificial stones.

## 2.13 DRESSING OF STONES

- The *art of cutting the stones to shape required for use in structures* is known as **dressing.**
- The dressing of stone is done at the *quarry itself.* The *quarry dressing* entails the following **advantages** (in comparison to *site dressing*):

(*i*) Freshly quarried stones contain some moisture called *quarry sap* and hence, are comparatively soft and can be easily dressed.

(*ii*) Cheap labour is available at the quarry site.

(*iii*) The natural bed surface of the stone can be made promiment during quarrying dressing.

(*iv*) It is possible to sort out stones for different works.

(*v*) Irregular and rough portions of the stones can be removed there and then, thus reducing the weight of the stones as well as facilitating easy transportation of the stones.

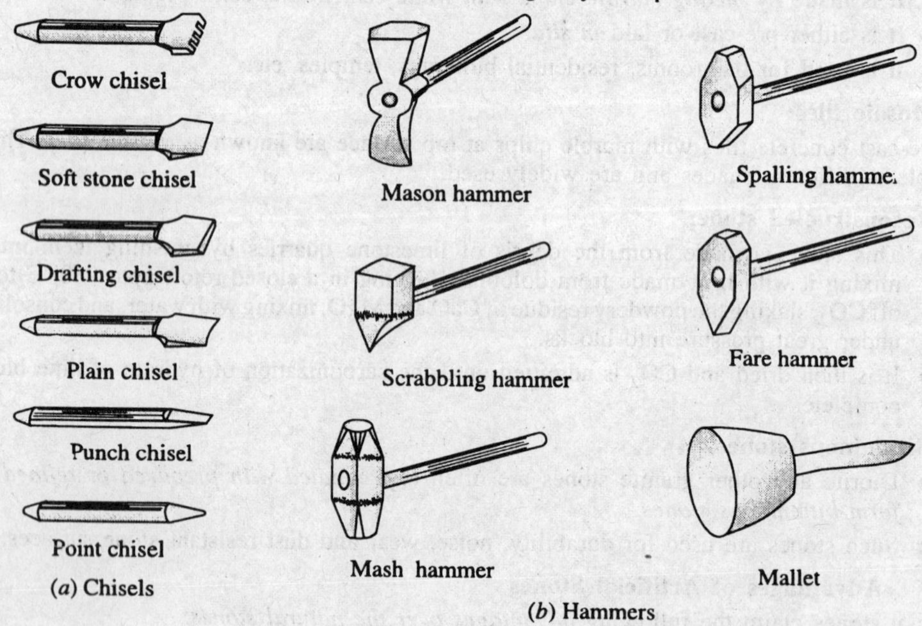


Fig. 2.4 Tools for cutting and dressing stones.

Figure 2.4 shows the commonly used tools and implements for the cutting and dressing of stone blocks.

Stones for different types of stone masonry are dressed as follows:

**1. Rubble masonry:**

- Stones, for such construction are roughly shaped by means of hammer. Lower and upper surfaces of stones are almost dressed flat.
- In such works, stones which have front face almost rectangular and remaining irregular are also used.

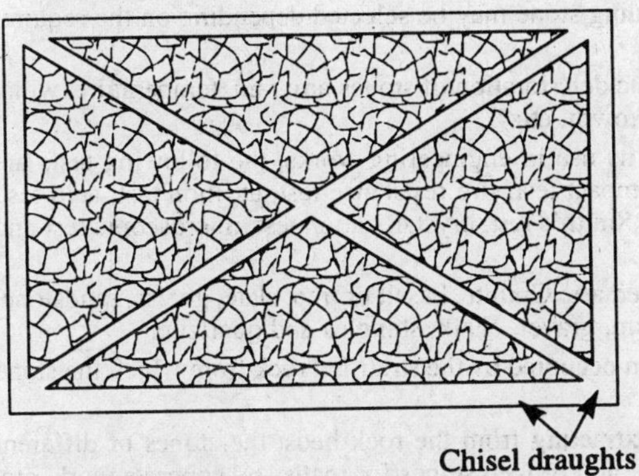


Fig. 2.5 Stone block with diagonals.

Chisel draught

Fig. 2.6 Rock faced stone.

**2. Ashlar masonry:**

- In this type of work the faces of stones are finely dressed by means of chisels.
- Chiseled faces are sometimes rubbed to give a smooth surface in case of fine and monumental work.

Figure 2.5 and 2.6 show the stone dressed with chisel.

The different types of *surface finishes* are:

1. Rough tooled surface
2. Tooled surface
3. Cut stone surface
4. Hammer dressed surface
5. Published surface
6. Rubbed surface.

Different types of surface finishes are shown in Fig. 2.7.

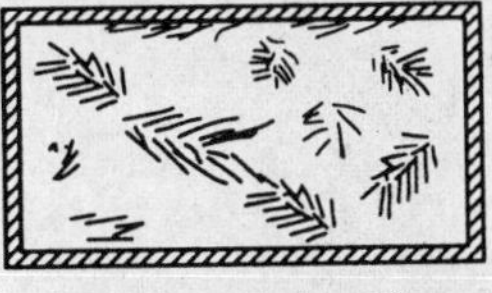

Rough tooled surface

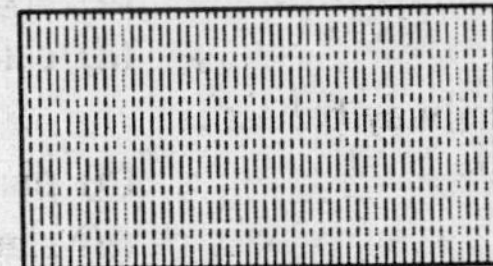

Hand tooled surface

Tooled surface

Cut stone surface

Machine tooled surface

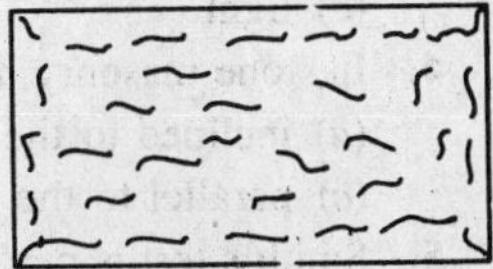

Hammer dressed surface

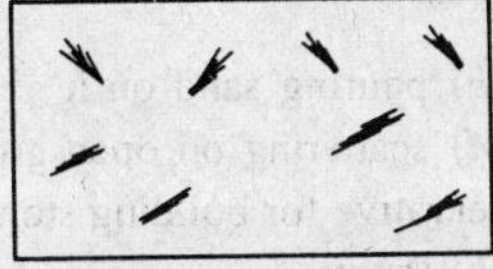

Polished surface

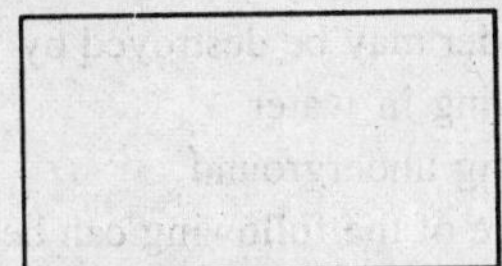

Rubbed surface

Fig. 2.7 Types of surface finishes.

## HIGHLIGHTS

1. The stones are derived from rocks which are classified according to geological, physical, chemical and practical classifications.
2. The common rock forming minerals are: silica, felspar, mica, hornblende, calcite and dolomite.
3. Use of a particular building stone will depend upon its various characteristics such as structure, durability, strength, etc. A particular building stone may be selected depending on the requirements of the structure to be built.
4. The various agencies which bring about the deterioration of stones are: rain, temperature, wind, frost, atmospheric impurities, vegetable growth, etc.
5. To determine the suitability of stone for its use in engineering works, the following tests are performed : hardness test, crushing test, impact test, fire resistance test, attrition test, acid test, electrical resistance/water absorption test; Smith's test, crystallization test, microscopic test and freezing and thawing test.
6. Some of the building stones commonly used are: Granite, basalt or trap, slate, gneiss, sandstone, limestone, marble, kankar, laterite, moorum, gravel, chalk, shingles and quartzite.
7. *Natural bed of stone* is the original position occupied by the stratified rock from which the stone is obtained.
8. *Quarrying* may be defined as an art of extracting from the rock beds, the stones of different varieties as used for general building work and broken stones for roads and concrete work, etc.
   The different methods of quarrying are:
   1. By digging 2. By heating 3. By wedging 4. By blasting
9. Important explosives are: Blasting powder, dynamite, gun cotton, blasting gelatine, cordile, rock-a-rock and gelignite.
10. Artificial stone cannot be used extensively in ordinary buildings due to its heavy cost, but it can be moulded into most intricate form easily and so it is most economical when it is substituted for the carvings and ornamentations in natural stone. For practical use it is formed into mortar or is used as concrete.
11. The art of cutting the stones to shape required for use is structures is known as *dressing*.

## OBJECTIVE TYPE QUESTIONS

**Choose the Correct Answers:**

1. Kaolin is chemically classified as
   (*a*) silicious rock (*b*) argillaceous rock
   (*c*) metamorphic rock (*d*) calcarious rock.
2. Stone generally preferred for railway ballast is
   (*a*) basalt or trap granite (*b*) marble
   (*c*) sand stone (*d*) slate.
3. For carving ornamental and architectural beauty, the stone should be
   (*a*) hard (*b*) soft
   (*c*) light (*d*) heavy.
4. In stone masonry, the direction of the pressure line is
   (*a*) inclined to the natural bed at 30° (*b*) inclined to the natural bed at 45°
   (*c*) parallel to the natural bed (*d*) perpendicular to the natural bed.
5. Smith's test is performed on stones for
   (*a*) durability (*b*) water absorption
   (*c*) soluble and clayey matter (*d*) hardness.
6. Gun powder may be destroyed by
   (*a*) throwing in water (*b*) pouring sand on it
   (*c*) burying underground (*d*) scattering on open ground.
7. Which one of the following can be used as a preservative for building stones?
   (*a*) Coal tar (*b*) Paints
   (*c*) Raw and boiled linsed oil (*d*) Alum-soap solution
   (*e*) Any of the above.

8. ....exhibits highest compressive strength.
(*a*) Granite (*b*) Gneiss
(*c*) Limestone (*d*) Laterite.

9. Stoneware products are usually
(*a*) hard (*b*) impervious to moisture
(*c*) compact (*d*) all of the above.

10. Laterite is a ...................... rock.
(*a*) calcarious (*b*) argillaceous
(*c*) volcanic (*d*) silicious.

11. Hard silicious rocks which could not be scratched by knife represent a hardness of
(*a*) 2 (*b*) 4
(*c*) 6 (*d*) 7.

12. Sandstones are generally weak in
(*a*) hardness (*b*) abrasion
(*c*) compression (*d*) all of the above.

13. Gun powder contains
(*a*) salt petre, sulphur and charcoal
(*b*) fine sand, potassium nitrate and charcoal
(*c*) fine glass, potassium chlorate and charcoal
(*d*) iron filings, sulphur and charcoal.

14. Which explosive is *not* preferred for exploding rocks under water?
(*a*) Dynamite (*b*) Gun cotton
(*c*) Cordite (*d*) Gun powder.

15. During storage an explosive should be protected from
(*a*) moisture (*b*) extreme cold
(*c*) extreme heat (*d*) all of the above.

16. Main component of mineral of granite is
(*a*) felspar (*b*) mica
(*c*) quartz (*d*) free silica.

## ANSWERS

| | | | | |
|---|---|---|---|---|
| **1.** (*b*) | **2.** (*a*) | **3.** (*a*) | **4.** (*d*) | **5.** (*c*) |
| **6.** (*a*) | **7.** (*e*) | **8.** (*b*) | **9.** (*d*) | **10.** (*b*) |
| **11.** (*d*) | **12.** (*b*) | **13.** (*a*) | **14.** (*d*) | **15.** (*d*) |
| **16.** (*a*). | | | | |

**B. Fill in the Blanks or Say Yes or No:**

1. The ...................... is a substance which is formed by the natural inorganic process and possesses a definite chemical composition and molecular structure.
2. The ...................... rocks are formed by cooling of the molten lava on or inside the earth's surface during the volcanic eruption.
3. The ....... rocks are formed by gradual deposition of disintegrated rocks, vegetable matter and clay at bottom of rivers, lakes or sea.
4. The ...................... rocks are those which exhibit distinct layers which can be separated.
5. The ...................... rocks do not show any sign of strata and cannot be easily split into slabs.
6. In case of silicious rocks silica is the main constituents.
7. Calcarious rocks have clay or alumina as the main constituent.
8. Quartz is pure or nearly pure silica and is a hard and glassy mineral.
9. Calcile is the leading constituent of the limestones and marble.
10. Mica is magnesium carbonate.
11. The art of shaping a stone is known as......................

12. ...................... is determined with the aid of the moh's scale.
13. ...................... of a stone can be determined by subjecting it to a hammer blow.
14. ...................... test is also known as abrasion test.
15. ...................... test indicates the presence of earthly matter.
16. Crystallization test determines the durability or weathering quality of a stone.
17. Granite is a metamorphic rock.
18. Basalt is an igneous rock.
19. Slate is an igneous rock.
20. The specific gravity of slate is 1.5.
21. Gneiss is an igneous rock.
22. Limestone is a sedimentary rock.
23. Laterite is basically igneous but not completely metamorphic.
24. Moorum is decomposed laterite.
25. Natural bed of stone is also known as the plane of cleavage.
26. ...................... may be defined as an art of extracting from the rock beds, the stones of different varieties as used for general building work and broken stones for roads and concrete work, etc.
27. The operation of breaking rocks by boring holes, filling with explosive charge and firing is called ......................
28. ...................... fuse is known as the best safety fuse.
29. A device containing a small amount of very sensitive and highly explosive blasting material used for initiating high explosions is called a ......................
30. An ...................... in a mechanical mixture or chemical compound of combustible and oxidising agents which release the energy when fired, almost instantaneously.
31. The filling of the hole containing explosive charge with clay or other non-inflammable material to enhance the force of expresion is known as ......................
32. The natural moisture formed in freshly quarried stone is known as quarry sap.
33. The line along which the explosion of the charge will meet with the greatest opposition to its passage into the air is called the line of least resistance.
34. Gun cotton is made by saturating cotton with nitric acid.
35. Dynamite does not necessitate hard tamping
36. Rock-a-rock contains 79% nitro-benzol and 21% potassium chlorate.
37. Gelignite is very powerful and can be used under water.
38. Artificial stones are also known as the cast stones.
39. ...................... is a mixture of marble chips and cement.
40. The pre-cast concrete tiles with marble clips at top surface are known as the ......................tiles.

## ANSWERS

| | | | | |
|---|---|---|---|---|
| 1. mineral | 2. igneous | 3. sedimentary | 4. stratified | 5. unstratified |
| 6. Yes | 7. No | 8. Yes | 9. Yes | 10. No |
| 11. dressing | 12. Hardness | 13. Toughness | 14. Attrition | 15. Smith's |
| 16. Yes | 17. No | 18. Yes | 19. No | 20. No 21.No |
| 22. Yes | 23. Yes | 24. Yes | 25. Yes | 26. Quarrying |
| 27. blasting | 28. Brickford's | 29. detonator | 30. explosive | 31. tamping |
| 32. Yes | 33. No | 34. Yes | 35. Yes | 36. No |
| 37. Yes | 38. Yes | 39. Terrazo | 40. mosaic. | |

## THEORETICAL QUESTIONS

1. What are the main classification of rocks? Explain briefly.
2. (*a*) Enumerate the characteristics of a good building stone.
   (*b*) What is natural bed of stone? What is the correct position to lay it in building work? Explain with reasons.

3. (*a*) Classify giving briefly the characteristics and uses of the following stones:
   (*i*) Marble (*ii*) Sandstone (*iii*) Slate (*iv*) Kankar.
   (*b*) Name the explosives generally used for quarrying stones.
4. Classify the different types of building stones and point out their suitability in building and paving works.
5. Enumerate the characteristics of a good building stone. What are different methods used in dressing it?
6. (*a*) What are the characteristics of good building stones?
   (*b*) Name the modern explosives for blasting stones on a large scale.
7. Give a short account of the classification of stones used for engineering works.
8. What are the factors affecting the durability of stones?
9. Name the principal varieties of building stones found in India and discuss their merits and demerits as materials of construction. Give the localities in which they are found.
10. Discuss the various tests you would carry out to determine the suitability of a building stone.
11. Discuss the geological classification of rocks.
12. What is natural bed of stone and what is its importance?
13. Explain how the following tests for stones are carried out:
    (*i*) Water absorption test (*ii*) Hardness test
    (*iii*) Attrition test (*iv*) Impact test.
14. Distinguish between the following:
    (*i*) Stratified rocks and foliated rocks (*ii*) Limestone and kankar
    (*iii*) Igneous rocks and metamorphic rocks.
15. What are the uses of various stones?
16. Define a quarry and mention the factors to be considered while making a selection for its site.
17. Enumerate the materials required in the process of blasting.
18. Describe in brief methods used for quarrying stones for building work.
19. How do natural stones decay?
20. Comment on the following as building stones and their specific use in building construction:
    (*i*) Granite; (*ii*) Sandstone; (*iii*) Marble.
21. State, with reasons, the kind of stone you would recommend for use in construction of:
    (*i*) External wall of a building situated on the seashore;
    (*ii*) Platform of a railway goods shed;
    (*iii*) Ornamental corner of a building.
22. What are the different tests applied to test the suitability of stones for structural use and what are the factors affecting the durability of stones?
23. What are the materials required in blasting? Describe the process of blasting rocks.
24. Give specifications of stone-dressing for different classes of stone masorry:
    (*i*) Ashlar, fine; (*ii*) Coursed rubble; (*iii*) Uncoursed rubble.
25. What are the agents that destroy stones?
26. Enumerate half a dozen important explosives used in blasting stones.
27. Describe briefly with neat sketches the process of blasting rock, mentioning the tools used in the operation and state how you would regulate the powder charge. What is the object of tamping?
28. Write a short note on "Preservation of stones".

# 3

# Bricks and Other Clay Products

*3.1 Introduction. 3.2 Comparison of stone and brick. 3.3 Constituents of brick earth. 3.4 Requirements of a good brick earth. 3.5 Field testing of brick earth. 3.6 Manufacture of bricks. 3.7 Qualities of good bricks. 3.8 Classification of bricks. 3.9 Tests for bricks. 3.10 Special bricks. 3.11 Building tiles. 3.12 Earthenware and stoneware. 3.13 Glazing. Highlights—Objective Type Questions—Theoretical Questions.*

## 3.1 INTRODUCTION

- In this chapter construction materials such as *bricks, tiles, refractory bricks, earthenwares* and *stonewares* will be discussed. All these materials are made from clay and are also known as **clay products.**
- The brick is the most commonly used building material, though tiles, refractory bricks, earthenwares and stonewares serve different construction purposes. The bricks are being commonly used for constructing *walls, columns, roofs, paving floors* and as *coarse aggregate for concretework in foundations, under-floors, etc.*

The bricks do not require dressing and the art of laying bricks is so simple that the brickwork can be carried out with the help of unskilled labourers. Thus, at places where stones are not easily available, but if there is plenty of caly suitable for the manufacture of bricks, the bricks replace stones.

## 3.2 COMPARISON OF STONE AND BRICK

Comparison of stone and brick is given in the Table 3.1:

**Tabole 3.1 Comparison of Stone and Brick**

| S. No. | Stone | Brick |
|---|---|---|
| 1. | It is heavier than brick. | It is lighter than stone. |
| 2. | It is obtained from rocks. | Made from clay. |
| 3. | Free form clay holes and flaws. | Free from lumps, flaws and cracks. |
| 4. | Hard and tough | Hard and sound. |
| 5. | It absorbs heat more than a brick. | It absorbs less heat comparatively. |
| 6. | Water absorption less than 5%. | Water absorption less than 16%. |

| | | |
|---|---|---|
| 7. | It is uniform in colour and can be shaped to the desired size. | Uniform in colour, shape and size. |
| 8. | It has high durability which depends upon its chemical composition and its physical structure. | Its durability is less than that of stone. |
| 9. | Particularly suitable for construction work in industrial area as it is acid and smoke proof. | Acid and smoke resistance is good but less than that of stone. |
| 10. | Quarrying dressing and transportation costly. | Overall cost of manufacture less than stone. |
| 11. | Labour cost for laying is high. | Easily laid to form walls of required uniform thickness, labour cost for brick masonry much less. |

## 3.3 CONSTITUENTS OF BRICK EARTH

- Brick earth is derived by the *disintegration of igneous rocks.* Potash felspars, orthoclase or microcline are mainly responsible for yielding clay minerals in the earth. This mineral decomposes to yield kaolinite a silicate of alumina which on hydration gives a clay deposit $Al_2O_3.2H_2O$ known as *kaolin*.
- A good brick earth should be such that it can be *easily moulded and dried* without cracking and wraping.

### 3.3.1 Chemical Composition of Brick Earth

A good brick earth should have the following composition:

1. *Alumina* or *clay* = 20–30 per cent by weight
2. *Silica* or *sand* = 35–50 per cent by weight.
3. *Silt* = 20–25 per cent by weight.

The total content of clay and silt may preferably be not less than 50 per cent by weight.

4. Remaining ingredients, which include
   (*i*) Iron oxide
   (*ii*) Magnesia (MgO)
   (*iii*) Lime (CaO)
   (*iv*) Sodium potash, etc.
   } 1–2 percent by weight

— The total lime (CaO) and magnesia (MgO) in case of *alluvial soil* should not be more than one per cent and in other cases should not be preferably more than 15 per cent.

— The lime should be in finely divided form.

— The total water-soluble material should not be more than one percent by weight.

### 3.3.2 Function of Constituents of Brick Earth

Functions of various constituents/ingredients of brick earth are described below:

**1. Alumina or clay**

- It is the *principal constituent* of every kind of brick earth.
- It provides *plasticity to the earth* and *makes the brick hard.*
- If not mixed with sand shrinkage and warping results.

**2. Silica or sand**

- It exists in brick earth either in chemical combination with alumina as silicate of alumina or mechanically mixed with caly as sand.
- If silica in *suitable proportion* is added to clay it *imparts hardness to the brick and checks it shrinkage, cracking and warping on drying.* But if added in *greater proportions* makes the brick *brittle.*
- Addition of silica also increases hardness, durability and resistance to heat.

**3. Lime**

- It helps silica to *melt at lower temperature* and *binds the particles of the bricks together.*
- It also reduces shrinkage of bricks.
- It should be present in *finely divided state* otherwise the *lumps on burning will change to quick lime and disintegrate the brick on absorbing moisture.*

**4. Iron oxide**

- A small quantity of iron oxide present in brick earth, acts as a flux, helps grains of sand to melt and binds the particles of clay together.
- It imparts pleasing red colour to the brick. Its excess percentage (8 to 11) makes the brick dark blue.

**5. Magnesia**

*Small quantities* of magnesia in brick earth make the bricks of *yellowish colour and reduce shrinkage.* But *excess* of magnesia leads to the *decay of bricks.*

**6. Manganese**

Manganese in small proportions along with iron makes the brick darker (or even black).

### 3.3.3 Harmful Ingredients in Brick Earth

**1. Lime stone and kankar nodules:**

The presence of limestone and kankar nodules in the brick earth is harmful and crops up serious troubles. On heating limestone is converted into lime which on contact with water swells and causes the brick to *split and crumble to pieces.* But however, a certain quantity of limestone is desirable in brick earth as it binds the particles of brick together and reduces shrinkage on drying. It is essential that limestone must be present in a finely divided state and not in lumps.

**2. Alkalies:**

If alkalies are present in the brick earth, they lower the fusion point of clays and cause the *brick to fuse, twist and warp during burning.* Alkaline salts if present in finished bricks absorb moisture from the atmosphere and create damp conditions which are detrimental to health. Further when the moisture dries up a greyish white deposit is left behind which spoils the appearance of the structure; the effect being known as *efflorescence.*

**3. Pebbles of stone and gravel:**

They do not exercise any harmful effect chemically, but do not permit the caly to be throughly mixed and thus impair the uniformity of a brick.

**4. Iron pyrites:**

Iron pyrites, if present in the earth decompose and oxidise in the brick and cause the brick to *split.*

**5. Kallar or reh**

It consists of sulphate of soda mixed with common salt and carbonate of soda. These salts prevent the bricks from being *poperly baked.* Its presence in the earth has the effect as that of alkalies.

**6. Vegetation and organic matter:**

The presence of vegetation and organic matter in brick assists in burning. But if such matter is not completely burnt, the bricks become porous (This is due to the fact that the gases will be evolved during the burning of the carbonaceous matter and it will result in the formation of small pores).

## 3.4 REQUIREMENTS OF GOOD BRICK EARTH

A good brick earth should fulfil the following *requirements:*

1. It must have proper proportions of sand, silt and clay.

3. It should have sufficient plasticity (so that the bricks can be properly and conveniently moulded with sharp and well defined edges).
4. It must be free from lumps of lime or nodules of kankar.
5. It must be free from earth containing alkaline salts, kankar or reh, etc.
6. It must be free from pebbles, grits and lumps of earth.
7. It must not contain vegetable and organic matter.
8. It should not be mixed with salty water.

## 3.5 FIELD TESTING OF BRICK EARTH

In order to determine the suitability of the brick earth, the following field tests are generally performed:

**1. Consistency test:**

The consistency test is performed to know *the proportions of clay and sand in the soil to be used for brick earth so that the soil can be modified, if required.*

This test is performed as follows:

- A handful of soil sample is taken and formed into a ball. The ball is dried in the sun.
- When dried, the ball is examined. If deformed on drying and the ball crumbles easily, it may be due to excess of sand. On the other hand, if the ball is hard and have cracks on the surface, it shows that the sand is less than the required quantity. If *no deformation takes place and cracks are very little on the surface of the ball, then the soil is suitable for brick making.*

If the soil is not of suitable composition, it can be modified by mixing different proportions of clay or sand as the case may be; the process is repeated until the ball on drying shows least deformity and cracks.

**2. Test for moulding properties:**

The main object of performing this test is to *fix the soil and water ratio so that the bricks can be moulded with sharp and well formed edges.*

This test is performed as follows:

- A small quantity of the original or modified soil used in the first test is taken and ground well.
- Then a required quantity of water is added in the soil so that soil mixture can be kneaded well to be rolled to form threads of about 3 mm diameter.
- The above kneaded soil is then put into the mould and a brick is moulded out of it. If the edges and corners of the moulded brick are not well formed or sharp, this test is repeated with varying water contents till a satisfactory result is obtained. Finally an approximate soil and water ratio may be found when the correct moisture content is obtained.

**3. Test for deformation and shrinkage on burning:**

The primary purpose of performing this test is to know the *shrinkage and deformation on burning of bricks, moulded and dried out of the soil under test,* so that any shortcoming in the composition of soil can be detected and corrected for achieving a satisfactory result.

This test is performed as follows:

- The sample of bricks moulded as above are burnt in the ordinary potter's clamp for three days and four nights until the bricks become red hot.
- The brick are then allowed to cool and removed from the clamp. These burnt bricks should then be examined for shrinkage and deformation.
- The bricks which have shrunk evenly and do not show any bad effect due to burning, should be *sorted out separately* and those which have lost their shape or deformed, should be *rejected.*

- The sorted out bricks should be compared with the total number of bricks. If they are less, then something is still missing from the soil to be used for brick making.
- Any shortcoming detected in the composition of the soil should be corrected and the process is repeated till a satisfactory result is obtained.

## 3.6 MANUFACTURE OF BRICKS

he manufacture of bricks involves the following *steps*:

1. Preparation of clay
2. Moulding of bricks
3. Drying of bricks
4. Burning of bricks.

### 3.6.1 Preparation of Clay

(*i*) *Selection of site and unsoiling*:

- The site selected for the manufacture of bricks must have suitable soil available in sufficient quantity; otherwise unnecessary labour and cost of digging and transportation of the soil would be involved.

— Generally, for making soil suitable for brick making, certain materials are required to be mixed with the soil. Therefore, the availability of such materials near the site of brick-making is amply important.

— It is also necessary that water and fuel like coal and wood are easily available in adequate quantities.

- The site should be selected after giving due consideration to *suitability of soil* and *location of the water table*. Site should be so selected that the water table during breaking session is at least 1 meter below the floor of the kiln.
- After selecting the site, the *top layer of soil, about 200 mm in depth is taken out and thrown away.* The clay in top soil is full of impurities, (*e.g.* roots of grass, vegetation etc.), and hence it is to be *rejected* for the purpose of preparing bricks.

(*ii*) *Digging and cleaning:* The soil below 200 mm is dug out preferably before rains. It is then cleaned off stones, pebbles, vegetable matter etc. All the lumps of soil should be broken into powder form.

(*iii*) *Weathering and blending:* The earth is then spread out to a depth of 600 to 1200 mm and left to the atmospheric action; this process being known as **weathering**. It extends from a few weeks to some months and imparts plasticity and strength to the clay. It is this stage during which different types of earths such as sandy or calcarious clays could be mixed in suitable proportions with little quantities of coke, ash, etc., to develop the essential qualities needed for moulding, drying and burning of bricks. The whole mass is mixed uniformly and proportionate quantity of water is added. The earth is covered up with cloth or mats. This process of mixing ingredients is called **blending.**

(*iv*) *Tempering:* Tempering process consists in *kneading the earth under feet of men or cattle to make it stiff and plastic*. Addition of water is made from time to time and tempering is continued till homogenous mass of earth is obtained. A soft plastic clay could be prepared by about 25 to 30 percent of water. This method is adopted for the majority of common hand made bricks but in case of superior bricks needed on large scale tempering is carried out with help of a *pug mill.*

**Pug mill**

Refer Fig. 3.1. A pug mill consists of a iron cylinder in the form of bucket with top and bottom diameters 120 cm and 75 cm respectively and height about 180 cm. Centrally placed is a long vertical iron shaft which is provided with number of horizontal arms furnished with knives for breaking clay lumps. A long horizontal arm is fitted at the top of vertical shaft which is driven round by a pair of bullocks or sometimes by electric power. Clay and water in the required ratio (generally 1 : $\frac{1}{4}$ ) are charged in the cylinder from the top and the tempered clay leaves the pug mill through an opening provided near the bottom of the iron tub or cylinder. The pug mill of the mentioned size can knead nearly 30 m$^3$ of earth per day.

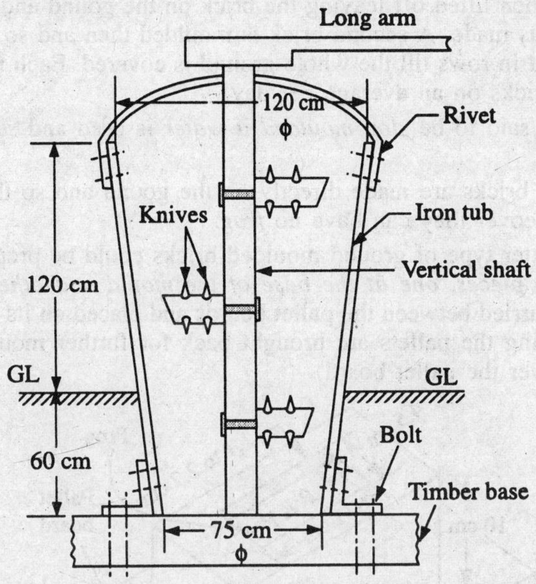


Fig. 3.1 Pug mill.

### 3.6.2 Moulding of Bricks

After the earth or clay is prepared the moulding of the bricks is carried out either by *hand* or by *machine*.

#### 3.5.2.1 Hand moulding

Hand moulding usually adopted in India is of the following two types:

(*i*) Ground moulding

(*ii*) Table moulding.

(*i*) *Ground moulding:* The process of moulding bricks on the ground by manual labour is known as ground moulding and it is carried out as follows:

- The moulder places his mould (wooden or steel box as shown in Fig. 3.2*a, b*) flat in one corner of the ground which is already levelled, plastered and sprinkled with sand.
- He then lifts a lumps of prepared clay over his head with both hands, dashes it into the mould and forces it with hands in order to fill the corners of the mould. He then strikes

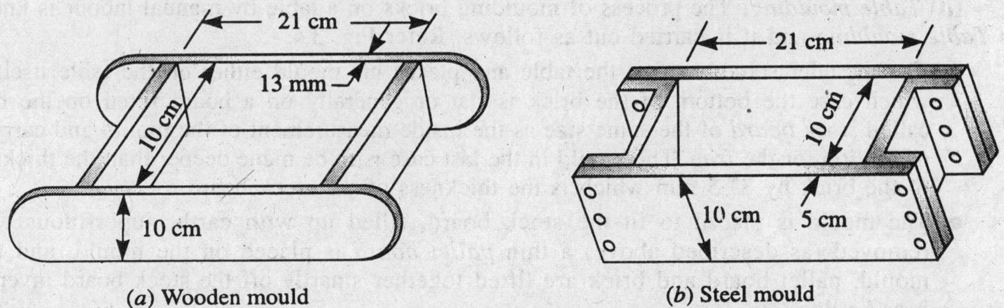


(*a*) Wooden mould (*b*) Steel mould

Fig. 3.2 Brick moulds.

off the superfluous clay above the level of the mould with a piece of wood or iron called a *strike* (Fig. 3.4) or a wire on a bow and throws the cut off clay on the ground to be mixed again with earth to be prepared.

- The mould is then lifted off leaving the brick on the gound and again placed by the side of the brick thus made. A second brick is moulded then and so on. The moulder moulds bricks lying flat in rows till the whole ground is covered. Each moulder can mould about 600 to 1000 bricks on an average per day.
- The bricks are said to be *slop moulded* if *water* is used and *sand moulded* if fine *sand* is used.
- In this process bricks are made directly on the gound and so their lower faces are very rough and moreover they can have no *frog*.

Comparatively better type of ground moulded bricks could be prepared by using *two pallet boards of thin wooden pieces, one at the base of the mould and other on top of it.* After the brick is moulded it is carried between the pallet boards and placed on its longer side face. Leaving the brick there for drying the pallets are brought back for further moulding (Fig. 3.3 shows a moulded brick lying over the pallet board).

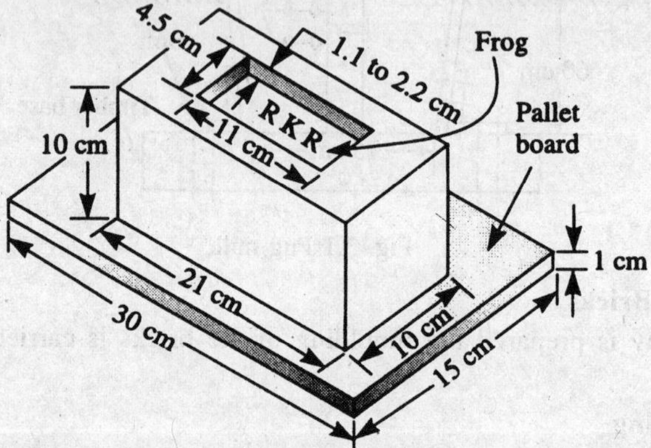


Fig. 3.3 A moulded brick lying over the pallet board.

**Frog or impression**

- It is an indentation mark left on the face of a brick during moulding process. It serves the following *two purposes:*
  (*i*) It indicates the *name* of the person or the firm who manufactures the bricks.
  (*ii*) It provides a *key for mortar for developing a structural grip when used.*
- Frog in the brick is formed by providing a projection or a fillet on the corresponding face of mould (usually of the lower pallet board).

(*ii*) *Table moulding:* The process of moulding bricks on a table by manual labour is known as *Table moulding* and it is carried out as follows: Refer Fig. 3.4.

- The moulder stands behind the table and places his mould either on the table itself, in which case the bottom of the brick is flat or generally on a board fixed on the table called *stock board* of the same size as the inside measurement of the mould and carrying projection for the *frog*. The mould in the last case is to be made deeper than the thickness of the brick by 37.5 mm which is the thickness of the stock board.
- The mould is placed to fit the stock board, filled up with earth, superfluous earth removed (as described above) a thin *pallet board* is placed on the mould, and then mould, pallet board and brick are lifted together smartly off the stock board inverting it as he does so.
- Then he (moulder) places the whole on the page (Fig. 3.4) on his left side with pallet board downwards and lifts up the mould leaving the brick on the pallet board. A boy

places another pallet board on the brick and carries it away between two boards to the drying ground, where it is placed on its edge. Pallet boards are removed.

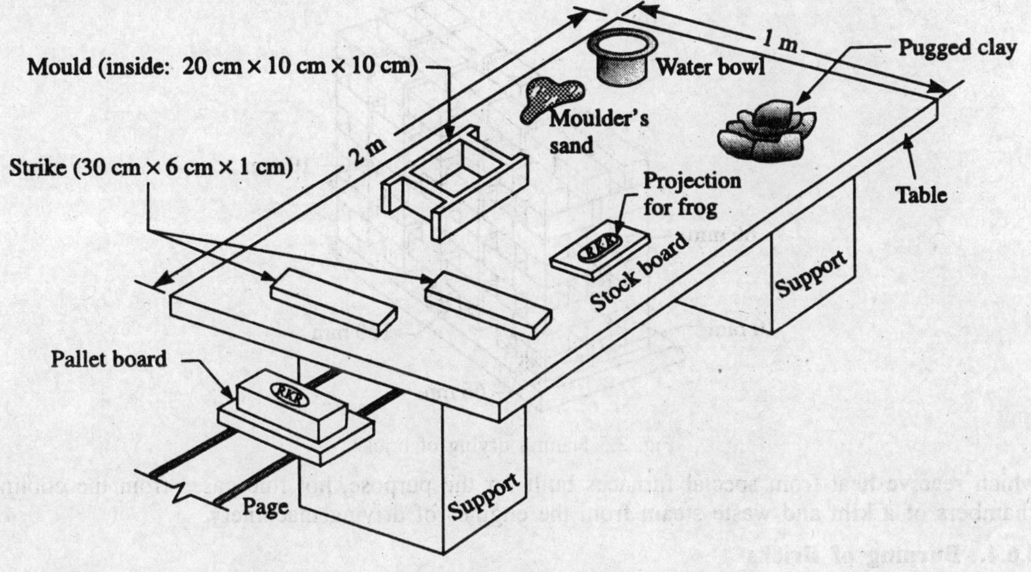


Fig. 3.4 Table moulding.

- The second brick moulded is placed next to the first and this is repeated until the whole space is covered.

**3.2.2.2 Machine moulding**

- This brick making method proves quite useful when a *large number of bricks have to be manufactured quickly.*
- *Machine moulding* can be performed by the following *two methods*:

(*i*) *Plastic method:* The clay in plastic condition is formed through an opening of dimensions equal to the length and the breadth of the bricks in a rectangular plastic band from which the bricks of required thickness are *cut with wires or knives fixed in a frame.* This is *quick* press of manufacturing bricks. It proves economical when bricks are required on a large scale.

(*ii*) *Dry process method:* Superior bricks which are to withstand heavy pressure are made by *compressing under great pressure moist powdered clay into well shaped bricks* called **pressed bricks.** The bricks made by this method are dense, smooth and more uniform in the shape. The bricks produced by this method are *much stronger* than ordinary bricks and *do not necessitate drying and can be directly fed into the kiln for burning.*

- Due to heavy initial cost of machine moulding method, it is not presently pupular in our country.

**3.6.3. Drying of Bricks**

After the bricks are moulded, they are dried by the following two methods:

(*i*) Natural drying (*ii*) Artificial drying.

(*i*) *Natural drying :* Refer Fig. 3.5. It is also called *hack drying* and consists in arranging the wet bricks in rows on their edges on a slightly raised ground called *hacks*. A little space for circulation of air is kept for each brick. It should be ensured that no sudden drying is caused by direct exposure to sun and wind and a portable cover is provided to protect the bricks from rain. In certain places where temperature variations are too much temporary sheds are erected for drying purpose. *The air and sun dried bricks are adequately strong and durable and produce good results when employed for construction of simple structure.*

(*ii*) *Artificial drying:* This method is used when bricks are required on a *large scale and at a rapid rate*. The heating of the bricks is carried out in special driers such as tunnels or chambers

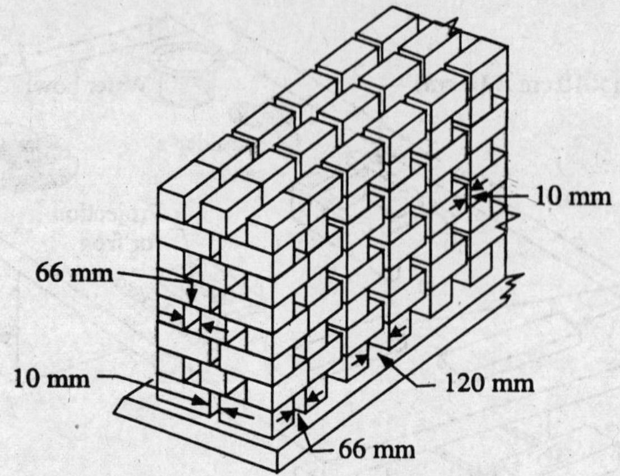


Fig. 3.5 Natural drying of bricks.

which receive heat from special furnaces built for the purpose, hot flue gases from the cooling chambers of a kiln and waste steam from the engines of driving machinery.

### 3.6.4. Burning of Bricks

Bricks after being moulded and dried are burnt in the **kilns.** The purpose of burning is *two-fold*: (*i*) *To impart hardness and strength to the bricks;* (*ii*) *To increase the density of the bricks so that they will absorb less quantity of water.*

Types of kilns:

The bricks may be burnt in any one of the following kilns :

1. Clamp or open kilns
2. Intermittent kilns
3. Continuous kilns.

**1. Clamp or open kiln:**

- It is also known as *Pazawah.*
- It is suitable for burning only ordinary bricks, *i.e.* ground moulded bricks of inferior type.
- This method of burning is usually adopted in *rural areas* where transport charges of bricks from urban areas are high. Furthermore as the number of bricks required for construction of ordinary building is small, it is not felt necessary to set up bigger kilns which entail a huge expenditure.

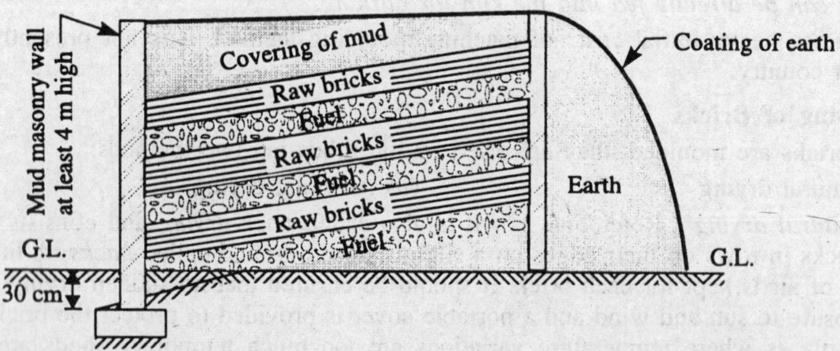


Fig. 3.6 Clamp burning.

- The size of the clamp depends on the number of bricks required. For one lakh bricks its size should be 15 m (length) × 13 m (breadth) × 5 m (height).
- These clamps sometimes are also made trapezoidal in form. A slight slope of 15° is unsually given to the ground lengthwise.
- In a brick clamp, fuel and bricks are arranged in alternate layers as shown in Fig. 3.6. On the sides and on the top of the clamp, some protection is provided by a coating of earth. The air dried bricks are arranged in layers in criss-cross way on their edges with about 5 cm air space between them to ensure proper circulation and upward draught of the hot gases from below. After the kiln is started it takes two to six months for burning and cooling the bricks (20000 to 100000). This clamp is a temporary structure for burning and is built again for the next burning of rawbricks.

**Advantages:**

1. Burning of bricks is cheap and economical.
2. Considerable saving of fuel.
3. Bricks produced are tough and strong (since the burning and cooling of bricks are gradual).
4. Skilled labour and supervision are not required for the construction and making of clamps.
5. The clamp is not liable to injury from high wind or rain.

**Disadvantages:**

1. Very slow process.
2. The quality of bricks is not uniform.
3. The bricks are not of regular shape (due to the settlement of bricks when fuel near bottom is burnt and turned to ashes).
4. The bricks are liable to burn unevenly (since it is not possible to regulate the fire).

**2. Intermittent kilns (overground kilns):**

These kilns are permanent one as opposed to clamps which are temporary structures, and can be put to use again for burning the bricks. Here the burning is not continuous. These are made rectangular or circular in shape and may be of 'upward draught' or 'down draught' type.

**3. Continuous kilns:**

The kilns falling under this category ensure an uninterrupted operation of burning. Following are the three types of countinous kilns used these days.

(*i*) Bull's trench kiln
(*ii*) Hoffman's kiln
(*iii*) Tunnel kiln.

**(*i*) Bull's trench kiln:**

*Construction*: This type of kiln is usually oval is plan and is made in a trench excavated below ground surface, the depth of trench being about 2 metres. Depending upon the type of soil, sides of the kiln are lined with bricks or two brick walls are erected outside and inside (where the soil does not slip, it is simply lined with bricks). In general oval shaped bench is divided into 8 to 12 zones there being an opening in each zone. Figure 3.7 shows a kiln having 12 openings $D_1$ to $D_{12}$. These openings are employed for loading and unloading of the bricks. In the compartments the bricks are so arranged that continuous horizontal and vertical flues are obtained. For even distribution of heat inside the kiln two chimneys provided are placed near the outer and inner walls. These chimneys are moved from zone to zone. To regulare the heat when required, small iron sheets are provided near the base of the chimney.

*Working:* Fire is started say is zone 7 and chimneys are arranged at the top near the side walls of zone 8. Damper (vertical iron sheet) is incorporated at a certain distance along the width. Small holes 36 to 48 in number are left at the top through which coal in the form of powder or small pieces is dropped, the holes being covered with iron discs. To check the loss of heat from inside, the top of stacked bricks is provided with a cover of earth and ashes. Hot gases from the burning zone 7 heat the dry bricks stacked in zone 8 before escaping to atmosphere through the

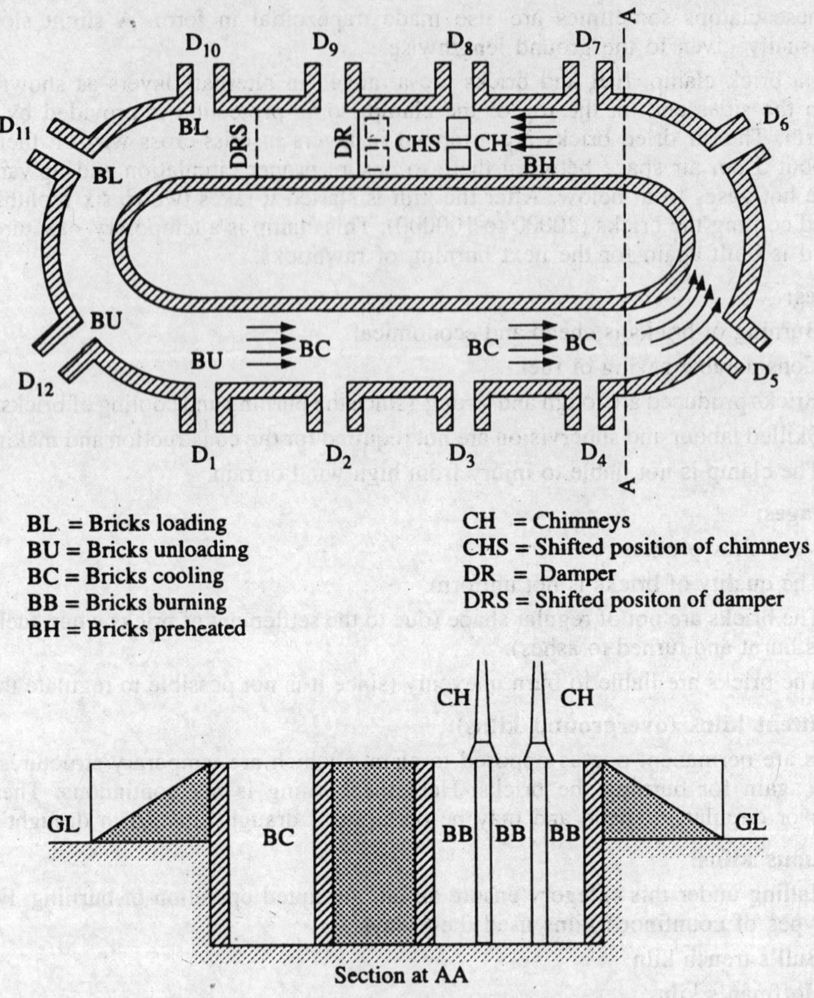

Fig. 3.7 Bull's trench kiln.

chimneys. As the bricks go one burning previous holes (from where the coal was dropped) are closed and new holes are made in forward direction for introducing the fuel. The positions of chimneys and damper are altered accordingly.

Each zone takes about 24 hours to burn. By the time the fire moves through a number of successive zones, the zone which is burnt first cools, is unloaded and reloaded with dry and raw bricks which are first warmed up by the hot gases from the previous zones and are burnt when the advancing fire reaches them. Thus the operation of the kiln is continuous.

The yield per day is 25000 bricks and depending upon the type of clay, 4 to 5 tonnes of coal is consumed. This kiln is widely used in India.

**Advantages:**

1. Cost of fuel is less as heat of hot gases escaping the rough the chimney is fully utilised for warming the bricks.
2. Supply of bricks is continuous.
3. High percentages of first class bricks.

4. As raw bricks are loaded in the kiln, the drying space is saved.

**Disadvantages:**

1. High initial cost.
2. Constant skilled supervision is essential.

***(ii)* Hoffman's kiln:** Refer Fig. 3.8.

This kiln is built *above the ground* and is circular in plan. Centrally placed is a chimney around which 12 chambers in a ring are arranged. For communication with the outside each chamber is provided with a door. This door can be closed when required with the help of dry bricks. There is a suitable arrangement for opening and closing the flue. In each chamber, small holes are provided at the top through which the coal can be dropped for burning bricks.

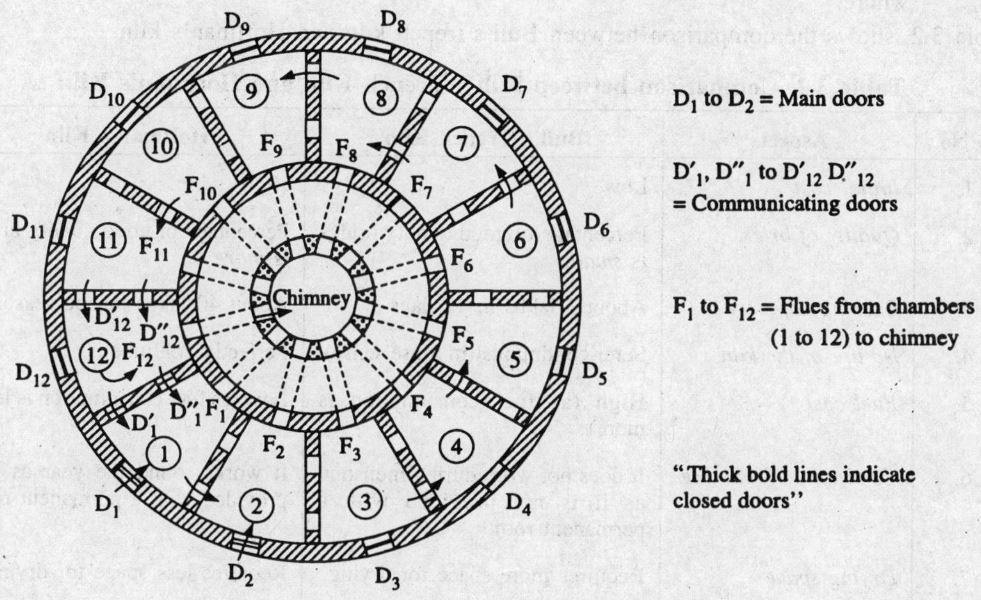


Fig. 3.8 Hoffman's kiln.

*Working:* To have an idea of working of this kiln let us consider following series of operation.

Chamber No. 1 is loaded with raw bricks, No. 2 is emptied of cooled down burnt bricks. The bricks in the chambers 3, 4, 5 and 6 are cooling and bricks stacked in chambers 7 and 8 are being supplied with fuel from fire holes provided for the purpose. Raw bricks in chamber 9, 10, 11 and 12 are being heated by the hot gases of combustion before they escape through the chimney.

As regards the circulation of flue gases through the kiln, cold air enters the chambers 1 and 2 through the doors and after passing through the cooling chambers 3, 4, 5 and 6 it enters in a heated condition into the burning chamber 7 and 8. The hot gases of combustion then travel through the chambers 9, 10, 11 and 12 and perheat the raw bricks.

In a continuous kiln the progress of work is so arranged that one chamber is loaded, another is unloaded and still another is freshly fired. The common size of a chamber is 11 m (length) × 4.6 m (mean width) × 2.4 m (height). Daily out-turn of the bricks is 25,000.

**Advantages:**

1. High percentages of first blass bricks.
2. Regular out-turn of bricks.
3. As the bricks can be loaded in the kiln, there is a saving in drying space.
4. Bricks are evenly burnt.
5. Economy in fuel since all the heat of combustion is utilised.

**Disadvantages:**

1. High initial cost of construction.
2. Constant and skilled supervision is required.
3. Economy demands regular consumption of its output, which may not be possible everywhere.

Table 3.2. shows the comparison between Bull's trench kiln and Hoffman's kiln.

**Table 3.2. Comparison between Bull's Trench Kiln and Hoffman's Kiln**

| S. No. | Aspects | Bull's Trench Kiln | Hoffman's Kiln |
|---|---|---|---|
| 1. | *Initial cost* | Low | High |
| 2. | *Quality of bricks* | Percentage of good quality bricks is *small.* | Percentage of good quality bricks is *more.* |
| 3. | *Burning capacity* | About 3 lakhs in 12 days. | About 40 lakhs in one season. |
| 4. | *Nature of the kiln* | Semi-continuous (in loose sense). | Perfectly continuous. |
| 5. | *Fuel cost* | High (as fuel consumption is more) | Low (as fuel consumption is less). |
| 6. | *Continuity of working* | It does not work during monsoon as it is not provided with a permanent roof. | It works round the year as it is provided with a permanent roof. |
| 7. | *Drying space* | Requires more space for drying. | Requires less space for drying. |
| 8. | *Popularity* | More popular (due to less initial cost). | Less popular (due to high initial cost). |

(*iii*) **Tunnel kiln**

- It is in the form of the tunnel which may be straight, circular or oval in plan.
- It contains a stationary zone of fire.
- The raw bricks are loaded on the trolleys which are then moved from one end to the other end of the tunnel.
- As the bricks approach the zone of fire, they get dried and preheated.
- These bricks are then burnt to the required degree in the zone of fire and, thereafter these are pushed for cooling.
- After getting adequately cooled, the bricks are unloaded.
- In this kiln uniform bricks of better quality are produced since the temperature can be controlled.
- The kiln proves to be economical when bricks are to be manufactured on a large scale.

Table 3.3 shows the comparison between clamp burning and kiln bearing of bricks.

**Table 3.3 Comparison between Clamp Burning and Kiln Burning**

| S. No. | Aspects | Clamp burning | Kiln burning |
|---|---|---|---|
| 1. | *Initial cost* | The initial cost is low. | The initial cost is high. |
| 2. | *Fuel cost* | Fuel cost is low; straw, cow dung and litter can be used. | Fuel cost is usually high as coal is used to burn the bricks. |
| 3. | *Structure* | It is a temporary structure. | It is a permanent structure. |
| 4. | *Wastage of heat* | A huge amount of heat is wasted from the top. Heat of hot gases cannot be utilised as there are no arrangements to do so. | No wastage of heat from sides or the top. Heat of hot gases is used for warming the bricks before burning. |
| 5. | *Out-turn* | It is used when a small number of bricks are required. About 20,000 to 1 lakh bricks are burnt by each clamp at a time. | It is employed when a continuous supply of bricks is required. The number of bricks burnt per day is about 25,000. |
| 6. | *Fire regulation* | The fire once started cannot be regulated. Only perfectly dry bricks should be put in the clamp otherwise sudden heat will warp them. | Fire can be regulated and even raw bricks can be burnt with safety. |
| 7. | *Labour* | Once the clamp is fired, no supervision is necessary. | Skilled supervision throughout the operation of the kiln is necessary. |
| 8. | *Time for burning and cooling* | It takes 2 to 6 months to burn and cool the bricks. | Time of burning any section or compartment is 24 hours and time of cooling about 12 days. |
| 9. | *Percentage of first class bricks* | Mostly the out-turn is of poor quality and on average 60% of bricks are of first class. | Bricks are generally of good quality and 90% of them are of first class. |
| 10. | *Wastage of heat* | Considerable wastage of heat. | Wastage of heat is the least. |

## 3.7. QUALITIES OF GOOD BRICKS

The good bricks which are to be used for the construction of important engineering structures should possess the following qualities/characteristics.

1. *Size and shape:* A good brick should be uniform in size (standard) and should have rectangular plane surfaces with parallel sides and sharp straight right-angled edges. Its surfaces should not be too smooth or, otherwise, the mortar will not stick to it.

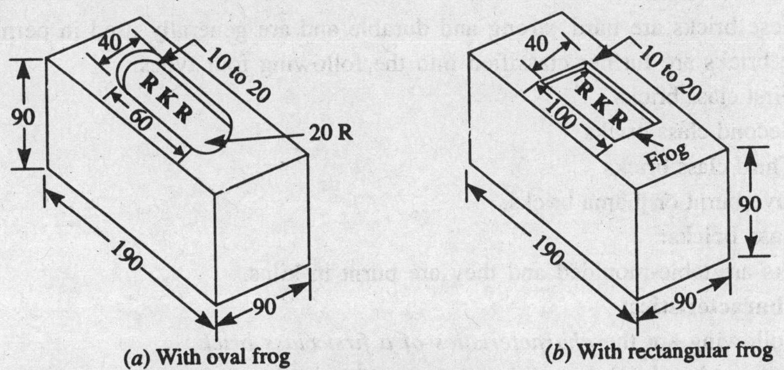


(*a*) With oval frog

(*b*) With rectangular frog

All dimensions in mm

Fig. 3.9 Standard bricks.

Figure 3.9 shows details of a standard brick. According to ISI specification, size of the brick should be 190 mm × 90 mm × 90 mm and the depth of the frog should be between 10 to 20 mm. Length and width of frog [Fig. 3.9(*b*)] should be 100 mm and 40 mm respectively. Thickness of mortar should be 5 mm on each surface of each brick. So the dimensions of the each brick including mortar joint will be 200 mm × 100 mm × 100 mm. *A standard brick weighs nearly 30 N.*

2. *Colour:* The bricks should be table-moulded, well burnt in kilns, copper coloured, free from cracks and with sharp and square edges. The colour should be uniform and bright.

3. *Structure:* A good brick should show fine, compact and uniform structure in broken form.

4. *Soundness:* A good brick should give a clear metallic ringing sound when struck with another brick.

5. *Hardness:* A good brick should be so hard that finger nails should not be able to make any impression on its surface when scratched.

6. *Porosity:* A good brick should not absorb more than 20 per cent of its dry weight of water when kept immersed for 24 hours.

7. *Strength:* A good brick should have minimum crushing strength of 10.5 N/mm$^2$. Ordinarily it should not break when dropped flat on hard ground from a height of about 1 m.

8. *Resistance to fire:* A good brick should have adequate resistance to fire. Ordinay bricks can resist temperature upto 1200 °C. Bricks to be used for lining of furnaces, etc., should be of special quality (known as *fire bricks*).

9. *Efflorescence:* A good brick should not contain much alkaline salts, which may cause efforscence on its surface and decay the brick.

10. *Durability:* A good brick should be able to resist the effects of weathering agencies like temperature variations, rain, frost action, etc.

## 3.8 CLASSIFICATION OF BRICKS

Conventionally, the bricks are classified in the following two types:

1. Sun dried or katcha bricks
2. Burnt or pacca bricks.

**1. Sun dried or katcha bricks:**

- These bricks are dried with the help of heat received from the Sun after the process of moulding. These bricks are not burnt in camp or kiln.
- These bricks can only be used in the construction of temporary and cheap structures.
- Such bricks should not be used at places exposed to heavy rains.

**2. Burnt or pacca bricks:**

- The bricks which are burnt in a clamp or kiln after moulding and drying are called burnt or pacca bricks.
- These bricks are hard, strong and durable and are generally used in permanent works.

These bricks are further classified into the following four types:

(*i*) First class bricks
(*ii*) Second class bricks
(*iii*) Third class bricks
(*iv*) Overburnt or jhama bricks.

(*i*) **First class bricks:**

These bricks are table-moulded and they are burnt in kilns.

(*a*) **Characteristics:**

The following are the *characteristics of a first class brick :*

(*i*) A first class brick is well burnt and regular in shape.
(*ii*) Edges are sharp and well defined.

(*iii*) It is sufficiently hard and sound.

(*iv*) Its surface is smooth, clean and free from cracks.

(*v*) A fractured surface shows a uniform compact structure (free from lumps or grits and holes).

(*vi*) *Water absorption* shall not be more than one sixth of its weight of water when kept immersed for sixteen hours.

(*vii*) When two bricks are struck against each other, a metallic ringing sound is produced.

(*viii*) When soaked in water for 24 hours and dried in shade does not show efflorescence.

(*ix*) It should *not* break when dropped from a height of 1 to 2 metres.

(*x*) It has specific gravity of 1.8.

(*xi*) It is so hard that no mark can be produced on it with finger nails.

(*xii*) It has a minimum crushing strength of 10.5 N/mm$^2$.

**Uses:** (*i*) In R. B. work.

(*ii*) For flooring and walling purposes.

(*iii*) As ballast for R.C. work.

(*iv*) Shaped bricks are used in carvings, arches and copings.

**(*ii*) Second class bricks:**

There bricks are ground-moulded and they are burnt in kilns.

(*a*) **Characteristics:**

(*i*) A second class brick is well-burnt but irregular in shape and size.

(*ii*) It has a rough surface.

(*iii*) Its edges are neither straight nor well defined.

(*iv*) It is not free from lumps and cracks.

(*v*) It does not have a uniform colour.

(*vi*) It has fine, compact and uniform texture.

(*vii*) It shall produce a ringing sound when struck with one another.

(*viii*) It shall have a minimum crushing strength of 7 N/mm$^2$.

**Uses:** Such types of bricks are used in unimportant situation and for internal walls (where the bricks are generally hidden from view.

**(*iii*) Third class bricks:**

The bricks which are underburnt and are not uniform in shape and size are called *third class,* or *pilla* bricks. These bricks are ground-moulded and then are burnt in clamps.

(*a*) **Characteristics:**

(*i*) A third class brick has a light yellowish colour.

(*ii*) It is not well-burnt.

(*iii*) It is soft.

(*iv*) When two bricks are struck a dull sound is emitted.

(*v*) It is associated with flaws and cracks.

(*vi*) Its edges are irregular and surface quite rough.

(*b*) **Uses:** Third class bricks are used in inferior and temporary building.

**(*iv*) Overburnt or jhama bricks:**

These are overburnt bricks with irregular shape and dark colour.

(*a*) **Characteristics:**

(*i*) They are overburnt, vitrified and distorted bricks.

(*ii*) They are irregular in shape and size and are available in the form of lumps.

(*iii*) They are dark in lumps.

(*iv*) They are hard and strong.

(*v*) They produce metallic ringing sound.

(*b*) **Uses:** These bricks are used as aggregates for making concrete to be used in *foundations, floors*, etc., and as road *metal* for soling coat.

## 3.9 TESTS FOR BRICKS

In order to find out the suitability of a brick for the construction work, it is generally subjected to the following *tests*:

1. Water absorption test
2. Compressive strength test
3. Efflorescence test
4. Structure test
5. Shape and size test
6. Soundness test
7. Hardness test.

**1. Water absorption test:**

- The specimen brick is first weighed dry.
- It is then immersed in water for a period of 16 hours.
- It is weighed again. The difference in weight indicates the amount of water absorbed by the brick. It should not in any case, exceed 20% of weight of dry brick.

**2. Compressive strength test:**

- The specimen brick is immersed in water for 24 hours.
- The frog of the brick is filled flush with 1 : 3 mortar and the brick is stored under damp jute bags for 24 hours followed by immersion in clean water for three days.
- The specimen is then placed between the plates of the *compression testing machine.*
- Load is applied axially at a uniform rate (of 14 N/mm$^2$) and maximum load at which the specimen fails is noted for determination of compressive strength of brick given by

$$\text{Compressive strength} = \frac{\text{maximum load at failure}}{\text{loaded area of brick}}$$

The crushing or compressive strength of common building bricks should not be less than 3.5 N/mm$^2$. Bricks of high quality do not have strength less than 14 N/mm$^2$.

If no testing machine is available, the strength can be roughly judged from the force required to break the brick with a hammer. The brick should not break when fallen on its end on their ground from a height of 2 metres.

The strength of the bricks is affected by the following *factors*:

(*i*) The composition of brick earth used.

(*ii*) The preparation of clay and blending of ingredients/constituents.

(*iii*) Type of moulding adopted.

(*iv*) Care exercised in drying and stacking of raw/green bricks.

(*v*) Type of kiln used (including type of fuel and its feeding).

(*vi*) Burning and cooling process.

(*vii*) Care exercised in unloading the bricks.

**3. Effloresence test**

- The soluble salts, if present in the bricks, cause efflorescence on the surface of bricks.
- In order to find out the presence of soluble salt in a brick, it is immersed in water for 24 hours. It is then taken out and allowed to dry in shade. The absence of grey or white deposits on its surface indicates absence of soluble salts.
- If the white deposits cover about 10% surface, the efflorescence is said to be slight and it is considered as moderate when the white deposits cover about 50% of surface. When

white (or grey) deposits cover more than 50% of surface the efflorescence becomes heavy and it is considered as serious.

**4. Structure test:**

- A specimen brick is broken and its structure is examined. It should be homogeneous, compact and free from any defects, *e.g.*, lumps, holes, etc.

**5. Shape and size test:**

- In this test a specimen brick should be closely inspected. It should be of standard size and its shape should be *truly rectangular* with *sharp edges*.
- Twenty bricks of standard size (190 mm × 90 mm × 90 mm) are randomly selected and stacked length-wise, along the width and along the height. For good quality bricks, the results should be within the following permissible limits: *Length*: 3680 mm to 3920 mm; *Width*: 1740 mm to 1860 mm; *Height*: 1740 mm to 1860 mm.

**6. Soundness test:**

This test is performed by striking two specimen bricks with each other; the bricks should not break and a clear ringing sound should be produced.

**7. Hardness test:**

This test is conducted by *making a scratch on brick surface* with the help of a finger nail; if no *impression* is left on the surface, the brick is treated to be *sufficiently hard.*

## 3.10 SPECIAL BRICKS

These bricks differ from the commonly used building bricks with respect to their shape, specifications and special purpose for which they are made. The special bricks may be of the following types:

1. Specially shaped bricks
2. Burnt clay facing bricks
3. Heavy duty bricks
4. Perforated bricks
5. Burnt clay hollow bricks
6. Sandlime bricks
7. Sewer bricks
8. Acid resitant bricks

**1. Specially shaped bricks:**

These are bricks of special shapes to suit different situations in which they are used. Specially

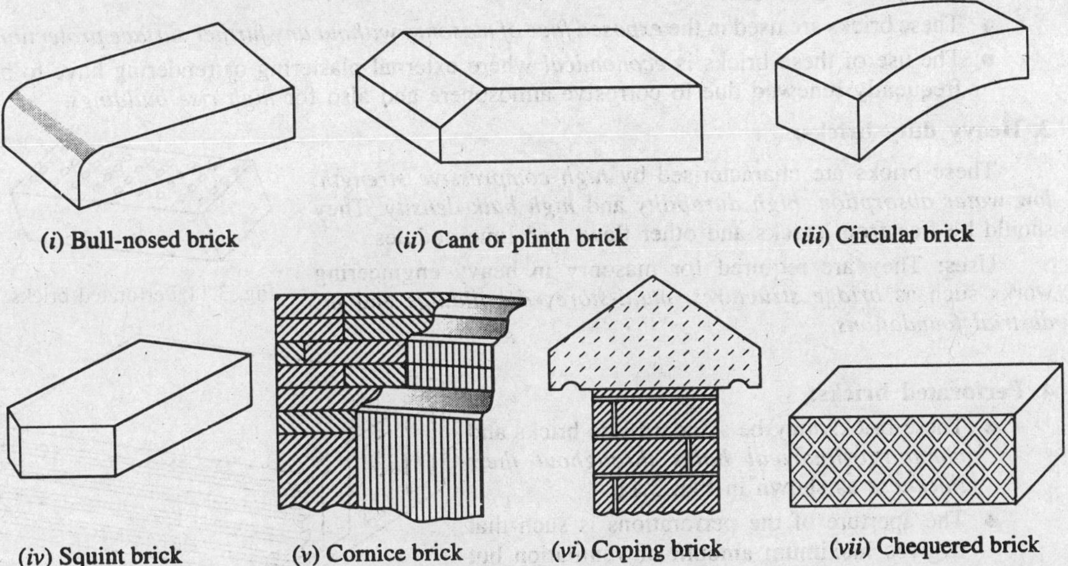

(*i*) Bull-nosed brick (*ii*) Cant or plinth brick (*iii*) Circular brick

(*iv*) Squint brick (*v*) Cornice brick (*vi*) Coping brick (*vii*) Chequered brick

Fig. 3.10. Specially shaped bricks.

moulded bricks avoid the cumbersome process of cutting and rounding the rectangular bricks to the desired shape. Some of the specially shaped bricks commonly used are given below:

(*i*) *Bull-nosed bricks:* Refer Fig. 3.10(*i*).

- A brick moulded with a rounded angle is termed as a bull nose.
- It is used for a rounded quoin (it is a connection which is formed when a wall takes a turn). The centre of the curved portion is situated on the long centre-line of brick.

(*ii*) *Cant or plinth bricks:* Refer Fig. 3.10(*ii*). These bricks have a bevel taken off one side and may be used in plinths or in doors and window joints.

(*iii*) *Circular bricks:* Refer Fig. 3.10(*iii*). These bricks have internal external faces curved to meet the requirement of the particular curve and radius of the wall.

- They are used for towers, wells etc.
- When ordering for such bricks, the radius should always be given.

(*iv*) *Squint bricks:* Refer Fig. 3.10(*iv*). These bricks are made in a variety of different shapes and are used in the construction of acute and obtuse squint quoins.

(*v*) *Cornice bricks:* Refer Fig. 3.10(*v*). These bricks are often made in different shapes and sizes. They are used for constructing cornice

(*vi*) *Coping bricks:* Refer Fig. 3.10(*vi*). These bricks are manufactured in a variety of shapes to suit the thickness of the wall and are throated on the underside to throw off rain water.

(*vii*) *Paving bricks:*

- These bricks are usually made from clay containing a higher percentage of iron. The excess iron vitrifies the bricks at a low temperature.
- They are machine-moulded and are burnt in a continuous kiln to ensure high degree of vitrification.
- They are unaffected by weather and ordinary traffic wear.
- They are laid on a bed of sand which in turn rests on a foundation of stone or concrete. The bricks are laid by grouting with cement mortar or asphalt.
- The paving bricks may be plain or chequered. Figure 3.10(*vii*) shows a *chequered brick.*

These bricks are extensively used for *street pavements, stable floors*, garden walks, etc. These bricks also render the floor *less slippery.*

**2. Burnt clay facing bricks:**

- These bricks are used in the *exposed face of masonry without any further surface protection.*
- The use of these bricks is *economical* where external plastering or rendering have to be frequently renewed due to corrosive atmosphere and also for *high rise buildings.*

**3. Heavy duty bricks:**

These bricks are characterised by *high compressive strength, low water absorption, high durability* and *high bulk density*. They should be free from cracks and other flaws and lime nodules.

**Uses:** They are required for masonry in heavy engineering works such as *bridge structures, multi-storeyed buildings* and *industrial foundations.*

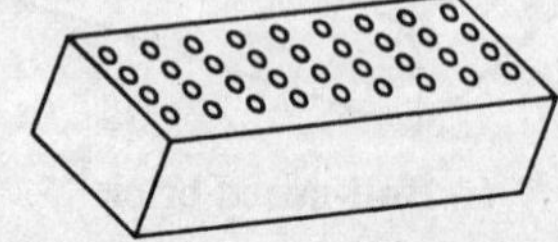

Fig. 3.11 Perforated bricks

**4. Perforated bricks:**

- These bricks may be standard size bricks and *contain cylindrical holes throughout their thickness* as shown in Fig. 3.11.
- The aperture of the perforations is such that it gives maximum amount of ventilation but does not permit the entry of rats or mice.

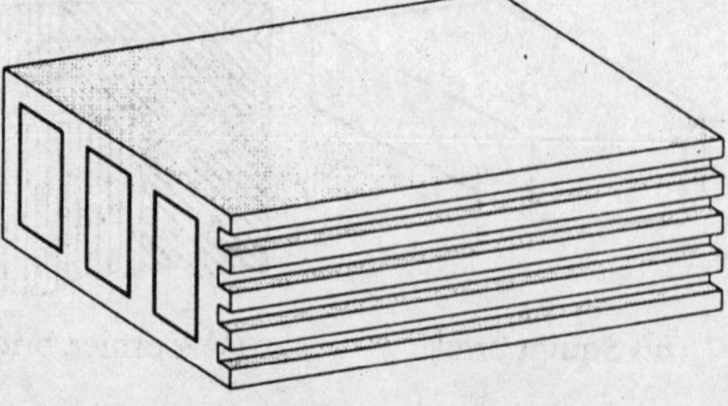

Fig. 3.12 Burnt clay hollow brick.

- They are light in weight and require less quantity of clay for their preparation.
- These bricks are easy to burn and their light weight makes it possible to cut down the weight of the structure and effect economy in foundations.

**Uses:** These bricks are used for constructing *panel walls for multi-storeyed buildings* and for *providing partition walls.*

**5. Burnt clay hollow bricks:** Refer Fig. 3.12.

- They are prepared from special homogeneous clay.
- They should be of uniform colour and have a fine, compact and uniform texture.
- These bricks/blocks are light in weight for masonry construction and also *being hollow impart insulation against heat, sound and dampness to the building.*

**Uses:** They are used for construction of *load bearing walls, partition walls or panel walls in multi-storeyed buildings.*

**6. Sand lime bricks:**

- These bricks are also called *calcium silicate bricks.*
- They consist of an intimate and uniform mixture of siliceous sand or crushed siliceous rock and lime combined by the action of saturated steam under pressure.

**Uses:** They are employed for *masonry construction* just like burnt clay bricks.

**7. Sewer bricks:**

These bricks are manufactured from surface clay, fire clay or shale or a combination of these materials.

**Uses:** They are used for *lining of walls, roofs* and *floors of sewers* used for *ordinary sanitary* (domestic) *sewage*. These bricks, however, are not suitable for sewers dealing with industrial effluent/sewage for which acid-resistant bricks may be considered.

**8. Acid-resistant bricks:**

These bricks are made of raw materials, such as clay or shale of suitable composition with low lime and iron content, feldspar, flint or sand and vitrified at high temperatures in a ceramic kiln.

**Uses.** They are used for masonry constructions, flooring subject to acid attacks, lining of chambers and towers in chemical plants, lining of sewers carrying industrial effluents, etc.

**9. Refractory bricks:**

The term **"Refractories"** *embraces all materials used in the arts for the construction of heat resisting containers, using the word in its broadest sense*, whether it be to afford space for evolution of gases in combustion processes or the holding of molten charges or of solids undergoing heat-treatment. The two principal functions involved in the use of refactory materials are those of *thermal insulation* and *conduction.* In the outside walls of a furnace, refractories serve the purpose of *confining the heat and preventing excessive loss to the atmosphere*; in a muffle or retort, they serve to *conduct the heat through the walls to the charge.*

A *good refractory material* should have the following *properties*:

1. It should be able to withstand high temperatures generated in the furnace.
2. It should be able to withstand sudden alternating, heating and cooling, *i.e.* thermal shocks.
3. It should be able to withstand abrasion and rough usage.
4. Its contraction and expansion due to the inevitable temperature variation should be minimum possible.
5. It should be able to withstand fluxing action of the slags and the corrosive action of gases.
6. It should have good heat insulating properties.
8. It should be chemically inactive at elevated temperatures.
9. It should be impermeable to gases and liquids as far as possible.
10. If used in electric furnaces, it must have low electrical conductivity.

*Refractory bricks* are made from *refractory clays which can withstand very high temperatures without becoming soft on melting.* The refractory clay is composed of heat resisting material such as silica, alumina, bauxite, magnesite, chromite, etc., and does *not* contain metallic oxides such as oxide of iron, lime and alkalies owing to their action as fluxes.

As regards *manufacture* of these bricks, the preparation of clay, moulding, drying, burning in kilns and annealing of refractory bricks are the same as detailed for ordinary bricks.

Following are the three varieties of refractory bricks:

1. *Acid bricks*
   (*i*) Fire bricks
   (*ii*) Silica bricks.
2. *Basic bricks*
   (*i*) Magnesite bricks
   (*ii*) Dolomite bricks
   (*iii*) Bauxite bricks
3. *Neutral bricks.*
   (*i*) Chrome bricks
   (*ii*) Chrome-magnesite bricks
   (*iii*) Spinel
   (*iv*) Forsterite bricks.

**1. Acid Bricks:**

These bricks are *used for acidic lining.*

(*i*) **Fire bricks:**

- These are made from fire clays (silica: 50 to 75%, aluminia: 20 to 40%, fluxes: less than 10%) which are generally mixed with a small percentage of sand to reduce shrinkage during burning.
- They are moulded either by stiff mud and dry press process or repressed after partial drying.

**Properties:**

1. Coefficient of thermal expansion is low.
2. Sufficient strength at high temperature.
3. Resist spalling.
4. Not so refractory as silica bricks but much cheaper.

**Uses:**

1. Linings of blast furnaces for the melting of iron, copper, lead ores, etc.
2. The lining of flues and stocks.
3. Heat treatment furnaces.
4. Glass furnaces.
5. Checker-work of regenerative furnaces.

(*ii*) **Silica bricks:**

- These are made from quartzite and sandstone or silica sand which contains more than 95% of silica. A small quantity of lime, about 1 to 2%, is added to work as binding material.
- These bricks are moulded by soft mud process and are burnt at a temperature of about 1480°C (slightly higher than that of fire bricks).
- The compressive strength of such bricks is about 15 N/mm$^2$.
- These bricks have a large coefficient of expansion and laid with a wide joint.

**Properties:**

1. Low porosity and free from air pockets.

2. Good thermal expansion and conductivity.
3. Remarkable load bearing capacity especially at high temperatures.
4. Not susceptible to thermal spalling at temperatures above 400 °C.
5. Ability to withstand thermal shock poor.

**Uses:** Silica bricks are extremly suitable for those parts of a furnace which are subjected to uniformly high temperature, *e.g.* fire bridges, roofs of acid and basic open hearth furnaces, electric furnaces, copper smelting and refining furnaces, etc.

**2. Basic Bricks:**

These bricks are used for *basic lining* and manufactured from basic refractory materials.

(*i*) **Magnesite brick:**

It contains magnesia = 85% (min.), calcium oxide = 25% (max.) and silica = 5.5% (max.)

**Manufacture:** The magnesite is first calcined at 800 °C and the carbon dioxide driven out. A large portion of this calcined magnesia is sintered at 1800 °C. It is then mixed with 1 to 5% of calcined magnesia and the mass is tempered with a small amount of water into bricks. A little magnesium chloride is added to act as binder. The bricks are then hand moulded, partially dried and repressed. After further drying the bricks are burnt.

**Properties:**

1. Thermal conductivity greater than that for fire-clays and silica.
2. Highly resistant to the action of basic slags and iron oxide.
3. Starts losing strength at temperature above 1500 °C.

Magnesite bricks are not capable of resisting sudden changes of temperature and show a tendency to spall under such conditions. They are costlier than silica and fireclay bricks and used in the construction of those parts of a furnace which are required to *withstand the corrosive action of basic slags.*

**Uses:** They are used for :

1. Hearths of basic open-hearth and copper reverberatory furnaces.
2. Electric-arc and Induction furnaces.
3. Lining of L.D. crucible (in the manufacture of steel).

(*ii*) **Dolomite bricks :**

- These are made of calcined dolomite, the *carbonate of magnesia* and *lime.*
- These bricks have low conductivity and refractoriness as compared to magnesite bricks.
- They are cheaper than magnesite bricks but are poor in performance.

These bricks are extensively used for repairing the banks and the bottoms of the Basic Open-hearth furnaces.

(*iii*) **Bauxite bricks :**

- Bauxite (containing 50 per cent or more of aluminia and iron oxide and silica as impurities) is calcined, crushes to a fine powder, mixed with 15 to 30 per cent of fire clay and with water and then moulded.
- Although these bricks are highly refractory (when properly burnt), they are not commonly used because of high shrinkage at high temperature.

**3. Neutral bricks**

These bricks are used for *neutral lining.* As compared to the basic bricks, the neutral bricks are more *inert to the slags.*

(*i*) **Chrome bricks:**

- These bricks contain 30 to 40 per cent of chrome oxide, the remainder being mainly oxide of magnesium, aluminium and silicon.

- They are less refractory and weaker than magnesia bricks.

**Properties:**

1. Low porosity.
2. High resistance to both acid and basic slags.
3. High thermal conductivity.
4. Low resistance to spalling.

**Uses:** Chrome bricks are extensively used in steel industry for lining open hearths and also in the bottom of soaking pits.

(*ii*) **Chrome-magnesite bricks:**

- These bricks are made from a mixture of chrome ore and magnesite with a predominating percent of chrome ore.
- They possess good mechanical properties.

(*iii*) **Spinel:**

- It is a special refractory consisting of magnesia and alumina.
- It is produced in an electric furnace.

(*iv*) **Forsterite bricks:**

- These bricks are made from the mineral fosterite.
- They have less shrinkage and possess good mechanical properties as high temperature.

In Table 3.4 are shown the common refractory materials with their chemical composition, melting points, etc.

**Table 3.4. Common Refractory Materials**

| Name | Formula | Composition | Melting point | Class |
|---|---|---|---|---|
| *Silica* | $SiO_2$ | 100% $SiO_2$ | 1713 °C | Acid |
| *Kaolin (Fire-clay)* | $Al_2O_3.2SiO_2\ 2H_2O$ | 4.50% $Al_2O_3$ | About 1775 °C | Acid (when pure) |
| *Sillimanite* | $Al_2O_3.SiO_2$ | 54.1% $SiO_2$ | About 1800 °C | |
| | | 63% $Al_2O_3$ | | Acid |
| | | 37% $SiO_2$ | | |
| *Mulite* | $3Al_2O_3.2SiO_2$ | 72% $Al_2O_3$ | | |
| | | 28% $SiO_2$ | 1900 °C | Acid |
| *Alumina* | | | | |
| *(Corrundum)* | $Al_2O_3$ | 100% $Al_2O_3$ | 2050 °C | Acid |
| *Lime* | $CaO$ | 100% $CaO$ | 2570 °C | Basic |
| *Magnesia* | $MgO$ | 100% $MgO$ | 2800 °C | Basic |
| *Chromite* | $FeO.Cr_2O_3$ | 32% $FeO$ | | Neutral |
| | | 68% $Cr_2O_3$ | 2180 °C | Neutral |
| *Zirconia* | $ZrO_2$ | 100% $ZrO_2$ | 2700 °C | Neutral |
| *Graphite* | $C$ | 100% $C$ | 3000 °C | Neutral |

## 3.11 BUILDING TILES

- *Tiles are thin slabs of brick earth, burnt in kiln.*
- These are thinner than bricks and have a greater tendency to crack and warp in drying and burning than ordinary bricks and are more liable to breakage. Therefore, a great care is needed in their manufacture. They should be dried in the shade, burnt and cooled gradually in specially made kilns.

### 3.11.1 Manufacture of Tiles

The manufacture of the common tiles includes the following operations:

1. Preparation of clay
2. Moulding of tiles
3. Drying of tiles
4. Burning of tiles.

**1. Preparation of clay:**

- The selected clay (the best blue clay for tiles is generally found a few feet below brick earth) is made free from any impurity such as grit, pebbles.
- The clay is ground into powder in *crushing roller* or *mills* and then thoroughly pugged. For the manufacture of *superior tiles*, the powdered clay is mixed with a large quantity of water in tanks, well stirred up and then mixture is allowed to settle. The heavier and coarse particles sink to the bottom of the tank and the finer particles are run off into the lower tank, where the mater is allowed to dry leaving the fine clay ready for mixing. The method is known as **blunging process**.
- To make the tiles *hard and impervious*, a *mixture of ground glass* and *potteryware may be added* in required quantity to the clay of tiles.

**2. Moulding of tiles:**

The following three methods are used for the moulding of tiles:

(*i*) *Wooden patterns:*

The tiles which are not uniform in section throughout their entire length are moulded in patterns (made of well-seasoned wood)

Flat cakes of clay are made and when partly dry, they are moulded, pressed and turned round a wooden pattern to give them the required shape.

(*ii*) *Potter's Wheel:*

- This method is similar to one that is adopted by a potter in the manufacture of earthenware vessels.
- This method is adopted when tile is of perfectly circular shape when on the wheel. It may have diameter varying along its length.

(*iii*) *Mechanical method:*

- This method of moulding is adopted for tiles having *a uniform section throughout their length.*
- The pugged clay is prepared by mechanical means through openings/orifices of required shape and size. The moulded clay coming out of the opening/orifice is received on a platform (wooden or iron) and the tiles are cut off to the required length by means of wire frames.

**3. Drying of tiles:**

- Two days after moulding, tiles are slightly beaten with a flat wooden mallet to correct the irregularity in shape due to warping.
- On the next day they are lifted up when hand-hard and the edges and under surfaces are scraped. They are then placed on edge and left to dry for about 2 days *under shade* to prevent warping and cracking when they are ready for loading in kiln.

**4. Burning of tiles :**

- The tiles are burnt in a typical kiln, known as *Sialkote kiln* (Refer Fig. 3.13); it can accommodate 30000 to 40000 tiles. Such kilns are partly underground and are generally provided with a roof.
- A layer of bricks is laid flat on rows of long narrow flues. The burning is effected by

firing wood placed in these flues. The bricks are arranged in such a fashion that open spaces are left in between them. Above the layer of bricks, the dried tiles are placed on edge layer by layer, until the kiln is completely loaded with raw tiles.

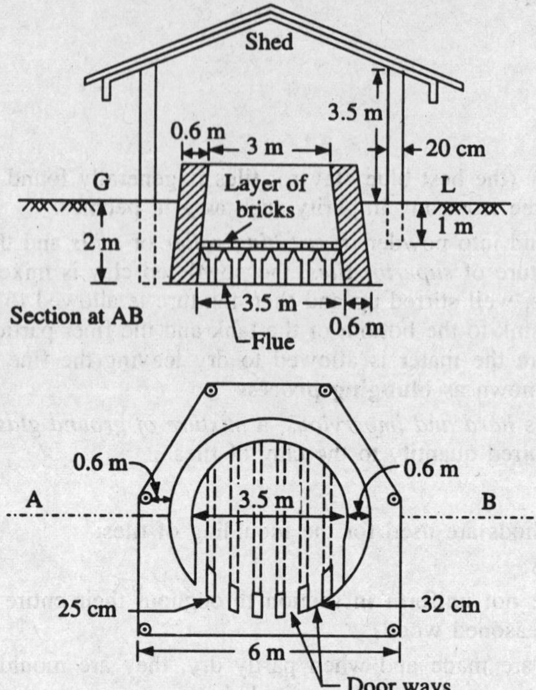


Fig. 3.13 Circular kiln for burning tiles.

- The doorways are then closed with bricks in mud plaster and top is covered with old tiles and ash.
- The firing of the kiln is started and the fire is kept low till the gases are driven out. The temperature is then raised gradually till the inner portions of the flues become red hot (at about 800 °C). It is then slackened for about 6 hours, raised again, till the flues are white hot (at about 1300 °C) and kept as such for about 3 hours. The fire is again slackened for 6 hours, then increased once more to white hot and kept steady for about 3 to 4 hours. Finally the flues are filled with fuel and doorways are closed by brickwork in mud. The kiln is then gradually allowed to cool down. The process of the burning the tiles is completed in about 72 hours.

  "*Single firing technology*"-a new *automatic process*, entails the following advantages:

(*i*) Drastic reduction of the firing cycle from 72 hours to an average of just one hour.

(*ii*) Reduced fuel consumption

(*iii*) Reduced total cost of production

(*iv*) Improved quality design, etc., of the burnt tiles.

### 3.11.2. Characteristics of a Good Tile

A good tile should possess the following *characteristics:*

1. It should possess uniform colour.
2. It should be properly burnt.
3. It should be free from cracks, flaws or bends.
4. It should be hard and durable.
5. It should have proper shape and size.

6. When placed in position, it should fit well.
7. Its broken surface should exhibit even and compact structure.
8. It should give a clear ringing sound when struck with light hammer or with another tile.

### 3.11.3 Types of Tiles

The tiles, according to their use, are divided into the following *five* types:

1. Roofing tiles
2. Flooring tiles
3. Wall tiles
4. Drain tiles
5. Glazed earthenware tiles.

**1. Roofing tiles:**

These tiles should be *strong, durable* and *perfectly leakproof.*

Although *expensive in first cost,* they need less maintenance cost.

Roofing tiles may be common tiles, encaustic tiles, manglore tiles and country tiles. The commonly used roofing tiles are described below:

(*i*) *Flat tiles:*

These tiles are rectangular in shape and are of various dimensions. They are laid in cement or lime mortar. The different types of flat tiles are discussed below:

(*a*) *Slate tiles:*

- The sizes available are: 60 cm × 30 cm × 15 mm and 50 cm × 25 cm × 10 mm.
- These tiles should be reasonably straight, of uniform colour, texture, free from veins, cracks, fissures, white patches, etc.
- The water absorption after 24 hours immersion in cold water should be maximum 21 per cent by weight.

(*b*) *Burnt clay flat terracing tiles:*

- These tiles should be uniform in shape, size and be tree from irregularities, (*e.g.* bends, twists, cracks etc.)
- The water absorption should not exceed 20 percent by weight.
- The compressive strength should not be less than 7.5 N/mm$^2$.

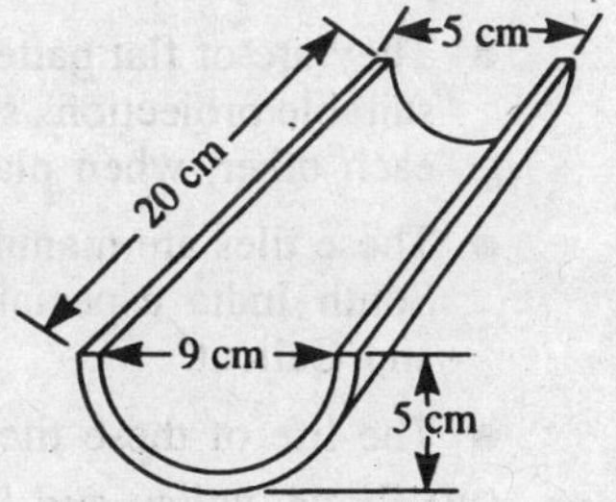

Fig. 3.14 Pot tile.

(*ii*) *Pot tiles or country tiles:*

- These tiles are also sometimes called *pan tiles*. They are hand moulded, first into a flat tile, then to the required shape on wooden pattern and burnt in a kiln after drying.
- These tiles are semi-circular in section (refer Fig 3.14). These are used either alone or with flat tiles.
- These tiles are less liable to be displaced by the birds.
- These tiles may be used as sole covering to the roof.
- These tiles are laid on sloped roofs along with the concave side up and longer end towards the ridge. Then another row of same tiles with convex side up and small end towards the ridge is laid covering the adjoing edges of every pair of tiles below.
- There tiles find extensively use in the rural areas.

Fig. 3.15 Allahabad tiles.

(*iii*) *Allahabad tiles: (Refer Fig. 3.15)*

- These tiles are of different shapes. They are generally laid side by side and joints are then covered with half round tile.
- They should not absorb water more than 20 per cent by weight.
- They are used for making good and pleasing roofs.

(*iv*) *Corrugated tiles:* Refer Fig. 3.16.

- These tiles have corrugations and when they are placed in position, a side lap of *one* or *two* corrugations is formed.

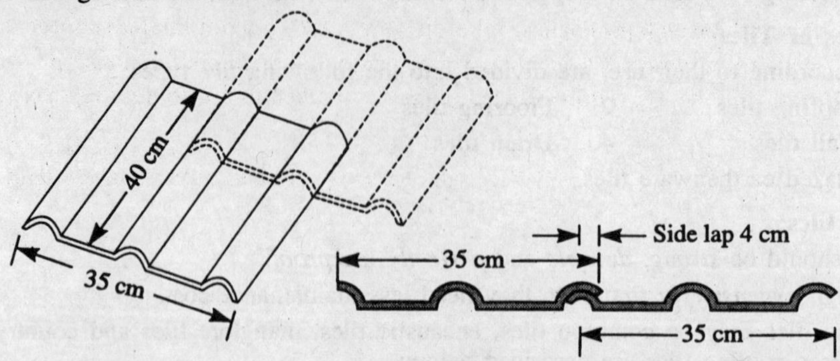


Fig. 3.16 Corrugated tiles.

- The placing of such tiles on a roof gives an appearance of corrugated galvanized iron sheets.
- They are very good in appearance, but can be easily blown away by a violent wind.

(*v*) *Manglore tile :*

- These tiles are red in colour and made of double chanelied Basel Mission Manglore pattern.
- They are of flat pattern and are provided with suitable projections so that they interlock with each other, when placed in position.
- These tiles are manufactured on large scale in South India especially in Manglore, Cochin and Calicut.

Fig. 3.17 Arrangement of two tiles, length-wise

- The life of these tiles is about 25 years with replacement of about 5 per cent per year.

(*vi*) *Ridge, valley and lip tiles:* (Refer Fig. 3.18).

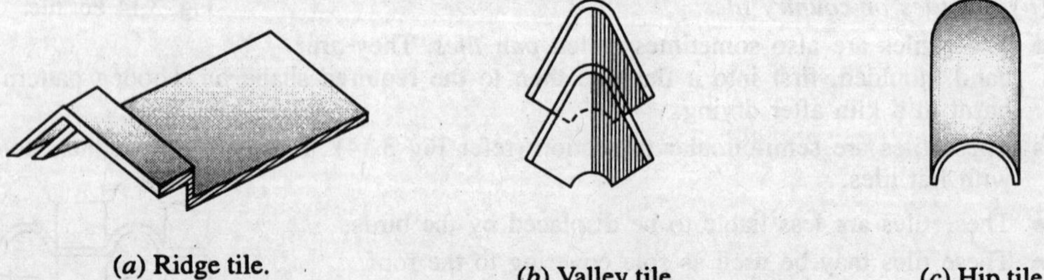

(*a*) Ridge tile. (*b*) Valley tile. (*c*) Hip tile.

Fig. 3.18 Purpose made tiles, for sloped roofs.

- These tiles are made from good earth, and are specially shaped.

(*vii*) *Encaustic tiles:*

These tiles consist of three layers:

(*i*) The face is a thin coat of pure clay of the colour required.

(*ii*) The body is of coarse clay; and

(*iii*) The back is formed with a thin layer of clay different from the body as it is to prevent warping.

- These tiles are employed for decorative purposes.

**2. Flooring tiles:**

- These tiles are flat and usually square or rectangular in shape. These can be made in any colour and of any geometrical shape.
- They should give a ringing sound when struck with each other.
- They should not absorb water more than 24 per cent by their weight.
- The fractured surface of the tile should be clean, dense and sharp at the edges.
- They should show maximum resistance to impact.
- Common sizes of flooring tiles are: 15 cm (length) × 15 cm (width) × 18 mm (thickness); 20 cm × 20 cm × 20 mm; 22.5 cm × 22.5 cm × 22 mm.

**3. Wall ties:**

- These are similar to the floor tiles except for their design and degree of burning. They are burnt comparatively at a low temperature, glazed and burnt again at a still lower temperature.
- They can be made in different designs and colours and be built to any size.

**Uses:** These are used on face work; to some extent on arches and for architectural ceilings, etc.

**4. Drain tiles:**

- These tiles are usually long curved sections of various shapes and sizes such as semi-circular, a segment of a circle or circular.

**Uses:** They are generally used for *draining waste water*. If they are to be employed for carrying sewage, glazed tiles should be used.

**5. Glazed earthenware tiles:**

- These tiles are of earthenware, having top surface glazed and underside unglazed so that tile may adhere properly to the base.
- These tiles are made in two sizes: 149 mm × 149 mm and 99 mm × 99 mm, having thicknesses 5 mm, 6 mm and 7 mm. The joint thickness is 1 mm.
- They should not absorb more than 18 percent of water.

**Uses:** These tiles and associated fittings are generally used for finishing the surfaces of walls and floors of water closets, bathrooms, kitchens, hospitals etc., (where cleanliness is important).

### 3.11.4 Specifications of Building Tiles

**1. Dimensions and tolerance:**

Dimensions and tolerance for various building tiles are given below:

| Type of tiles | Dimensions | Tolerance |
|---|---|---|
| *Roofing tiles* | 20 cm × 20 cm or<br>20 cm × 10 cm or<br>15 cm × 15 cm<br>Thickness: 15 mm or 20 mm | ± 5 cm for length and width<br>± 2 cm for thickness |
| *Floor tiles* | 15 cm × 15 cm × 1.5 cm<br>15 cm × 15 cm × 2.0 cm<br>20 cm × 20 cm × 2.0 cm<br>20 cm × 20 cm × 2.5 cm<br>25 cm × 25 cm × 3.0 cm | ± 5 cm for length and width<br>± 2 cm for thickness |

| | | |
|---|---|---|
| *Terracing tiles* | Length : 25 cm to 15 cm in stages of 2.5 cm<br>Width : 20 cm to 10 cm in stages of 2.5 cm<br>Thickness : 20 mm and 15 mm | ± 2% for length<br>± 2% for width<br>± 2% for thickness |

**2. General quality:**

- The building tiles should be made from good clay of even texture.
- They should be well-burnt.
- They should be uniform in size and shape.
- They should be free from irregularities such as bends, twists, cracks, etc.

**3. Warpage test:**

- In case of *flooring tiles*, warpage should not exceed 2 percent along the edges and 1.5 percent along the diagonal.
- In case of *terracing tiles*, maximum warpage should not exceed in any direction by 1 percent.

**4. Water absorption test:**

Water absorption test for roofing, flooring and tracing tiles is carried out as follows:

- Take six tiles and dry them in an oven at a temperature of 100 °C to 110 °C till they attain a constant weight and then cool.
- Weigh them and then immerse the dry specimen completely in clean water at 24 °C to 30 °C for 24 hours.
- Remove each specimen, wipe off the surface water carefully with a damp cloth and weigh the specimen within 3 minutes after removing the specimen of water.

$$\%\text{age of water absorption} = \frac{W_2 - W_1}{W_1} \times 100$$

where $W_1$ = weight of the dry specimen, and
$W_2$ = weight of the specimen after 24 hours immersion in cold water.

## 3.12 EARTHENWARE AND STONEWARE

### 3.12.1 Earthenware

- The clay product which is manufactured from *ordinary clay mixed with sand, crushed pottery*, etc., by burning *at low temperature* is known as *earthenware.*
- The earthenwares are generally *soft* and *porous.* They are liable to damage by atmospheric action.
- They should not absorb water more than 18% by weight.
- When glazed, the earthenwares become impervious to the water and they are not affected by acids or atmospheric agencies.

**Uses:** Earthenwares are used for making *ordinary drain pipes, partition blocks* and *electrical cable conduits.* Glazed earthenware *tiles* are used in finishing floors and walls of kitchens, bathrooms, water closets, hospitals, etc., where cleanliness is an important criterion.

#### 3.12.1.1 Terra-cotta

Terra-cotta is a kind of *high quality earthenware* which is used as a substitute for stone for *ornamentation of buildings.*

*Manufacture*:

— It is made from a *finely ground mixture of refractory brick clay* and shale or refractory clay and impure clay which is mixed with ground brick or other burnt clay product to reduce shrinkage.

— It is then ground to a fine powder, put into a large water tub and thoroughly mixed with a spade. Water is then run off from the tub.

— This clay is thoroughly pugged when dry enough to make it fit for moulding.

— Moulding is done by pressing the clay carefully in moulds in a manner similar to brick moulding.

— After being moulded the terra-cotta pieces are slowly dried to avoid warping and burnt (in special kilns without coming in contact with fire) with utmost case to prevent distortion or discolouration by flames and gases. The maximum temperatures attained are 1100 °C to 1200 °C.

— Porous terra-cotta is made by mixing sawdust in the clay.

*Advantages:*

(*i*) Can be cleaned easily.
(*ii*) Light in weight.
(*iii*) Strong and durable material.
(*iv*) Available in several colours.
(*v*) Can be moulded into desired shapes easily.
(*vi*) Being fire-proof, it can be used with R.C.C work conveniently.
(*vii*) Not affected by acids and atmospheric agencies.
(*viii*) Cheaper than ordinary finely dressed stones.

*Uses:* (*i*) Used for all sorts of ornamental work.
(*ii*) Used as building material in the form of shaped blocks.

#### 3.12.1.2 Poreclain

- The term porcelain is used to indicate fine earthenware which is white, thin and semi-transparent; it is sometimes referred to as *whiteware*.
- It is prepared from clay, felspar, quartz and minerals. The constituents are finely ground and then are thoroughly mixed in liquid state. The mixture is given the desired shape and is burnt at high temperature.
- The various types of porcelains available are used for various uses such as sanitary-wares, electric insulators, crucibles, reactor chambers, etc.
- Low voltage porcelain is prepared by *dry process* and is mainly used for lamp sockets, switch blocks, etc. *High voltage porcelain* is prepared by *wet process*.

### 3.12.2 Stoneware

- It is the clay product which is manufactured from *refractory clay mixed with crushed pottery, powdered stone, etc.*, by burning at *high temperature*.
- The clay is prepared carefully and moulding is done accurately to the required shape. It is then dried and the material burnt in a kiln, raising the heat gradually to a high temperature and maintaining this for one to three days depending upon the size of kiln and the articles in it. The burnt stoneware is allowed to cool down gradually.
- It is impervious to moisture, hard, closed grained and durable.
- The stoneware is more compact and dense than earthenware.
- The stonewares can be kept clean easily.
- When glazed, the stonewares become impervious to the water and they are not affected by acids or atmospheric agencies.

*Uses:* These are employed for:

(*i*) Drain and sewer pipes.
(*ii*) Sanitary wares
(*iii*) Vessels or jars to hold chemicals, etc.

## 3.13 GLAZING

- It is the *process of forming some transparent film over the surfaces of bricks, tiles, earthenware or stoneware to improve upon their appearance and also to protect them from the action of weather, sewage, chemicals or other destroying agents.*

- A *glaze* is a glassy coat of thickness about 0.1 to 0.2 mm applied on the surface of an item and then fused into place by burning at high temperature.
- Glazing should be such that it does *not craze* (formation of fine cracks on the surface of clay) when the article is exposed to any change of temperature.
- Glazing may be *transparent* like glass or opaque like enamels. For obtaining coloured glazes, the oxides and salts of various metals or special refractory colouring agents are added.

### 2.13.1 Transparent Glazing

Transparent glazing is of two types

1. Salt glazing
2. Lead glazing.

1. *Salt glazing:*

- In this method, common salt (sodium chloride) is thrown into the kiln, containing the articles to be glazed at the close of the burning operation when the temperature reaches to about 1300 °C. The salt vapourizes and attaches itself chemically to all exposed surfaces of the clay products, thus forming a thin layer of transparent glass over the surfaces. The quantity of salt and throwing at proper time should be done with extreme care.
- This method is useful for *sanitary pipes* and *chemical stonewares.*

2. *Lead glazing:*

- Lead glazing is preferred to salt glazing for getting articles of better quality.
- In this method, hot article to be glazed is *dipped in a bath containing oxide of lead and tin when particles of these materials adhere to the article dipped. The article is then withdrawn and reburnt when high temperature of the finance melts it forming a film of glass over the exposed surface.*
- In this method of glazing, the glaze does not penetrate into the body of the ware and as a matter of fact, it can be easily detached from the surface of the ware.

**Uses:** This method is used for terra-colta, earthenwares and fire-clay wares.

### 3.13.2 Opaque Glazing or Ship Glazing

- The glazing materials consist of quartz, felspar, boric oxides, oxide of tin, zinc, lead, china clay etc. A mixture of these materials is prepared and ground to fine paste (known as slip) in the presence of water.
- The clay product to be glazed is first dried and dipped into the slip (paste) and then fired into the furnace at a temperature of about 120°C.During the process of firing the composition of glaze gets completely vitrified and forms uniform glaze of desired colour over the surface of the burnt clay product.
- In this glazing process, burning and glazing of the clay products are simultaneously achieved.

**Uses:** Opaque glazing is used where it is desired to give the whole or any part of the article a better appearance than that presented by the burnt material. Basins, dishes, cups, etc., are opaque glazed.

## HIGHLIGHTS

1. Brick earth is obtained from the disintegration of igneous rocks. The main constituents of brick earth are silica (or sand) and alumina (or clay). Lime and oxides of iron act as fluxes helping the grains of sand to melt and bind the particles of clay together.
2. Harmful ingredients in brick earth are: Limestone and kankar moulds, alkalies, pebbles of stone and gravel, iron pyrites, kallar or reh, vegetation and organic matter.
3. The manufacture of bricks involves the following steps:

(*i*) Preparation of clay (*ii*) Moulding of bricks
(*iii*) Drying of bricks (*iv*) Burning of bricks.

4. Bricks are burnt in clamps or kilns. The clamp is a temporary structure for burning and is built again for the next burning of raw bricks. Continous kilns are of the following three types:
(*i*) Bull's french kiln (*ii*) Hoffman's kiln
(*iii*) Tunnel kiln.

5. Bricks are classified as: (*i*) Sun dried or kutcha bricks, (*ii*) Burnt or pacca bricks. Burnt bricks are further classified as:
(*a*) First class bricks, (*b*) Second class bricks,
(*c*) Third class bricks, and (*d*) Overburnt or Jhama bricks.

6. Refractory bricks are of the following types:
(*i*) Acid bricks, (*ii*) Basic bricks, and
(*iii*) Neutral bricks.

7. Tiles are thin slabs of brick earth, burnt in kiln.
There are principally three types of building tiles depending upon their use:
(*a*) Roofing tiles
   (*i*) Common tiles
   (*ii*) Encaustic tiles
   (*iii*) Manglore tiles
   (*iv*) Country tiles
(*b*) Flooring tiles
(*c*) Wall tiles.

8. Earthenware and stoneware are made from mixed earths or clays mixed with sand, crushed pottery, etc. Generally the softer, porous variety, burnt at low temperature, is called *earthenware* while the more compact variety made from refractory clay mixed with crushed pottery and stone and burnt at a high temperature is called *stoneware*.
*Uses:* Sanitary wares, drain and sewer pipes, vessels or jars to hold chemicals, etc.
*Terra-cotta* is a kind of high quality earthenware which is used as substitute for stone for the ornamentation of buildings.

9. *Glazing* is the process of forming some transparent film over the surfaces of the bricks, tiles, earthenware or stoneware to improve upon their appearance and also to protect them from the action of weather, sewage, chemicals or other destroying agents. Glazing may be transparent like glass or opaque like enamels.

## OBJECTIVE TYPE QUESTIONS

### A. Choose the Correct Answers:

1. Excess of alumina in clay in bricks
(*a*) makes the bricks crack and warp on drying
(*b*) makes the bricks dense and sound
(*c*) improves impermeability and durability of the bricks
(*d*) imparts yellow colour to bricks.

2. Hollows bricks are used for
(*a*) ornamental designs (*b*) increasing the bearing area
(*c*) resistance against heat flow (*d*) earthquake proof buildings.

3. Excess of silica in clay for bricks
(*a*) makes bricks hard and sound (*b*) makes the bricks crack and wrap on drying.
(*c*) results in deep red colour of bricks (*d*) improves impermeability and durability of bricks.

4. Bricks containing a relatively large proportion of soluble salts are liable to become discoloured by the formation of a whitish deposit, known as
(*a*) chuffs (*b*) efflorescence
(*c*) bloating (*d*) underburning.

5. The indentation marks left on bricks during the process of moulding are known as

(a) fillets (b) frogs
(c) marks (d) projections

6. If rain water falls on hot brick then the shape of the brick gets deformed, the defect being known as
(a) chuffs (b) efflorescence
(c) nodules (d) lamination.

7. A bull nose brick is *not* used in
(a) walls (b) arches
(c) pillars (d) rounding off sharp corners.

8. Fire bricks are always set in a mortar of
(a) fire clay (b) cement
(c) lime (d) any of the above.

9. Which of the following types of bricks are prepared for panel walls in multi-storeyed buildings?
(a) Fine bricks (b) Hollow bricks
(c) Squint bricks (d) None of the above.

10. The crushing strength of a first class brick is
(a) 3 N/mm$^2$ (b) 5.5 N/mm$^2$
(c) 7.5 N/mm$^2$ (d) 10.5 N/mm$^2$.

11. Water absorption by a first class brick in 24 hours immersion in cold water should not exceed
(a) 5 percent (b) 10 percent
(c) 15 percent (d) 20 percent.

12. Which of the following bricks are expected to have highest percentage of water absorption (by dry weight)?
(a) Common building bricks (b) Engineering bricks
(c) Pressed bricks (d) Fire bricks.

13. Glazing of clay products is done to
(a) make the products of uniform size (b) increase strength
(c) provide a surface for painting (d) improve appearance and protect from corrsive action.

14. The dry brick is usually expected to have moisture content of about
(a) 2 percent (b) 6 percent
(c) 10 percent (d) 20 percent.

15. Dolomite bricks are
(a) acid refactory bricks (b) basic refractory bricks
(c) neutral refractory bricks (d) ordinary bricks.

16. The major constituent of fire clays is
(a) hydrocarbon (b) lime
(c) iron oxide (d) hydrated aliminium silicate.

17. Refractory bricks resist
(a) high temperature (b) dampness
(c) chemical action (d) shocks and vibrations.

18. The advantage of kiln drying is
(a) volatile constituents are driven off
(b) reduced time for seasoning
(c) reduction for moisture level below that is possible in air seasoning
(d) all of the above.

19. In chamber drying, the range of temperature of air is around
(a) 10 °C to 25 °C (b) 25 °C to 40 °C
(c) 40 °C to 100 °C (d) 100 °C to 200 °C.

20. Efflorescence is caused by
(a) low silica content of bricks (b) the presence of alkaline salts

(*c*) the high pH content of water used during pugging (*d*) all of the above.

21. Which of the following is the harmful constituent of brick clays?
(*a*) Limestone (*b*) Iron pyrites
(*c*) Vegetation and organic matter (*d*) Pebbles of stone
(*e*) All of the above.

22. Swelling of bricks takes place due to the presence of excess of carbonaceous matter and any gas forming material in the clay and due to bad burning or very rapid firing. This defect is known as
(*a*) chuffs (*b*) lamination
(*c*) nodules (*d*) bloating.

## ANSWERS

| | | | | |
|---|---|---|---|---|
| **1.** (*a*) | **2.** (*c*) | **3.** (*b*) | **4.** (*b*) | **5.** (*b*) |
| **6.** (*a*) | **7.** (*b*) | **8.** (*a*) | **9.** (*b*) | **10.** (*d*) |
| **11.** (*d*) | **12.** (*a*) | **13.** (*d*) | **14.** (*b*) | **15.** (*b*) |
| **16.** (*d*) | **17.** (*a*) | **18.** (*d*) | **19.** (*c*) | **20.** (*b*) |
| **21.** (*e*) | **22.** (*d*) | | | |

## B. Fill in the Blanks/Say 'Yes' or 'No'.

1. Brick earth is derived by the disintegration of .................... rocks.
2. Alumina clay is the principal constituent of every kind of brick earth.
3. .................... helps silica to melt at lower temperature and binds the particle of the bricks together.
4. Natural drying (of bricks) is also called hack drying.
5. Clamp or open kiln is also known as ....................
6. A first class brick when soaked in water for 24 hours and dried in shade should not show ....................
7. A brick moulded with a rounded angle is termed as ....................
8. A good refractory material should be chemically active at elevated temperature.
9. .................... bricks are made of calcined dolomite, the carbonate of magnesia and lime.
10. ....................bricks are used for neutral lining.
11. .................... are thin slabs of brick earth, burnt in kiln.
12. The earthenwares are generally soft and porous.
13. Terra-cotta is a kind of high quality earthenware.
14. Porous terra-cotta is made by mixing .................... in the clay
15. In opaque glazing process, burning and glazing of the clay products are simultaneously achieved.

## ANSWERS

| | | | | |
|---|---|---|---|---|
| **1.** Igneous | **2.** Yes | **3.** Lime | **4.** Yes | **5.** Pazawah |
| **6.** Efflorescence | **7.** Bull nose | **8.** No | **9.** Dolomite | **10.** Neutral |
| **11.** Tiles | **12.** Yes | **13.** Yes | **14.** Sawdust | **15.** Yes. |

## THEORETICAL QUESTIONS

1. What are the characteristics of good brick earth?
2. Explain briefly the harmful ingredients in good brick earth, stating their effects on the properties of the brick.
3. What are the essential constituents of good brick earth? How is the quality of brick affected by the excess of one or other of the constituents?
4. Give briefly the characteristics of good bricks.
5. Discuss the relative advantages and disadvantages of bricks as compared with stone as a building material.
6. What is sand moulding?

7. Compare the merits and demerits of burning bricks in clamp and kilns.
8. What is the principle of continuous brick kiln? Show by a line diagram how it is followed in a Hoffman's kiln.
9. What are the characteristics of a 'First class brick'? How will you test the quality of brick?
10. What is a pug mill? What are its uses and advantages? How is it set up? Describe its working with dimensioned sketches.
11. Describe with sketches the manufacture of bricks by Bull's trench kiln.
12. Describe with sketches the different stages of brick manufacture. How does the manufacture of bricks differ from that of the tiles.
13. Explain briefly the terms: (*a*) Country bricks, (*b*) Table-moulded bricks, (*c*) Wire-cut bricks. Describe briefly the process of their manufacture, mentioning their relative merits.
14. What are the chief points to be observed in selecting bricks for walling?
15. What are the tests usually carried out on bricks required for a large construction?
16. Give the characteristics of different kinds of bricks generally turned out from brick kilns, and their uses for construction purposes.
17. What is good brick-earth? Presence of what material in excess in the clay makes the brick brittle, and what is pure clay?
18. What are the requisite properties of earth that is to be used for the manufacture of tiles and fine bricks?
19. Describe fireclay and its properties and uses in engineering industries.
20. What are stoneware products and how are they manufactured and used?
21. Explain briefly the following: Earthenware, terra-cotta, stoneware, porcelain.
22. What are the encaustic tiles and for what purpose are they used?
23. What materials are necessary for making firebricks, terra-cotta ware and roof tiles?
24. Write a short note on "Glazing".

# 4

# Lime

> *4.1 Introduction. 4.2 Sources of lime. 4.3 Properties of lime. 4.4 Uses of lime. 4.5 Important technical terms. 4.6 Constituents of limestones. 4.7 Classification of limes. 4.8 Manufacture of lime. 4.9 Determining the slaking nature of lime. 4.10 Precautions in handling of lime. 4.11 Storage of lime. 4.12 Testing of limestones-physical tests – chemical tests. 4.13 Pozzolanic materials – introduction - properties and uses – types of pozzolanic materials – Highlights – Objective Type Question – Theoretical Questions.*

## 4.1 INTRODUCTION

- *Lime* is produced by *heating limestone which is more or less pure calcium carbonate.* During heating, carbon-dioxide is driven off and non-volatile impurities such as oxides of silicon, magnesium, aluminium, iron and manganese are left behind in the lime.
- *Lime* or *quicklime* as it is sometimes called is an essentially calcium oxide, CaO, although the term is also applied to mixtures in which calcium oxide is the principal ingredient. Pure calcium oxide is a white solid which melts at 2570 °C. When the molten material is cooled, it solidifies in cubic crystals, but as ordinarily prepared it is non-crystalline, easily powdered and has a specific gravity varying from 3.08 to 3.30. The specific gravity of lime depends on the temperature to which it has been heated, *higher temperatures giving a denser product.*
- Lime has been used as a material of construction from ancient times. Egyptians used lime for plastering, while Romans used this material extensively for making mortar, plaster and concrete. In India, the historical structures such as palaces, massive forts, dams and bridges, etc., were built with lime as cementing material. Even today, it is used in states like Rajasthan, Punjab and Haryana where limestone is abundantly available. Although the cement has replaced lime to a great extent, yet in view of the rising cost it is quite likely that lime will come into prominence again.

## 4.2 SOURCES OF LIME

The lime is not generally found in nature in the *free state* but it is obtained by *burning* one of the following materials:

1. Limestone found in limestone hills,
2. Limestone builders found in the beds of old rivers;
3. Kankar found below ground;
4. Shells of sea animals.

**White chalk** is *pure limestone* and *kankar* is *an impure limestone.*

## 4.3 PROPERTIES OF LIME

Following are the properties of good quality lime, which makes it suitable for use as an engineering material:

1. Easily workable.
2. Provides strength to the masonry.
3. Possesses good plasticity.
4. Offers good resistance to moisture.
5. Stiffens early.
6. An excellent cement and adheres to the mesonry units perfectly.
7. Lime masonry proves durable due to low shrinkage in drying.

## 4.4 USES OF LIME

Following are the various uses of lime:

1. It is used as a matrix for concrete.
2. It is used as a *binding material* in mortars for stoneware and also in *bedding* and *jointing* brickwork of low strength.
3. It is used for *plastering* walls, ceilings, etc.
4. It is employed for *white washing* and as a *base coat for distempers.*
5. It is used for knotting of timber work before painting.
6. It is used for production of *artificial stone*, *lime-sand bricks*, *foam-silicate products* etc.
7. When mixed with Portland cement, the lime-cement mortar attains such valuable properties that it replaces the costly cement plaster and serves as a plasticiser.
8. It is used as a flux in the manufacture of steel.
9. Eminently hydraulic lime can be used for masonry work below ground level.
10. It is used in the manufacture of paints.
11. It is used for stabilizing the soils.
12. It is employed for creating good sanitary conditions in foul, damp and filthy places.

## 4.5 IMPORTANT TECHNICAL TERMS

**1. Calcination and Quick lime**

- *Calcination* is the process of heating limestone to redness in contact with air. Due to calcination, the moisture and carbon dioxide are removed from the limestone and the remaining product is known as *Lime*; its chemical composition is oxide of calcium (CaO). The chemical reaction is as follows:

$$\underset{\text{(Lime stone)}}{CaCO_3} = \underset{\text{(Lime)}}{CaO} + CO_2$$

- The *lime obtained by calcining pure limestone is called* **Quick lime** (CaO). It has *great affinity for moisture*. It is *amorphous*, (*i.e.* not crystalline) and highly caustic having no *affinity for the carbonic acid.*

**2. Lump lime:** It is the lime obtained in lumps after burning or calcination in kilns.

**3. Fat lime:** It is that lime which has high calcium oxide content and can set and become hard only in the presence of carbon dioxide (from atmosphere). This type of lime is perfectly white in colour.

**4. Hydraulic lime :** Lime containing small quantities of silica, alumina and iron oxide, which are in chemical combination with calcium oxide and can set and become hard even in the absence of $CO_2$ and can set under water.

**5. Slaked lime/hydrate of lime:** It is formed *by the absorption of water by quick lime* and is the hydrated oxide of calcium, $Ca(OH)_2$. The chemical reaction is as follows:

$$\underset{\text{(Quick lime)}}{CaO} + H_2O \xrightarrow{\text{slaking}} \underset{\text{(Hydrated lime)}}{Ca(OH)_2} + \text{heat}$$

- A *dry powder obtained by treating quick lime with water, just sufficient to convert the lime into calcium hydroxide* is called **hydrated lime.**
- A *plastic mass of lime which results from the slaked lime in the presence of a sufficient quantity of water* is called **putty**.
- A *thin pourable suspension of slaked lime in water* is called **milk of lime.**

6. **Slaking:** It is the *process of chemical combination of quick lime with required quantity of water for killing the heat and getting into powder form for use.*

$$\underset{\text{(Quick lime)}}{CaO} + H_2O = \underset{\text{(Hydrate of lime)}}{Ca(OH)_2}$$

7. **Seting:** It is the *hardening of lime which has been mixed to a paste with water.* It is quite different from mere drying. In case of drying, the water evaporates only and no setting action takes place.

**8. Hydraulicity**

- It *is the property due to which lime (or cement) will set in damp places or under wat*[illegible] *or in thick masonry walls there is no free access of air.*
- Hydraulicity is due to the crystallizing energy of the aluminate and silicate of lime.
- Pure limestone (such as white chalk and marble) and also a mechanical mixture of limestone and sand do not possess hydraulic property.

**9. Pozzolana**

- The *volcanic dust containing 80 per cent clay with lime, magnesia and iron oxide varying proportions* is called pozzolana.
- When pozzolana is added in powdered form with lime, it imparts hydraulic properties, to the lime.
- Surkhi is considered as artificial pozzolana.

## 4.6 CONSTITUENTS OF LIMESTONES

Following are the constituents of limestones (main source of getting lime) :

**1. Clay:**

- It is responsible for producing hydraulicity in lime.
- It retards slaking when present in small quantity, arrests slaking when it is in excess.
- It makes lime insoluble in water.
- In order to make a good lime it is desirable to have 8 to 30% of clay.

**2. Soluble silica:**

- In order to develop hydraulicity it is imperative to have silica and alumina present in chemical combination with limestone. Hydraulicity is caused due to silicates of calcium, magnesium and aluminium. These silicates are inert at low temperatures but they become active and combine with lime at high temperatures.

**3. Magnesium carbonate:**

- It *increases the setting process but reduces slaking.*
- Lime countaining *not more* than 5% of magnesium carbonate gives *better results and imparts hydraulic properties* to the limes. But limes containing *large proportions* of magnesium carbonate are *liable to crack* (which ultimately disintegrate).
- When the content of magnesium carbonate is about 30%, the hydraulicity is rendered to the lime even in absence of clay.

**4. Alkalies and metallic oxides:**

They tend to become soluble silicates at low temperatures only and thus cause hydralulicity.

**5. Sulphates:**

They tend to *reduce slaking but increase setting action only when they are not in excess.*

6. **Iron compounds:**

The iron compounds, if present, lower the temperature of calcination of limestone.

7. **Carbonaceous matter:**

The presence of carbonaceous matter is harmful; it produces very poor quality of lime.

## 4.7 CLASSIFICATION OF LIMES

Limes obtained by *calcining* limestones are broadly classified as follows :

1. Fat lime
2. Hydraulic lime
3. Poor lime.

**1. Fat lime:**

- It is also know as *rich, fat, common, high calcium, air* or *pure lime.*
- It is so called (fat lime) as it slakes vigorously and its volume is increased to about 2 to $2\frac{1}{2}$ times the volume of quick lime.
- It is prepared by *calcining comparatively pure carbonate of lime* which contains about 95% of calcium oxide.

**Properties:**

(*i*) Slake vigorously.
(*ii*) Possesses high degree of plasticity
(*iii*) Hardens very slowly.
(*iv*) Possesses perfectly white colour.
(*v*) Sets slowly in presence of air.
(*vi*) Soluble in water which is changed frequently.

**Uses:** (*i*) It is used in whitewashing and plastering walls.
(*ii*) Lime mortar, made of *lime* and *sand,* is used for thin joints of brickwork and stonework.
(*iii*) Lime mortar, made of *lime* and *surkhi* (powder obtained by grinding of burnt bricks) may be used for thick masonry walls, foundations, etc.

**2. Hydraulic lime:**

- It is also known as *water lime.* The hydraulic lime, depending upon the *percentage of clay* is divided into the following three types:

(*i*) *Feebly hydraulic lime* (*ii*) *Moderately hydraulic lime* (*iii*) *Eminently hydraulic lime.*

The increase in percentage of clay makes the slaking difficult and increases the hydraulic property.

- It can set under water and in thick walls where there is no free circulation of air.
- It forms a thin paste with water.
- It is not perfectly white in colour, as such it appears *less* sanitary than the fat lime.
- The hydraulic lime containing about 30% of clay resembles natural cement.

Table 4.1 shows the comparison between various types of hydraulic limes.

**Table 4.1. Comparison between Different Types of Hydraulic Limes**

| S. No. | Aspects | Feebly hydraulic lime | Moderately hydraulic lime | Eminently hydraulic lime |
|---|---|---|---|---|
| 1. | *Percentage of clay* | 5–10 | 11–20 | 21–30 |
| 2. | *Slaking time* | Few minutes. | One or two hours. | Slakes with difficulty. |

| | | | | |
|---|---|---|---|---|
| 3. | *Setting time (in water)* | Three weeks or so. | One week or so. | A day or so. |
| 4. | *Hydraulicity* | Feeble. | Moderate. | Eminent. |
| 5. | *Uses* | Ordinary masonry work (lime mortar-reasonably strong) | Superior type of masonry work (lime mortar-quite strong) | Used for damp places, (lime mortar-similar to the ordinary cement mortar) |

**3. Poor lime:**

- It is also known as *lean* or *meagre lime.*
- It contains more than 30% of clay.
- Its colour is muddy white.
- It slakes very slowly.
- It forms a thin paste with water.
- It sets/hardens very slowly.
- It has poor binding properties.

**Uses:** As the mortar made from this type of lime is of poor quality, therefore, such mortar is employed for inferior type of work.

**Classification of Building Limes according to I.S.I. Specifications:**

According to Indian standard specifications, limes are classified as follows:

1. **Class *A***: *Eminently hydraulic lime* - used for *structural purposes.*
2. **Class *B***: *Semi-hydraulic lime* - used for *masonry mortars.*
3. **Class *C***: *Fat lime* - used for finishing coat of *plastering, whitewashing*, etc., and with addition of *pozzolanic material for masonry mortar.*
4. **Class *D***: *Magnesium lime* - used for finishing coat in *plastering, whitewashing*, etc.
5. **Class *F***: *Kankar lime* - used for *masonry mortars.*

The comparison between class *A*, class *B*, class *C* limes is given in Table 4.2.

**Table 4.2. Comparison between Class *A*, Class *B* and Class *C* Limes**

| S. N. | Characteristics | Class A (*Eminently hydraulic lime*) | Class B (*Semi-hydraulic lime*) | Class C (*Fat lime*) |
|---|---|---|---|---|
| 1. | *Composition* | Calcium oxide = 60 to 70%, clay = 20 to 30%. | Calcium oxide = 70%, clay = 15 to 20%. | Calcium oxide = 85%, clay = Nil. |
| 2. | *Slaking* | • It slakes sluggishly and does not produce any heat or noise. | • It slakes slowly producing little heat. | • It slakes vigorously producing sufficient heat with hissing noise. |
| | | • It undergoes slight increase in volume. | • There is slight increase in volume. | • The increase in volume is from 2 to 3 times. |
| 3. | *Setting time* | It sets without the presence of $CO_2$. Initial set takes place in not less than 2 hours and final set within 48 hours after the addition of water. | Sets without the presence of $CO_2$. | Sets in the presence of $CO_2$. |
| 4. | *Compressive strength* | 1.75 $N/mm^2$ after 14 days and 2.8 $N/mm^2$ after 28 days. | 1.25 $N/mm^2$ after 14 days and 1.75 $N/mm^2$ after 28 days. | ................ |

| | | | | |
|---|---|---|---|---|
| 5. | *Transverse strength* | Modulus of rupture not less than 1.05 N/mm² at 28 days. | Modulus of rupture not less than 0.7 N/mm² at 28 days | ................ |
| 6. | *Soundness* | Expansion should not be more than 10 mm. (Lechatelier apparatus). | Expansion should not be more than 10 mm (Lechatelier apparatus) | ............. |
| 7. | *Colour* | Slight greyish tinge. | Slight greyish tinge. | White. |
| 8. | *Uses* | Used in wet foundations and under water. | Employed as mortar in the interior of thick walls. | Extensively used for white washing and plastering. |

## 4.8 MANUFACTURE OF LIME

Following operations are involved in the manufacturing of lime:

1. Collection of raw materials
2. Burning of limestone
3. Slaking of burnt lime.

**1. Collection of raw materials:**

- The limestones of required quality are collected and stacked in sufficient quantity near the site of manufacture of lime so as to produce lime continuously. Pure limestone is used for manufacturing quick lime and its impure forms may be used for producing other types of lime. For *fat lime,* the percentage of impurities in limestone should not exceed 5%. *Kankar limestone* (containing about 30% clay in the form of alumina and silica and greyish in colour) is used for the preparation of *hydraulic lime.*
- The following kinds of fuels may be used for calcining limestones: Wood, coal, coke, coal gas, charcoal, cinder, oil.

**2. Burning of limestone:**

Limestone may be burnt in *clamps* or in *kilns.* Fuel generally used for burning limestone is *coal, charcoal* or *firewood.*

**Clamp burning:**

- This method of burning limestone is used only in places where limestone and the fuel is available in abundance and quantity of burnt lime required is small.
- In this method a *clamp is formed by stacking limestone and fuel in alternate layers.* The height of the clamp is usually not more than 3.6 m and its sides are sloped. The clamp is covered with mud plaster to prevent the escape of heat during the process of burning. A *fire place* is kept at the bottom of the clamp for firing.

*Demerits/Shortcomings:*

1. It is a *wasteful method* as the burnt product often contains high percentage of unburnt and overburnt lime. (The lime near the outside faces of the clamp, being in contact with air, is liable to remain unburnt or underburnt while the lime stacked in the heart of the clamp may overburnt).

2. Clamp burning is *uneconomical* as much heat is lost through the mud plaster which cracks by the heat from inside and gives an inferior quality of lime (as lime produced partly slacks by the entry of moist air through cracks in mud plaster and falls into powder and also gets mixed with the fuel ashes which cannot be separated).

3. It is *not a continuous process of burning and cannot be performed in rainy season.*

**Kiln burning:**

For burning of limestones, kilns are built in varying sizes and shapes. In general, the kilns may be of the following two types:

(*i*) Intermittent kilns

(*ii*) Continuous kilns.

**(*i*) Intermittent kilns:**

Figure 4.1 shows an intermittent kiln for burning limestone.

It is built by the side of a hill or river bank. An arch over the furnace is formed with large pieces of limestone and small pieces are then arranged on the top of arch. Here the *charge does not come in direct contact with the fuel* and consequently the *outturn is least associated with impurities* (ashes, etc.). Each charge takes 2 days for burning and 1 day for cooling before the kiln can be recharged. About 40 kg of wood fuel is required for burning a charge of 3 cubic metre.

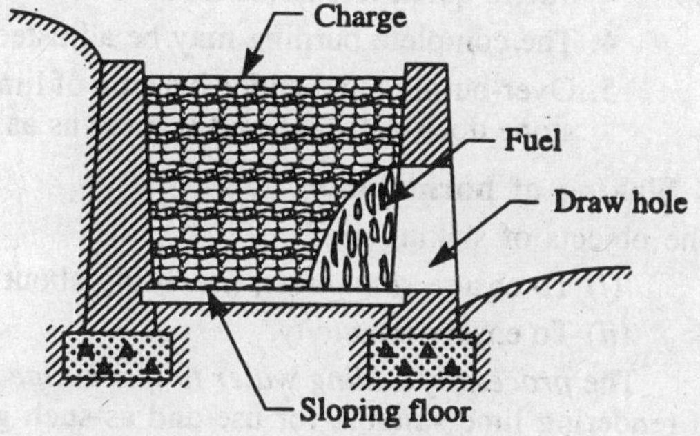


Fig. 4.1 Intermittent lime kiln.

**(*ii*) Continuous kilns:**

These kilns are used when lime is to be manufactured on *large scale.* The kiln is a vertical structure lined with fireclay bricks. It is usually built on a slope to facilitate continuous removal of lime. Alternate layers of limestone and coal (sometimes wood) are charged from the top and lime is removed from an opening at the ground level. Fire is lighted in the kiln and as the calcination progresses lime is continuously withdrawn and fresh change is fed from the top. Air required for combustion enters the same opening through which lime is removed; thus it gets heateo while the lime is cooled. *Lime obtained from such klin is not of high purity as it gets mixed with ash of fuel.* Its *thermal efficiency is however higher than of other furnaces.*

When lime of high purity is required *continuous separate feed kilns* are used. The kiln is similar to the kiln described above except that two or four fireplaces are provided by the side of the kiln near the bottom (Fig. 4.2). In the fireplaces coal or wood are burnt and the hot gases of combustion move up the shaft of the kiln.The advantage of using wood as a fuel is that long flames are produced and the *temperature of limestone is not raised too much.* When coal is burnt *some steam is introduced which maintains a low temperature and forms water gas. The water gas so produced burns and heats the kiln upto a greater height.* Limestone is supplied to kiln from the top and the lime removed from the bottom.

Fig. 4.2 Continuous lime kiln.

The *Rotary kiln* used for burning lime is similar to that used for cement manufacture but is much *smaller* in size and is worked at a *lower temperature.* It consists of an inclined steel cylinder 1.5 to 2.5 metres in diameter and 15 to 45 metres long, lined with refractory bricks. Limestone crushed to much smaller pieces is charged at the upper end. Kiln is fired with coal, oil or producer gas and daily outturn is nearly 70 to 80 tonnes of lime. Due to better control on the operation product obtained is of *superior quality*. Rotary kilns are widely used in *highly industrialised countries and for making lime for chemical industries*.

**Precautions in burning of limestone:**

The following *precautions* should be taken while burning limestone in kilns:

1. Before burning the limestones, they should be broken into suitable sizes.
2. Quantity of fuel to be used should be carefully worked out.
3. The limestones should be gradually heated (sudden heating may blow the stones to pieces due to quick release of moisture and carbon dioxide).
4. The complete burning may be adjusted by the bright red colour of the flame.
5. Over-burning and under-burning of limestone should be avoided (overburnt or underburnt stone does not slake and is knowns as **dead lime**).

**3. Slaking of burnt lime:**

The objects of slaking are:

(*i*) To ensure *soundness*, (*i.e.* bring about volumetric stability)

(*ii*) To ensure *plasticity*.

The *process of adding water to quick lime* is called *slaking of lime*. It is an important process in rendering lime suitable for use and as such great care should be taken to slake the lime. Lime should be slaked soon after it is drawn from the kiln, preferably within one week of its calcination.

The burnt lime or quick lime can be slaked in the form of **dry powder** or in the form of a cream paste called *putty*.

The common methods employed for slaking are enumerated and described below:

1. Air slaking
2. Basket slaking
3. Platform slaking
4. Tank slaking.

**1. Air slaking:**

- Lime is spread over a non-porous platform, it absorbs moisture from the air and slakes slowly into a powder form.
- This method is *unsuitable* since *complete slaking is not obtained.*

**2. Basket slaking:**

- In this method, the quick lime is broken into pieces (size not more than 50 mm) and filled in a basket.
- The basket is then immersed in clean water for a few seconds. It is then taken out and thrown on a pacca platform in the form of a heap; the quck lime crumbles and falls into **powder form**.
- This method is employed for *slaking semi-hydraulic limes.*

This method is also not very suitable as complete slaking is not obtained.

**3. Platform slaking:** Refer Fig. 4.3.

- It is usually carried out on a *clear, dry, masonry platform.*
- Large lumps of quick lime are broken to less than 50 mm size. They are spread in a 150 mm thick layer on the platform and water is then sprinkled over it from a weathering can fitted with a rose. The lime slakes and gets reduced to powder form. During the process of sprinkling water, the heap of lime is turned over and over again with a phaora and no more water is used than that required for the lime to retain its powder form. Slaked lime is then screened through a sieve of 3 meshes to a cm to remove unburnt lumps, unslaked material or other foreign matter.
- The screened lime can be used in the *preparation of mortar.*

Fig. 4.3 Platform slaking.

**4. Tank slaking:**

This method is used when lime is required in the *form of paste*, known as **lime putty**. In this process sufficiently large slaking vessels or tanks (water tight) are needed.

- Two brick-lined tanks are normally required to be constructed. One of the tanks is generally built at *ground level* and is 45 cm deep, the other tank is made adjacent to first one but at a lower level *below the ground* (Refer Fig. 4.4); usually 60 to 75 cm deep.
- The tank at the ground level is partly filled with water and sufficient quicklime is added gradually to fill up the tank to about half the level of water. It may be noted that **lime should be always added to water** and **not water to lime.**
- The lime is then stirred properly and *no part of lime* is allowed to get *exposed above water.* As the lime slakes with the evolution of heat, the temperature of water will begin to increase and water begins to boil. Thus, when the required temperature (85 °C to 98 °C) for slaking is reached, additions of lime and water are made in small quantities with constant stirring so as to maintain the required temperature. The operation is continued until the required quantity of lime or whole of the lime has been slaked.

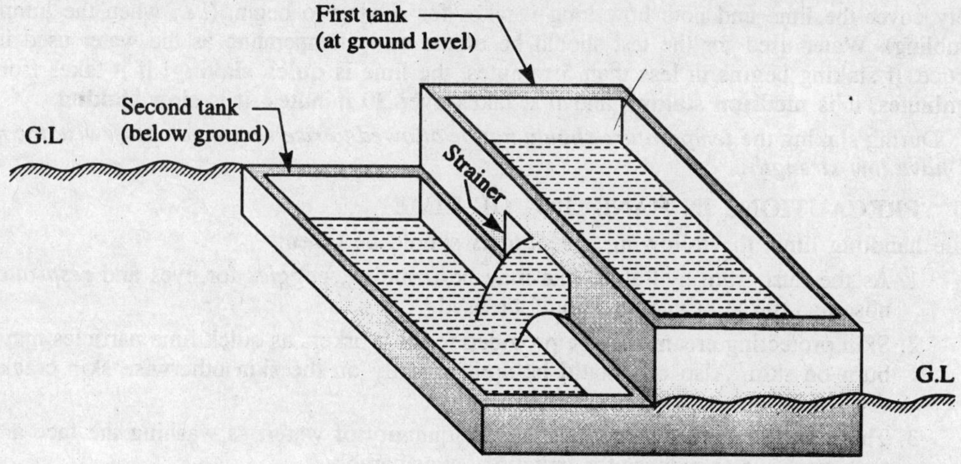


Fig. 4.4 Tank slaking.

- The lime in the state of suspension (or the milk of lime) is then allowed to pass through a sieve of 3 meshes to a cm and flow into the second tank built at lower level. The particles of lime settle down and water gets partly evaporated and partly absorbed in ground and the remaining surplus water is removed.
- The putty, prior to use, is allowed to mature in the tank for 2 to 3 days; by doing so, the *complete slaking is ensured* and *the workability of the putty improves.*

**Manufacture of Hydraulic Lime:**

Following two methods are used to manufacture hydraulic lime:

1. By natural method. 2. By artificial method.

**1. By natural method:**

In this method impure kankar (limestone) is used. Kankar containing 30 per cent of clay gives lime *additional hydraulic properties* and *ability to withstand exposure to sun and rain without disintegrating.*

- Kankar is broken in lumps of size not exceeding 50 mm, stacked in the kiln in alternate layers of fuel, (*e.g.* coal, charcoal or firewood) and burnt.
- The burnt kankar is slaked with water and screened before use.

**2. By artificial method:**

The process of manufacture, in this method, depends upon the class of lime to be used.

- In case the limestone to be used is *soft* (much as chalk, etc.), it is first ground into a fine powder. It is then mixed with correct proportion of clay and the mixture is burnt in a kiln. The burnt product has all the qualities of hydraulic lime and can be used after slaking and screening.
- In case the limestone available is **hard,** *it is first burnt and slaked.* The slaked lime is mixed with pre-determined proportion of clay and the mixture is then burnt in a kiln. In this case, as the lime is burnt *twice*, it is called "*Twice kilned lime*". The burnt product is hydraulic lime and is slaked and screened before use.
- A good hydraulic lime mortar can be made out of such a lime by mixing it with *surkhi* and *sand.* The ratio of 1 : 1 : 2 or 1 : 2 : 1 for lime, surkhi and sand respectively may be used for this purpose.

## 4.9 DETERMINING THE SLAKING NATURE OF LIME

Following is the procedure of finding slaking nature of lime. In a bucket, put two or three lumps (about the size of a man's fist) or an equal amount in case of granular lime. Add sufficient water to barely cover the lime and note how long it takes for slaking to begin, (*i.e.*, when the lumps start crumbling). Water used for the test should be of the same temperature as the water used in field practice. If slaking begins in less than 5 minutes, the lime is **quick slaking;** if it takes from 5 to 30 minutes, it is **medium slaking** and if it takes over 30 minutes, it is **slow slaking**.

During slaking the *temperature should not be allowed to rise very much otherwise the mortar will have low strength.*

## 4.10 PRECAUTIONS IN HANDLING OF LIME

While handling lime, the following *precautions* should be taken:

1. As the dust from hydrated lime may be irritating, *goggles* for eyes and *respirators* for nose and throat protection should be used.
2. Skin protecting cream may be provided to the workers as quick lime particles may cause burn on skin. Also oil should be applied daily on the skin otherwise skin cracks may develop.
3. There should be provision of adequate quantity of water, as washing the face and skin with fresh water reduces the irritations considerably.

## 4.11 STORAGE OF LIME

- The quicklime, soon after delivery, should be slaked or run into putty otherwise it will deteriorate when stored. The deterioration is caused due to the action of the atmospheric moisture on the unslaked limes, which causes it air slaked. The lime thus attacked loses a great deal of its useful properties and is considered unsuitable for sound construction.
- Lime can be safely stored in the form of lime putty for a maximum of 14 days and must be consumed within this period.
- If unslaked lime is to be stored for a short time, it should be piled up and covered with a blanket of lime powder to exclude moist air.
- In case storing of lime for a long period cannot be avoided, it should be stored in closed store houses properly insulated against the penetration of moisture from the atmosphere and from the ground below.
- *Hydrated lime* can be safely stored for considerable time without deterioration if the bags are kept in a dry place.

## 4.12 TESTING OF LIMESTONES

The following tests are generally performed on limestones:

1. Physical tests

2. Chemical Tests.

### 4.12.1 Physical Tests

These tests are based on the *texture, appearance, colour* and *odour* of the limestones.

- Limestones except crystalline variety used in the lime and cement trades are amorphous in texture.
- Deposits of chalk and fat limes have white or whitish brown or grey colour.
- Magnesian limestones are very heavy and hard.
- A good hydraulic limestone has white (rarely) bluish grey; yellow or brown colour. It has a compact texture with an irregular dull fracture. The freshly fractured surface of hydraulic limestones gives a clayey taste and an earthly smell.

### 4.12.2 Chemical Tests

The chemical tests are classified as:

1. Heat test
2. Hydraulic test-slaking
3. Acid test.

**1. Heat test :**

This test is usually carried out in open and temperature maintained is 1500 °C approximately. The limestone disintegrates giving $CO_2$ as per the equation given below:

$$CaCO_3 \rightarrow CaO + CO_2$$

Due to removal of $CO_2$ from sample it loses its weight and this *loss in weight gives an indication of the presence of $CaCO_3$ in the sample.* For every loss in weight of 44 parts there must be 56 parts of lime by weight.

**2. Hydraulic test–slaking:**

When quick lime (anhydrous product of calcination) is brought into contact with water it hydrates as follows:

$$CaO + H_2O = Ca(OH)_2$$

The process is known as **slaking**. During slaking considerable amount of heat is released due to which quick lime crumbles to powder. If the *slaking is vigorous it indicates that the quantity of calcium content is high.* In hydraulic limes the slaking action is very slow.

**3. Acid test:**

The purpose of this test is to assess the *classification* and *calcium-carbonate content of lime.*

- Place a teaspoon of powdered lime in a test tube and pour dilute hydrochloric acid into it.
- Stir the contents well with a glass rod and leave the test tube with the contents for 24 hours; then:

— If effervescence is in abundance, it indicates high percentage of calcium carbonate.

— If the content of calcium carbonate is less, there will be less effervescence and more formation of residue; this indicates that the limestone is impure or hydraulic.

— The *formation of a thick jel*, which will not flow even when the test tube is turned upside down shows **eminently hydraulic lime**.

— If the jel is *not quite thick* and *tends to flow when the test tube is tilted*, it indicates **feebly hydraulic lime.**

— *Absence of jel indicates non-hydraulic* or **fat lime.**

## 4.13 POZZOLANIC MATERIALS

### 4.13.1 Introduction

- **Pozzolanic materials** are *the silicious and aluminous mineral substances which do not* *possess cementitious properties themselves but form cementitious compounds by reacting with lime in the presence of water.*

- Pozzolanic is a *volcanic ash* containing about 80 per cent clay.
- The Greeks and Romans used calcined limestone and later developed the *pozzolanic cement* by grinding together *lime* and *pozzolana* a (this name is derived from Pozzouli town in Italy near Vesuvius; the sand available around this town when mixed with hydrated lime in the presence of water was found to form cementitious compounds).

### 4.13.2 Properties and Uses

The properties and uses of pozzolonic materials or pozzolans are given below:

**Properties:**

1. Improve workability of mortar and concrete.
2. Economical to use.
3. Do not possess cementitious properties themselves.
4. Lower the heat of hydration.
5. Assist in early setting and hardening of the mortar (as they make the mortar porous, thereby permitting the air to penetrate for providing $CO_2$, necessary for this purpose).
6. Reduce shrinkage of the mortar or concrete.
7. Enhance hydraulic properties of the lime mortar and concrete.
8. Enhance resistance of mortar or concrete to chemical attacks.

**Uses:** (*i*) Useful in mortars and concretes to be used for hydraulic structures and for *structures exposed to sea water and alkali soils.*

(*ii*) Also useful for producing lean mixes of concrete to improve their workability.

### 4.13.3 Types of Pozzolanic Materials

Various types of pozzolanic materials are given below:

**1. Naturally occurring pozzolans:**

(*i*) Clay and shales, *viz.* kaolinite, etc.

(*ii*) Opaline materials, *viz.* opaline shales, diatomaceous earth, etc.

(*iii*) Volcanic tuffs and pumicities, *viz.* Rhyolite, Andesite and Basalitic types.

**2. Artificial pozzolanas:**

(*i*) Fly ash

(*ii*) Burnt clay or ground bricks

(*iii*) Slags (*viz.*, blast furnace, water quenched slags)

The description of artificial pozzolanas given below:

(*i*) **Fly ash:**

- Fly-ash or cinder is obtained as a residue from the burning of coal in furnaces and locomotives in the formof powder or small nodules respectively.
- When ash or cinder are to be used as pozzolanic materials, they should be thoroughly cleaned of unburnt contents and finely powdered so that they should pass through an IS sieve No. 320.
- Fly-ash is used in making lime mortars (by adding one part of lime to three parts of flyash, in general), known as *black mortars*. Such a mortar is hard setting and is generally suitable for internal walls and for masonry work where more importance is given to strength than appearance.

(*ii*) **Burnt clay or ground bricks:**

- Burnt clay balls or well-burnt bricks (specially made for the purpose) when ground properly (ground material termed as *surkhi*) and added to lime provide hydraulic properties.

- Widely employed in preparing lime mortars which can be used in *joints of walls, for plastering the inside surface and for hydraulic works.* They are, however, *not* used on works exposed to weathering actions (because they disintegrate after sometime).

(*iii*) **Slags:**

The combination of slags such as blast furnace slag etc. (these *cannot* be classified as true *pozzolanas* because they possess cementitious properties) with cement produces a binding material, known as *blast furnace slag cement* which is economical and quite useful.

## HIGHLIGHTS

**1.** Lime is produced by heating limestone which is more or less pure calcium carbonate. The constituents of limestones are: Clay, soluble silica, magnesium carbonate, alkalies and metallic oxides, sulphates, iron compounds and carbonaceous matter.

**2.** Limes obtained by calcining limestones are broadly classifies as: Fat lime, Hydraulic lime and Poor lime.

**3.** Operations involved in the manufracture of lime are:
(*i*) Collection of raw materials,
(*ii*) Burning of limestone, and
(*iii*) Slaking of burnt lime.

**4.** In general, the kilns may be of the following two types:
(*i*) Intermittent kilns;
(*ii*) Continuous kilns.

**5.** The process of adding water to quick lime is called *slaking of lime.* The common methods employed for slaking are:
(*i*) Air slaking,
(*ii*) Basket slaking,
(*iii*) Platform slaking, and
(*iv*) Tank slaking.

**6.** The following tests are generally performed on limestones.
1. Physical tests
2. Chemical tests
(*i*) Heat test
(*ii*) Hydraulic test-slaking
(*iii*) Acid test.

**7.** *Pozzolanic materials* are the silicious and aluminous mineral substances which do not possess cementitious properties themselves but form cementitious compounds by reacting with lime in the presence of water.

## OBJECTIVE TYPE QUESTIONS

**Choose the Correct Answer:**

**1.** The fat lime swells ........... times of its original volume during slaking.
(*a*) two to three (*b*) three to four
(*c*) four to five (*d*) five to ten.

**2.** Quick line if left exposed to air
(*a*) will absorb carbon dioxide (*b*) will absorb moisture
(*c*) will become inert powder of calcium carbonate (*d*) all of the above.

**3.** The heating of limestone to redness in contact with air is known as
(*a*) hydration (*b*) calcination
(*c*) burning (*d*) oxidation.

**4.** Lime is used in all of the following except.
(*a*) in the manufacture of glass (*b*) as a base for cellulose paints

(c) as a matrix for concrete (d) for plastering walls, ceilings etc.

5. Which of the following is produced as a result of calcination of pure lime?
   (a) Hydraulic lime (b) Quick lime
   (c) Fat lime (d) Hydrated lime.
6. In case of hydraulic lime, the constituent mainly responsible for hydraulicity is
   (a) silica (b) clay
   (c) calcium oxide (d) carbon dioxide
7. For construction of structures under water, the lime used in
   (a) pure lime (b) quick lime
   (c) fat lime (d) hydraulic lime
8. The advantage of adding puzzolanas to lime is to
   (a) increase resistance to cracking
   (b) reduce shrinkage
   (c) increase resistance to chemical attack
   (d) all of the above.
9. Lime having high calcium oxide content, which can set and become hard only in the presence of carbon dioxide from the atmosphere, is known as
   (a) slaked lime (b) fat lime
   (c) lump line (d) calcined lime.
10. Lime is considered to be hydraulic lime when it sets under water within
    (a) 3 to 5 days (b) 7 to 30 days
    (c) 2 to 3 months (d) 3 to 6 months.
11. The process of adding water to lime to convert it into hydrated lime is known as
    (a) quenching (b) hydration
    (c) slaking (d) calcination.
12. The setting of hydraulic lime and cement is facilitated by presence of
    (a) water (b) oxygen
    (c) air (d) all of the above.

## ANSWERS

| | | | | |
|---|---|---|---|---|
| 1. (a) | 2 (d) | 3. (b) | 4. (b) | 5. (b) |
| 6. (b) | 7. (d) | 8. (d) | 9. (b) | 10. (b) |
| 11. (c) | 12. (a) | | | |

## THEORETICAL QUESTIONS

1. Mention the uses of lime in engineering works.
2. Compare and contrast the advantages and disadvantage of using lime and cement in engineering works.
3. Explain, with sketches, the process of manufacture of lime.
4. Differentiate between ordinary and hydraulic limes and mortars.
5. Describe briefly a method of burning limestone for lime manufacture.
6. Describe, giving sketches, the process of manufacture of lime.
7. Give the classifications and properties of different types of lime.
8. Explain the terms:
   (a) Quick lime,
   (b) Slaked lime,
   (c) Fat lime,
   (d) Hydraulic lime,
   (e) Artificial hydraulic lime, and
   (f) Poor lime.
9. What are the different kinds of lime available for use on works? Mention the uses of each variety.

Indicate the chemical reactions taking place during the slaking and setting of lime.

10. Describe with sketches the process of lime burning in India. What limes are used for (*a*) foundation in damp soil, (*b*) superstructuring and roofing of a building. If they differ give reasons for your selection.
11. Describe with sketches the manufacture of lime. State the chemical changes involved in the burning, slaking and setting of lime. Indicate the influence upon the setting properties of lime, of the impurities contained in the limestone and describe briefly a method of obtaining hydraulicity with a lime made from a pure limestone.
12. Describe the manufacturing of hydraulic lime and give its standard specifications.
13. How would you classify limes as per IS standards and give briefly their main properties.
14. How are limestones tested?
15. How will you determine the slaking nature of lime?
16. Explain the difference between quicklime, hydraulic lime and fat lime.
17. Describe in detail the method of burning and slaking lime.
18. How are limes classified?
19. Describe with a neat sketch an intermittent lime kiln.
20. Write short notes on:
    (*i*) Hydraulic lime; (*ii*) Hydrated lime; (*iii*) Storage of lime; (*iv*) Slaking lime to powder.
21. Compare clamp with kiln for burning.
22. What do you understand by pozzolanic materials? Explain briefly their properties and uses.

# 5

# Cement



## 5.1. INTRODUCTION

- **Cement** may be prescribed as a *material with adhesive and cohesive properties which make it capable of bonding mineral fragments into a compact whole.* The definition embraces a large variety of cementing materials.
- For constructional purposes the meaning of the term cement is restricted to the *bonding materials used with stones, sand, bricks, building blocks etc.* The principal constituents of this type of cement are *compounds of lime*, so that in building and civil engineering we are concerned with calcareous *cement.* The cements of interest in the making of *concrete* have *property* of *setting* and of interest in *under water* by *virtue of a chemical reaction with it* and are therefore called *hydraulic cement.*

## 5.2. CLASSIFICATION

*Cements* can be broadly **classified** as :

1. Natural cement
2. Artificial cement.

**1. Natural cement:**

- It is manufactured from stones containing 20 to 40 per cent of clay, the remainder being carbonate of lime mixed with carbonate of magnesia. The stones are first burnt and then crushed.
- It possesses a brown colour and sets rapidly when mixed with water.
- It does not find much use in India.

**2. Artificial cement:**

- It may be *Portland cement* or *special cement.*
- *Portland cement is so named because a paste of cement with water, after it sets hard, resembles in colour and hardness a Portland stone, a limestone quarried in Dorset.*
- It is prepared in different varieties.

## 5.3. PROPERTIES OF CEMENT

A *good cement* possesses the following properties (which depend upon its chemical composition, thoroughness of burning and fineness of grinding).

1. Provides strength to masonry.
2. Stiffens or hardens early.
3. Possesses good plasticity.
4. An excellent building material.
5. Easily workable.
6. Good moisture-resistant.

## 5.4. USES OF CEMENT

Following are the various *uses* of cement :

1. It is used in cement mortar for *masonry work, plastering, pointing,* etc.
2. It is used for making *joints for pipes, drains* etc.
3. It is used in concrete for laying floors, roofs and constructing lintels, beams, stairs, pillars etc.
4. It is employed for manufacturing *precast pipes, piles, fencing posts* etc.
5. It is used in the construction of important engineering structures such as *bridges, culverts, dams, tunnels, light houses* etc.
6. It is used in the preparation of *foundations, water tight floors, footpaths* etc.
7. It is employed for the construction of wells, water tanks, tennis courts, lamp posts, telephone cabins, roads etc.

## 5.5. COMPARISON BETWEEN CEMENT AND LIME

Table 5.1 gives the comparison between cement and lime.

**Table 5.1 Comparison between Cement and Lime**

| S.No. | Aspects | Cement | Lime |
|---|---|---|---|
| 1. | *Colour* | Greenish grey. | White or greyish. |
| 2. | *Slaking* | Does not slake when wetted with water. | Slakes when wetted with water. |
| 3. | *Setting* | Sets rapidly when mixed with water. | Sets slowly when mixed with water. |
| 4. | *Strength* | Artificial cement possesses more strength. | Possesses less strength. |
| 5. | *Suitability* | *Can* be used for important and heavy engineering structures. | *Cannot* be used for important and heavy engineering structures. |

## 5.6. CHEMICAL CONSTITUENTS OF CEMENT

- Percentage of various ingredients for the manufacture of Portland cement should be as follows :

| Ingredient | | Proportion |
|---|---|---|
| *Lime* (CaO) | ........... | 63.0% |
| *Silica* ($SiO_2$) | ........... | 22.0% |
| *Alumina* ($Al_2O_3$) | ........... | 6.0% |
| *Iron oxide* ($Fe_2O_3$) | ........... | 3.0% |
| *Magnesium oxide* (MgO) | ........... | 2.50% |
| *Sulphur trioxide* ($SO_3$) | ........... | 1.75% |
| *Loss on ignition* | ........... | 1.50% |
| *Insoluble residue* | ........... | 0.25% |

- Besides the above ingredients, calcium sulphate ($CaSO_4$), commonly known as *gypsum* is also added by 3 to 4% during the grinding process for *controlling the initial setting time of cement* (larger the proportion of gypsum, more will be the initial setting time and vice versa).

## 5.7. FUNCTIONS OF INGREDIENTS OF CEMENT

The functions of various ingredients of an ordinary Portland cement are given below :

**1. Lime (CaO):**

- It is the major constituent of cement and its proportion needs to be maintained carefully.
- The lime in *excess* makes the cement *unsound* and causes the cement to expand and disintegrate.
- If lime is in deficiency, the strength of cement is *decreased* and cement *sets quickly*.
- If lime is in *right proportion,* it makes the cement *sound* and *strong.*

**2. Silica ($SiO_2$):**

- It is an important ingredient of cement and *imparts strength to the cement due to the formation of dicalcium and tricalcium silicates.*
- Silica in *excess provides greater strength to the cement but at the same time prolongs its setting time.*

**3. Alumina ($Al_2O_3$):**

- It imparts *quick setting quality* to the cement.
- It acts as a flux and lowers the clinkering temperature.
- Alumina in excess reduces the strength of cement.

**4. Calcium sulphate ($CaSO_4$):**

- It is present in the form of gypsum.
- It helps in *increasing the initial setting time of cement.*

**5. Iron oxide ($Fe_2O_3$):**

- It provides *colour, hardness* and *strength to cement.*
- It also *helps the fusion of raw materials* during manufacture of cement.

**6. Magnesium oxide (MgO):**

- This ingredient, if present in small amount, imparts hardness and colour to the cement.
- Magnesium oxide *in excess* makes the cement *unsound.*

**7. Sulphur trioxide ($SO_3$):**

- If present in very small quantity, it makes the cement *sound.*
- If present in excess, it causes the cement to become *unsound.*

**8. Alkalies:**

- They should be present in small quantities.
- Alkalies in excess will cause efflorescence.

**Harmful constituents of cement:**

The presence of the following *two* oxides affects the quality of cement *adversely* :

1. *Alkali oxides ($K_2O$ and $Na_2O$):*

If the amount of alkali oxides *exceeds 1%,* it *leads to the failure of concrete made from that cement.*

2. *Magnesium oxide (MgO) :*

If the content of magnesium oxide *exceeds 5%,* it causes *cracks after mortar or concrete hardness.*

## 5.8. SITE FOR CEMENT FACTORY

The following *points* should be considered while making selection of a suitable site for a cement factory :

1. The *climatic conditions* of the site should be *favourable* for the manufacturing process of cement.
2. The *raw materials* should be *easily and continuously available* in *large quantity* near the site.
3. *Labour* should be *easily and cheaply available.*
4. As far as possible, the site selected should be *connected by roads and railway* with the important towns.
5. *Electric power and fuel* should be available at *economic rates.*
6. The site for cement factory, as far as possible, should be near to the market of sale.
7. General facilities such as medical, recreation centres, schools and colleges etc. should be available near to the proposed site.
8. The bearing capacity of the soil at the site must be adequate to take the thrust of the machines safely.
9. A sufficient quantity of drinking water should be available at the proposed site.

## 5.9. MANUFACTURE OF PORTLAND CEMENT

The process of manufacture of cement consists essentially of *grinding the raw materials* (calcareous and argillaceous stones containing silica, alumina and iron oxide), *mixing them* intimately in a certain proportion and *burning in a large rotary kiln* at a temperature of approximately 1300°C to 1900°C, when the material *sinters* and partially fuses into balls known as **clinker.** The clinker is cooled and ground to fine powder, with *some gypsum added*, and the resulting product is the commercial Portland cement so widely used throughout the world.

### Details of Manufacture

There are two processes employed for the manufacture of the cement :

1. Wet process
2. Dry process

Both the processes involve the following **steps** in the manufacture of Portland cement:

(*i*) Collection of raw materials.

(*ii*) Crushing, grinding and mixing of raw materials.

(*iii*) Burning.

(*iv*) Grinding of clinker.

### 1. Wet Process :

(*i*) **Collection of raw materials :** Calcareous materials such as limestone, marl, chalk etc. are quarried by blasting. *Argillaceous materials* such as clay, slate etc. are transported to the site by wine ropeways or dumping trucks.

(*ii*) **Crushing, grinding and mixing of raw materials** : When *chalk* is used it is finely broken up and dispersed in water in a wash mill; this is a circular pit with revolving radial arms carrying rakes which break up the lumps of solid matter. The clay is also broken up and mixed with water, usually in a similar wash mill. The two mixtures are now pumped so as to mix in predetermined proportions and pass through a series of screens. The resulting cement slurry flows into storage tanks.

When *limestone* is used it has to be blasted and then crushed usually in two progressively smaller crushers, and then fed into a ball mill with the clay dispersed in water. There the grinding of the limestones (to the fineness of flour) is completed, and the resultant slurry is pumped to storage tanks.

The slurry is a liquid of creamy consistency, with a water content between 35 to 40 per-

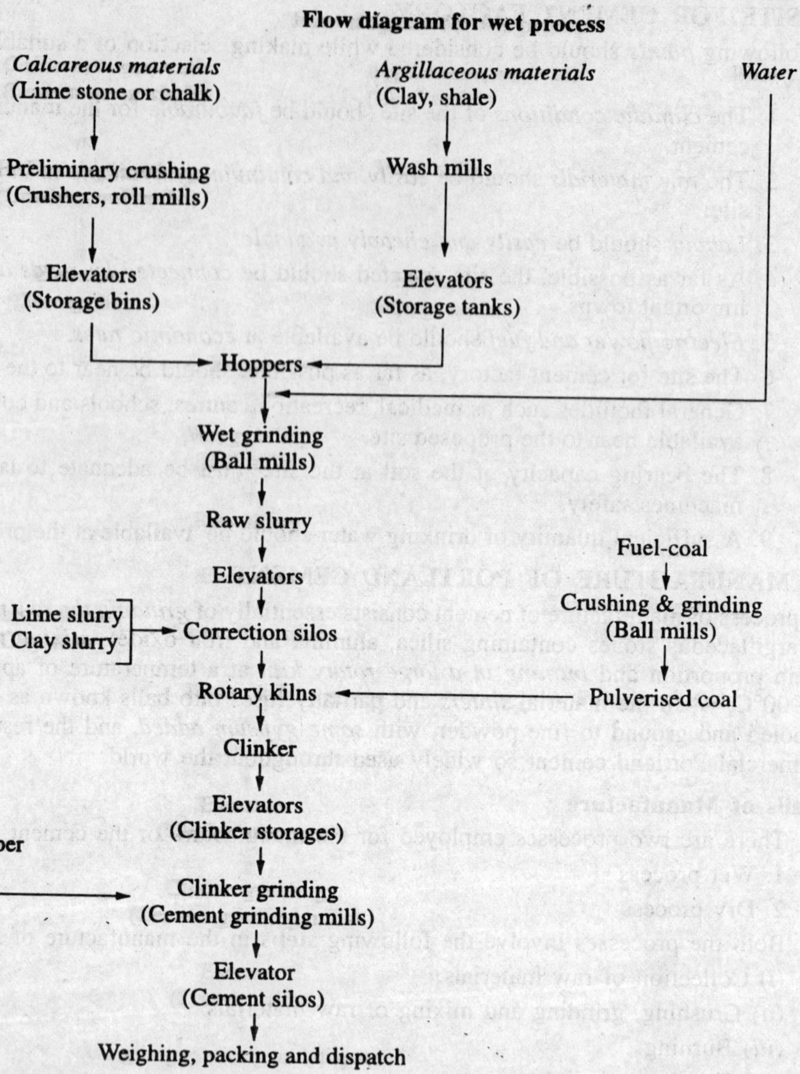

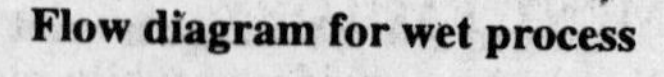

**Flow diagram for wet process**

*Calcareous materials* (Lime stone or chalk) → Preliminary crushing (Crushers, roll mills) → Elevators (Storage bins) → Hoppers

*Argillaceous materials* (Clay, shale) → Wash mills → Elevators (Storage tanks) → Hoppers

*Water* → Wet grinding

Hoppers → Wet grinding (Ball mills) → Raw slurry → Elevators → Correction silos

Lime slurry, Clay slurry → Correction silos

Correction silos → Rotary kilns

Fuel-coal → Crushing & grinding (Ball mills) → Pulverised coal → Rotary kilns

Rotary kilns → Clinker → Elevators (Clinker storages) → Clinker grinding (Cement grinding mills)

Gypsum → Gypsum hopper → Clinker grinding

Clinker grinding (Cement grinding mills) → Elevator (Cement silos) → Weighing, packing and dispatch

cent. There is usually a number of storage tanks in which slurry is kept, the sedimentation of the suspended solid being prevented by mechanical stirrers or bubbling by compressed air. The slurry then passes into *silos* where the proportioning is finely adjusted to ensure the correct chemical composition. From these silos the slurry is fed into the *kilns*.

(*iii*) **Burning** : The fine slurry is pumped into the upper end of the *rotary kiln* **set at a slight gradient.** The rotary kiln (Fig. 5.1) is a large, refractory-lined steel cylinder upto 4 metres in diameter sometimes as long as 150 metres, slowly rotating about its axis. The slurry is fed in at the upper end while pulverised coal is thrown in by an air blast at the lower end of the kiln, where the temperature reaches 1400 to 1500°C. The coal used must not have too high an ash content. Oil and natural gas can be used instead of coal.

The slurry, in its movement down the kiln, encounters progressively higher temperature. At first the water is driven off and $CO_2$ is liberated; further on, the dry material undergoes a series of chemical reactions until finally, in the hottest part of the kiln, 20 to 30 per cent of the material becomes liquid, and lime, silica and alumina recombine. The mass then fuses into balls, 0.3 cm to 2.5 cms diameter, known as *clinker*. The clinker drops into coolers, which are of various types and often provide means for exchange of heat with the air subsequently used for the combustion of the pulverised coal. A large kiln can produce 700 tonnes of cement a day.

(*iv*) **Grinding** : The cool clinker which is characteristically black, glistering and hard is **interground with 3 to 4% *gypsum* in order to prevent flash setting of cement**. The grinding is done in a *ball mill consisting of several compartments with progressively smaller steel balls.* In some plants a closed-circuit grinding system is used; the cement discharged by the mill is

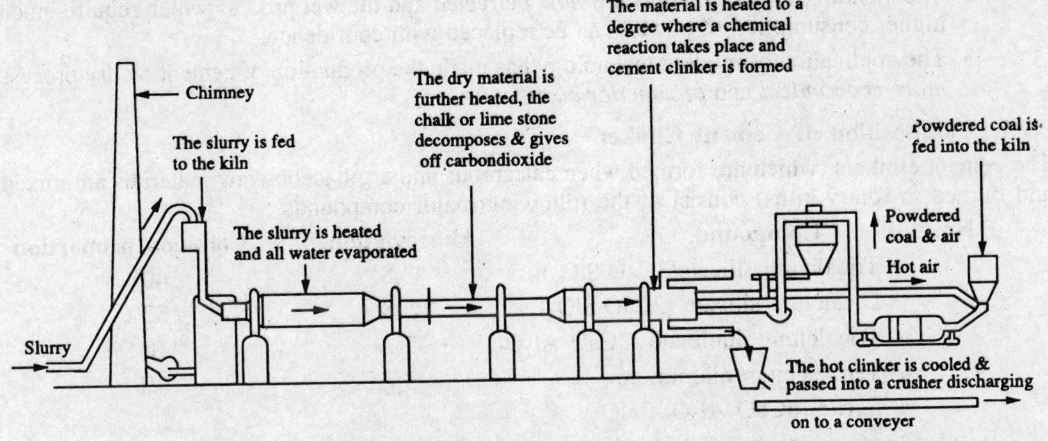


Fig. 5.1 : Rotary kiln.

passed through a separator, fine particles being removed to the storage silo by an air current, while the coarser particles are passed through the mill once again. Close circuit grinding avoids the production of a large amount of excessively fine material or of a small amount of too coarse material, faults often encountered with open circuit grinding.

Once the cement has been satisfactorily ground it is ready for packing in bags. Each bag contains 50 kg of cement.

**2. Dry Process :**

In the dry or semi-dry processes the materials are crushed and fed in the correct proportions into a grinding mill, where they are dried and reduced in size to fine powder. The dry powder, called, raw meal, is then pumped to a blending silo, and final adjustment is now made in the proportions of the material required for the manufacture of the cement. To obtain uniform and intimate mixture the raw meal is blended, usually by means of compressed air inducing an upward movement of the powder and decreasing its apparent density. The air is pumped over one quadrant of the silo at a time and, this permits the apparently heavier material from the unaerated quardant to move laterally into the aerated quadrant. Thus the aerated material tends to behave almost like a liquid, and by aerating all quadrants in turn for a total period of about one hour a uniform mixture is obtained. In some work continuous blending is used.

The blended meal is now sieved and fed into a rotating dish called a granulator, water weighing about 12 per cent of the meal being added at the same time. In this manner hard-pallets about 1.25 cm diameter are formed. This is necessary, as powder fed direct into a kiln would not permit the air flow and exchange of heat necessary for the chemical reactions of formation of cement clinker.

The pallets are baked hard in a pre-heating grate by means of hot gases from the kiln. The pallets then enter the kiln, and subsequent operations are the same as in wet the process of manufacture. Since, however, the moisture content of the pallets is only 12 per cent as compared with the 40 per cent moisture content of slurry used in the wet process, the *dry-process kiln is considerably smaller.* The amount of heat required is also very much lower as only some 12 per cent of moisture has to be driven off additional, but additional heat has already been used in removing the original moisture content of the raw materials (usually 6 to 10 per cent). The process is thus *quite economical, but only when the raw materials are comparatively dry.*

**Dry process** (modern technology) is *now preferred and adopted in most of the cement industries* due to the following *reasons :*

1. For dry process the *heat required per kg of clinker produced is less.*
2. The *blending of dry powders has now perfected* and the wet process which require much higher consumption of power, can be replaced with confidence.
3. The application of modern technology has made the production of cement by dry process *more economical* and *of superior quality.*

### 5.9.1. Composition of Cement Clinker

The cement clinkers (which are formed when calcareous and argillaceous raw materials are mixed and burned in rotary kilns) consist of the following major compounds :

| S.No. | Compound | Abbreviation | Common proportion |
|---|---|---|---|
| 1. | Tricalcium silicate(3CaO $SiO_2$) | $C_3S$ | 40% |
| 2. | Dicalcium silicate (2CaO $SiO_2$) | $C_2S$ | 32% |
| 3. | Tricalcium aluminate (3CaO $Al_2O_3$) | $C_3A$ | 10.5% |
| 4. | Tetracalcium aluminium ferrite (4CaO $Al_2O_3$ $Fe_2O_3$) | $C_4AF$ | 9% |

Besides these major compounds, the proportion of other constituents varies from 0 to 5% and that of gypsum ($CaSO_4$) 3 to 4% which is added while grinding the cinders (*Gypsum is added to retard or to slow down the initial setting action of cement*).

The properties of the compounds of clinker are discussed below :

**1. Tricalcium silicate ($C_3S$) :**

(*i*) Generates heat more rapidly.
(*ii*) Hydrates more rapidly.
(*iii*) Possesses less resistance to chemical attack
(*iv*) Develops early strength..

**2. Dicalcium silicate ($C_2S$) :**

(*i*) Imparts ultimate strength to the cement.
(*ii*) Offers more resistance to chemical attack.
(*iii*) Hardens more slowly.
(*iv*) Hydrates slowly.

**3. Tricalcium aluminate ($C_3A$) :**

(*i*) Weak against sulphate attack.
(*ii*) Reacts fastly, generating a large amount of heat.
(*iii*) Does not contribute to develop strength.
(*iv*) Causes initial setting of cement.

**4. Tetra calcium aluminium ferrite ($C_4AF$) :**

(*i*) Poor cementing value
(*ii*) Reacts slowly, generating small amount of heat
(*iii*) Comparatively inactive

It has been observed/analysed that most of the strength developing properties of cement are *controlled by $C_3S$ and $C_2S$* (the sum of their percentages varies from 70 to 80 per cent).

- High percentage of $C_3S$ and low $C_2S$ result in
  — rapid hardening,
  — high early strength with high heat generation, and
  — less resistance to chemical attack.

- Low percentages of $C_3S$ and high $C_2S$ result in
  - — slow hardening,
  - — much more ultimate strength with less heat generation, and
  - — greater resistance to chemical attack.

Thus, by changing the relative proportions of these compounds, various types of cements are manufactured.

## 5.10. SETTING AND HARDENING OF CEMENT

- The *chemical reaction between cement and water* is called **hydration of cement** or simply **cement hydration.**
- When cement is mixed water (varying from 25 to 35 per cent by weight), a *stiff* and *sticky paste* is formed which remains plastic for a short period. With the passage of time, the plasticity gradually disappears and the cement paste becomes stiff *due to initial hydration of cement. This phenomenon by virtue of which the plastic cement changes into a solid mass* is known as **setting of cement**. On setting, the cement binds the aggregates into a solid mass which gains strength as the time lapses, till the hydration of cement is complete. Thus, the *phenomenon by virtue of which the cement paste*, which is finally set, *develops strength* is known as **hardening of cement.**
- As mentioned above, cement during its reaction with water sets and then hardens. This *reaction evolves heat*. The rate of setting and hardening of cement, the rate of evolution of heat and the resistance to sulphate attack are *affected by the proportions of different cement compounds* as discussed below :
  - — $C_3S$ and $C_2S$ constitute about 70 to 80% of all Portland cements. $C_3S$ hydrates more rapidly than $C_2S$ and develops strength in concrete for the first 28 days. It also generates more heat. $C_2S$ is next to hydrate but it *hydrates slowly and is responsible for the ultimate strength.* $C_2S$ takes 2 to 3 years for its complete hydration which contributes towards ultimate strength of cement mortar on concrete. It is more resistant to sulphate attack.
  - — When cement reacts with water, $C_3A$ is the first to react with water and *causes the initial set*. It generates great amount of heat and is easily affected by sulphates. It *contributes little to the strength of concrete*. It is *rendered ineffective by addition of gypsum during the grinding of clinkers.* Gypsum reacts with $C_3A$ and turns it into calcium sulpho-aluminate which causes expansion during setting. $C_3A$ is therefore not available in substantial quantity for reaction when water is added to cement.

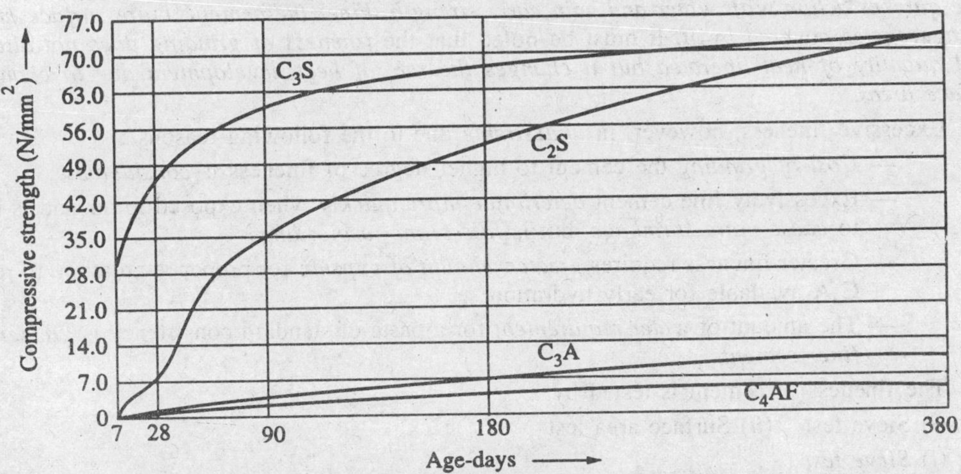


Fig. 5.2: Compressive strength of cement compounds with age.

— $C_3AF$ is comparatively *inactive* and *contributes little to the strength of concrete and the heat of hydration.*

Fig. 5.2 shows the contribution of each cement compound to the development of strength.

- The rate of hardening of cement governs the time upto which the formwork should be kept in position and the time after which the structure should be loaded.

## 5.11. TESTING OF PORTLAND CEMENT

For normal setting of ordinary Portland cement the following tests (laboratory) are carried out :

1. Chemical composition.
2. Fineness test.
3. Consistency test.
4. Setting times test.
5. Soundness test.
6. Tensile strength test.
7. Compressive strength test.

**1. Chemical composition:**

The various tests are carried out to determine the chemical constituents of cement. Following are the chmical requirements of ordinary cement as per IS 269-1967/1975 :

(*i*) Ratio of percentage of lime to those of silica, alumina and ironoxide known as *lime saturation factor* (LSF) when calculated by the formula

$$\frac{CaO - 0.7\ SO_3}{2.8\ SiO_2 + 1.2\ Al_2O_3 + 0.65\ Fe_2O_3}$$ ... Not greater than 1.02 and and not less than 0.66.

(*ii*) Ratio of percentage of alumina to that of iron oxide ... not less than 0.66

(*iii*) Weight of insoluble residue ... not less than 2%

(*iv*) Weight of magnesia ... not more than 5%

(*v*) Total sulphur contents calculated as sulphuric anhydride ($SO_3$) ... not more than 2.75%

(*vi*) Total loss on ignition ... not more than 4%.

**2. Fineness test:**

The object of conducting this test is to *check the proper grinding of cement as finer cements have quicker action with water and gain early strength. Finer the cement, more surface area of the aggregates covered by it.* It must be noted that the *fineness of grinding does not alter the total quantity of heat liberated but it changes the rate of heat development due to change in surface areas.*

Excessive fineness, however, in *undesirable* due to the following reasons :

— *Cost of grinding* the cement to higher degree of fineness is *considerable*;

— Excessively fine cement *deteriorate more quickly* when exposed to air and is likely to *cause more shrinkage*, but is *less prone to bleeding.*

— Greater fineness requires *greater amount of gypsum* for proper retardation, to reduce $C_3A$ available for early hydration.

— The amount of *water requirement* for a paste of standard consistency is *greater with finer cement.*

The fineness of cement is tested by :

(*i*) Sieve test (*ii*) Surface area test

(*i*) *Sieve test* :

The maximum residue after sieving through a 90-micron IS sieve should be limited to 10%

by weight for ordinary Portland cement and 5% by weight for rapid hardening Portland cement. This test, however, *does not give any idea of the smaller grains of the particles which in fact play a great role in the process of hydration and development of strength.*

(*ii*) *Surface area test* :

- This test is also known as *specific surface test*. It is *better* than sieve test. *Specific surface* is the *total surface of all the particles of cement per unit weight*. It can be determined by *Air pemeability method* or *Wagner's turbidimeter.*
- Properly ground cement of the ordinary type should not have specific surface less than 2250 $cm^2/g$ of cement or 1600 $cm^2/g$ of cement if found by "Air permeability" and Wagners turbidimeter" methods respectively.

**3. Consistency test:**

This test is conducted to *determine the percentage of water required for preparing cement pastes of standard consistency for other tests* (*e.g.*, setting time, soundness and compressive strength tests).

This test is performed with the help of Vicat's apparatus (Fig. 5.3) as described below :

**Vicat's apparatus :** It consists of a metal frame to which is attached a movable rod weighing 300 g (along with cap and attachment) and having diameter and length as 10 mm and 50 mm respectively. The movable rod is provided with a releasing pin to let the rod free and is attached with an indicator to take readings on a vertical scale which is graduated from 0 to 40 mm in either direction which gives the penetration. The Vicat mould is in the form of a cylinder (it can be split into two halves) and is placed on a non-porous plate. There are following three attachments :

(*i*) *Square needle* ... used for *initial setting time test* ;

(*ii*) *Plunger* .... used for *consistency test;*

(*iii*) *Annular collar* ... used for *final setting time test.*

The whole apparatus may be made of gun metal, aluminium or steel.

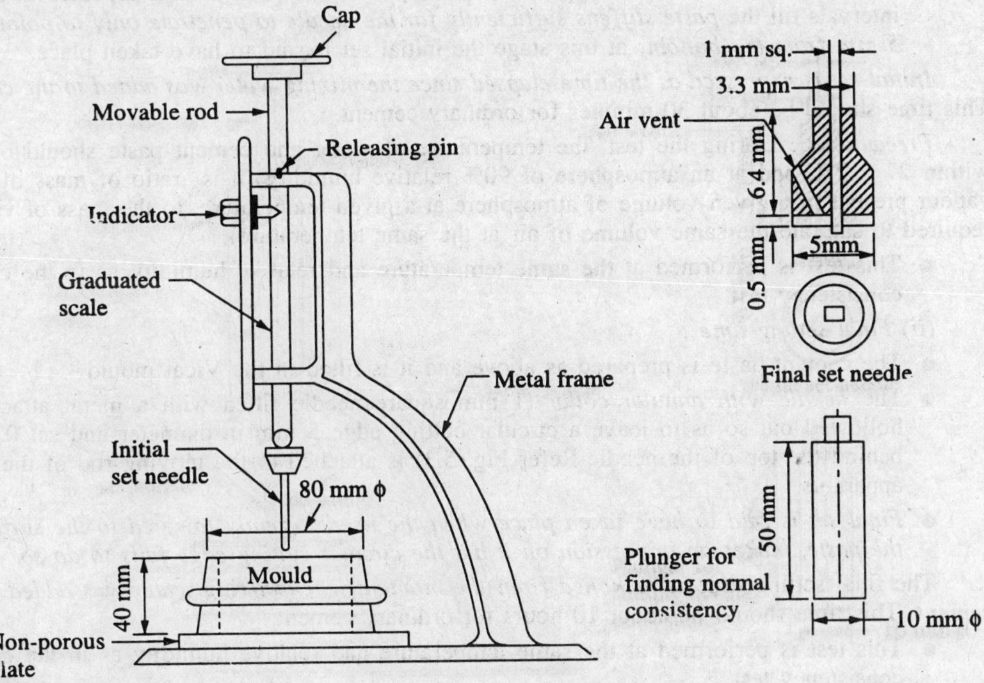


Fig. 5.3 : Vicat's apparatus.

**Method :**

The consistency is measured by the Vicat apparatus by using a *10 mm diameter plunger* fitted into the needle holder.

- A trial paste of cement (300 g) and water (30% by weight or 90 g) is mixed and placed in the mould.
- The plunger is then brought into contact with the top surface of the paste and released. Under the action of its weight the plunger will penetrate the paste, the depth of penetration depending on the consistency of the paste. This is considered to be the *standard when the plunger penetrates the paste to a point 5 to 7 mm from the bottom of the mould.*
- The water content of the standard paste is expressed as a percentage by weight of the dry cement, the usual range of values being between 26 and 33 per cent.

**4. Setting times test:**

— The *object* of this test is to *check the initial and final setting times of the cement.*

— The *initial setting time* is determined as to give *sufficient time* for *various operations* such as *mixing, transportation, placing* and *compaction of cement mortar or concrete.*

— The *final setting time* is determined to find that after laying the mortar or concrete, the hardening should be rapid so that the *structure may be used as early as possible.*

**Method:**

(*i*) *Initial setting time :*

- The cement paste is prepared (as mentioned in the consistency test) and is filled in the Vicat mould.
- A round or square needle with a cross-sectional area of 1 mm$^2$ is attached to the moving rod. The needle is then quickely released and it is allowed to penetrate the cement paste. It is then taken out and dropped at a fresh place. The procedure is repeated at regular intervals till the *paste stiffens sufficiently for the needle to penetrate only to point about 5 mm from the bottom;* at this stage the initial set is said to have taken place.

*Initial set is expressed as the time elapsed since the mixing water was added to the cement.* This time should be about 30 minutes for ordinary cement.

*Precautions :* During the test, the temperature of water and cement paste should be kept within 27 ± 2°C and at an atmosphere of 90% relative humidity (it is ratio of mass of water vapour present in a given volume of atmosphere at a given temperature, to the mass of vapours required to saturate the same volume of air at the same temperature).

- This test is performed at **the same** temperature and relative humidity as in the case of consistency test.

(*ii*) *Final setting time :*

- The cement paste is prepared as above and it is filled in the Vicat mould.
- The *needle with annular collar* (1 mm square needle fitted with a metal attachment hollowed out so as to leave a circular cutting edge 5 mm in diameter and set 0.5 mm behind the top of the needle-Refer Fig. 5.3) is attached to the moving rod of the Vicat apparatus.
- *Final set is said to have taken place when the needle, gently lowered to the surface of the paste, makes an impression on it but the circular cutting edge fails to do so.*

The final setting time is *reckoned from the moment when mixing water was added to the cement.* This time should be about 10 hours for ordinary cement.

- This test is performed at the same temperature and relative humidity as in the case of consistency test.
- According to Indian standard specifications the setting times for *ordinary, rapid hardening* and *low heat Portland cements* are :

| | Ordinary | Rapid hardening | Low heat |
|---|---|---|---|
| (*i*) Initial setting time | | | |
| *not less than* | 30 minutes | 30 minutes | 60 minutes |
| (*ii*) Final setting time | | | |
| *not more than* | 10 hours | 10 hours | 10 hours. |

It should be remembered that *speed of setting and rapidity of hardening i.e., gain of strength, are entirely independent of one another.* For instance, the prescribed setting times of rapid hardening cement are no different from those for ordinary Portland cement, although two cements harden at different rates.

**5. Soundness test:**

This test is performed to *detect the presence of uncombined lime and magnesia in cement.*

Soundness of cement is tested with 'Le Chatelier apparatus' (Fig. 5.4). It consists of a small brass cylinder (30 mm dia., 30 mm high and 0.5 mm thick) split along its generatix (split does not exceed 0.5 mm). Two indicators with pointed ends are attached to the cylinder on either side of the split.

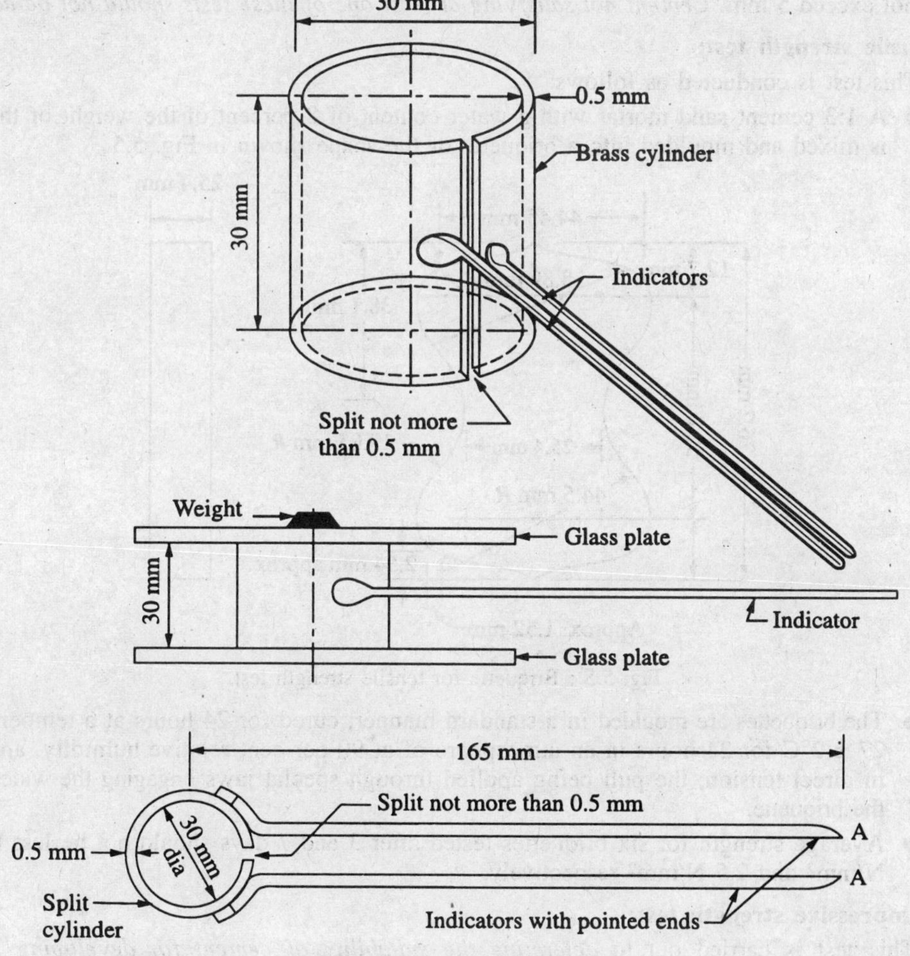


Fig. 5.4 : Le Chatelier apparatus.

**Method:**

- The cement paste is prepared; the percentage of water is taken as determined in the consistency test.
- The cylinder is placed on a glass plate and it is filled with the cement paste. It is covered at the top with another glass plate. A small weight is placed on the top.
- The whole assembly is immersed in water at 24°C to 35°C for 24 hours. At the end of that period the *distance between the indicators is measured.*
- The mould is immersed in water again and brought to boil in 30 minutes. After boiling for one hour the mould is removed, and after cooling, the *distance between the indicators is again measured.* The increase in this distance represents the expansion of the cement and according to Indian standard specification **it should not exceed 10 mm** *for any type of Portland cement.*

If the expansion *exceeds* the above value, the cement is *not rejected.* A *further test* is made after the *cement has been spread and aerated for 7 days,* during which time, some of the lime and magnesia may further hydrate or carbonate and a physical breakdown in size may take place. At the end of 7 days, the Le Chatelier test is repeated and the expansion of the aerated cement must not exceed 5 mm. *Cement not satisfying at least one of these tests should not be accepted.*

**6. Tensile strength test:**

This test is conducted as follows:

- A 1:3 cement sand mortar with a water content of 8 percent of the weight of the solids is mixed and moulded into a briquette of the shape shown in Fig. 5.5.

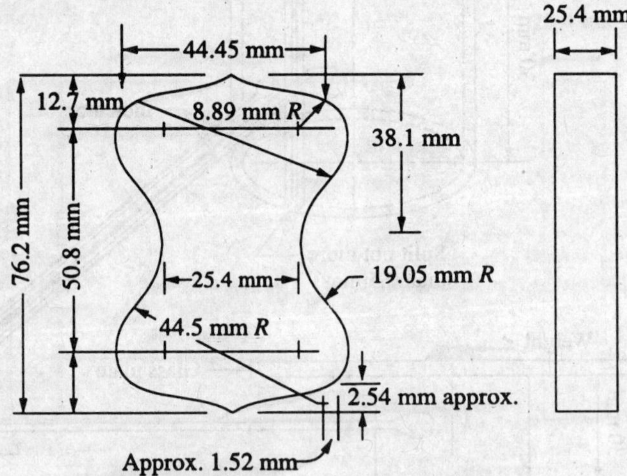


Fig. 5.5 : Briquette for tensile strength test.

- The briquettes are moulded in a standard manner, cured for 24 hours at a temperature of 27 ± 2°C for 24 hours in an atmosphere of at 90 per cent relative humidity, and tested in direct tension, the pull being applied through special jaws engaging the wide ends of the briquette.
- Average strength for six briquettes tested after 3 and 7 days should not be less than 2.0 N/mm$^2$ and 2.5 N/mm$^2$ respectively.

**7. Compressive strength test:**

This test is carried out to *determine the suitability of cement for developing required compressive strength of concrete and mortar.*

- The test is performed on 1:3 cement mortar cubes made by gauging 185 g of cement 555 g of standard sand and 74 g of water.

- As per Indian standard specifications *average compressive strength* for three cubes should not be less than 11.5 N/mm$^2$ and 17.5 N/mm$^2$ after *2 and 7 days respectively*.

**Field tests for cement:**

In order to ascertain the quality of cement roughly, the following *four field tests* are carried out :

1. *Colour* : The colour of the cement should be uniformly greenish grey.

2. *Physical properties :*

- It should feel smooth when rubbed in between fingers. If it is felt rough, it indicates adulteration with sand. If the cement contains too much of powder clay and silt as adulterant, the paste will give an earthy smell.
- When hand is thrusted into a bag of cement it should feel cool and not warm.
- A handful of cement thrown into a bucket of water, should float in case the cement is of good quality and it sinks, if it contains some impurities.

3. *Presence of lumps :* Any lump found in the cement bag should be powdered by pressing between the thumb and the forefinger. If it does not turn into powder form, the cement is considered to be spoiled by air setting and should be rejected.

4. *Strength :* The strength of cement can be determined/ascertained in the following ways:

- A thick paste of cement with water is prepared on a piece of thick glass and is kept under water for 24 hours; the paste should set and *not* crack.
- The briquettes (75 mm × 25 mm × 12 mm) are prepared with a lean or weak mortar (1:6) and are immersed in water for 3 days. If the cement is of sound quality, these briquettes will not be broken easily; also it will be difficult to convert them into powder form.
- A cement block measuring 25 mm × 25 mm × 200 mm (length) is prepared and immersed in water for 7 days. The block should not show signs of failure when placed on supports 150 mm apart and loaded with a load of 340 N.

## 5.12. TYPES OF CEMENT

Following are the various types of cement available in the market:

1. *Portland cements:*

| | |
|---|---|
| (*i*) Ordinary Portland cement | (*ii*) Modified Portland cement |
| (*iii*) Rapid hardening Portland cement. | (*ix*) Low heat Portland cement |
| (*v*) Sulphate resisting Portland cement | (*vi*) Water-repellent Portland cement |
| (*viii*) Water-proof Portland cement. | |

2. *Other varieties of cement:*

| | |
|---|---|
| (*i*) High alumina cement | (*ii*) Quick setting cement |
| (*iii*) Blast furnace slag cement | (*iv*) White cement |
| (*v*) Coloured cements | (*vi*) Acid resistance cement |
| (*vii*) Expanding cement | (*viii*) Hydrophobic cement |
| (*ix*) Portland Pozalana cement | (*x*) Supersulphated cement |
| (*xi*) Masorny cement. | |

### 5.12.1. Portland Cements

**1. Ordinary Portland cement:**

- This is by far the most common cement in use.
- It has medium rate of strength development and heat generation.
- It has adequate resistance to dry shrinkage and cracking, but has less resistance to chemical attack.

**Uses** : (*i*) It is admirably suitable for use in general concrete construction when there is no exposure to sulphates in the soil or in ground water.

(*ii*) It is used in small structures where heat of hydration will not cause any defect.

**2. Modified Portland cement:**

This cement, on setting, develops less heat of hydration than ordinary Portland cement.

**Uses:** Due to lower heat of hydration it can be employed in hot climates and for construction of heavy abutments, large piers, retaining walls etc; where *sulphate content is not high.*

**3. Rapid hardening Portland cement:**

- It is also known as *High early strength cement.*
- It contains large proportion of lime and other constituents are the same as for ordinary Portland cement.
- It is prepared more carefully and burnt at a higher temperature than that of an ordinary Portland cement.
- The strength developed at the age of 3 days is of the same order as the 7 days strength of ordinary Portland cement with same water-cement ratio. The *increased rate of gain of strength of this cement is achieved by a higher $C_3S$ content and by finer grinding of the cement clinker.* It is ground finer and has specific surface not less than 3250 $cm^2/g$.
- It is ligher than ordinary Portland cement.
- The curing period is short, hence it is economical.

**Uses** *:* It is used where a rapid strength development is required; *e.g.*, when frame work is to be removed early for re-use, or where sufficient strength for further construction is wanted as quickly as practicable.

**Extra rapid hardening cement**

- This type of cement is obtained by intergrinding calcium chloride with rapid hardening Portland cement. The quantity of calcium chloride should not exceed 3 percent.
- Its strength is about 25 percent higher than that of rapid hardening cement at 1 or 2 days and 10 to 20 percent higher at 7 days.

*Uses* : It is suitable for *cold weather concreting*, or when a very high early strength is required but when it is inadvisable to use aluminous cement.

**4. Low heat Portland cement:**

- It contains a low percentage (about 5%) of tricalcium silicate ($C_3S$) which hydrates quickly and a higher percentage (about 46%) of dicalcium silicate ($C_2S$) which hydrates slowly.
- It contains less lime than ordinary cement; other materials remain the same as in the case of ordinary cement.
- It possesses less compressive strength.
- The initial setting time is about one hour and final setting time is about 10 hours.

**Uses** : During the setting action of cement, a considerable amount of heat is produced, in *order to reduce the amount of heat, this type of cements is used.* It is mainly used for *mass concrete work.*

**5. Sulphate resisting Portland cement:**

- In this cement, the percentage of tricalcium aluminate ($C_3A$) is kept *below* 5 percent and it results in the increase in resisting power against sulphates.
- The heat developed by this type of cement is not much higher than that of low heat cement. Theoretically it is an ideal cement but because of the special requirement for the composition of the raw materials used in its manufacture, sulphate-resisting cement cannot be easily and cheaply made.

**Uses :** It is used at places *where sulphate action is severe.* It is employed for structures which are likely to be damaged by severe alkaline conditions such as *canal linings, culverts, syphons* etc.

**6. Water-repellent Portland cement:**

- It contains a small percentage of water-proofing material uniformly mixed with the cement and is manufactured under the name "*Aquacrete*". The cement is prepared with ordinary or rapid hardening cements and white cement.
- If such a cement is used in concrete, a considerable care is needed to avoid reduction in strength.

**Uses :** It is chiefly used in water-tight concrete and water-tight renderings to check moisture penetration in basements etc. and for coloured renderings and stucco.

**7. Water-proof Portland cement:**

- These cements are prepared by mixing with ordinary or rapid hardening cements, a small percentage of some metal stearate (Ca, Al etc.) at the time of grinding.
- Concrete made with such a cement is more resistant to penetration by water and some oils than that made from ordinary Portland cement.
- It is also adequately resistant to the corrosive action of acids and alkalies or other harmful salts usually present in industrial waters.

**Uses :** It is used for the construction of water-retaining structures like tanks, reservoirs, retaining walls, swimming pools, dams, bridges, piers etc.

**5.12.2. Other Varieties of Cements**

**1. High alumina cement:**

- It is a rapid hardening cement of chocolate colour.
- It is manufactured from bauxite and limestone in special reverberatory furnaces. It contains nearly 35 per cent of alumina and the ratio of alumina to lime lies between 0.55 and 1.3.

**Advantages:**

(*i*) It resists the action of acid and high temperature and does not expand on setting.

(ii) Its initial and final setting times are $3\frac{1}{4}$ hours and 4 to 5 hours respectively. It therefore allows more time for mixing and placing operations.

(*iii*) It sets quickly and attains higher ultimate strength in a short period. Its strength after 1 day is about 40 N/mm$^2$ and after 3 days is about 50 N/mm$^2$.

(*iv*) It is completely resistant to the action of sulphates.

(*v*) It offers excellent resistance to fire.

(*vi*) It is *not effected by frost* since it evolves great heat during setting.

(*vii*) It is not necessary to grind it to fine powder since its setting action mainly depends on the chemical reactions.

**Disadvantages:**

(*i*) It is unsuitable for mass concrete construction as it develops considerable heat on setting.

(*ii*) It is much costlier than ordinary Portland cement.

(*iii*) Extreme care has to be taken to ensure that it does not come in contact with even traces of lime or ordinary cement.

**Uses:**

(*i*) Its field of application includes works in chemical plants and furnaces.

(*ii*) It is employed in colder regions having temperatures 18°C or below and during war time emergency.

(*iii*) It is used in under-water construction.

**2. Quick setting cement:**

- It contains less *percentage of gypsum* (retarder) and is ground much finer than ordinary Portland cement. The *setting action* is accelerated by adding a *small percentage of aluminium sulphate, during grinding.*
- It is costlier than ordinary Portland cement.
- Its initial and final setting times are 5 minutes and 30 minutes respectively.

The setting action of such a cement starts within 5 minutes and it becomes stone-hard in less than half an hour.

**Uses :** Due to its quick setting property it is used in works where concrete is to be placed under water or in running water.

**3. Blast furnace slag cement:**

- It is made by *intergrinding Portland cement clinker and granulated blast-furnace slag,* the proportion of the latter not exceeding 65 per cent of the weight of the mixture.
- It has lower evolution of heat.
- It is more resistant to attack of weathering agencies.
- It is cheaper than ordinary Portland cement.
- Its colour is blackish grey.
- Its initial setting time is not less than 30 minutes. Its final setting tine is not more than 10 hours.

**Uses :** (*i*) It can be used in *mass concrete structures* (since its heat of hydration is lower than that of ordinary Portland cement). However, in cold weather, the low heat of hydration of blast-furnace slag cement, coupled with a moderately low rate of strength development can lead to frost damage.

This cement should not be used in thin R.C.C. structures since its early strength is less.

(*ii*) Because of its fairly high sulphate resistance, this type of cement is *frequently used in sea water construction.*

**4. White cement:**

- White Portland cement is made from raw materials containing *very little iron oxide and manganese oxide.* China clay is generally used, together with chalk or limestone free from specified impurities. Oil is used as fuel for the kiln in order to avoid contamination by coal ash. Since iron acts as flux in clinkering, its absence necessitates high kiln temperatures but sometimes cryolite (sodium aluminium fluoride) is aded as a flux. Contamination of the cement with iron during grinding has also to be avoided. For this reason, the rather inefficient pebble grinding is used instead of a usual ball mill, although nickel and molybdenum alloy balls have been introduced. The *cost of grinding* is thus higher, and this, coupled with the more *expensive raw materials*, makes white cement rather *expensive*.
- It dries quickly
- It possesses high strength.
- It has superior aesthetic values.
- It should not set earlier than 30 minutes.
- It should be carefully transported and stored in closed container only.

**Uses:** (*i*) It is used for *floor finish, plaster work, ornamental work* etc.

(*ii*) It is also used as mortars for marbles and tiles.

(*iii*) Miscellaneous applications of white cement include *swimming pools* (where it replaces the use of glazed tiles with coloured shades usable under water), moulding sculptures and statues, painting garden furniture etc.

(*iv*) It is also employed for ready mixed concrete and precast concrete blocks.

**5. Coloured cements:**

- Coloured cements are prepared by adding 5 to 15 percent of a suitable colouring pigment before the cement is finally ground.
  - — Iron oxide is added to give red and yellow colours.
  - — Chromium oxide provides green colour.
  - — Cobalt oxide provides blue colour.
- Coloured cements are also known as "*Colourcrete*".
- These are much costlier than the ordinary cement.

**Uses :** These cements are widely used for finishing of floors, external surfaces, artificial marble, stair treads, textured panel faces, window sill slabs etc.

**6. Acid resistance cement:**

An acid resistance cement contains : (*i*) acid-resistant aggregates such as quartz, quartzites etc., (*ii*) additive such as sodium fluosilicate ($Na_2SiF_6$), and (*iii*) aqueous solution of sodium silicate or soluble glass.

- — The addition of additive sodium fluosilicate accelerates the hardening process of soluble glass and also increases the resistance of cement to acid.
- — The soluble glass works as the binding material of acid-resistant cement.

- It *cannot* resist the action of water well. Its resistance to water can be increased by adding 0.5% of linseed oil or 2% of cresit and it is then known as *acid and water resistant cement.*

**Uses :** It is used for *acid-resistant and heat-resistant coatings of installations of chemical industry.*

**7. Expanding cement:**

- It is produced by *adding an expanding medium like sulpho-aluminate and a stabilising agent to the ordinary cement.*
- This cement expands whereas other cements shrink.

**Uses :** (*i*) It is used for the construction of *water retaining structures.*

(*ii*) It is also employed for repairing the *damaged concrete surfaces.*

**8. Hydrophobic cement:**

- It contains *admixtures* (*e.g.*, acidol, naphthene soap, oxidized petrolatum, etc.) *which form a thin film around cement grains and decrease the melting ability of cement grains.*
- In this type of cement, in initial stage, the gain in strength is *less* as hydrophobic films on cement grains prevent the interaction with water. However, after 28 days, its strength equals that of ordinary Portland cement.

**Uses :** When this cement is used in the preparation of concrete, the fine pores in concrete are uniformly distributed and thus the frost resistance and water resistance of such concrete are increased considerably.

**9. Portland pozzolana cement:**

- It is an interground blended mixture of *cement* and *pozzolana.* **Pozzolana** is a *natural or artificial material containing silica and alumina in a reactive form.* Pozzolanic materials most commonly met with are : Volcanic ash, pumic, opaline shales and cherts, burnt clay, flyash etc.
- In the manufacture of cement, about 25 per cent of prozzolanic material is added to the ordinary cement clinkers and *mix is thoroughly ground.*

**Advantages :** (*i*) Possesses higher tensile strength.

(*ii*) Evolves less heat during setting.

(*iii*) Attains compressive strength with age.

(*iv*) Offers great resistance to expansion.

(*v*) Imparts higher degree of water-tightness.

(*vi*) Imparts plasticity and workability to mortar and concrete prepared from this type of cement.

(*vii*) Offers higher resistance to chemical attack and to the action of sea water.

(*viii*) Not costly.

**Disadvantages :** (*i*) Less compressive strength in early days.

(*ii*) Less resistance to erosion and weathering action.

*Uses :* (*i*) It is widely used for *hydraulic structures* (mass concrete works) such as *dams, wiers etc.*

(*ii*) It is also used in *sewage works* and for *laying concrete under water.*

**10. Supersulphate cement**

- It is made by intergrinding a mixture of 80 to 85% of granulated slag with 10 to 15% of calcium sulphate and about 5% Portland cement clinker and ground to a fineness of 4000 to 5000 $cm^2/g$.
- It is highly resistant to sea-water.
- It can withstand the highest concentration of sulphates normally found in soil or ground water.
- It also offers resistance to peaty acids and oils.
- The heat of hydration of this type of cement is *low*.
- Its compressive strength should *not* be *less* than 15 $N/mm^2$, 22 $N/mm^2$ and 30 $N/mm^2$ after 3 days, 7 days and 28 days respectively.
- This type of cement should not be mixed with other cements.
- It combines chemically with more water than is required for the hydration of Portland cement, so that concrete with water-cement ratio less than 0.5 should *not* be made and mixes richer than 1:6 (cement : total aggregate) are not recommended.

*Uses :* (*i*) Supersulphated cement can be used in all cases where normal cements are used *except in very hot weather.*

(*ii*) It is used in variety of agressive conditions, for example, marine works, mass concrete jobs to resist the attack of aggressive water; reinforced concrete pipes in ground water, concrete construction in sulphate bearing soils.

- It has been used for the undersides of bridges, over railways tracks and for concrete sewers, carrying industrial effluents.

**11. Masonry cement:**

- It is prepared by intergrinding a mixture of Portland cement clinker with inert materials (non-pozzolanic) such as limestone, dolomite limestone and dolomite gypsum and an air-entraining plasticizer in suitable proportion.
- Its initial and final setting times are 90 minutes and 24 hours respectively.
- Its compressive strength is 2.5 $N/mm^2$ for 7 days.
- Because of its property of producing a smooth, plastic, cohesive, strong yet workable mortar when mixed with fine aggregate, masonry cement is superior to lime-mortar, lime-cement mortar and cement mortar.

## 5.13. STORAGE OF CEMENT

Cement has great affinity for moisture and hence it should be stored well shielded from moisture-laiden current of air. The exposed cement is attacked by air setting which gradually spreads; it gives rise to lumps. If the lumps formed are so hard that they cannot be powdered by pressing in between thumb and the forefinger, the cement should not be used in any sound construction.

The following **precautions** should be taken in the storage of cement:

1. Cement should be stored in special water-tight shed with a dry damp-proof floor, water-proof walls and leak-proof roof on the top.
2. The cement bags should be placed at a distance of 30 cm from the walls. Stacks of bags should be of such a height as can be easily handled.
3. Cement bags should be so arranged that they can be used on the principle of "first come first served".
4. Cement bags should be stacked close together to avoid free circulation of air.
5. Storing of cement during rainy season should be avoided.
6. Cement bags should not be placed directly on floors; planks or iron sheets may be placed underweath to guard against dampness.
7. As far as possible, the storing of cement for a long period should be avoided.
8. When cement is to be stored in bulk form, it should be stored in air tight cylindrical containers, known as **silos** or **bins**.

## 5.14. CEMENT WATER PROOFERS

- Cement mortar or concrete, rich in cement, becomes fairly water-tight when set. However, addition of certain water proofers to cement mortar or concrete, during mixing, makes them absolutely water-tight on setting. These water proofers render mortar or concrete water-tight either by filling the pores physically or reacting chemically.
- The water proofers may be in *powder, paste* or *liquid forms*. The amount to be added must be in accordance with the instructions of the manufactures; generally the following proportions are used :

— 2 to 5 per cent ... when in *powder form*

— 1 part paste and 10 parts water ... when in *paste form*

— 1 litre liquid and 15 litres water... when in *liquid form*

Indian water proofers are : **"Cico"** and **"Acco Proof"**.

Foreign water proofers are : **"Pudlo"** and **"Impermo"**.

The water proofers are required for all water retaining structures especially for :

(*i*) Swimming pools (*ii*) Basements

(*iii*) Hospitals (*iv*) Refrigeration rooms

(*v*) Cold storages (*v*) Water supply and sewage works

(*vii*) Exterior plaster (*viii*) Bath rooms and kitchens

(*ix*) Reservoirs.

## 5.15. ADMIXTURES

**Admixtures** are *the materials which are added in cement mortar or concrete to improve upon their quality*. Various admixtures under differ trade names are available in the market.

The admixtures serve the following *purposes* :

1. Improve the workability.
2. Retard setting action of the mortar and concrete.
3. Increase the bond strength between reinforcement and concrete.
4. Improve the water proofing properties of the cement mortar or concrete.
5. Reduce shrinkage during setting of mortar or concrete.
6. Reduce bleeding and segregation effect of concrete.

## 5.16. VARIOUS CEMENT PRODUCTS — THEIR TRADE NAMES, USES AND DISTRIBUTORS

In the table 5.2. are given the various cement products along with their tradenames, uses and distributors.

**Table 5.2. Various Cement Products-Their Trade Names, Uses and Distributors**

| S. No. | Trade names | Uses/Purpose | Distributors |
|---|---|---|---|
| 1. | *Accocid* | Acid proof cement–resists the corrosive action of acids and chemicals. | A.C.C. |
| 2. | *Acco-proof* | Cement water proofing additive recommended for all types of constructions. | A.C.C. |
| 3. | *Acco Set-50* | A high refractory cement suitable for all types of furnaces. | A.C.C. |
| 4. | *Acco Set super* | A high temperature refractory cement used in high temperature zones. | A.C.C. |
| 5. | *Aquacem* | For total protection of concrete from inner most to outermost. | Bird & Co. Private Ltd (Kanpur, Mumbai, Madras) |
| 6. | *Aquacrete* | A concentrated coloured cement for coloured floors, cast stone mosaic flooring etc. | – do – |
| 7. | *Calundum* | High alumina refractory cement for furnaces, kilns, boilers etc. It possesses good hydraulic setting and rapid hardening properties. | A.C.C. |
| 8. | *Cema Plast* | A plasticizing admixture for high strength of concrete. | Construction Materials Agency (25- Parsee bazar street Mumbai-1) |
| 9. | *Cema Rapid* | An accelerating admixture for obtaining high early strength in concrete. | –do– |
| 10. | *Cemix* | New mortar plasticizer | A.C.C. |
| 11. | *Cemprover* | Cement primer | Snowcem India Ltd. (Mumbai-1) |
| 12. | *Colocrete* | Coloured Portland cement | –do– |
| 13. | *Feronite* | A metallic compound used for concrete hardening and heavy duty floors. | Bird & Co. Private Ltd. |
| 14. | *Firecrete and Finecrete super* | Castable refractories suitable for construction of furnace arches, furnace walls, hearths, lintels etc. recommended where temperature requirements are more stringent. | A.C.C. |
| 15. | *Glamour* | A polish for floors, furniture, leather goods, tiles etc. | Bird & Co. Private Ltd. |
| 16. | Hardcrete | Concrete hardening and water proofing liquid. | Snowcem India Ltd. |
| 17. | *Heco Aea* | An air entraining agent for all classes of concrete | Construction Materials Agency |
| 18. | *Impermo* | Cement water proofing compound | Snowcein India Ltd. |
| 19. | *Sand tax* | New long life decorative finish | A.C.C. |
| 20. | *Silvicrete* | White cement, excellent for flooring, interior and exterior finishes. | A.C.C. & Snowcem India Ltd. |
| 21. | *Snowsal* | Stabilising cement | A.C.C. |
| 22. | *Snowcem* | Decorative water-proof cement coating | Snowcem India Ltd. |
| 23. | *Snow filla* | New expansion filler | A.C.C. |
| 24. | *Surfacit* | Concrete surface hardening liquid | Snowcem India Ltd. |
| 25. | *Water seal* | Most satisfactory water proofing cement compound. | Bird & Co. Private Ltd. |
| 26. | *Wet seal* | Water proofing compound for cement; ideal for water bearing structures such as dams etc. (only 20 per cent needed to impart water tightness) | Jayat Colour & Chemical Industries (Mumbai - 10) |

## HIGHLIGHTS

1. Cement may be prescribed as a material with adhesive and cohesive properties which must be capable of bonding mineral fragments into a compact whole.

   Cement may be classified as : (*i*) Natural cement (*ii*) Artificial cement.

2. Percentage of various ingredients for the manufacture of Portland cement should be as follows:

   Lime = 63%; Silica = 22%; Alumina = 6%, Iron oxide = 3%, Magnesium oxide = 2.5%, Sulphur trioxide = 1.75%; Loss on ignition = 1.5%; Insoluble residue = 0.25%.

   Besides there ingredients 3 to 4% gypsum is also added during the grinding process for controlling the initial setting of cement.

3. Cement consists of four basic compounds $C_3S$, $C_2S$, $C_3A$ and $C_4AF$ which combine with water to form a gel on hydration. The strength gain of cement mainly depends on $C_3S$ during the early stage and on $C_2S$ during the later stage. Hydration of $C_3A$ is mainly responsible for setting and initial heat evolution while $C_4AF$ acts as inert compound.

4. The process of manufacture of cement consists essentially of *grinding the raw materials* (calcareous and orgillaceous stones containing silica, alumina and iron oxide), *mixing* them intimately in a certain proportion and *burning* in a *large rotary kiln* at a temperature of approximately 1300°C to 1900°C, when the material *sinters* and partially fuses into balls known as **clinker**. The clinker is cooled and ground to lime powder, with some *gypsum added* and the resulting product is the commercial Portland cement.

   Cement may be manufactured by wet process or dry process. Dry process (modern technology) is now preferred and adopted in most of the cement industries.

5. The chemical reaction between cement and water is called **hydration of cement**. The phenomenon by virtue of which the cement paste, which is finally set, develops strength is known as **hardening of cement.**

6. For normal setting of ordinary Portland cement, the following tests (laboratory) are carried out:

   (*i*) Chemical composition (*ii*) Fineness test
   (*ii*) Consistency test (*iv*) Setting times test
   (*v*) Soundness test (*vi*) Tensile strength test
   (*vii*) Compressive strength test.

7. Cement should be stored in special water-tight shed with a dry damp-proof floor, water-proof walls and leak-proof roof on the top. Cement bags should be stacked close together to avoid free circulation of air.

8. Addition of certain water-proofers to cement mortar or concrete, during mixing, makes them absolutely water tight on setting.

## OBJECTIVE TYPE QUESTIONS

### Choose the Correct Answer :

1. The soundness of cement can be tested by

   (*a*) Vicat needle (*b*) Specific surface analysis
   (*c*) Le Chatelier's apparatus (*d*) Sieve analysis.

2. Which of the following cements is expected to have the highest compressive strength after 72 hours?

   (*a*) High alumina cement (*b*) Rapid hardening cement
   (*c*) Supersulphated cement (*d*) Ordinary Protland cement.

3. In an asbestos cement sheet the percentage of asbestos is around

   (*a*) 2 per cent (*b*) 6 per cent
   (*c*) 10 per cent (*d*) 15 per cent.

4. White cement should have least percentage of

   (*a*) aluminium oxide (*b*) iron oxide
   (*c*) silica (*d*) magnesium oxide.

5. Initial setting time of quick setting cement is nearly

   (*a*) 5 minutes (*b*) 10 minutes
   (*c*) 30 minute (*d*) 50 minutes.

6. Bulking is
   (*a*) segregating sand of particular quality
   (*b*) compacting of sand
   (*c*) increase in volume of sand due to moisture which keeps sand particles apart
   (*d*) ramming of sand so that it occupies minimum volume.
7. The constituents of cement which act as binders are
   (*a*) carbon and silica
   (*b*) silica and sand
   (*c*) tricalcium silicate, dicalcium silicate and carbon
   (*d*) tricalcium silicate, dicalcium silicate, and tricalcium aluminate.
8. Bulking of sand is caused due to
   (*a*) surface moisture
   (*b*) clay content
   (*c*) air voids
   (*d*) viscosity.
9. Basically the apparatus used for the determination of heat of hydration is
   (*a*) calorimeter
   (*b*) hygrometer
   (*c*) photometer
   (*d*) hydrometer.
10. The insoluble residue in cement is
    (*a*) less than 1.5 percent
    (*b*) less than 10 percent
    (*c*) less than 20 percent
    (*d*) between 10 and 20 percent.
11. In quick setting cement the compound added is
    (*a*) gypsum
    (*b*) aluminium sulphate
    (*c*) calcium sulphate
    (*d*) aluminium silicate.
12. The ultimate strength of cement is provided by
    (*a*) dicalcium silicate
    (*b*) tricalcium aluminate
    (*c*) silica
    (*d*) tricalcium silicate.
13. A slow setting cement will have higher prcentage of
    (*a*) gypsum
    (*b*) free lime
    (*c*) dicalcium silicate
    (*d*) tricalcium silicate.
14. The compound first to settle in cement is
    (*a*) dicalcium silicate
    (*b*) tricalcium silicate
    (*c*) tricalcium aluminate
    (*d*) terra-calcium alumino-ferrite.
15. The raw materials used in cement manufacturing, having more than 10 per cent proportion are
    (*a*) sand and coke
    (*b*) silicon oxide and ferrous oxide
    (*c*) calcium oxide and magnesium oxide
    (*d*) silicon oxide and magnesium oxide
16. A basic water content of 1.25 means
    (*a*) 1.25 litres of water have been added per litre of cement
    (*b*) 1.25 litres of water have been added per litre of mortar
    (*c*) 12.5% more water has been added than the basic water content requirements
    (*d*) 25% more water has been added than the basic water content requirements
    (*e*) none of the above.
17. The type of cement generally used for the construction of road pavements is
    (*a*) rapid hardening cement
    (*b*) low heat cement
    (*b*) blast furnace slag cement
    (*d*) ordinary Portland cement.
18. A high value of basic water content means
    (*a*) normal fluidity
    (*b*) more fluidity
    (*c*) less fluidity
    (*d*) excess cement proportion.
19. In low heat cement the compound which is kept to the minimum is
    (*a*) free lime
    (*b*) dicalcium silicate
    (*c*) tricalcium aluminate
    (*d*) tricalcium silicate.

20. The number of bags of cement required per cubic metre of concrete (1:2:4) would be roughly
(*a*) 2 to 3 (*b*) 3 to 4
(*c*) 4 to 5 (*d*) 5 to 6.

21. In case of Portland cement the permissible loss on ignition is
(*a*) 5 per cent (*b*) 10 per cent
(*c*) 15 per cent (*d*) 20 per cent.

22. The maximum percentage of ingredient in cement is that of
(*a*) lime (*b*) alumina
(*c*) silica (*d*) magnesium oxide.

23. Le Chatelier's method can be used to determine
(*a*) fineness of aggregates (*b*) soundness of cement
(*c*) compressive strength of cement (*d*) all of the above.

24. Four bags of cement (50 kg each) will require .... litres of water.
(*a*) 20 (*b*) 40
(*c*) 90 (*d*) 150.

25. A good quality cement should have higher percentage of
(*a*) free lime (*b*) silica
(*c*) dicalcium silicate (*d*) tri-calcium silicate.

26. For ordinary cement, the maximum extension by a Le Chatelier's method should not be more than
(*a*) 2 mm (*b*) 4 mm
(*c*) 6 mm (*d*) 10 mm.

27. An ordinary Portland cement when tested for fineness of sieve No. 9, will leave residue to
(*a*) 28 per cent (*b*) less than 20 per cent
(*c*) less than 15 per cent (*d*) less than 10 per cent.

28. Aerated cement is produced by the addition of
(*a*) copper sulphate (*b*) zinc sulphate
(*c*) powdered aluminium (*d*) sodium silicate.

29. The cubes for testing cement in compression are kept at
(*a*) 17 ± 2°C and 100% humidity (*b*) 27 ± 2°C and 90% humidity
(*c*) 37 ± 2°C and 70% humidity (*d*) none of the above.

30. Addition of pozzolana to cement causes all of the following *except*
(*a*) increased heat of hydration (*b*) increased curing time
(*c*) reduction in bleeding (*d*) reduction in permeability.

31. A sample of cement is said to be sound when it does not contain free
(*a*) silica (*b*) lime
(*c*) iron oxide (*d*) alumina.

32. Vicat's apparatus is used to determine all of the following *except*.
(*a*) fineness modulus of cement (*b*) normal consistency of cement
(*c*) initial setting time of cement (*d*) final setting time of cement.

33. The specific surface of an ordinary cements is determined by ..........
(*a*) autoclave method (*b*) Le-Chaterlier's method
(c) air permeability method (*d*) photo-calorimeter method.

34. The rate of hydration and hydrolysis of cement depends upon its
(*a*) soundness (*b*) fineness
(*c*) setting lime (*d*) none of the above.

35. For rapid hardening cement the minimum cube strength in a work test is taken as strength at
(*a*) 2 days (*b*) 7 days
(*c*) 15 days (*d*) 20 days

36. A cement bag stored for 24 months is likely to result in
(*a*) loss of strength by 50% (*b*) swelling by 10%

(*c*) change in colour of cement (*d*) increase in strength by 20%.

**37.** Water-cement ratio by weight is
(*a*) same as that by volume (*b*) lower as compared to that by volume
(*c*) higher as compared to that by volume (*d*) none of the above.

**38.** As the cement sets and hardens it generates heat. This is called
(*a*) heat of vapourisation (*b*) sensible heat
(*c*) latent heat (*d*) heat of hydration.

**39.** Which of the following cements is used for lining work of deep tube wells?
(*a*) Oil well cement (*b*) High alumina cement
(*c*) Blast furnace slag cement (*d*) Sulphate resisting Portland cement.

**40.** Low heat cement is used in
(*a*) thin structures (*b*) thick structures
(*c*) reinforcement free structures (*d*) underwater structures.

## ANSWERS

| | | | | |
|---|---|---|---|---|
| **1.** (*c*) | **2.** (*a*) | **3.** (*d*) | **4.** (*b*) | **5.** (*a*) |
| **6.** (*c*) | **7.** (*d*) | **8.** (*a*) | **9.** (*a*) | **10.** (*a*) |
| **11.** (*b*) | **12.** (*a*) | **13.** (*c*) | **14.** (*c*) | **15.** (*d*) |
| **16.** (*d*) | **17.** (*a*) | **18.** (*b*) | **19.** (*c*) | **20.** (*d*) |
| **21.** (*d*) | **22.** (*a*) | **23.** (*b*) | **24.** (*c*) | **25.** (*d*) |
| **26.** (*d*) | **27.** (*d*) | **28.** (*c*) | **29.** (*b*) | **30.** (*a*) |
| **31.** (*b*) | **32.** (*a*) | **33.** (*c*) | **34.** (*b*) | **35.** (*b*) |
| **36.** (*a*) | **37.** (*c*) | **38.** (*d*) | **39.** (*a*) | **40.** (*b*) |

## THEORETICAL QUESTIONS

1. What are the various ingredients of Portland cement? Discuss the function played by each in imparting specific properties to it.
2. Describe briefly with sketches the process of manufacture of good Portland cement.
3. State the various tests of ordinary Portland cement and give their recommended values. Describe how the tensile strength is tested.
4. What is Portland cement? What are its important properties and how are they tested?
5. Give an outline of a method of its manufacture, state why gypsum is added to it.
6. Compare the setting of cement with that of lime mortar.
7. What are the important precautions necessary in storing cement?
8. Why is sand used with cement? What kind of sand do you consider the best to be mixed with cement?
9. What are the different sources of obtaining sand? Which of them is the best? Give the characteristics of good sand.
10. What precautions will you take in storing Portland cement, and why?
11. What is meant by rapid hardening cement? What are the conditions under which it is used?
12. What is a plaster and why is it used? Enumerate the different kinds of plaster commonly used, mentioning their composition and typical applications.
13. Define Portland cement and mention the important raw materials used in manufacture of this cement.
14. Describe the important uses of cement in the construction industry.
15. Distinguish between setting and hardening of cement.
16. Explain the significance of testing the initial and final setting time of cement and how is it done? What is the effect of adding gypsum on setting time?
17. Explain the meaning of normal consistency of cement paste.
18. What is the importance of testing the fineness of cement? Why is the surface area method of testing fineness more reliable than the sieve test?
19. Explain where the following types of cements are most suitable:

(*i*) High alumina cement;
(*ii*) Blast furnace slag cement;
(*iii*) Ordinary Portland cement;
(*iv*) Sulphate resisting cement;
(*v*) Rapid hardening Portland cement.

**20.** Enumerate the laboratory tests for cement and describe any two of them.

**21.** What are the harmful constituents of cement?

**22.** Write short notes on :

(*i*) Admixtures
(*ii*) Compressive strength test
(*iii*) Soundness test
(*iv*) Hydrophobic cement
(*v*) Acid-resistant cement
(*vi*) White cement.

# 6

# Mortar

*6.1. Introduction. 6.2. Functions of mortar. 6.3. Properties of a good building mortar. 6.4. Uses of mortar. 6.5. Ingredients of mortars. 6.6. Types of mortars – Cement mortars-lime mortars – lime-cement mortars – special mortars: mud mortar-cement clay mortar – light weight and heavy mortars – decorative mortars – air entrained mortar – gypsum mortar – fire-resisting mortars – packing mortar – sound absorbing mortar – X-ray shielding mortar. 6.7. Precautions in the use of mortars. 6.8. Tests for mortars – adhesiveness test – crushing strength test – tensile strength test – setting time test. 6.9. Selection of mortars for different engineering works. 6.10. Plastering – objects of plastering – lime plastering – cement plastering – stucco plastering. – special type of finishing – special materials used in Plastering – defects in plaster work. 6.11. Pointing – procedure of pointing – forms of pointing. Highlights – Objective Type Questions – Theoretical Questions.*

## 6.1. INTRODUCTION

- **Mortar** is a *mechanical mixture in varying proportions of a binding material* like cement or lime and an *inert material or fine aggregate* like sand, surkhi, etc. The binding material and fine aggregate are sometimes referred to as *matrix* and *adulterant* respectively.
- Mortars are usually named according to the *binding material* used in their preparation *e.g., cement-mortar, lime-mortar,* etc.
- Mortar is used as a binding material in stone or brick masonry, or concrete, as a covering material to walls in the form of plaster to provide, a smooth, hard, and decorative surface.
- On account of its plastic nature, when green, the mortar forms an excellent, even bed for irregular surface of stone and this results in equal distribution of pressure over it.

## 6.2. FUNCTIONS OF MORTAR

Following are the functions of a mortar :

1. It binds together stones or bricks properly (so as to provide strength to the structure).
2. In any concrete, it holds coarse aggregates together (so as to form a solid mass).
3. In stone masonry and brick masonry, it fills up empty joints; a thin liquid mortar used for such purposes is termed as *grout*.
4. It provides a durable/weather resisting layer between different courses of masonry in the structure.

5. It form a homogeneous mass of the structure so that it may resist all the loads coming over it and transfer the same uniformly to its foundation.
6. It does pointing or plastering to the structure.

## 6.3. PROPERTIES OF A GOOD BUILDING MORTAR

Following *properties* are required of a good building mortar :

1. It should be *easily workable.*
2. It should develop adequate strength in tension, compression and bond for the work for which it is used.
3. It should be durable and should not affect the durability of other materials.
4. It should set quickly so that the speed of construction is ensured.
5. It should bind the bricks or stones to give a tight joint through which water cannot penetrate.
6. It should be cheap.
7. It should be capable of developing the designed stresses.
8. The joints formed by mortar should not develop cracks and they should be able to maintain their appearance for quite a long period.

## 6.4. USES OF MORTAR

Following are the *uses* of the mortars:

1. It is used to fill up the spaces between bricks or stones for making wall weather-tight.
2. It is used to bind together the bricks or stones in brick or stone masonry.
3. It is employed to give a soft even bed between the different layers of brick or stone masonry for equal distribution of pressure over the bed.
4. It is used in concrete as a matrix.
5. It is used in pointing and plastering to protect the joints and surfaces of brick and stone masonry.
6. It is employed for moulding purposes.
7. It is used to form joints of pipes.
8. It is used to improve general appearance of structure.
9. It is used to hide open joints of a brickwork and stonework.

## 6.5. INGREDIENTS OF MORTARS

The **ingredients** of various mortars used for different engineering purposes are enumerated and described below :

1. Binding or cementing materials ................ such as *cement* and *lime*;
2. Fine aggregates ......................... such as *sand, surkhi, ashes, cinder etc.;*
3. Water.

### 1. Binding or Cementing Materials:

(*i*) **Cement** : It is used as a binding material for the preparation of mortars required for different engineering applications where *strength and durability are the primary requirements*. For the purpose, *ordinary portland cement* (also called normal setting cement) is mostly used; however for special works rapid hardening cement, low heat cement or any other type of cement may also be suitable used.

*Functions:* In cement mortar, cement performs the following *functions* :

(*i*) It makes the mortar impermeable by filling up the voids existing in the fine aggregate.
(*ii*) It imparts strength to the mortar on setting and hardening.
(*iii*) It binds the aggregates into a solid mass by virtue of its setting and hardening properties when mixed with water.

(*ii*) **Lime:** It is also used as a binding material in preparing the mortars tor various purposes but its *strength is less than that of cement.* The different types of limes *viz.*, eminently hydraulic, semi-hydraulic and fat limes are used for preparing the lime mortars for different purposes. For preparing the mortar, the lime should be slaked before mixing it with sand, surkhi, etc.

*Functions:* In lime mortar, lime performs the following functions :

(*i*) It imparts strength to the mortar on setting and hardening.

(*ii*) It binds the fine aggregate into a solid mass by virtue of its setting and hardening properties when mixed with water.

**2. Fine Aggregates:**

(*i*) **Sand:**

- It is commonly used as fine aggregate in preparing the cement as well as lime mortars.
- Sand consists of small angular or rounded grains of silica ($SiO_2$).
- It is formed by decomposition of sandstone under the effect of weathering agencies.
- It may be either natural sand (*e.g., river sand, nala sand, pit sand* and *sea sand*) or artificial sand (prepared by crushing stones and gravels to powder). Mortars which are to be exposed to view should not be prepared with sea sand. The sea sand should be washed with fresh water (to remove maximum of its salt contents) before use.

**Classification of sand:**

The sand is classified as fine, coarse and gravelly, according to the *size of grains.*

— *Fine sand* will pass through a screen with clear openings of 1.5875 mm and is *mainly used for plastering.*

— *Coarse sand* will pass through a screen with clear openings of 3.175 mm and is *generally used for masonry work.*

— *Gravelly sand* will pass through a screen with clear openings of 7.62 mm and is generally used for *concrete work.*

**Requirements/characteristics or properties of good sand:**

A good sand should possess the following *characteristics*:

(*i*) Its grains should coarse, angular, sharp and hard.

(*ii*) It must be strong and durable.

(*iii*) It should be chemically inert.

(*iv*) It should not contain salts which attract moisture from the atmosphere.

(*v*) It should be free from any organic or vegetable matter.

(*vi*) It should be well graded. The fineness modulus of sand should be between 2 and 3.

(*vii*) it must be clean and free from coatings of clay and silt.

**Functions of sand in mortar:**

Following are the *functions of sand* in mortar:

(*i*) It does not increase the strength of mortar but is used as an adulterant for *economy.*

(*ii*) It *reduces the shrinkage of the binding* material and hence cracking of mortar during setting is avoided.

(*iii*) It *increases the resistance of mortar against crushing.*

(*iv*) It helps in *setting/hardening of fat lime by making it porous* ($CO_2$ is absorbed through the voids of sand which cause fat lime to set/harden effectively).

(*v*) It subdivides the paste of the binding material into a thin film and then *more surface is offered for its spreading and adhering.*

(*vi*) It helps in *making mortars of any desired strength by varying its proportion with the binding material.*

(*ii*) **Surkhi:**

- Surkhi (brick dust) is also used as a fine aggregate in the preparation of lime mortar.
- *It is finely powdered burnt clay and generally made from slightly underburnt or pilla bricks.*
- It should be free form admixture of foreign substances.
- *It plays the same functions as those of sand but in addition, it* **gives strength and improves hydraulic property of the mortar.**
- Surkhi if used wholly in place of sand particularly in exposed situations, it *causes the mortar to disintegrate.*

(*iii*) **Ashes or cinders:**

- These are in the form of small nodules (obtained from furnaces or locomotives) which are ground and used as fine aggregate in lime mortar.
- They *impart strength and hydraulic properties to the lime mortar.*
- These act as *adulterant,* hence the bulk or volume of mortar is increased which results in *reduction of cost.*

**3. Water:**

- Water plays a significant role in *mixing, transporting* and *proper laying of the mortar.*
- *The strength of the mortar depends upon the quality and quantity of water used in the mix.*
- The water to be used for preparing the mortar should meet with the following *requirements*:
  - — Should be *clean and fresh.*
  - — Should be free from *organic impurities, hygroscopic, greasy* and *oily substances.*
  - — Should preferably be drinkable.

*Functions:* Following functions are performed by water in the preparation of mortars :

(*i*) It *wets the surfaces of fine aggregates.*

(*ii*) It *facilitates spreading of cement/lime over the fine aggregate.*

(*iii*) It makes the mortar *workable* (by acting as a lubricant for the aggregate).

(*iv*) It causes hydration of cement and eminently hydraulic lime due to which setting and hardening of cement and lime take place.

## 6.6. TYPES OF MORTARS

Mortars are classified is several ways, *i.e.,* based on the *binding agent, properties of binding agent and aggregate* etc.

In general, mortars are classified as follows :

1. Cement mortars
2. Lime mortars
3. Lime-cement mortars (or composite or gauge mortar)
4. Special mortars
   (*i*) Mud mortar
   (*ii*) Cement-clay mortar
   (*iii*) Light and heavy mortars
   (*iv*) Decorative mortars
   (*v*) Air-entrained (plasticized) mortar
   (*vi*) Gypsum mortar
   (*vii*) Fire-resistant mortar
   (*viii*) Packing mortar
   (*ix*) Sound absorbing mortar

(*x*) *X*-ray shielding mortar.

### 6.6.1. Cement Mortars

- Cement mortar consists of *cement as binding material, sand in different proportions and water.*
- It should be noted that surkhi and cinder are not chemically inert substances and hence they cannot be used as adulterants with matrix as cement. Thus the *sand only* can be used to form cement mortar.
- It is the strongest type of mortar and is therefore preferred for use in the construction of structures subjected to heavy loading.
- The ratio of cement to sand should normally be 1 : 3 to 1 : 6 by weight for the mortar to be *workable and strong*. The use of mortars of lower cement content is not satisfactory since any notable reduction in cement content leads to reduction in workability and less cohesion and will produce *porous joints with a tendency for low frost resistance.*

  1 : 8 cement mortar is nearly twice as strong as 1 : 3 lime mortar.
- Cement mortar when used for protective plaster provides a *water proof layer and protects the elements covered from weathering effects.*
- Cement mortars should be *immediately used* after mixing with water as they *set and harden very quickly*.

**Preparation :** Cement mortars are prepared by the following methods :

(*i*) Hand mixing (*ii*) Machine mixing.

(*i*) **Hand mixing:** This method is generally adopted when a small quantity of mortar is required.

- In this method first, clean dry sand is spread in a uniform layer on a pucca platform. On it the requisite quantity of cement is uniformly spread.
- The whole mass is then mixed dry by working with spades till the whole mass becomes uniform in colour.
- A depression is then formed in the centre of the mortar mix and required quantity of water is added to this dry mix.
- Dry material from sides is placed on the edge of the depression containing water. It is done gradually till the water is completely absorbed by the dry mass. Care is taken not to let the water break the banks and flow out.
- The wet mass of mortar is then worked with spades to have a mortar of uniform consistency. Nearly *28 litres of water per bag of cement is required for making the mortar of suitable consistency.*
- Cement mortar should be prepared is small quantities which can be used before the initial setting time of the cement.

(*ii*) **Machine mixing**: When *large quantity* of mortar is required *continuously at a fast rate*, it is prepared by mixing of the ingredients in *mechanical mixers.*

- Cement and sand in specified proportion are put into the drum of the mixer and then required quantity of water is immediately added before revolving the drum.
- The drum is then revolved for a sufficient period to form uniform mixture of required consistency.
- The mixed mortar is then poured out for use.

**Uses:** (*i*) Cement mortar is used where high strength is required and in structures which are subjected to *wet condition* such as *piers, dams, deep foundations* etc.

(*ii*) It is also used for *external plastering* and *pointing work*.

(*iii*) It is used for damp-proof course below the ground level, exposed work such as exterior free standing walls and parapets, engineering construction with bricks of high strength.

(*iv*) In concrete, cement mortar binds the particles of coarse aggregate into one solid mass.

**Precautions:**

Following *precautions* should be taken while using cement mortar :

1. It should be prepared by mixing it *uniformly* to a workable consistency.
2. It should be used immediately after preparing and should be consumed within half an hour after adding water.
3. Before laying the bricks or stones in structure, they should be fully saturated with water, otherwise they may absorb greater part of water from the mortar, thus making the mortar joint weaker due to lack of water required for hydration.
4. The masonry or plastered surface should be continuously kept moist by sprinking water for *at least seven days.*

**6.6.2. Lime Mortars**

These are further classified as:

(*i*) Non-hydraulic lime mortars

(*ii*) Hydraulic lime mortars

(*iii*) Black mortars.

(*i*) **Non-hydraulic lime mortars:**

- These mortars are prepared by mixing *fat lime* (*well slaked before use*) with sand. The usual proportions of lime and sand are 1 : 2 or 1 : 3 by volume.
- They have a light colour.
- They do not cause efflorescence.
- They are **unsuitable** *for damp situations, foundations, thick walls, as their setting action depends on the presence of* $CO_2$.
- They are **suitable only** for *thin joints in brickwork.*
- When used in moderately thick sections, their quality can be improved by the addition of surkhi or cement.

(*ii*) **Hydraulic lime mortars:**

- These mortars are made from class *A* and class *B* limes.
- Mortars prepared from class *A* lime is *next to cement mortar* in strength and so is *used in all heavy and important engineering works, in foundations and for external plastering and pointing works.*

Mortar made with 1 part of fat lime and 2 parts of surkhi or 1 part of the lime plus 1 part of surkhi and 1 part of sand is mostly used in the foundations and superstructures of ordinary buildings.

In *works where more strength or hydraulicity is needed, pure hydraulic lime without sand is preferred.*

(*iii*) **Black mortars:**

- These mortars are so called because of their colour.
- In these mortars, the usual proportions are 1 part lime to 3 parts of ash or cinder (1 : 3).
- They become hard on setting.
- Due to *their less pleasing* colour, they are *used in internal walls.*

**Preparation of lime mortar:**

The lime mortar is prepared by the following methods :

1. By bullock-driven mortar mill or ghani.
2. By power-driven mortar mill.

**1. By bullock-driven mortar mill or ghani:**

When mortar is required on a small scale, it is ground in bullock-driven, circular trench mill (Fig. 6.1)

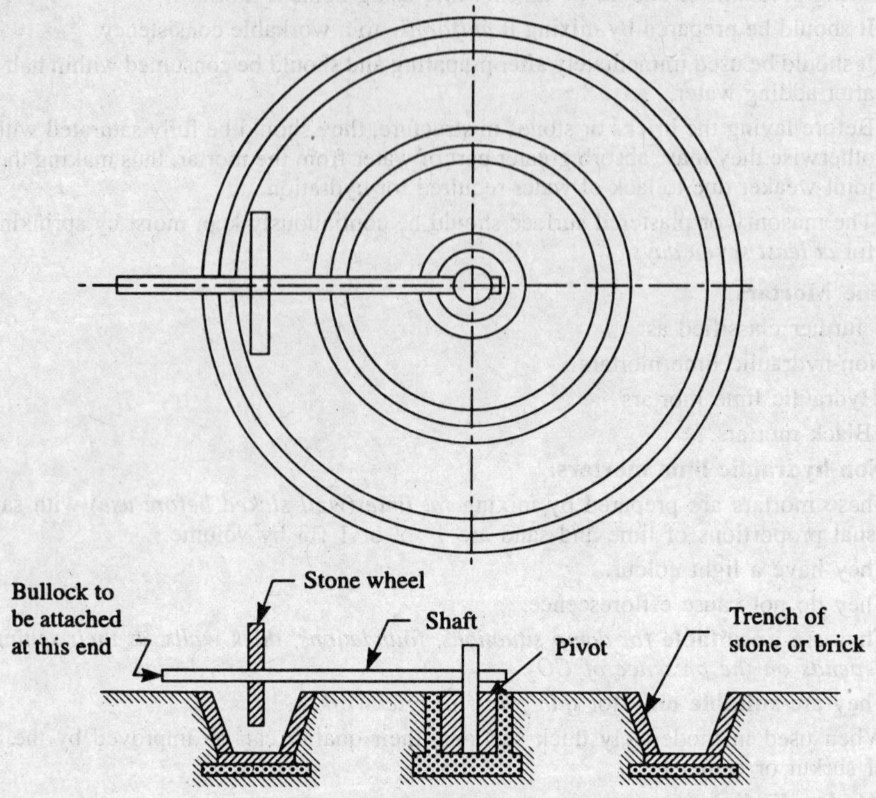


Fig. 6.1 : Bullock-driven mortar mill.

- Slaked lime and sand in specified proportions are mixed with (or without) the addition of water on a dry platform or in a mixer and the mix is then fed to the mortar mill.
- As the object is of crushing the lime and thoroughly mixing it with sand, the mortar must be continually turned over by means of a 'spoons', *i.e.*, an iron rod with a flat end attached to a *ghani* axle by a rope and used in a vertical position for scraping and turning over the contents of the trench. Water if needed is adding during grinding so as to bring the material to consistency of stiff paste.
- Mixing is continued till every part of aggregate is coated uniformly with the cementing material.

**2. By power-driven mortar mill:**

- This method is employed when a steady and a continuous supply of lime mortar is required on a large scale.
- In this method, a power driven pan type mortar mill (Fig. 6.2) is generally used for grinding the mortar.

**Properties of lime mortars:**

Following are the *properties* of lime mortars :

1. They are *quite plastic* and *workable when wet.*
2. They have good working qualities, if made from high-calcium limes.
3. They *develop strength very slowly but gain continuous strength over long periods.*

4. They do not *'set' but stiffen* as water is lost by absorption by the units or blocks in contact, and by evaporation. Further, the gain in strength is acquired by very slow reaction of lime with carbondioxide from air.
5. They provide a fairly strong surface when used for plastering.
6. They provide enough bond between masonry blocks or bricks when used for masonry joints.

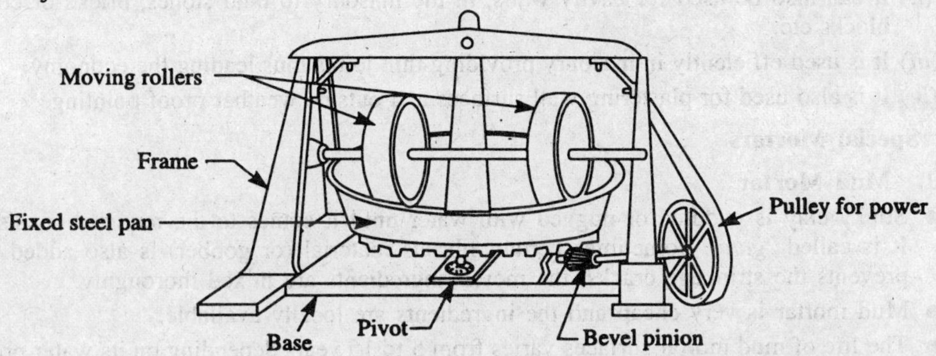


Fig. 6.2 : Power driven mortar mill (pan type).

**Uses:**

(*i*) Lime mortar is used in masonry to *bind* stones, bricks or concrete blocks together.

(*ii*) It is suitable for *chimneys* as one of their characteristics is the continued strength development over a long period.

(*iii*) It is also suitable for *masonry* and *plastering* in *cheap and light load bearing wall construction.*

(*iv*) It is also used for internal work with very thin mortar joints, or external walls in sheltered conditions where the mortar is protected by a frost-resistant pointing.

**Precautions:**

While using lime mortar, following *precautions* should be taken :

1. When hydraulic lime is *used* as the main binding material, the mortar should be used *within 4 hours after grinding.*
2. Lime motar made with *surkhi* or other *pozzolana* is hydraulic in nature and should be used within 24 hours after grinding.
3. It is not advisable to allow the mortar surface dry suddenly; it is essential to sprinkle water mildly on masony or plastered surface.
4. *After grinding,* all lime mortars *must be kept wet* and should never be allowed to dry during maturing and storage (to do so, make a small hollow in the middle of the heap and keep it filled with water all the time).
5. The workers who are to handle lime mortar should protect their skin by using oil rubber gloves etc.

### 6.6.3. Lime-Cement Mortars

- Lime-cement mortar is also known as '*compo*' mortar or *gauged mortar.*
- It is prepared by mixing cement with lime mortar in suitable proportions. *Addition of cement increases the hydraulicity and the strength.* The process of adding cement is known as *gauging*. It makes the lime mortar *economical, strong* and *dense.*
- *Non-hydraulic, semi-hydraulic* or *hydrated lime* are generally used for preparing these mortars.

- The usual proportions of cement, lime and sand are :
  1 : 1 : 6 ; 1 : 2 : 9 and 1 : 3 : 12.
- These mortars should be used within two hours after the cement is added.

**Uses:**

(*i*) It is used for bedding and for thick brick walls.

(*ii*) It can also be used for cavity walls, in the masonry to bind stones, bricks or concrete blocks etc.

(*iii*) It is used efficiently in masonry providing thin joints thus leading the economy.

(*iv*) It is also used for plastering wall surfaces and outside weather proof pointing.

### 6.6.4. Special Mortars

#### 6.6.4.1. Mud Mortar

- Sticky clay is puddled or pugged with water until it comes to the required consistency. It is called *'gara'*. Sometimes certain fibrous material (or gobber) is also added which prevents the strinkage cracks; the mortar ingredients are mixed thoroughly.
- Mud mortar is very cheap and the ingredients are locally available.
- The life of mud mortar surfaces varies from 5 to 15 years depending on its water proofing treatment and weather conditions.

**Uses:**

(*i*) These mortars are also used for *brickwork of ordinary buildings.*

(*ii*) It is used for *surfacing floors* and plastering the *internal and also external surfaces.* In order to improve the weather resistance of mud mortar on the outside walls which are exposed to rain, the *surface is sprayed or painted with bituminous material* (sometimes, for providing weather proof properties, the bituminous material is mixed during the preparation stage).

#### 6.6.4.2. Cement-clay Mortar

- Here clay is introduced as an effective finely ground additive in quantities ensuring a cement-clay proportion of not over 1 : 1. The additon of clay improves the grain composition, the water retaining ability and the workability of mortar and also increases the density of mortar.
- This type of mortar has better covering power and can be used in thin layers.

**Uses :** It is used for *masonry joints and plastering.*

#### 6.6.4.3. Light-weight and Heavy Mortars

**Light weight mortars :**

- These are prepared from light porous sands from pumice and other fine aggregates. They are also prepared by mixing wood powder, wood shavings or saw dust with cement mortar or lime mortar.
- In such mortars, fibres of jute coir and hair, cut into pieces of suitable size, or asbestos fibres can also be used.
- These mortars have bulk density less than 15 kN/m$^3$.

**Uses :** They are used in cases where a *reduction in the conductivity of structure is desirable* and *also structure is to be kept lightly loaded.*

**Heavy mortars:**

- These are prepared from heavy quartz or other sands.
- They have bulk density of 15 kN/m$^3$ or more.

**Uses :** They are used in load bearing constructions.

### 6.6.4.4. Decorative Mortars

These mortars are obtained by using :

— Colour cements or pigments, and

— Fine aggregate of appropriate colour, texture and surface.

**Uses :** They are used to impart a *pleasant outer appearance* to the surfaces of structures.

### 6.6.4.5. Air-entrained (plasticized) Mortar

- The working qualities of lean cement-sand mortar can be improved by entraining air in it (air serves as a *plasticizer* producing minute air bubbles which help in flow characteristics and workability).
- The air bubbles increase the volume of the binder paste and help to fill the voids in the sand.
- The air entraining also makes the mortar light weight and a better heat and sound insulator.

### 6.6.4.6. Gypsum Mortar

These mortars are prepared from gypsum binding materials such as building gypsum and anhydrite binding materials.

### 6.6.4.7. Fire-resistant Mortar

- It is prepared by *adding aluminous cement to a finely crushed powder of fire-bricks* (usual proportion being *one part* of aluminous cement to *two parts* of powder of fire-bricks).

**Uses :** This mortar, being fire-resistant, is used with fire-bricks for lining furnaces, fire places, ovens etc.

### 6.6.4.8. Packing Mortar

- For packing oils, special mortars are required which should possess the following *properties.*

(*i*) Water resistance
(*ii*) High homogenity
(*iii*) Predetermined setting time
(*iv*) Resistance to subsoil water pressure
(*v*) Ability to form solid water-proof plugs in cracks and voids of rocks etc.

- The composition of this type of mortar depends on hydrogeologic conditions, type of timebering and packing methods.
- The varieties of packing mortars are :
  Cement-sand, cement-loam and cement-sand-loam packing mortars.

### 6.6.4.9. Sound Absorbing Mortar

- Thsre mortars may have binding materials such as cement, lime, gypsum, slag etc. and aggregate (lightweight porous materials) such as pumice, cinders etc.
- The bulk density of such a mortar varies from 6 to 12 kN/m$^3$.
- Noise level can be reduced by using sound-absorbing plaster formed with the help of sound-absorbing mortar.

### 6.6.4.10. X-ray Shielding Mortar

- The aggregates for this type of mortar are obtained from heavy rock. In order to enhance the *protective property* of such a mortar, *suitable admixtures are added.*
- It has a bulk density of 22 kN/m$^3$

**Uses:** It is used for providing the plastering coat to walls and ceiling of X-ray cabinets.

## 6.7. PRECAUTIONS IN THE USE OF MORTARS

1. All mortars prepared for masonry works should be *used* as soon as possible with the maximum limits given below :

(*i*) *Cement mortars* — immediately after the addition of water, before the initial setting time of the cement.

| | | |
|---|---|---|
| (*ii*) | *Eminently hydraulic lime mortar* | within 4 hours after grinding. |
| (*iii*) | *Lime surkhi mortars* | within 24 hours of grinding. |
| (*iv*) | *Gauged mortars* | within 2 hours of the addition of cement |

2. Only that much quantity of cement mortar should be prepared at a time which is likely to be used before the initial setting time of cement used in mortar.
3. All lime mortars after grinding, should we kept damp by covering them with wet sacks. They shall never be allowed to go dry.
4. Partly set and dried mortar should not be retempered and used.
5. The bricks or stones should be thoroughly soaked in water till all the air bubbles cease, other wise they might absorb moisture from mortar and prevent it from setting.
6. The mortar should be used as stiff as possible consistent with workability since excess water weakens the mortar. Grout (a thin liquid mortar) should only be used in joints which cannot be filled by mortar of proper consistency.
7. In case of masonry work with lime mortars, work after every height of 1.5 m or less should be allowed to set for at least 2 days before starting further construction over it.
8. After the work has been laid, it should be kept wet for a week or so to prevent the rapid drying of the mortar particularly in hot weather.

Fig. 6.3 : Test for adhesiveness.

## 6.8. TESTS FOR MORTARS

In order to test the quantity of mortar, the following tests are usually conducted :

1. Adhesivenss to building units test.
2. Crushing strength test.
3. Tensile strength test.
4. Setting time test.

### 6.8.1. Adhesiveness to Building Units Test

This test is conducted as follows :

- Two bricks of size 190 mm × 90 mm × 90 mm are jointed together and cemented with the mortar under test. This gives a common central sectional horizontal joint surface area of 90 mm × 90 mm forming a cross with 50 mm free projections on either side of both the bricks (Fig. 6.3).
- The upper brick is suspended from an overhead support and a board is hung from the ends of the lower brick.
- The board is loaded till the joint between the brick fails.
- The ultimate adhesive strength of mortar is obtained by dividing maximum load by area of of contact (*i.e.*, 8100 mm²).

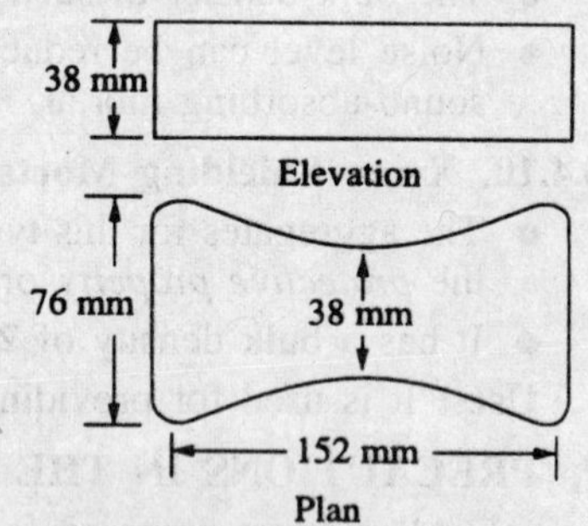

Fig. 6.4 : Briquette for tensile strength test of mortar.

### 6.8.2. Crushing Strength Test

The following procedure is used to conduct the test :

- The brickwork is carried out with the mortar to be tested.

- A sample of this brickwork is taken and it is gradually loaded in a compression testing machine till the failure occurs due to crushing.
- The ultimate load causing failure divided by the cross-sectional area gives the ultimate crushing strength of the mortar per sq. mm.

### 6.8.3. Tensile (or Cohesive) Strength Test

This test is conducted as follows :

- A briquette (Fig. 6.4) made with water is tested in a testing machine.
- The total applied tensile load at which the briquette breaks is divided by the area of cross-section of the briquette at the place of its breaking (central portion area = 38 × 38 = 1444 mm$^2$);' this gives the ultimate tensile strength of the mortar, per sq. mm.

### 6.8.4. Setting Time Test

This test is preformed as follows :

- Mortar is tested in Vicat apparatus for the setting time.
- Depending on the type of lime mortar, the test may be conducted after one day, three days or one week.
- In case of cement mortar, this test is conducted within 3 to 5 minutes of adding water.

## 6.9. SELECTION OF MORTARS FOR DIFFERENT ENGINEERING WORKS

The various engineering works along with the recommended mix of different mortars is given in table 6.1.

**Table 6.1. Selection of Mortars for Different Engineering Works**

| S. No. | Nature of Work | Mortar-type and Composition |
|---|---|---|
| 1. | Thick joints in stone masonry. | Hydraulic lime sand mortar (1:2 to 3) |
| 2. | Stone masonry in foundations and superstructure of ordinary buildings. | 1:2 fat lime surkhi mortar or 1 part lime, 1 part surkhi and 1 part sand. |
| 3. | Brickwork in arches, plastering inside of walls. | 1:5 to 1:6 cement and mortar, or lime surkhi mortar (1:2) or lime, surkhi and sand. (1 : 1 : 1) mortar. |
| 4. | Reinforced brickwork. | 1:3 cement sand mortar |
| 5. | Mass concrete in foundations, paving tiles, cavity walls, plastering of ceiling and external plastering work etc., where good finish is required. | 1:4 cement sand mortaror or 1:2 to 3 hydraulic lime mortar. |
| 6. | Massive work below ground level especially in water logged areas. | 1:3 cement sand mortar or 1:3 lime (eminently hydraulic) sand mortar. |
| 7. | Massive works, dams, retaining walls, damp proofing, flooring, etc. where very high finish is requried. | 1:3 cement sand mortar. |
| 8. | Pointing work. | 1:1 to 1:2 cement sand mortar |
| 9. | General R.C.C. works such as slabs, beams and columns, cement concrete flooring etc. | 1:2 cement sand mortar. |
| 10. | Damp proof courses and cement concrete roads | 1:2 cement sand mortar |
| 11. | R.C.C. tanks and other retaining structures etc. | $1:1\frac{1}{2}$ cement sand mortar |
| 12. | Highly stressed numbers of structure | 1:1 cement sand mortar |
| 13. | Laying fire-bricks | Fire-resisting, mortar consisting of 1 part of aluminous cement to 2 parts of finely crushed powder of fire-bricks. |

## 6.10. PLASTERING

**Plastering** *is an art of covering rough walls and uneven surfaces in the construction of houses and other structures with a plastic composition, called* **plaster**.

Plastering, in general, is not required for good brickwork, made of well burnt bricks or neat stone masonry constructed of durable stones and good mortar.

### 6.10.1. Objects of Plastering

The *principal objects* of plastering are :

1. To provide a true, even, smooth and finished surface to the work and improve the appearance.
2. To protect the surface from atmospheric influences.
3. To cover defective workmanship.
4. To provide a covering for the unsound and porous materials.

### 6.10.2. Tools of Plastering

For plastering, the following tools are used :

1. Gauging trowel
2. Float
3. Floating rule
4. Plumb-bob.

**1. Gauging trowel:**

- It is an ordinary trowel used by masons in general work.
- At the time of masonry work and plastering work it is used for taking mortar from the mortar pan which is generally kept mixed near the working mason.
  Dashing of mortar against the prepared walls and then spreading and pressing is done with the help of trowels.

**2. Float:**

- It is used to spread the mortar evenly over the surface.
- It may be made of timber or tempered steel. Steel float is also known as *lay trowel* whereas that of timber *skimming float.*
- Floats are also made in which nail remain projecting about 3 mm from the surface; such a float is known as *devil float*. This float is *used to make zig-zag lines on the plastered surface so as to form a key for the coats to be subsequently applied.*

**3. Floating rule:**

- It is used to check the level of the plastered surface between successive screeds.
- It is made of timber and its length varies from 1 m to 2 m. Its one face in made perfectly plain and smooth.

**4. Plumb-bob:**

This tool is very much used in forming vertical screeds on the surface to be plastered.

Some commonly used tools of plastering are shown in Fig. 6.5

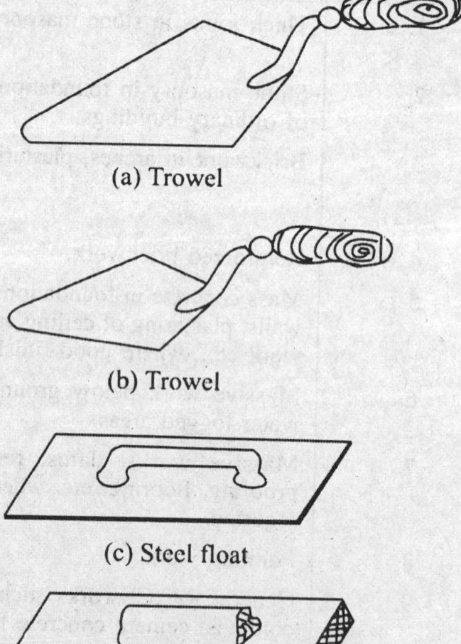
(a) Trowel

(b) Trowel

(c) Steel float

Wooden float

**Fig. 6.5** : Some commonly used tools of plastering.

### 6.10.3 Lime Plastering

Lime and sand in different proportions (depending on the type of work and number of coats to be applied) go to form a lime plaster. *Fat or pure lime is best for preparing lime plaster.* If hydraulic lime is to be used, it should be either *(i)* ground dry to a fine powder and then mixed with sand, water and used; or *(ii)* it should be ground twice, *i.e.,* ground for first time, then left for two or three weeks in a heap to slake, after that period reground and used.

The usual proportions for external or internal works are :

*(a) "Lime plaster if applied in one coat"*: I part lime and $1\frac{1}{2}$ cm parts river sand.

*(b) "Lime plaster if applied on two coats"* : Ist coat : same as *(a)* ; 2nd coat: 2 parts lime and 1 part white sand.

*(c) "Lime plaster if applied in three coats"* : Ist coat : same as *(a)* or *(b)*; 2nd coat: 1 part lime and 1 part lime sand; 3rd coat : 4 parts lime and 1 part fine white sand.

**Procedure of plastering :**

*(i) Preparation of surface :* All the joints and surface of the masonry are well cleaned with a wire brush and care is taken to see that they are free from oil, grease, soot etc. Holes and hollows, if any, are properly filled and projections more than $1\frac{1}{2}$ cm beyond the general surface of the wall are knocked off to make the surface of the wall uniform. In case of old brick walls, the bricks are pricked over so as to expose a new rough surface. Finally the mortar joints and surface of the walls are well washed, wetted with clean water and kept wet for at least 6 hours prior to the application of plaster.

*(ii) Ground work for plaster* : To act as a guide for the thickness of plaster, vertical strips or ledges, of plaster, called screed or bands are formed at the ends of the surface to be covered and at intervals of about 1.5 metres across it, and levelled upto strings stretched at the distance from the masonry. The screeds are usually 15 cm in width and extend from the top of the wall to the bottom. The screed thus formed form the ground work for the application of plaster.

*(iii) Application of plaster coats* :

*(a) First Coat :* It is also called *rough coat of plastering.* It is generally $1\frac{1}{4}$ cm on brick or ashlar masonry and nearly $1\frac{3}{4}$ cm on rubble masonry. The first coat of plaster is usually applied by dashing it against the wall surface between the screeds. It is frequently sprinkled with sufficient quantity of water and rubbed well with floats.

*(b) Second Coat :* It is also called *floating coat of plastering.* Before applying second coat the surface of the first coat is left exposed to air to set and harden for about 2 days, but not to dry, and after this period, the surface (of the first coat) is sprinkled with water and either freely scored all over with the edge of the trowel or well beaten with a thin cane. The second coat is applied with mason's trowels, fairly spread out, tested, pressed and rubbed with the straight edge. The plastered surface is sprinkled with water and rubbed well with floats. The thickness of this coat is usually $\frac{5}{8}$ cm to $1\frac{3}{4}$ cm.

*(c) Third coat :* It is also known as *final or finishing coat.* It is applied about 5 days after the application of second coat and is usually of cream of white or fat lime, called *nerru* or *plaster's putty.* It is applied in a thin coat of 0.3 cm thick, with a perfectly straight plane, and the paste well rubbed with a straight edge. It is then floated with a wooden flat and polished with a trowel till no trowel marks are visible on the surface. To get a fine polished surface, a soap stone or polishing stone is oftenly used.

### 6.10.4. Cement Plastering

The cement plaster may be applied either in one coat or two coats.

- The first or rough coat of plaster usually comprises of 1 part of Portland cement and 3 or 4 parts of sand by volume. The surface of the wall is first prepared as described for lime plastering and vertical screeds, 15 cms in width are formed on the wall surface at suitable intervals. The spaces or bays between the screeds are applied with cement plaster with mason's trowels and the plaster is levelled by means of flat wooden floats and wooden straight edges. Finally the surface is polished without sprinkling water.

- If second or fine or flusting coat is to be applied, the surface of first coat is not polished but roughened to give a key to the second coat of plaster. Second coat consists of a paste of pure Portland cement mixed with sufficient quantity of water and is applied in a thin layer over the moist and damp first coat, well trowelled and rubbed perfectly smooth.
- To *develop adequate strength and hardness the finished plastered surface* (whether one or two coats) is *kept wet by sprinkling water for at least a week.*

### 6.10.5. Stucco Plastering

- Stucco is the name given to a *decorative type* of plaster *which gives as excellent finish.* The surface plastered with stucco compares very much with marble finish both in smoothness and elegance.
- Stucco plaster is *mixture of lime, white stone, gypsum and oxides for colouring.*
- Stucco plaster can be used for interior as well as exterior surfaces. It is usually laid in *three coats* making the total thickness of plaster of about 25 mm.
  — The *first coat is called the scratch coat*;
  — The *second* a finer coat, also known as *brown coat*;
  — The *third coat* is called the *white coat* or *finishing coat.*

Each coat should be permitted to dry thoroughly before the next coat is applied.

**Stucco for interior walls :**

— In this case the *first coat or scratch coat* consists of a coat of lime plaster 12 mm in thickness.

— The *second coat* or the *brown coat* consists of lime plaster which is made richer in lime and is 10 mm in thickness.

— The *third coat* or *finishing coat* is 3 mm in thickness and consists of a mixture of finest lime and well-ground white stone (which may be marble or quartz). This coat is carefully polished with a bag of linen containing moist chalk in first instance and then oil and chalk. The surface is finally rubbed with oil only and finished smooth and bright.

**Stucco for exterior walls :**

— In this case the mortar for the rough coat is prepared by mixing cement and sand in the ratio of 1 : 3 and then adding 10% by weight of hydrated lime to the mixture. The thickness of the *first coat* in this case is also kept 12 mm.

— The *second coat* has the same composition as the rough coat but is 10 mm in thickness.

— The finishing coat which may be 3 to 6 mm in thickness consists of a mixture of cement and sand taken in proportions of 1 : 2 to 1 : $2\frac{1}{2}$. White or coloured cement may be used in finishing coat to obtain the desired shade.

### 6.10.6. Special Types of Finishing

The following types of finishing treatments are given to the external plastered surfaces:

1. Smooth cast
2. Rough cast
3. Pebble dash
4. Scrapped finish
5. Textured finish.

1. **Smooth cast :**

- This type of finish presents levelled and smooth surface.
- The mortar for such a finish is made by mixing cement and fine sand in the ratio of 1 : 3.
- The mortar may be applied in one or two coats and finished by working with a float. Steel floats are not recommended for external rendering as they give a very smooth finish which is liable to crack and craze due to exposure to atmospheric conditions.

2. **Rough cast :**

- In this type of finish the mortar for final coat contains a proportion of fairly big size coarse aggregates. The wet plastic mix is thrown on the surface by means of a scoop or trowel, and is left in rough condition. The texture desired is regulated by the size of the coarse aggregate which generally varies between 3 to 12 mm. Rough cast finish is normally 10 mm thick and it is applied over 13 mm thick under layer of plaster.

- The top layer is applied after one or two days after the underlayer is laid. The mortar for the finish is made by mixing cement, fine sand and coarse aggregate in the ratio of 1 : $1\frac{1}{2}$ : 3.
- Rough cast finish is *water-proof, durable and resistant to cracking and crazing.*
- This type of finish is particularly considered suitable *for buildings subjected to heavy rainfall and strong winds.*

3. **Pebble dash :**

- In pebble dash finish small pebbles or crushed stones of suitable size (usually varying from 10 to 20 mm) are thrown on to a freshly applied final coat of mortar and left exposed. The mortar of the final neat is usually made by mixing cement and coarse sand in the ratio of 1 : 3 and is laid in about 1 mm thickness.
- This finish possesses all the advantages of the rough cast finish and is recommended in similar condition.

4. **Scrapped finish :**

- In this type of finish the final coat after being levelled and allowed to stiffen for a few hours, is *scrapped* with a steel straight edge, old saw blade or other such tool to remove the surface skin. In this case, the final coat in usually 6 to 12 mm thick of which about 3 mm is removed in the scrapping process. The scrapping is done after the setting has taken place.
- Scrapped finish is less liable to crack than plain finishes.

5. **Textured finish :**

- In textured finish, ornamental patterns or textured surfaces are produced by working with various tools or the freshly applied final coat.
- This type of finish possesses all the advantages of the rough cast finish.

### 6.10.7. Special Materials used in Plastering

The following special materials are used in plastering:

1. **Plaster of paris.** It is also known as calcium sulphate plaster and is *obtained by heating gypsum.*
2. **Keene's cement.** This is hardest and densest form of gypsum.
3. **Barium plaster.** This is essentially made of barium sulphate.
4. **Accoustical plaster.** This is gypsum mixture, which is employed as a final coat to serve the purpose of a sound-repellent finish.

- The first coat is very coarse, the second is finer and forms a smooth even surface, while the final is the finishing coat. The usual proportions in the first two coats are : Cement 1 part, sand 3 parts, hydrated lime 10 per cent and in the final coat, cement 1 part, sand 2 parts, coloured cement 1 part or any metallic oxide for colour.
- It is an excellent decorative plaster used on both interior and exterior walls.

### 6.10.8. Defects in Plaster work

Some of the *defects* which usually occur is a plaster are:

1. Cracks
2. Falling of plaster
3. Blistering or blowing of plaster
4. Efflorescence.

**1. Cracks.** The cracks which develop in a plastered surface may be *thin hair cracks* or *wide open cracks*. The cracks which are most easily noticeable are not considered as defects. Development of thin hair cracks is known as *crazing*.

The cracks may develop due to the following *reasons*:

(i) Excessive thickness of plaster–causes excessive shrinkage.

(ii) Very smooth finished surface– Rough surfaces crack less than smooth finished surfaces.

(iii) Expansion or shrinkage of plaster itself during drying.

(iv) Movement in the backing due to expansion or shrinkage of backing material.

(v) Improper plastering of old surface.

**2. Falling of plaster:**

(i) Imperfect bonding of plaster with the background.

(ii) Too much thermal changes either in backing or plaster itself.

(iii) Imperfect bond between successive coats of plaster.

(iv) Non-uniform suction by the backing material from plaster.

**3. Blistering or blowing of plaster.** In this defect small patches of plaster swell out beyond the plastered surface. *This defect occurs due to slaking of unslaked lime particles present in the plaster.*

**4. Efflorescence.** Some soluble salts are always present in all the materials making plaster and backing materials. Even water used in construction may contain soluble salts. When newly constructed wall or nearly plastered wall dries out, the soluble salts are brought to the surface and they appear in the form of a whitish crystalline substance. Such a growth is known as *efflorescence*. The surface which may be naked wall or plastered wall gets *disfigured and concentrated stains present an ugly appearance*.

**Remedies for plaster defects :**

The plaster defects can be minimised by using the following *measures* :

1. The surface to be plastered should be properly prepared and well-watered before plaster coat is laid over it. This will cause *non-absorption of water from the plaster*.
2. The finished surface of the plaster should *not be trowlled excessively*.
3. Superior quality of bricks should only be used for brickwork.
4. *Water* used for construction should be *free from soluble salts*.
5. Efflorescence if any should be removed by rubbing the surface with brushes.
   – Alternatively, a solution of one part of hydrochloric acid or sulphuric acid with four to five parts of clean water may be prepared and applied in the affected surface with brushes. The surface is lastly washed with water.
6. Water should not be used for washing the surface to remove efflorescence. In that case some soluble salts again get absorbed with water in the surface and reappear again in the form of efflorescence.
7. The surfaces should be painted only when efflorescence has fully ceased.
8. Proper damp proof should be laid correctly.

### 6.10.9. Plaster boards

These days ready made plaster boards are available in the market. They can directly be fixed on the surface plastered. The boards may be fixed to the walls and ceilings directly or may be fixed through a framework of timber batten.

The boards are manufactured by combining the plaster, convass and lath together, *the form of suitable precast slabs or mouldings*, as follows:

- On a pucca levelled platform wooden frames or forms, as per the size of the panels to be manufactured are laid and a thin layer of plaster is spread on them.
- On thin layer of plaster, convass and lath are laid and finally plaster in required amount is spread to cover the lath and convass, completely.
- When plaster boards have become sufficiently hard, they are taken out of the forms and stacked on edge to harden.

The boards may have it surface smooth or rough. *Smooth surface boards are used, where finishing coat is not to be applied and board's surface is to present the final finishing.* If final finishing coat of plastic plaster is to be applied, rough surface boards should be used.

## 6.11. Pointing

***Pointing** is the art of finishing the mortar joints of the walls or similar structure with either cement mortar or lime mortar in order to protect the joints from atmospheric agencies and also to improve the apperance of the structure.*

Pointing is restored to :

(i) Where a uniform and smooth surface is not required.

(ii) Where the exhibition of the natural beauty of the materials such as stone blocks, bricks etc., used in construction is to be made.

(iii) Where the materials used in construction can withstand effects of weather.

(iv) Where the workmanship is quite good.

### 6.11.1. Procedure of Pointing

- Prior to pointing, all the mortar joints on the face are raked out by a special pointing tool to a depth of about $1\frac{3}{4}$ cm to provide an adequate key for the fresh mortar used for pointing.
- All the loose mortar and dust are then brushed out of joints and the wall surface is well washed, wetted with clean water and kept wet for a few hours.
- The joints after being prepared, are filled with lime or cement mortar with a small trowel.

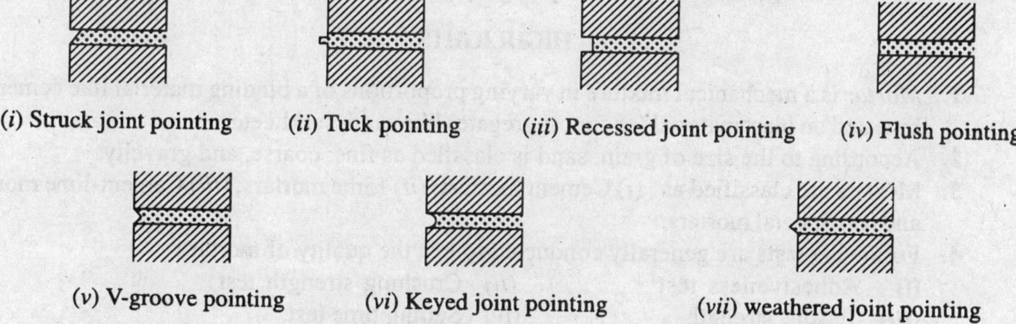

(*i*) Struck joint pointing (*ii*) Tuck pointing (*iii*) Recessed joint pointing (*iv*) Flush pointing (*v*) V-groove pointing (*vi*) Keyed joint pointing (*vii*) weathered joint pointing

**Fig. 6.6** : Forms of pointing.

The mortar is well pressed into the joints to ensure a solid contact with the internal old mortar joints. Excess mortar sticking to the sides is scraped away carefully.

- The finished pointing is kept wet for about 4 days for lime pointing and 10 days for cement pointing.
  - — *Lime pointing* : It is done with *lime mortar* prepared with equal parts of lime and fine white sand, carefully ground in a mortar mill.
  - — *Cement pointing* : It is done with cement mortar prepared by mixing *equal parts of cement and fine sand using it quickly before it starts setting*.
  - — *Surkhi pointing* : In this case the mortar may consist of 1 part of lime and 2 parts of surkhi.

### 6.11.2. Forms of Pointing

(*i*) *Struck joint pointing* : Refer Fig. 6.6 (*i*). The upper portion is inside the face of masonry by 12 mm while the bottom portion of the pointing is flush with masonry.

(*ii*) *Tuck pointing* : Refer Fig. 6.6 (*ii*) :

- The joint, after having been raked, is filled with cement mortar.
- A groove 6 mm wide and 3 mm deep is formed along the centre of the joint while the previously applied cement mortar is soft.

- The groove is filled with lime putty which projects by 3 mm. Lime putty is sometimes repalced by cement.
- The pointing gives a pleasing appearance.

(*iii*) *Recessed joint pointing* : Refer Fig. 6.6 (*iii*)

- Mortar is pressed behind the walls.
- It is used for high class masonry work.

(*iv*) *Flush pointing* : Refer Fig. 6.6 (*iv*)

- It is suitable for brick as well as stone masonry.
- The raked portion of the joint is filled with mortar and surface made flush to the masonry for trowelling.

(*v*) *V-groove jointing* : Refer Fig. 6.6 (*v*).

- It derives its name for its shape.
- It is suitable for rubble and ashlar masonry work.

(*vi*) *Keyed joint pointing* : Refer Fig. 6.6 (*vi*)

- After filling the joint with mortar a semi-circle is formed inside the pointing by some tool.
- The pointing gives an elegant appearance.

(*vii*) *Weathered joint pointing*: Refer Fig. 6.6 (*vii*)

As the name suggests, this type of joint gives adequate protection against weathering. However, it requires comparatively large quantity of water.

## HIGHLIGHTS

1. *Mortar* is a mechanical mixture in varying proportions of a binding material like cement or lime and an inert material (or fine aggregate) like sand, surkhi etc.
2. According to the size of grain, sand is classfied as fine, coarse, and gravelly.
3. Mortars are classified as : (*i*) Cement mortars; (*ii*) Lime mortars; (*iii*) Cement-lime mortars and (*iv*) Special mortars.
4. Following tests are generally conducted to test the quality of mortar :
   (*i*) Adhesiveness test (*ii*) Crushing strength test
   (*iii*) Tensile strength (*iv*) Setting time test.
5. *Plastering* is an art of covering rough walls and uneven surfaces in the construction of houses and other structures with a plastic composition, called *plaster*.
6. Pointing is the art of finishing mortar joints of the walls or similar structures with either cement mortar or lime mortar in order to protect the joints from atmospheric agencies and also to improve the appearance of the strucuture. There are verious types of pointings such as *flush joint pointing, recessed joint point pointing, keyed joint tuck pointing,* etc.

## OBJECTIVE TYPE QUESTIONS

**A. Choose the Correct Answers :**

1. Water mixed in a mortar should be always
   (a) less than basic water content (b) equal to the basic water content
   (c) more than the basic water content (d) none of the above.
2. The sand in mortar
   (a) promotes cracking (b) reduces the volume of mortar
   (c) increases the shrinkage (d) helps the pure lime to set by allowing penetration of air which provides the needed carbondioxide.
3. Which proportions of cement mortar is used for pointing work ?
   (a) 1 : 2 (b) 1 : 4
   (c) 1 : 5 (d) 1 : 6.

4. Which proportion of cement mortar is used for plastering work ?
   (a) 1:2 (b) 1:3
   (c) 1:5 (d) 1:7.
5. Sand is mixed with lime mortar to
   (a) reduce setting time (b) reduce cost
   (c) prevent shrinkage and cracking (d) improve strength.

### ANSWERS

1. (c) 2. (d) 3. (a) 4. (b) 5. (c)

**B. Fill in the Blanks/Say Yes or No :**

1. The binding material in a mortor is sometimes referred to as .............. .
2. Mortar should not set quickly.
3. Mortar may be used to form joints of pipes.
4. .............. sand is mainly used for plastering.
5. ............... sand is generally used for masonry work.
6. ............... sand is generally used for concrete work
7. A good sand should be chemically inert.
8. Surkhi gives strength and improves hydraulic property of the mortar.
9. Water to be used in mortar should preferably be drinkable.
10. .................. motars should be immediately used after mixing with water.
11. Cement mortars set and harden very slowly.
12. ................ mortars develop strength very slowly but gain continuous strength over long periods.
13. Lime mortars do not set but stiffen.
14. Partly set and dried mortar should not be retempered and used.
15. ................ is an art of covering rough walls and uneven surfaces in the construction of houses and other structures with a plastic omposition, called plaster.

### ANSWERS

| | | | | |
|---|---|---|---|---|
| 1. matrix | 2. No | 3. Yes | 4. Fine | 5. coarse |
| 6. Gravelly | 7. Yes | 8. Yes | 9. Yes | 10. Cement |
| 11. No | 12. Lime | 13. Yes | 14. Yes | 15. Plastering |

### THEORETICAL QUESTIONS"

1. What is mortar ?
2. What are the functions of mortar ?
3. State the properties of a good mortar.
4. List the uses of mortar.
5. Enumerate and describe briefly the ingredients of mortars.
6. Give the functions of sand in a mortar ?
7. How are sands classified ?
8. List the charateristics / properties of good sand.
9. How are mortars classified?
10. Explain briefly the following types of mortars :
    Cement mortars, lime mortars and lime-cement mortars.
11. How is a lime mortar prepared ?
12. Give the properties of lime mortars.
13. Explain briefly any three of following special mortars :

(i) Cement-clay mortar; (ii) Decorative mortars
(iii) Gypsum mortars; (iv) Packing mortar;
(v) Sound absorbing mortar.

14. Explain briefly the following tests for mortars :
    (i) Adhesiveness test; (ii) Crushing strength test; and
    (iii) Tensile strength test.
15. What is "plastering"?
16. What are the objects of plastering?
17. Explain briefly the following:
    (i) Lime plastering; (ii) Cement plastering;"
    (iii) Stucco plastering.
18. What is "pointing" ?
19. Explain briefly the procedure of "pointing".
20. Given the various forms of pointing

# 7

# Concrete

7.1. Introduction. 7.2. Characteristics of good concrete. 7.3. Advantage and disadvantages of concrete. 7.4. Uses of concrete. 7.5. Classification of concretes. 7.6. Plain cement concrete – ingredients–bulking of sand – water cement ratio – grading of aggregates – methods of proportioning concrete – mixes used in concrete works – workability – preparation of plain cement concrete. 7.7. Properties of concrete. 7.8. Water-proof concrete. 7.9. Precast concrete. 7.10. Reinforced concrete. 7.11. Prestressed concrete. 7.12. Special concretes – Ferrocement – fibre-reinforced concrete – polymer concrete – super plasticized concrete – ultra high strength concrete – sulphur-impregnated concrete – gap graded concrete. 7.13. Concreting under water. 7.14. Placing concrete in cold weather. 7.15. Placing concrete in hot weather. 7.16. Deterioration of concrete and its prevention. 7.17. Admixtures. 7.18. Joints in concrete. 7.19. Formwork. 7.20. General precautions in cement concrete construction. Highlights – Objective type Questions – Theoretical Questions.

## 7.1. INTRODUCTION

- **Concrete** is *mixture of cement (or lime), sand, brick or stone ballast and water, which when placed in forms and allowed to cure, becomes hard like stone. The hardening is caused by the chemical reaction between the cement and water*. The cement and water form a paste which, upon hardening, binds the aggregates to a permanent mass. Cement is called the "*binding material*". The stone or brick ballast is called the "*coarse aggregate*" as distinguished from the "*fine aggregate*" which is sand. The mortar used in concrete is called "*matrix*". Cement concrete when used by itself is known as "*mass concrete*".

- Concrete is *much stronger in compression than tension*. In order to enable it to resist tensile stresses it is reinforced or strengthened with steel in the form of steel bars or wire netting etc The concrete so obtained is called "*Reinforced concrete*".

## 7.2. CHARACTERISTICS OF GOOD CONCRETE

A good concrete should possess the following *characteristics* :

1. It should have *high compressive* strength (so that it can resist the heavy loads of (the structures). The compressive strength should not be less than 15.5 N/mm$^2$.
2. On hardening, it should exhibit *minimum shrinkage*.
3. It must be adequately dense. The density of a good concrete should be about 24 kN/m$^3$.

4. It should be *economical* for the desired strength.
5. It should *sufficiently hard and provide enough resistance to abrasion*. This property is of paramount importance when concrete is to be used for making *steps of stairs* and *road pavements*.
6. It must be *adequately durable* to resist the effects of weathering agencies (e.g., rain, frost action, variations in temperatures etc.).
7. It should have *minimum thermal expansion* so as to provide good resistance to fire.
8. It should provide the *required finish* to the concrete structure.
9. It should have *sufficient impermeability* or *water tightness* (so that the moisture does not find its access inside the structure, corrode the steel and eventually cause the structure to fail).
10. It should have *minimum creep* (*creep, time yield* or *plastic flow* is the continuous strain with time which the concrete undergoes due to application of external load).

## 7.3. ADVANTAGES AND DISADVANTAGES OF CONCRETE

**Advantages :**

Concrete as a construction material entails the following *advantages* :

1. It has a *high compressive strength,* and the *corrosive and weathering effects are minimal.* Its strength equals that of a hard material stone when properly prepared.
2. As compared to other materials, concrete is *economical in the long run.*
3. It is *durable and fire resistant and requires very little maintenance.*
4. The green concrete can be *easily handled and moulded* into any shape or size according to specifications.
5. The concrete can be *pumped* and hence it can be *laid in the difficult positions also.*
6. Concrete can even be *sprayed on and filled into* the *cracks for repairs by the grinding process.*
7. Being strong in compression, it has unlimited structural applications in combination with steel reinforcement.

**Disadvantages :**

Following are the *disadvantages* of concrete :

1. Concrete is to be *reinforced with steel bars or meshes* since it has *low tensile strength* and hence *cracks easily.*
2. *Provision for contraction joints has to be made* to avoid the development of cracks due to drying shrinkage and moisture movement in fresh concrete.
3. In order *to avoid the formation of cracks due to thermal movement* (concrete expands and contracts with the changes in temperature), *expansion joints have to be provided.*
4. Concrete as a material lacks in ductility and this factor proves disadvantageous with respect to earthquake resistance design.
5. Concrete is *liable to disintegrate by alkali* and *sulphate attack.*
6. Concrete undergoes *creep* under sustained loading resulting *in reduction of prestress in the prestressed concrete construction.*

## 7.4. USES OF CONCRETE

The concrete is *used in the construction of the following* :

| | |
|---|---|
| (*i*) Foundations | (*vi*) Docks and harbours |
| (*ii*) Buildings | (*vii*) Dams |
| (*iii*) Roads | (*viii*) Bridges |
| (*iv*) Arifields | (*ix*) Bunkers |
| (*v*) Water retaining structures | (*x*) Silos, etc. |

## 7.5. CLASSIFICATION OF CONCRETES

The concretes are classified as follows:

I. According to binding material
II. According to design
III. According to purpose.

### 7.5.1. According to Binding Material

The concretes, according to binding material, are classified as follows:

1. Cement concrete
2. Lime concrete

**1. Cement concrete**

- Cement concrete consists of *cement, sand* and a *coarse aggregate, mixed in suitable proportions in addition to water*. Here cement is used as a *binding material*, sand as *fine aggregate* and shingle, gravel, broken brick or crushed stone are used as *coarse aggregates*.
- In cement concrete, usual proportions of its ingredients are : 1 part *Portland cement* : 1½ to 8 parts *clean sand* : 3 to 16 parts *coarse aggregate*.
- The bulk density of ordinary concrete is 23 kN/m$^3$.

*Uses :* Cement concrete/ordinary concrete is used in :

(*i*) Walls of houses in continuous mass or in blocks
(*ii*) Retaining walls and reservoirs
(*iii*) Roofing and floorings
(*iv*) Foundation of engineering structure
(*v*) Stair and steps
(*vi*) Ornamental moulding and pipes.
(*vii*) Lintels, arches etc.

**2. Lime concrete**

- *Lime concrete* consists of lime, a fine aggregate and a coarse aggregate, mixed in suitable proportions in addition to water. Here, *hydraulic lime* is generally used as a binding material; sand, surkhi and cinder are used as *fine aggregates* and broken bricks, broken stone etc. as coarse *aggregates.*
- It is cheaper and has less strength than cement concrete.
- It provides good *bases capable of taking sufficient loads and a certain degree of flexibility specially when in moist condition* and hence *adjusts very well with the surface in contact.*
- It also exhibits a certain degree of *water-proofing property* and thus prevents sub-soil water from causing dampness in floors or walls etc. It also eliminates the movement of dampness from roof to ceiling.
- It also *exhibits volumetric stability* when matured lime is used and lime concrete is cured suitably over a long period.
- It can be *prepared easily* and made available at much *cheaper rates.*
- It also *resists weathering effects* very well and is *quite durable.*
- *For achieving good quality lime concrete for specific purposes certain mixtures such a gur water, finugreek powder, fibres etc. should be sprinkled during final ramming and finishing.*

**Preparation of lime concrete**

For preparing a good lime concrete, *hydraulic lime* is generally used. The *fat lime is not suitable for use in large masses.* The sand to be used should be free from impurities. Well burnt broken bricks or broken stones are used as coarse aggregate. The water to be used should be clean and free from impurities.

Usual proportions of lime concrete are :

1 part lime : 2 to 3 parts sand : 3 to 4 parts coarse aggregate.

— 1 : 2 : 3 concrete mix. is used *under floors*.

— 1 : 2 : 4 concrete mix. is used for *foundation work*.

The following *procedure* is adopted to *prepare lime concrete:*

- The measured quantity of coarse aggregate is stacked and adequately soaked with water. Then the measured quantity of sand and lime, in turn are spread evenly over the stack of coarse-aggregate. The whole mass is then thoroughly mixed dry by overturing with kassies or shovels.
- After dry mixing, sufficient quantity of water is added and the whole mass is again mixed thoroughly to have the desired consistency of the concrete.
- The concrete is then laid in uniform layers not exceeding 15 cm when compacted. The compaction is done with light rammers. When the mortar comes over the surface, the concrete should be allowed to set for about 24 hours. Then the surface of the concrete is watered and rammed with heavy rammers till the concrete is thoroughly compacted.

**Uses :** (*i*) It is widely used for foundation bases of load bearing walls, columns and underlayers of floors.

(*ii*) It is also used for *filling haunches over masonry arches*.

(*iii*) It also forms good terrace finishing material due to its characteristics of volumetric stability, water proofing and smooth surface finishing.

**Precautions :** (*i*) As lime causes rashes on human skin, hence persons working with lime concrete must be provided with protecting rubber gloves and rubber gum boots.

(*ii*) The persons engaged in lime concrete work must also apply oil on their skin to avoid rashes and cracking of their skin due to reaction of lime.

### 7.5.2. According to Design

The concretes, *according to design*, are classified as follows :

1. Plain cement concrete
2. Reinforced cement concrete (R.C.C.)
3. Pre-stressed cement concrete (P.C.C.)

**1. Plain cement concrete**

- The cement concrete in which no reinforcement is provided is called a *plain cement concrete* or *mass cement concrete.*
- It is considerably stronger in compression but weak in tension or shear.

*Uses:* It is commonly used in the construction of columns foundation, bed blocks, massive gravity dams, heavy foundations etc.

**2. Reinforced cement concrete (R.C.C.)**

- It is the cement concrete in which *reinforcement is embedded for taking tensile, excessive compressive or shear stresses.*
- The usual proportions of ingredients are :
  1 part Portland cement : 1 to 2 parts clean sand : 2 to 4 parts shingle or crushed stone. The steel reinforcement used is generally in the form of round bars. 6 mm to 32 mm dia. Expanded metal or steel fabric in the form of a mesh of steel rods, welded together, are also used as reinforcement in concrete walls etc.

**Uses :** It is commonly used for the construction of *beams, slabs, lintels, raft or mat foundations etc.*

**3. Pre-stressed cement concrete (P.C.C.)**

- It is the cement concrete in which *high compressive stresses are artificially induced before its actual use*. The high compressive stresses are induced by *pre-tensioning the reinforcement before placing the concrete and the reinforcement is released when final setting of the concrete takes place.*

- Here, high tensile steel is used as reinforcement.
- This type of concrete can take up high tensile and compressive stresses without development of cracks.
- By the use of this concrete, the quantity of concrete reinforcement can be considerably reduced.

*Uses :* It is used where high stresses develop and it is uneconomical to use ordinary reinforced cement concrete.

### 7.5.3. According to Purpose

The concrete, according to purpose, is classified into various types as discussed below :

1. Light-weight concrete
2. Cellular or aerated concrete
3. Sawdust concrete
4. Vacuum concrete
5. White and coloured concrete
6. High early strength concrete
7. No-fines concrete.

**1. Light-weight concrete:**

- It is prepared by using the following materials :

(*i*) *Binding material* ... Ordinary portland cement and its varieties (lime-slag, lime-cinder etc. may also be used, if locally available)

(*ii*) *Aggregates* ...... The loose porous materials such as crushed light-weight rocks (*natural porous* aggregates); the *artificial porous* aggregates can be obtained from *industrial wastes*.

(*iii*) Steel ......... The light weight concrete should be made adequately dense when used for R.C.C. work. Sometimes the reinforcement is coated with anticorrosive compounds (when light weight concrete is used).

(*iv*) *Water* ........... Pure drinking water should be used for preparing the conerete. The water-cement ratio should be carefully worked out for achieving optimum strength of the concrete.

- It is light in weight.
- Its bulk density varies from 5 to 18 kN/m$^3$.
- It possesses high insulating property.

**Advantages :** (*i*) Its fire resistance is greater as compared to the ordinary concrete (since it has comparatively less tendency to spall).

(*ii*) As compared to ordinary concrete, it has a lower thermal expansion.

(*iii*) The local industrial waste, if found suitable, can be economically utilized for the preparation of light weight concrete.

(*iv*) By the use of light weight concrete, the cost is reduced to the extent of about 30 to 40 percent or so.

(*v*) It does not present problems relating to freezing and thawing.

(*vi*) Due to reduced weight of light weight concrete, it is easy to remove, transport and erect precast products.

(*vii*) Its sound absorption coefficient is nearly *twice* than that of the ordinary concrete.

**Disadvantage :** Special care needs to be taken to provide sufficient cover to the reinforcement of the light weight structures to provide protection against corrosion.

**Uses :** It is employed for making *precast structural units for partitions and wall-lining.*

**2. Cellular or aerated concrete:**

- It is made by *adding aluminium powder to a rather wet mix of cement, sand, lime and water*. The mass begins to rise due to the hydrogen bubbles evolved by the chemical action between cement and mixture and thereby *cells are formed* in concrete. It is also called *air-entrained concrete*.

- It is light in weight and spongy in structure.

**Uses :** It is useful for roof slab and precast units in partitions etc. for heat and sound insulation.

**3. Saw dust concrete:**

- It is prepared by mixing *Portland cement with sawdust* in specified proportion in addition to water.
- It expands when becomes wet and contracts when dry.

**Uses :** It is used as a *heat and sound insulating material.*

**4. Vacuum concrete:**

It is the concrete in position from which *air and excess water are removed with a vacuum pump*. Water-cement ratio is reduced and thereby it is 20 per cent *stronger* than ordinary concrete.

**Uses :** 1. It is employed for all reinforced concrete works.

2. It is also used in the construction of horizontal and sloping concrete slabs, such as floor slabs, road and airport pavements, their load-beaming and partition walls.

**5. White and coloured concrete:**

- The concrete in which white cement is used as a binding material is called *white concrete.*
- In *coloured concrete*, coloured cement is used as a binding material alongwith coloured aggregates.

**Uses :** (*i*) Ornamental finishes in buildings;
(*ii*) Underground pedestrian crossings etc.;
(*iii*) Preparing park lanes;
(*iv*) Manufacture of items for public welfare.

**6. High early strength concrete:**

- In this type of concrete high *early strength cement is used as a binding material.*
- As compared to ordinary concrete, it *sets and hardens quickly.*

**Uses :** Employed for construction work, especially in cold weather (since it sets earlier and thus the formwork can be re-used quickly).

**7. No-fines concrete:**

It consists of *cement, coarse aggregate and water; fine aggregate or sand is eliminated.*

**Advantages :**

(*i*) Drying shrinkage is relatively low (as compared to ordinary concretes).
(*ii*) Considerable saving of cement per $m^3$ of concrete (due to the absence of sand).
(*iii*) Due to the presence of large voids, the insulating characteristics of this type of concrete are better than ordinary concrete.
(*iv*) As it does not segregate, it can be dropped from a considerable height and placed in very high lifts.
(*v*) No transmission of water by capillary action (since capillary passages are absent)
(*vi*) Its formwork need *not-be watertight*; it is possible to use *cheap formwork* (its unit weight being two-thirds that of ordinary concrete).
(*vii*) Being a type of lightweight concrete, it grants the advantages associated with the light-weight concrete construction.

**Disadvantages/Limitations :**

(*i*) As compared to ordinary concrete, its compressive, bond and flexural strengths are *considerably lower*.
(*ii*) As there is little or no cohesion in its fresh state, this type of concrete, requires a long time for the removal of forms.

## 7.6. PLAIN CEMENT CONCRETE

### 7.6.1. Ingredients

The various ingredients of plain cement concrete are enumerated and discussed below:

1. Cement
2. Sand (fine aggregate)
3. Coarse aggregate (broken stone, gravel, broken bricks, etc.
4. Water.

**1. Cement:**

- Cement is used as a *binding material* in the concrete required for different engineering works where strength and durability are of significant importance. For this purpose ordinary Portland cement is commonly used. However, when work needs to be carried out quickly, rapid hardening cement is also used.
- The cement to be used should be as fresh as possible and should satisfy the latest specifications of Indian standard Institute.

**Functions :** (*i*) It makes the concrete impermeable by filling up the voids existing in the fine aggregate.

(*ii*) It binds the aggregates into a solid mass (due to its setting and hardening properties, when treated with water).

(*iii*) It imparts strength to the concrete.

**2. Sand:**

- It consists of small angular or rounded grains of silica ($SiO_2$) and is formed by decomposition of sand stone under the effect of weathering agencies.
- It may be either natural sand (*e.g.*, river sand, nalah sand, pit sand and sea sand) or *artificial* sand (prepared by crushing stones and gravels to powder form).
- Sand is commonly used as a fine aggregate in the plain cement concrete.
- Sea sand is not recommended for its use for concrete work which is to be *exposed to view*. It should be washed with fresh water before use, to remove its salts contents.

**Characteristics/requirements**

Fine aggregate *should* be :

(*i*) Clean, sharp and angular;

(*ii*) Chemically an inert material;

(*iii*) Highly silicious and free from impurities such as clay, loam, dust , coal particles and organic matter;

(*iv*) Hard strong and durable.

**Functions**

(*i*) It fills the voids present in the coarse aggregate.

(*ii*) It assists in hardening of cement by allowing water to penetrate through its voids.

(*iii*) It minimizes shrinking and cracking of concrete.

(*iv*) It prepares concrete economically, of any required strength, by varying its proportions.

**3. Coarse aggregate:**

- It acts as main filler, and forms the main bulk of concrete, around the surfaces of which the binding materials adhere in the form of film.
- *Broken stone, broken brick* and gravel are generally used as coarse aggregates. *Granite, trap and basalt give excellent coarse aggregates.*
- The crushing strength and watertightness of concrete and its resistance to wear and tear depend upon the aggregates. Further the aggregates *balance the shrinkage* and volume changes of concrete.

**Characteristics/requirements**

A good coarse aggregate should be :

(*i*) hard and tough;
(*ii*) angular or cubical in shape;
(*iii*) absolutely clean;
(*iv*) sound, fire resisting and durable;
(*v*) free from chemicals or coating of clay;
(*vi*) free from organic matter;
(*vii*) pass through 75 mm IS sieve and entirely retain on 4.75 mm IS sieve.

**Functions**

(*i*) It imparts greater volumetric stability and durability to concrete (as compared to the cement paste alone).
(*ii*) It makes a solid and hard mass of concrete (with cement and sand).
(*iii*) It increases the crushing strength of concrete.
(*iv*) It is cheaper than cement and hence directly helps in achieving economy in concrete manufacture.

The properties of concrete are greatly influenced by aggregate characteristics given below :

(*i*) Size and grading of aggregate;
(*ii*) Shape and surface texure;
(*iii*) Specific gravity and bulk density;
(*iv*) Strength;
(*v*) Water absorption and surface moisture;
(*vi*) Bulking of sand;
(*vii*) Deleterious substances;
(*viii*) Soundness, durability including alkali-aggregate reactivity.

**4. Water:**

- It plays an important role in *mixing, laying, compaction, setting* and *hardening of concrete*.
- Water *influences the strength development and durability of concrete.*
- Ordinary drinking water can be used for preparing concrete. It should be free from certain chemicals injurious to cement.
- The quantity of water used should be *just sufficient for chemical reaction and suitable workability of the concrete.*

**Functions**

(*i*) It wets the surface of the aggregates.
(*ii*) It facilitates spreading of cement over the fine aggregate.
(*iii*) It *lubricates the surface of the aggregate* which influences the workability and compaction of concrete during plastic stage (leading to the influence on the strength and durability of concrete).
(*iv*) It chemically combines with various compounds in cement to form a binding medium for pieces of aggregates.

### 7.6.2. Bulking of Sand

The *increase in volume of sand due to presence of surface moisture upto some extent* is called **bulking of sand.**

- The presence of moisture in aggregate necessitates the correction of the actual mix properties; the *weight of water added to the mix has to be decreased by the weight of the free moisture in the aggregate, and the weight of aggregate must be increased by a like amount.* In the case of sand there is a second effect of the presence of moisture : *bulking*. This is the *increase in the volume of a given weight of sand caused by the films of water pushing the sand particles apart.* While bulking does not effect the proportioning of materials by weight, in case of volume batching bulking results in a smaller weight of sand occupying the fixed volume of the measuring box. For this reason the mix becomes deficient in sand and appears "*stony*" and the concrete may be prone to *segregation* and *honeycombing*. Also the yield of the concrete is *reduced*. The remedy, of course, lies in increasing the apparent volume of sand to allow for bulking.

- The *extent of bulking depends on the percentage of moisture present in the sand and on its fineness.* The increase in volume relative to that occupied by a saturated and surface-dry sand increases with an increase in the moisture content of the sand up to a value of some 5 to 8 per cent, when bulking of 20 to 30 percent occurs. Upon further addition of water the films merge and the water moves into voids between the particles so that the total volume is approximately, the same as the volume of dry sand for the same method of filling the container. This is apparent from the (Fig. 7.1), which also shows that *finer sand bulks considerably more and reaches maximum bulking at a higher water content than does coarse sand.* Extremely fine sand has been known to bulk as much as 40 per cent at a moisture content of 10 per cent; but such a sand is in any case *unsuitable* for the manufacture of good quality concrete.

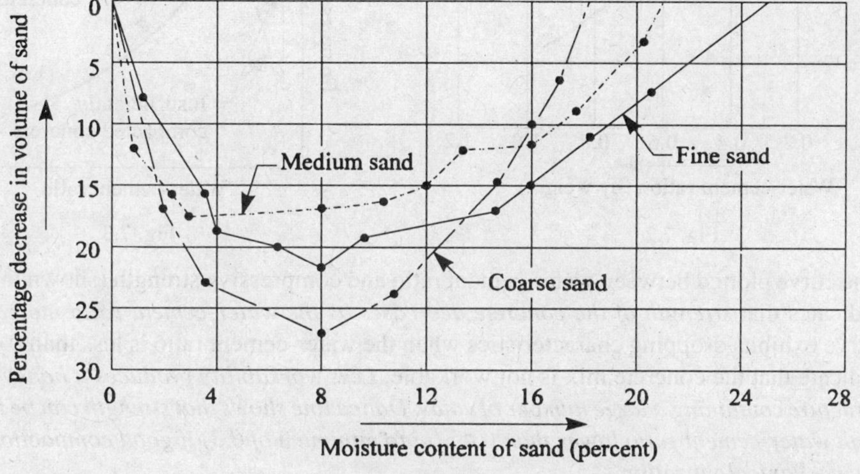


Fig. 7.1. Bulking of sand.

- *Coarse aggregate shows only a negligible increase in volume due to the presence of free water, as the thickness of moisture films is very small compared with the particle size.*

**Method of determining bulking of sand:**

The following simple test is performed to determine the percentage of bulking of sand :

- Take a measuring cylinder of about 250 c.c. capacity.
- Take damp sand (to be tested) and pour it into the cylinder and shake it well.
- Note the level of the sample of damp sand, say it is $h_1$.
- Add sufficient quantity of water into the cylinder to *saturate the sand completely* and stir the sample thoroughly (complete saturation can be judged by a film of water standing above the surface of sand). It will be seen that the level of saturated sand is *below the original level of damp sand.*
- Note the reading again, say it is $h_2$.
- Then calculate the bulking of sand by using the following relation :

$$\textit{Percent bulking of sand} = \frac{h_1 - h_2}{h_2} \times 100$$

**7.6.3. Water-cement Ratio**

- In engineering practice, the *strength of concrete at a given age and cured at a prescribed temperature* is assumed to depend primarily on two factors only, the **water-cement ratio** and the **degree of compaction.**

- **Water-cement ratio** *is the ratio of water to cement and is expressed as the ratio of the weight or volume of water to the weight or volume of cement in concrete mixture*. Generally it is expressed as so many litres of water per bag of cement (50 kg).

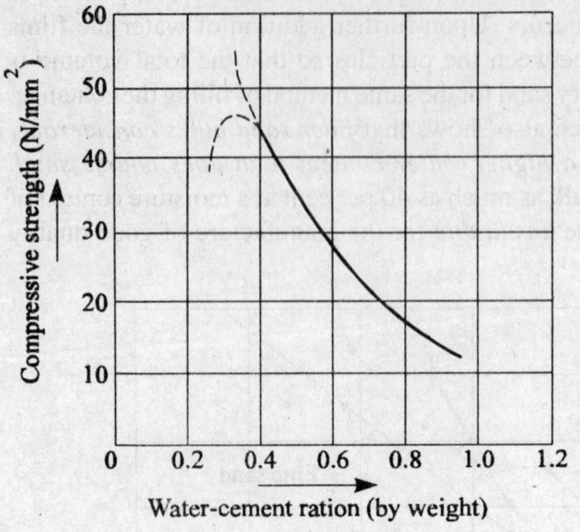


Fig. 7.2

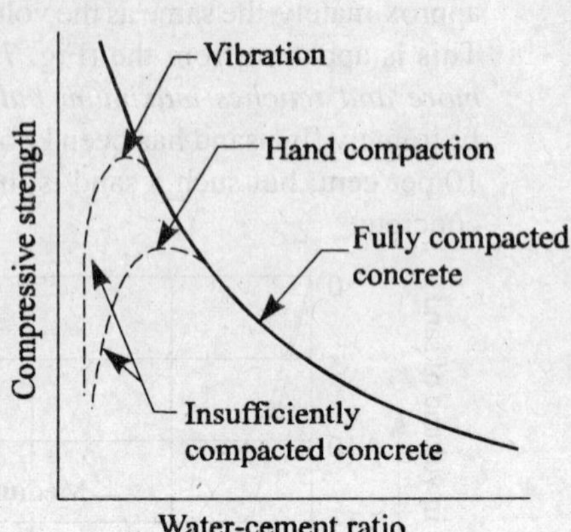


Fig. 7.3

- The curve plotted between water-cement ratio and compressive strength (shown in Fig. 7.2) indicates that *strength of the concrete decreases as the water-cement ratio increases*. The curve exhibits dropping characteristics when the water-cement ratio is less than 0.45 which indicate that the concrete mix is not workable. *Less workability produces a honey- combed structure containing a large number of voids. Dotted line shows that strength can be increased with water-cement ratio lower than 0.45 (upto minimum of 0.3) if good compaction is done by mechanical vibration.*
- Fig. 7.3 shows compressive strength vs water-cement ratio curves for fully compacted and insufficiently compacted concrete.

### 7.6.4. Grading of Aggregates

- The *grading of aggregates* is the *art of doing gradation* (particle size distribution) of *an aggregate as determined by sieve analysis.*
- Fine and coarse aggregates to be used in a concrete mix, should be well graded because grading of an aggregate has considerable effect on workability, uniformity and finishing qualities of plastic concrete (which ultimately affect the *economy, strength, durability* and other properties of hardened concrete).
- The grading of a fine aggregate has a *more effect* or workability, uniformity and finishing qualities of plastic concrete as compared to that of a coarse aggregate.

For grading of different types of aggregates, the following Indian standard test sieves are generally used :

*With fine mesh wire cloth :* 3.35 mm, 2.36 mm, 1.18 mm, 600 micron, 300 micron, 150 micron and 75 micron.

*With square hole perforated hole :* 80 mm, 63 mm, 50 mm, 40 mm, 31.6 mm, 25 mm, 20 mm., 16 mm, 12.5 mm, 10 mm, 6.3 mm, 4.75 mm.

### 7.6.5. Methods of Proportioning Concrete

- *The process of determining the proportion of cement to fine aggregate and coarse aggregate for concrete mix* is called **proportioning of concrete.**
- The *principle of proportioning of aggregates is that the smaller particles will fill up the*

*voids between large particles*. A well proportioned concrete mix *requires minimum amount of cement per unit volume.*

The various methods of proportioning concrete are :

1. Arbitrary method
2. Trial mixtures method
3. Minimum voids method
4. Fineness modulus method.

**1. Arbitrary method:**

- In this method one part of cement to *M* parts of fine aggregate and 2*M* parts of coarse aggregate are taken as the basis.
  The quantity of water required for mixing is determined by according to desired workability.
- By this method such mixes as $1 : 1\frac{1}{2} : 3$ (one part cement, $1\frac{1}{2}$ parts fine aggregate, 3 parts coarse aggregate), 1 : 2 : 4, 1 : 3 : 6, 1 : 4 : 8 are arbitrarily chosen by volume.

These days the ratio of fine to coarse aggregate is not taken as 1 : 2 as experiments have shown that the quantity of fine aggregate needed is *more* than 50% of the coarse aggregate; thus the ratio of coarse to fine aggregate; may be varied from $1\frac{1}{2}$ to $3\frac{1}{2}$

**Limitations :** The chief drawbacks of this method are :

(*i*) The lack of proper control over the quantity of water that goes into a batch of concrete tends to result in a concrete of *varying strength and denseness*.

(*ii*) The fixed specified proportions of cement, fine aggregate and coarse aggregate *do not permit taking advantage of favourable types and grading of aggregates to secure a workable mix at the least possible cost of materials, and without harm to the strength of the concrete.*

However, this method is *simple and if carefully considered gives fairly satisfactory results.*

**2. Trial mixtures method:**

The procedure is as follows:

- A certain weight of cement, fine aggregate, coarse aggregate and water are put into a graduated cylindrical vessel and tamped.
- The vessel is emptied after noting the height of the upper surface.
- The vessel is then cleaned carefully and the method repeated for several trial mixtures in varying proportions.

The *mixture that gives the smallest volume for the same weight i.e.*, one which is the densest is adopted.

This method is quite *suitable for impermeable construction.*

**3. Minimum voids methods**

- This method is based on the *principle that the concrete which has the minimum voids is the densest and strongest*. Amount of fine aggregate should be such as to fill the voids in coarse aggregate and the amount of cement should be such as to fill the voids in fine aggregate. Due to *wedging action* of fine aggregate with sides of coarse aggregate and that of cement with the sides of fine aggregate, 10% extra sand and 15% extra cement are required.
- Practical experience has shown that this method *gives no better results than Arbitrary method.*

**Example 7.1.** *Design a concrete mix if coarse aggregate and sand used have 40% and 30% voids respectively.*

**Solution :** Sand required for 1 m$^3$ of coarse aggregate

$$= 1 \times \frac{40}{100} + \frac{1 \times 40}{100} \times \frac{10}{100} = 0.4 + 0.04 = 0.44 \text{ m}^3$$

Cement required for 1 m$^3$ of coarse aggregate

$$= 0.44 \times \frac{30}{100} + 0.44 \times \frac{30}{100} \times \frac{15}{100} = 0.132 + 0.0198 \simeq 0.152 \text{ m}^3$$

∴ Cement : sand : aggregate : 0.152 : 0.44 : 1 = 1 : 2.9 : 6.6 **Ans.**

**Example 7.2 :** *In cement concrete the voids in coarse and fine aggregates amounted to 45% and 30% respectively. Find the proportions of ingredients in cement concrete when 12% cement and 10% fine aggregate in excess are allowed.*

**Solution :** Coarse aggregate = 1.00 m$^3$

$$\text{Fine aggregate } = 1.00 \times \frac{45}{100} + 1.00 \times \frac{45}{100} \times \frac{10}{100} = 0.495 \text{ m}^3$$

$$\text{Cement } = 0.495 \times \frac{30}{100} + 0.495 \times \frac{30}{100} \times \frac{12}{100} = 0.166 \text{ m}^3$$

∴ Cement : fine aggregate : coarse aggregate = 0.166 : 0.495 : 1 = 1 : 3 : 6 **Ans.**

**4. Fineness modulus method**

- This method is based on a factor known as "*fineness modulus*" of aggregates. The term fineness modulus is used to *indicate an index number which is roughly proportional to the average size of the particle in the entire quantity of aggregates.*
- In order to determine fineness modulus, the standard test are conducted using the following set of sieves (metric):

*For fine aggregates* : Nos. 480, 240, 120, 60, 30 and 15

*For coarse aggregates* : 80 mm, 40 mm, 20 mm, 10 mm and no. 480

*For all in aggregates* : 80 mm, 40 mm, 20 mm, 10 mm, nos. 480, 240, 120, 60, 30 and 15.

*The percentage of sample by weight retained on each of standard sieves is determined. The total of these percentages divided by 100 gives the fineness modulus of the sample.*

The *percentage of fine to coarse aggregate* ($x$) can be found out from the following relation:

$$x = \frac{F_c - F}{F - F_f}$$

where $F_c$ = fineness modulus of coarse aggregate,
$F$ = desired fineness modulus for a concrete mix of fine and coarse aggregates, and
$F_f$ = fineness modulus of fine aggregate.

**Example 7.3.** *The fineness modulus of coarse and fine aggregates are 6.0 and 3.0 respectively. If the mixture has a fineness modulus of 5.0 find the percentage of fine aggregate to be mixed with coarse aggregate.*

**Solution :** Given : $F_c = 6.0; F_f = 3.0; F = 5.0$

The percentage of fine aggregate to be mixed with coarse aggregate is given by the relation:

$$x = \frac{F_c - F}{F - F_f} \times 100$$

or

$$x = \frac{6 - 5}{5 - 3} \times 100 = 50\% \textbf{ Ans.}$$

Hence, in *order to prepare the suitable unix, one part of fine aggregate is to be mixed with two parts of coarse aggregate.*

### 7.6.6. Mixes Used in Concrete Works

The mixes used in various concrete works are given below :

| | Mixes | Uses |
|---|---|---|
| 1. | 1 : 1 : 2 | Long span arches, high-load columns, piles etc. |
| 2. | $1 : 1\frac{1}{2} : 3$ | Arches, beams, copings, sills, slabs, stains, landing etc. columns, culverts, engine beds, pipes, toilets, pillars, thin partitions, tanks, reservoir etc. |
| 3. | 1 : 2 : 4 | Beams, blocks, engine beds, conduits, columns, flat roofs, lintels, pavements, hearths, pillars, piers, surface channels, upper floors, landings etc. |
| 4. | $1 : 2\frac{1}{2} : 5$ | Abutments, bottom courses of reinforced concrete, roads, external steps, machinery foundations, sills etc. |
| 5. | 1 : 3 : 6 | Block masonry walls, concrete beds, under wooden floors, ground floors, foot paths, lintels, hearths, pavings, foundations etc. |
| 6. | 1 : 4 : 8 | Ordinary footings, mass concrete works etc. |

### 7.6.7. Workability

The strength of concrete of given mix proportions is very seriously affected by *degree of its compaction*; it is vital, therefore, that the consistence of the mix be such that the concrete can be transported, placed and finished sufficiently easily and without segregation. A concrete satisfying these conditions is said to workable, but to say merely that workability determines the case of placement and the resistance to segregation is too loose a description of this vital property of concrete. Furthermore, the desired workabilitv in any particular case would depend on the means of compaction available; likewise, a workability subitable for mass concrete is not necessarily sufficient for thin inaccessible or heavily reinforced sections. For these reasons *workability should be defined as a physical property of concrete alone without reference to circumstances of particular type of construction.*

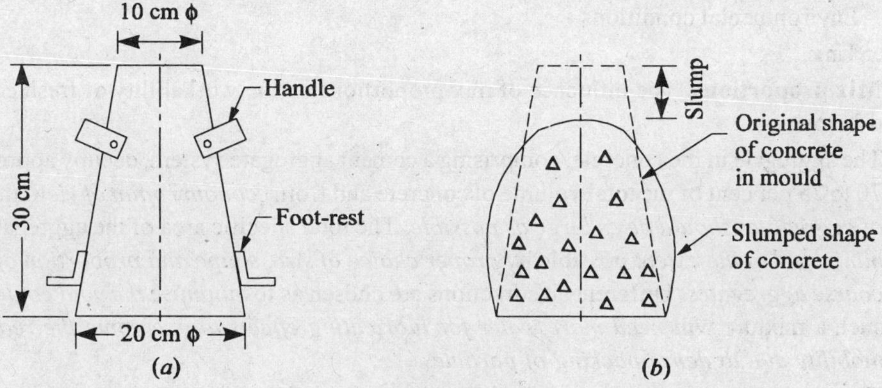


Fig. 7.4 : Slump test.

**Measurement of workability :** Unfortunately no test is known that will measure directly the workability. Numerous attempts have been made, however, to correlate workability with some easily determinable physical movement, but none of these is fully satisfactory although they may provide useful information within a range of variation in workability.

**Slump test.** Refer Fig. 7.4. This is a test used in site work all over the world. The *slump test does not measure the workability of concrete but is very useful in detecting variations in the uniformity of a mix of given nominal proportions.*

- The mould for this is a frustrum of a cone 30 cm high and top and bottom diameters 10 cm and 20 cms respectively. It is placed on a smooth surface with the smaller opening at the top, and filled with concrete in four layers (As per ISI specifications nominal maximum size of aggregate should not exceed 38 mm in slump test).
- Each layer is tamped 25 times with a standard bull nosed diameter steel rod, and the top surface is struck with a trowel. The mould must be firmly held against its base during the entire operation; that is facilitated by handles or foot-rests brazed to mould.
- Immediately after filling, the cone is slowly lifted, and the unsupported concrete will now slump - hence the name of the test. The *decrease in the height of the highest part of the slumped concrete is called* **slump.**
- In order to reduce the influence on the slump of the variation in the surface friction, the *inside of the mould and its base should be moistened* at the beginning of every set, and prior to lifting of the mould the area immediately around the base of the cone should be cleaned from the concrete which may have dropped accidently.

The following slumps are recommended for *different works*:

(i) *Vibrated concrete* ... 10 to 25 mm
(*ii*) *Mass concrete* ... 25 to 50 mm
(*iii*) *Road work* ... 25 to 50 mm
(*iv*) *Ordinary beams and slabs* ... 50 to 100 mm
(*v*) *Columns, retaining walls and thin vertical sections* ... 75 to 125 mm

**Factors Affecting workability:**

The following factors affect workability of fresh concrete:

1. Mix proportions
2. Aggregate properties
3. Environmental conditions
4. Time.

**1. Mix proportions.** The influence of mix proportions on the workability of fresh concrete is discussed below:

- The aggregate in the concrete, comprising a cement aggregate system, occupy approximately 70 to 75 per cent of the total volume of concrete and from *economy point of view,* the volume *of aggregates should be as large as possible.* The total specific area of the aggregate is to be *minimized to the extent* possible by *proper choice of size, shape and proportion of fine and coarse aggregates.* Different size fractions are chosen as to *minimise the void content* and as such a mixture will *need more water for lubricating effects to overcome the reduction in mobility due to dense packing of particles.*
- *The water-cement ratio* determines the intrinsic properties of cement paste and *requirements* of workability is such that there is sufficient paste to surround the aggregate particles as well as fill the voids in the aggregate.
  – An increase of water content results in monotonous increase in workability but eventually a stage is reached where segregation and bleeding occur, and *use of higher water content will result in the more serious problems of shrinkage and creep of hardened concrete.*

– The water content is *limited* to some volume given by *water-cement ratio* which is dependent on the *target design strength* of hardened concrete, making it essential to study the effect on the workability of other factors.

**2. Aggregate properties.** The aggregate properties affect the workability of fresh concrete as follows:

- The use of coarse aggregate of large size and /or rounded aggregate, for the same volume of aggregate in concrete, *gives higher workability because of reduction in total specific surface area and particle interference.*

– The use of *elongated aggregates* results in low *workability, primarily due to increase in particle interference.*

- The *use of finer sand, for the same water content, decreases workability.*

– In practice there is an optimum finers content for maximum workability, such that *either an increase or decrease of fines reduces workability.*

- Normally, the mixes with *higher water cement ratio* would require a *somewhat finer grading* and for mixes with low *water-cement-ratio* a *coarse grading* is preferable.
- A *rapid-hardening cement* will *reduce workability as compared to ordinary portland cement* because of its higher specific surface and the fact that it *hydrates more rapidly,* and also the fineness of cement has an influence on *bleeding.*
- The workability may be considerably affected by the presence and nature of admixtures or cement replacement.

**3. Environmental conditions.** The ambient temperature affects the workability of concrete. On a hot day it becomes necessary to increase the water content of the concrete mix in order to maintain the desired workability.

**4. Time:**

- The fresh concrete loses workability with time mainly because of loss of mixture due to evaporation.
- The loss of workability varies with the type of cement, the concrete mix proportions, the initial workability and the temperature of the concrete (on an average a 125 mm slump concrete may lose about a 50 mm slump in the first one hour).

For some particular total time after mixing, the loss in workability is small and initial level could be regained without loss in the strength of hardened concrete *simply by adding extra water.*

**Requirements of workability:**

The workability of the fresh concrete should be such that *it can be placed in the framework and compacted with minimum effort, without causing segregation and bleeding.*

- The choice of workability depends upon the type of *compacting equipment available, the size of the section and concentration of reinforcement.*

– Compaction by hand using rodding and tamping is not possible when compacting factor is less than 0.85.

– Ordinary techniques of vibrations are not applicable if the compacting factor falls below 0.70. In such cases techniques like "vibro-pressing" have to be adopted.

- *For heavily reinforced sections* or when the *sections* are *narrow or contain inaccessible parts* or when the spacing of reinforcement makes the placing and compaction difficult, the *workability should be high to achieve full compaction with a reasonable amount of effort.*

The workability should be assessed depending upon the situation at hand. The aim should be to have the maximum possible workability consistent with satisfactory placement and compaction of concrete.

An insufficient workability may result in incomplete compaction, thereby severely affecting the strength, durability and surface finish of concrete and may prove to be uneconomical in the long run.

**7.6.8. Preparation of Plain Cement Concrete**

The plain cement concrete of required strength and durability is *prepared* by the following *steps*:

1. Selection, storing and batching of ingredients.
2. Mixing of ingredients.
3. Transportation of concrete.
4. Placing of concrete.
5. Compaction of concrete.
6. Finishing of concrete.
7. Curing of concrete.

**7.6.8.1. Selection, storing and batching of ingredients**

- The various ingredients should be carefully selected according to the purpose and strength of the concrete desired and stored separately and properly.
- '*Batching of ingredients*' is the *process of proper and accurate measurement of all ingredients of concrete to ensure uniformity of proportions and aggregate grading*. In order to obtain full strength of the concrete, it must be ensured that quantities are correctly measured.

**7.6.8.2. Mixing of ingredients**

The various ingredients of concrete are mixed thoroughly to achieve the following *objectives*:

1. To impart uniform colour to concrete.
2. To distribute various sizes of particles uniformly/evenly.
3. To spread evenly the binding matrix over every particle of aggregates.
4. To impart required consistency to concrete.

Ingredients of concrete are mixed by the following methods:

(*a*) Hand mixing (*b*) Machine mixing

**(*a*) Hand mixing:**

"*Hand mixing*" is the *process of mixing the ingredients of concrete by manual labour*; the concrete thus produced is known as *hand mixed concrete.*

- The ingredients of mortar are measured out in a dry state by measuring boxes in the proportions decided upon, on a clean platform of concrete or brickwork, and then mixed by turning over two or three times for thorough mixing.
- Then the dry mixture of mortar is spread over well soaked aggregate.
- Necessary water is then sprinkled over the dry mixture from a water can with a rose to mix the whole very thoroughly.
- The mixture is again turned over two or three times for thorough and even mixing of materials. This is known as dry *method of mixing concrete.*
- The prepared mix should be consumed in 30 minutes after adding water.

The mixing by hand is allowed in case of *small works or unimportant works where small quantity of concrete is required.*

If hand mixing is to be adopted for important works, it is advisable to use 10% more cement than specified.

This method is less efficient and *requires more cement than that required in machine mixing for obtaining the same strength of concrete.*

**(*b*) Machine mixing:**

- "*Machine mixing*" is the *process of mixing the ingredients of concrete by a machine*; the concrete thus produced is known as *machine mixed concrete.*

- The machine used in this method is known as concrete mixer which can be operated electrically or mechanically. The capacity of the mixer depends upon the size of the drum. The revolving drum has *paddles inside to ensure thorough mixing* by 50 to 60 rotations for each batch of mix. It receives the raw materials on one side and delivers the mixed concrete on the other side. There are *other concrete mixers in which loading, mixing and delivery are kept in a continuous operation.*

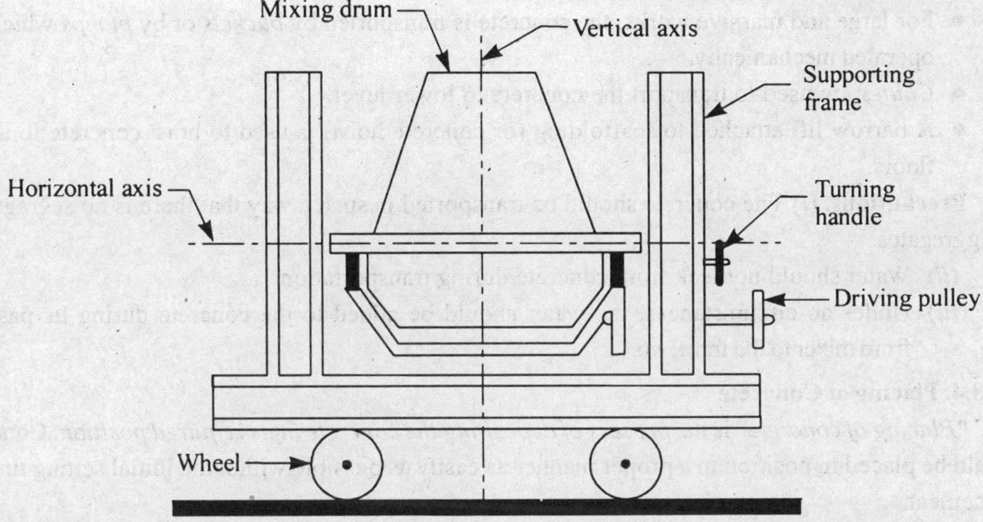


Fig. 7.5 : Concrete mixer.

Following points are worth noting :

- Too much water should not be used as it will wash away the lime or cement in the mixture. Generally 4.5 litres of fresh water is required for 0.03 $m^3$ of concrete. If hydraulic lime is to be used it should be ground in a mortar mill and carefully slaked. No unslaked particle should be allowed to mix with other ingredients.
- The water should enter the mixer at the same time or before the other materials are placed. This ensures even distribution of water.
- The mixing time should be at least one minute and preferably two-minutes (The time of mixing the materials in the mixer and the speed of the mixer are very important factors in deciding the strength of concrete).
- The concrete discharge by the mixer should be consumed within 30 minutes.
- The concrete mixer should be thoroughly washed and cleaned after use.
- There should be regular inspection of the inside portion of the mixer, the damaged/broken blades should be replaced promptly.

**Advantage.** This method is *superior* to handmixing as it *ensures more thorough mixing and requires less quantity of cement than that required in hand mixing for obtaining same strength of concrete.*

**Uses :** It is generally used where continuous supply of concrete is required for a big or important engineering project.

The various *types of concrete* mixers used are :

| | |
|---|---|
| 1. Tilting drum mixers | 2. Non-tilting |
| 3. Continuous mixers | 4. Pan mixer |
| 5. Truck mixers. | |

**7.6.8.3. Transportation of concrete**

The concrete can be transported either *normally* or *mechanically*.

- For small jobs the concrete is transported in *iron pans, wheel barrows or two wheel carts by manual labour*.
- For large and massive works, the concrete is transported by *buckets* or by *pumps* which are operated mechanically.
- *Chutes* are used to transport the concrete to lower level.
- A barrow lift attached to scaffolding (or concrete hoist) is used to hoist concrete to upper floors.

**Precautions.** (*i*) The concrete should be transported in such a way that there is no segregation of aggregates.

(*ii*) Water should not leak from concrete during transportation.

(*iii*) Under no circumstances, the water should be added to the concrete during its passage from mixer to the framework.

**7.6.8.4. Placing of Concrete**

"*Placing of concrete*" is the *process of depositing the concrete in its required position*. Concrete should be placed in position in a proper manner as easily as possible within the initial setting time of the cement.

**Precautions.** (*i*) Concrete should not be thrown from a height to avoid segregation.

(*ii*) The layers of concrete should be horizontal so as to prevent water running off and carrying some mortar with it.

(*iii*) Layers should follow one another so that each is laid before the previous one has had time to set otherwise the surface of previous one is to be roughened by crosslines when green so as to ensure a key being formed between the layers.

(*iv*) All "Laitance" (*i.e.*, squeezed out mortar), collecting at the top which becomes soft and chalky when hardened, must be removed before placing new concrete.

(*v*) The concrete should be thoroughly worked around the reinforcement and tapped in such a way that no honeycombed surface appears on removal of the framework.

(*vi*) In case placing of concrete is to be suspended for sometime, grooves must be made in the finished work, for joining future work, before initial setting time of the cement.

(*vii*) The placing of concrete should be carried out uninterrupted between predetermined construction joints.

(*viii*) During placing, it should be seen that all edges and corners of concrete surface remain unbroken, sharp and straight in line.

The concrete is *placed under water* by the use of **dump buckets** or **tremies** (Tremie is the name given to a steel pipe 15 to 30 cm in diameter and sufficiently long enough to reach the bed of water shed keeping its one end above the water level; the pipe is fitted with a hopper at its upper end for pouring concrete inside the pipe).

### 7.6.8.5. Compaction of concrete

"*Compaction of concrete*" is the *process of consolidating concrete after placing it in position.* The main aim of consolidation of concrete is to *eliminate air bubbles and thus to give maximum density to the concrete.* Proper consolidation ensures an intimate contact between concrete and reinforcement (5 per cent voids reduce 30 per cent strength of concrete).

Compaction of concrete can be carried out either *manually* or *mechanically.*

- The concrete is compacted **manually** by using handtools such as rammers, templates for tamping and spading rod.
  - — for *mass concrete work* such as foundation *heavy flat bottomed rammers* are used until a thin film of mortar or paste appears at the surface.
  - — In case of *flat surfaces* which are of comparatively smaller thickness *templates* corresponding to the profile of the finished surface are used and tamping is done till slurry or paste appears at the top.
  - — At or near the *vertical faces spading* is done for compacting.
- The concrete is compacted **mechanically** by jets of **compressed air** or by **vibrators.**

*Various type of vibrators used are* : (*i*) Internal or immersion vibrators; (*ii*) Surface vibrators; (*iii*) Form or shutter vibrators; (*iv*) Vibrating tables.

- — The concrete *laid in thin walls and columns* is compacted by *forcing a jet of compressed air* through it.
- — The concrete, on large scale and important works, is compacted by *internal or external vibrators* (**Vibrators** are the mechanical devices which are used to *compact concrete in the framework*).

In case of *mass concrete works*, the concrete is compacted by *internal vibrator* whereas *external* or *surface vibrators* are used for *compacting concrete in thin concrete slabs* such as R.C.C. roof slabs, road slabs etc.

*Advantages : Compaction by vibration has the following advantages :*

1. The *mechanically compacted concrete produces a denser and stronger concrete construction as compared to that compacted manually.*
2. Since by the use of vibrators the consolidation time is reduced, therefore vibrators are employed where the *rapid progress of work* is of significant importance.
3. By the use of vibrators it is possible to *deposit concrete in small openings or places* where it will be difficult to deposit concrete manually.
4. With the use of vibrators it is possible to make a harsh and stiff concrete mix, with a slump of about 40 mm or less, workable.
5. Compaction by vibration permits the use of *less water and a leaner mix and results in a stronger concrete.*
6. Forms may also be *removed much earlier* than hand-compacted concrete.
7. A large proportion of coarse aggregate and small proportion of fine aggregate can be used giving *less shrinkage.*

### 7.6.8.6. Finishing of concrete surfaces

Finishing of concrete surfaces is an important process from engineering point of view. The results of finishing are good if slump is about 50 mm. The choice of concrete finish depends upon the *ultimate use of completed job and desired effect.*

Finishing may be achieved by the following *operations* :

- *Screeding.* It is the levelling operation that removes humps and hollows and provides a true and uniform concrete surface.
- *Floating.* It is the process of removing irregularities from the surface of the concrete left after screeding. It also helps in levelling the surface and compacting concrete.
- *Trowelling.* It is final operation of finishing; it is performed where smooth and dense surface is required.

**7.6.8.7. Curing of concrete**

- **Curing** is *the process of keeping the set concrete damp for some days in order to enable the concrete gain more strength.* It has been established that strength of concrete *increases with age* provided it is kept damp. During the process of curing the *concrete absorbs the water necessary for its complete chemical action to reach its required strength. The strength of concrete increases more rapidly in the first few days after setting* and afterwards the rate of increase in strength *goes on retarding.*
- The period for which curing should be continued depends upon atmospheric condition such as *temperature, humidity* and *wind velocity.* It also depends on the type of cement and nature of work.
- For *Portland cement, the curing period is about 7 to 14 days.*
- If *rapid hardening cement* is used, the curing period *can be considerably reduced.*
- In cold weather, the concrete should be cured for at least 14 days.

**Methods of curing:**

While selecting any mode or method of curing, the following *two factors* are considered :

(*i*) The loss of water should be prevented.

(*ii*) The temperature should be kept minimum for dissipation of heat of hydration.

Depending upon the type of construction work, the concrete is cured by any of the following methods :

1. **By ponding.** This method is employed for *flat surfaces.*

Here small clay bunds or water retaining material is provided around the perimeter of the surface. The whole of the surface is divided into small portions by clay bunds and the enclosed areas are filled with water. The flat surfaces are also cured by covering them with a layer of wet sand or sand dust 2 to 3 cms thick. This method is generally used for curing *pavements, floors slabs* etc.

2. **By using wet sacks and sprinkling water frequently.** Columns and similar members are best cured by wrapping round them wet sacks and keeping pot full of water on thin tops with a pin hole at the bottom.

*Vertical members are also cured by spraying water on the surface at regular intervals.*

3. **By steam.** Precast concrete is often cured by steam under pressure.

4. **By alternating current.** Concrete is also cured by passing alternating current through freshly laid concrete for 24 hours.

5. **By sealing curing compounds.** In this method, the concrete surface is applied with a coat of sealing curing compound immediately after finishing it. The coat of such compounds *prevents the evaporation* of water from the concrete and then concrete remains moist for a longer period.

6. **By chemicals.** In this method of curing, water is sprinkled over the surface after adding certain quantity of some hygroscopic material (*e.g.*, sodium chloride or calcium chloride). The hygroscopic material further absorbs moisture from the atmosphere and thus keeps the surface damp. This method is employed where there is *scarcity of water.*

**Effects of improper curing:**

If the concrete is not properly cured, the following *disadvantages* result:

1. Due to higher permeability, the durability decreases.
2. Decrease in compressive and flexural strengths.
3. The weathering and fast resistances are decreased.
4. The chances of ingress of atmospheric chemicals and chlorides are increased.
5. Due to plastic shrinkage, drying shrinkage and thermal effects the cracks are formed.
6. The abrasion resistance is lowered on account of coating the surfaces with sand and dust.
7. The carbonation rate enhances.

### 7.6.8.8. Principle of hardening of cement concrete

When water reacts with the surface of cement grain it produces a gelatinous material or gel. The gel so produced serves as a cementing material between the grains, but it loses water to the hydrated layer underneath (below) and is hardened during the process.

In addition to *gel formation*, hardening process may take place due to *crystallisation* as well.

In *cement concrete*, the aggregate is mixed with a cement paste which binds the aggregate lumps together. The ingredients of cement paste set and then bond the aggregate through the action of its water of hydration. The basic operative materials are the calcium silicate $2CaO\ SiO_2$ and $3CaO\ SiO_2$ which when hydrated form a material called *tobermorite gel*. This gel consists of distorted crystalline layered structure with water of hydration dissolved between adjacent layers and has the formula $2CaOSiO_2 . x(H_2O)$ where $x$ may vary to a large extent depending on water-cement ratio.

On addition of water, while preparing, an initial fast reaction takes place as follows :

$$C_3A + 6H_2O \longrightarrow C_3A\ \ 6H_2O$$

During this reaction heat of hydration is involved.

The main subsequent hydration reactions are :

$$C_2S + x.H_2O \longrightarrow C_2S \,.\, x(H_2O)$$

and, $$C_3S + x.H_2O \longrightarrow C_2S \,.\, x(H_2O) + Ca(OH)_2$$

It takes about 30 days for $C_3S$ to reach 70% of its ultimate strength. The $C_2S$ reaches about 2/3rd of its final strength in about six months. In this fashion, *concrete may go on hardening for years together*.

## 7.7. PROPERTIES OF CEMENT CONCRETE

*Strength and durability* of concrete depend on the following *factors* :

(*i*) Quantity of cement;
(*ii*) Size, grading and moisture contents of aggregates and presence of foreign matters;
(*iii*) Water-cement ratio;
(*iv*) Proportioning of ingredients;
(*v*) Method of measuring materials–by volume or weight;
(*vi*) Hand mixing or machine mixing;
(*vii*) Methods adopted for placing and consolidation before initial setting; and
(*viii*) Method of curing.

In order to obtain quality concrete, its properties in *plastic as well as hardened stage play important roles*. These properties are enumerated and discussed below

1. Workability
2. Segregation
3. Bleeding
4. Harshness.

} ........... Plastic stage.

5. Strength
6. Durability
7. Impermeability
8. Dimensional changes.

} .................. Hardened stage.

1. **Workability:**
   - It is defined as the *ease with which concrete can be compacted fully without segregation and bleeding.*
   - It depends *on the quantity of water, grading of aggregate, and the percentage of fine materials in the mix.*

2. **Segregation:**
   - It implies the *separation of coarser particles from the mix which results in non-homogeneity of the concrete mix.* In order to avoid segregation, the concrete should *not* be thrown from a height.
   - Segregation results in *honeycombing and decrease in density and ultimately loss of strength of hardened concrete.*

3. **Bleeding:**
   - Bleeding refers to the *appearance of water along with cement particles on the surface of freshly laid concrete on compaction and finishing.* It occurs either due to *excessive compaction* or when there is *excess water* present in the mix.
   - It *causes the formation of pores in the concrete making it porous and weak.*
   - In order to make good quality concrete, bleeding must be *avoided* by :
     — controlling the quantity of water;
     — providing finer grading of fine aggregate;
     — using finely ground cement; and
     — performing suitable optimum compaction.

4. **Harshness:**
   - *"Harshness" is the resistance offered by concrete to its surface finishing.* It is due to the presence of *lesser fines, lesser cement mortar*, and use of properly graded *crushed or angular aggregate* and *insufficient water content* required for workability.
   - The surface of harsh concrete remains rough and porous.

5. **Strength:**
   - *"Strength"* of concrete is its *resistance to bear loads after it has hardened.*
   - The strength of concrete is commonly measured in terms of its compressive strength which is an important property of hardnened concrete; it generally varies from 5 $N/mm^2$ to 45 $N/mm^2$.
   - The *compressive strength of concrete depends on the type of cement, water-cement ratio, sizes, shape and grading of aggregate, degree of compaction, curing and other concreting operations.*
   - The resistance of concrete to tension is very low; resistance to tension under bending (flexural loading) is called its *flexural strength*. The flexural tensile strength of cement concrete generally varies from 10 to 12 percent of its compressive strength.

6. **Durability:**
   - *"Durability" of concrete refers to its resistance to deterioration under the force of environment* (*e.g.*, weathering, chemical attack, fire, and corrosion of steel etc).
   - Generally dense and strong concretes have better durability in extreme weather conditions.
   - Porous-concrete leads to corrosion of steel and hence to increase durability of reinforced cement concrete, it should form a cover of dense and impermeable concrete around steel.

7. **Impermeability:**

- *"Impermeability" is the resistance of the concrete to the flow of water through the pore-spaces in it.*
- In order to obtain impermeable concretes, the following steps should be taken :
  - — Adopt a low water-cement ratio;
  - — Use dense and well graded aggregate;
  - — Ensure full compaction; and
  - — Cure continuously under moist and low temperature conditions etc.
- It is important in exposed and water-retaining structures.

8. **Dimensional changes:**

- In concrete, dimensional changes are caused due to shrinkage (of concrete), thermal changes, elasticity and creep in concrete.
- The dimensional changes may *induce certain stresses in concrete which may lead to its cracking*. These changes in sample structural designs are generally ignored.

## 7.8. WATER-PROOF CONCRETE

The impermeability of concrete is absolutely essential in works like reservoirs, water storage tanks, roofs, basement walls etc. It can be stated in general, that if concrete is made dense and free from cracks, it is watertight. Concrete is made dense/waterproof in the following ways:

1. The best way of obtaining water-proof concrete is by *proper design of mix and proper compaction preferably using mechanical vibrators*.
2. By using more cement.
3. By using special water-proofing cements.
4. By using certain admixtures (*e.g.*, calcium chloride, ammonium chloride, zinc sulphate, aluminium sulphate) which fill the voids of the concrete and make it watertight.
5. By using patented products obtainable from the market under various trade names *e.g.*, *Pudlo, Cika, Pyrex, Nombad*. Usually 5 percent by weight of cement is added to the concrete mix to get water-proofing qualities.

Water-proofing compound of good quality entails the following *advantages* :

(*i*) It makes good concrete from the poor materials.
(*ii*) It corrects a badly proportioned concrete mixture.
(*iii*) It permits less rigid supervision of the workmanship.
(*iv*) It cures immature green concrete.

*Water-proof concrete* is specially useful *for liquid retaining structures*.

## 7.9. PRECAST CONCRETE

- Precast concrete differs from cast-in-situ concrete in that the former is a factory made product while the latter is prepared at site of work.
- The precast products may vary from simple structures such as pipes, fencing posts, paving slabs etc. to elaborate and complicated artificial concrete blocks.
- The precast products are prepared as follows :
  - — The moulds (of sand, timber or steel) are prepared according to the shape of the product.
  - — As per the design, the reinforcement, if any, is placed in the moulds.
  - — The concrete is mixed in the desired proportion and then placed in the mould; the finishing of the products is then carried out.
  - — The products are then adequately cured in specially constructed tanks.
  - — The products are then despatched for use at site of work.

**Advantages:**

1. No need to provide joints in the precast construction.
2. The labour required can be easily trained.

3. Desired shape and finish can be given with accuracy.
4. Concrete produced is of superior quality (due to better control on the concrete production in factory).
5. The precast structures can be dismantled, when required and they can be suitably used elsewhere.
6. Moulds used are accurate and more durable and can be used several times.
7. Reduced amount of scaffolding and formwork.
8. Less time required to complete the work.

**Disadvantages**

1. Lack of proper handling may damage the precast units during transport.
2. For lifting and moving of the precast unit special equipment is required.
3. It becomes difficult to produce satisfactory connections between the precast members.

## 7.10. REINFORCED CEMENT CONCRETE (R.C.C.)

### 7.10.1. Introduction

- The *cement concrete in which reinforcement is provided for taking tensile or excessive compressive and shear stresses is called* **reinforced cement concrete.**
- The reinforcing material for concrete should possess the following *characteristics* :
  (*i*) It should have a *high tensile strength, a high modulus of elasticity.*
  (*ii*) It should have *same temperature co-efficient of expansion and contraction as concrete so that thermal stresses do not develop.*
  (*iii*) It should be *cheap and easily available.*
  (*iv*) It should be able to *develop perfect bond with concrete.*
  (*v*) It should be able to *resist* very harmful effect, if any, produced by the concrete.
  (*vi*) It should be easy to cut, bend or weld.

Almost all the above requirements are fulfilled by ***steel*** and thus it is in common use as a reinforcing material. In early days attempts were made to use cheaper materials like glass, bamboo, asbestos etc. as reinforcing materials but due to some inherent defects in them they could not claim a wider field of application.

— The steel reinforcement to be used in generally in the form of round bars 6 mm to 32 mm diameter.

— Expanded metal or steel fabric in the form of a mesh of steel rods, moulded together, are also used as reinforcement in concrete walls etc.

— Flat bars are more useful in case of tanks and pipes etc.

— Usual proportions of ingredients in reinforced cement concrete are :
1 part Portland cement : 1 to 2 parts clean sand : 2 to 4 parts shingle or crushed stone.

### 7.10.2. Properties

The combination of steel and concrete becomes practicable for the following *three reasons* :

1. The co-efficient of linear expansion of concrete and steel are almost the same. Therefore, no internal stresses are set up within reinforced concrete due to change in temperature.
2. The coating of cement on the surface of steel bars protects them from corrosion and does not produce an adverse chemical effect.
3. The concrete while setting grips very fast the surface of steel bars. Therefore the concrete is able to transmit to the steel bars those stresses which it cannot resist itself.

### 7.10.3. Advantages Disadvantages and Uses

**Advantages :** Reinforced cement concrete entails the following *advantages* :

1. It does not rot or decay and is not attacked by termites.
2. It is fire resisting.
3. It is economical in cost.

4. It is almost impermeable to moisture.
5. Its maintenance cost is negligible.
6. Its constituents are easily available.
7. It does not develop cracks.
8. It can be easily moulded to any desired shape.
9. It is durable.
10. It provides monolithic (without joint) construction of structural members.
11. It provides good structural appearance to the structure.
12. The structural members can be made of any desired strength.

**Disadvantages :** The disadvantages of reinforced cement concrete are :

1. High initial cost.
2. Heavier than steel construction.
3. Skilled supervision and workmanship required.
4. Involves additional cost of form-work.

**Uses:** Reinforcement cement concrete is commonly used for the construction of the following:

(*i*) Slabs (*ii*) Lintels
(*iii*) Beams (*iv*) Columns and their footings
(*v*) Raft or mat foundations (*vi*) Precast or cast-in-situ concrete piles etc.

### 7.10.4. Basic Assumptions

The following *assumptions* are made in the reinforced cement concrete for *design purposes*:

1. The transverse section which is plain before bending remains plain after bending.

2. All tensile stresses are taken up by steel reinforcement only and the resistance of concrete to tension is nil.

3. Concrete is homogeneous and elastic. It is also of uniform quality and strength. The stress, therefore is proportional to strain. It also follows the stress diagram on the compression side which will be a triangle, the resultant compressive stress acting at the centre of the triangle.

The bond or adhesion between steel and concrete is perfect within certain limits and they act unitedly.

**General requirements of R.C.C. design:**

The mix should be carefully designed so that maximum compressive strength of the concrete is available and both steel and concrete are stressed to their maximum permissible stresses after taking into consideration all possible loads and various criteria of failure.

Some of the *important design requirements* are :

1. Main reinforcement should be sufficient to resist direct tensile stresses.
2. Area of concrete should be adequate to take up compressive stresses.
3. Adequate bend should be developed between joints of the rods.
4. Adequate provision should be made for shearing stresses.

### 7.10.5. Causes of Failure of Reinforced Cement Structures

Following are the causes of failure of reinforced cement structures:

1. When the member is subjected to excessive tension, so as to exceed the permissible stress in steel.
2. When the loading is such that the compressive stress in concrete exceeds its safe permissible value.
3. On account of the slipping of the steel bars from concrete.
4. When the concrete is subjected to excessive shear.
5. Due to bad quality of material used, shrinkage creep or thermal effects.

## 7.11. PRESTRESSED CONCRETE

### 7.11.1. Introduction

- It *is the concrete in which high compressive stresses are artificially induced before its use.* In reinforced concrete members; the *prestress is commonly introduced by tensioning the steel reinforcement.*
- The development of *early cracks in reinforced concrete due to incompatibility in the strains of steel and concrete* was perhaps the starting point in the development of a new material like "*prestressed concrete*". The application of permanent comprehensive stress to a material like concrete, which is strong in compression but weak in tension, increases the apparent tensile strength of that material, because the subsequent application of tensile stress must first nullify the compressive prestress.
- In the words of *Guyon*: "There is probably no structural problem to which prestress cannot provide a solution, and often a revolutionary one. Prestress is more than a technique, it is a general principle."
- In prestressed concrete high strength concrete is necessary, as the material offers high resistance in tension, shear, bond and bearing.

High strength concrete entails the following *advantages* :

(*i*) It is less liable to shrinkage cracks.

(*ii*) It has a higher modulus of elasticity and smaller ultimate creep strain, resulting in a smaller loss of prestress in steel.

(*iii*) Its use results in a reduction in the cross-sectional dimensions of prestressed concrete structural elements. With a reduced dead-weight of the material, longer spans become technically and economically practicable.

### 7.11.2. Advantages:

*Advantages* of prestressed concrete, in comparison with other forms of construction, such as reinforced concrete and steel are given below:

1. The cross-section of fully prestressed members, which are free from tensile stresses under working loads, is more efficiently utilised, when compared with a reinforced concrete sections which is cracked under working loads.
2. The members of prestressed concrete possess improved resistance to shearing forces, due to the effect of compressive prestress, which *reduces the principal tensile stress.*
3. Under working loads, a prestressed concrete flexural member is *stiffer than a reinforced concrete member of the same depth.*
4. The use of high-strength concrete and steel in prestressed members *results in higher and slender members than is possible with reinforced concrete.*
5. Prestressing of concrete *improves the ability of the material for energy absorption under impact loads.*
6. The prestressed concrete proves economical for long-span structures.
7. It has *considerable resilience* due to its capacity for completely recovering from substantial effects of overloading without undergoing any serious damage.
8. Its fatigue strength is comparatively better than that of other materials (chiefly due to the small variations in prestressing steel); as such it is recommended for dynamically loaded structures, such as railway bridges and machine foundations.
9. The use of prestressed concrete *leads to an overall economy,* as the decrease in dead weight reduces the design loads and the cost of foundations.

### 7.11.3. Applications

The following are the *applications* of prestressed concrete:

1. It is ideally suited for long-span bridge construction.
2. Its finds extensive applications in the *construction of* :
   — long-span folded plate roofs;
   — air craft hangers;
   — nuclear containment vessels;
   — pavements;
   — rail-road sleepers;
   — poles;
   — piles;
   — television towers and masts.
3. Prestressed concrete, in recent years, has found novel applications in a variety of *marine structures,* such as floating docks, off-shore oil-drilling platforms, giant floating liquified gas and oil-storing vessels.
4. It is used in the construction of tall towers to support restaurants, observation desks etc.

Notable examples of *prestressed concrete structures in India* include the following : Lubha bridge; Gomti aqueduct; Ball tank; Trombay (Maharashtra), The Boeing hanger at Santa Cruz airport; Ganga bridge at Patna; Silo at Nangal; Zuari bridge at Goa.

### 7.11.4. Materials for Prestressed Concrete

**1. High strength concrete:**

***(i) High strength concrete mixes:***

- Prestressed concrete requires concrete which has a *high compressive strength at a reasonably early age,* with *comparatively higher tensile strength than ordinary concrete.*
- Low *shrinkage, minimum creep characteristics and a high value of Young's modulus are generally deemed necessary for concrete used for prestressed members.*
- Several desirable properties such as durability, impermeability and abrasion resistance, are highly influenced by the strength of concrete.
- It has been indicated by the experimental investigations that in high-strength concrete mixes, workability, type and maximum size of aggregate, and the strength requirement influence the selection of the water-cement ratio. Crushed rock aggregates, being angular, generally produce stronger concrete at the same age in comparison with gravel aggregates.
- In high strength concrete mixes, *the water content should be as low as possible* with due regard to workability and the concrete should be suitable for compaction by means available at the site.

***(ii) Strength requirements:***

As per Indian standard code IS:1343-1980 the minimum 28-day cube compressive strength shall be:

*Pre-tensioned members* 40N/mm$^2$;
*Post-tensioned members* 30 N/mm$^2$.

***(iii) Light-weight aggregate prestressed concrete:***

- The use of light-weight aggregate concrete for prestressed concrete construction is well established since long. The main advantage of light weight concrete is that it reduces the self-weight of the structure, thus minimising the amount of concrete and steel required for carrying the load.
- The light-weight aggregates, generally used for prestressed concrete, are *foamed slag, lytag and aglite.*

- The modulus of elasticity of light-weight concrete is about 50 to 55 percent of modulus of the elasticity of normal weight concrete and hence the loss of prestress due to elastic deformation is higher and deflection of flexural members are comparatively higher due to the lower values of modulus of elasticity.
- The unit weight of light-weight concrete varies considerably between 1450 and 1750 kg/m$^3$.
- The shrinkage and creep of light-weight concrete is comparable, with marginal variations, to that of sand and gravel concrete.

**2. High tensile steel:**

- For prestressed concrete members, the high-tensile steel used generally consists of *wires, bars* or *strands.*
- The higher tensile strength is generally achieved by marginally increasing the carbon content in steel in comparison with mild steel. High tensile steel usually contains 0.6 to 0.85 per cent carbon, 0.7 to 1 per cent manganese, 0.05 per cent of sulphur and phosphorus with traces of silicon.
- The ultimate tensile strength of a plain hard-drawn steel wire varies with its diameter. The *tensile strength decreases with increase in the diameter of the wires.*
- The ultimate tensile strength of bars *does not vary appreciably with the diameter,* the high strength of bars is due to alloying rather than to cold working as in case of wires.

### 7.11.5. Prestressing Systems/Methods

- The most widely used method for prestressing of structural concrete element is *longitudinal tensioning of steel by different tensioning devices.* (The various types of devices used for tensioning steel are grouped with four principal categories namely: (*i*) Mechanical, (*ii*) hydraulic, (*iii*) electrical/thermal and (*iv*) chemical.)
  - Prestressing by the application of *direct forces* between abutments is generally used for *arches and pavements,* while *flat jacks* are invariably used to impart the desired forces.
  - For *circular structures,* such as tanks and pipes, it is common practice to impart pre-compression in concrete by *circular prestressing.*
  - With the development of expansive cements, prestress in concrete can be developed by *chemical prestressing.*
- In **pre-tensioning system,** the tendons (a *tendon* is a stretched element used in a concrete member of structure to impart prestress to the concrete), are first tensioned between rigid anchor-blocks cast on the ground or in a column or unit-mould type pretensioning bed, prior to the casting of concrete in the moulds. The tendons comprising, individual wires or strands are stretched with constant or variable eccentricity with tendon ancharage at one end and jacks at the other. With the forms in place, the concrete is cast around the stressed tendon.
- In **post-tensioning system**, the concrete units are first cast by incorporating ducts or grooves to house the tendons. When the concrete attains sufficient strength, the high tensile wires are tensioned by means of jack bearing on the end face of the member and anchored by wedges or nuts. Post-tensioning is ideally suited for medium to long-span in situ work where the *tensioning cost is only a small proportion of the cost of the whole job.*

Post-tensioning is invariably used for:

(*i*) Strengthening concrete dams;

(*ii*) Circular pre-stressing of large concrete tanks;

(*iii*) Biological shields of nuclear reactors;

(*iv*) Concrete construction work involving stage pre-stressing.

## 7.12. SPECIAL CONCRETES

Under this heading following special concretes will be dealt with:

1. Ferrocement
2. Fibre reinforced concrete
3. Polymer concrete
4. Superplasticized concrete
5. Ultra high strength concrete
6. Sulphur impregnated concrete
7. Gap-graded concrete.

### 7.12.1. Ferrocement

#### 7.12.1.1. Introduction

- **Ferrocement** may be considered as a *type of thin reinforced concrete construction where cement mortar-matrix is reinforced with many layers of continuous and relatively small diameter wire-meshes.* While the *mortar provides the mass, the wire-mesh imparts tensile strength and ductility to the material.*
- The ferrocement, in terms of structural behaviour, *exhibits very high tensile strength-to-weight ratio and superior cracking performance.* The distribution of a small diameter wire-mesh reinforcement over the entire surface, and sometimes over the entire volume of the matrix, *provides a very high resistance against cracking.* Moreover, many other engineering properties, such as *toughness, fatigue resistance, impermeability* etc. are *considerably improved.* In order to achieve a stiff reinforcing cage, sometimes, conventional bars in a skelton form are added to thin wire-meshes.

#### 7.12.1.2. Materials

**1. *Cement mortar matrix:***

- The ferrocement composite is a rich cement-mortar matrix of 10 to 60 mm thickness with a reinforcement volume of five to eight percent in the form of one or more layers of very thin wire-mesh and a skelton reinforcement consisting of either welded-mesh or mild steel bars.
- Normally *portland cement and fine aggregate (sand) mix is used in ferrocement.* The matrix constitutes about 95 per cent of the ferrocement and *governs the behaviour of the final product. Plasticizers* and *admixtures* may be added for achieving an improved *workability, water reduction for increase in strength and reduction in permeability, water-proofing* and *increase in durability.*
- The *mix proportions* in terms of *sand-cement* ratio (by mass) normally recommended are *1.5 to 2.5.* The *water-cement ratio,* by mass, may vary between *0.35 and 0.6;* it *must be kept below 0.4 in order to reduce permeability.*

**2. *Reinforcement:***

- The reinforcement used in ferrocement is of the following two types:
  (*i*) Skelton steel (*ii*) Wire mesh.
- The *skelton steel frame* is made conforming exactly to the geometry and shape of structure, and is *used for holding the wire-meshes in position and shape of the structure.*
  The skelton steel comprises of relatively large diameter (about 3 to 8 mm) steel rods typically spaced at 70 to 100 mm. It may be tied-reinforcement or welded wire fabric.
- The *wire-mesh* consisting of galvanized wire of diameter 0.5 to 1.5 mm spaced at 6 to 20 mm centre-to-centre, is formed by welding, twisting or weaving.
  Fig. 7.6 shows different types of welded wire meshes.
  — The *hexagonal wire-mesh* is *cheaper but structurally less efficient* than the mesh with square opening.

— The *rectangular meshes* have better rigidity when placed or tied over the skelton frame.
— The *welded wire-meshes* have a higher Young's modulus.
— The *woven-wire meshes* are a little more flexible and easy to work with than the welded meshes.

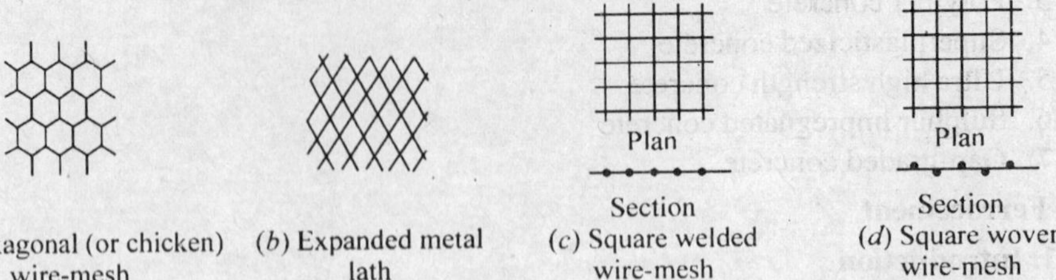


(*a*) Hexagonal (or chicken) wire-mesh (*b*) Expanded metal lath (*c*) Square welded wire-mesh (*d*) Square woven wire-mesh

Fig. 7.6 : Different types of wire-meshes.

### 7.12.1.3. Construction in ferrocement

The construction in ferrocement may be divided into the following *four* phases:

1. Fabrication of skelton framing system.
2. Fixing of bars and mesh
3. Application of mortar
4. Curing

### 7.12.1.4. Properties of ferrocement

- Its cracking resistance, ductility, impact and fatigue resistances are higher than those of concrete.
- It has a very high tensile strength-to-weigh ratio and superior cracking behaviour. *Tensile strength* of ferrocement depends mainly on the *volume of reinforcement in the direction of force and the tensile strength of the mesh.*
- Ferrocement has *tensile strength as high as its compressive strength i.e.*, 27 MPa, and the widths of cracks are very small even at failure (about 0.05 mm). Ferrocement structures can be designed to be watertight at service loads
- The ferrocement beams *show poor resistance to fatigue under cyclic loading. Impact tests* on ferrocement slabs show that the *impact resistance increases* almost linearly with the increase in *specific surface (volume traction) and ultimate strength of mesh reinforcement.
- The impermeability of ferrocement products is far superior to that of ordinary reinforced concrete products.

### 7.12.1.5. Advantages of ferrocement

The major *advantages* of ferrocement are listed below:

1. Ferrocement structures are *thin* and *light.*
2. The construction technique is *simple* and hence highly skilled labour is not required.
3. Partial or complete elimination of framework is possible.
4. The local damage due to abnormal loads (such as impact) can be easily repaired by ferrocement construction.
5. Ferrocement is suitable for manufacturing the precast units which can be easily transported.

* *Specific surface area* of reinforcement is defined as the total surface area of the wires in contact with cement mortar divided by the volume of the composite.

#### 7.12.1.6. Applications

The ferrocement is used in the *construction of the following:*

(*i*) Light and water-tight structure.
(*ii*) Portable structures such as mobile homes.
(*iii*) Water tanks.
(*iv*) Silos and bins.
(*v*) Boat hulls.
(*vi*) Biogas holders.
(*vii*) Pipes.
(*viii*) Folded plates and shell roofs.
(*ix*) Floor units.
(*x*) Service core units.
(*xi*) Wind tunnels.
(*xii*) Modular housing.
(*xiii*) Kiosks.
(*xiv*) Swimming pools.
(*xv*) Permanent form of concrete columns.

### 7.12.2. Fibre-reinforced Concrete (FRC)

#### 7.12.2.1. Introduction

- Plain cement concrete, due to its *low tensile strength* and *impact resistance is* considered to be a *brittle material.* However, marked improvement in these properties can be brought about by the addition of small diameter, short length, randomly distributed *fibres.*
- The fibres can be imagined as an aggregate with an extreme deviation in shape from the rounded smooth aggregate. The fibres inlock and entangle around aggregate particles and considerably *reduce the workability,* while the *mix becomes more cohesive and less prone to segregation.*
- The fibres suitable for reinforcing the concrete have been produced from *steel, glass* and *organic polymers.*
- The major factors affecting the characteristics of fibre-reinforced concrete are:
  (*i*) Water-cement ratio;
  (*ii*) Percentage (volume fraction) of fibres;
  (*iii*) Diameter and length of fibres.
  — The *location and extent of cracking under load will depend upon the orientation and number of fibres in the cross-section.*
  — The fibres *restrain the shrinkage and creep movements of unreinforced matrix.* However, *fibres have been found to be more effective in controlling compression creep than tensile creep of unreinforced matrix.*

#### 7.12.2.2. Types of fibre-reinforced concrete

Out of various possible types of fibre-reinforced concrete the following two types are mostly being recommended:

1. Steel-fibre reinforced concrete
2. Glass-fibre reinforced concrete

**1. Steel-fibre reinforced concrete:**

- This type of concrete is formed by adding steel fibres in the ingredients of concrete. A number of steel-fibre types are available as reinforcement. Round steel fibres, the commonly used type, are produced by cutting wires into short lengths. The typical diameters lie in the range of 0.25 to 0.75 mm. Steel fibres having a rectangular cross-section are produced by slitting the sheets about 0.25 mm thick.
- Steel fibres have a tendency to cluster together which creates difficulties in ensuring their uniform random distribution in the concrete. Their difficulty is overcome by using *fibre-bundles* (Fibre losely essembled in the form of bundles with the help of water soluble glue). The steel fibres in the fibre-bundles separate out during the process of mixing concrete and get distributed in a random fashion in the concrete mix.

- By addition of 2 to 3 per cent of fibre (by volume) it is possible to achieve two to three times increase in the flexural strength of concrete and substantial increase in explosion resistance, crack resistance and other properties of concrete.

**Applications.** Steel-fibre reinforced concrete is considered suitable for construction of *pavements, bridge decks, pressure vessels, tunnels lining* etc.

**2. Glass-fibre reinforced concrete:**

- Glass fibres are made up from 200 to 400 individual filaments which are highly bonded to make up a strand. These strands can be chopped into various lengths or combined to make cloth, mat or tape.
- The process of manufacture of glass-fibre cement products may involve spraying, premixing or incorporation of continuous rovings.
- It has been observed that *addition of 10 per cent of glass fibre* (by volume) brings almost *two folds increase in tensile strength* and *substantial increase in impact resistance of concrete.*

**Applications.** The glass-fibre reinforced cement finds use in *form work systems,* ducting, roofing elements, sewer lining, swimming pools, tanks etc. Sometimes it is used in combination with polymer impregnated in *situ* concrete.

**7.12.2.3. Properties of fibre-reinforced concrete**

Fibre-reinforced concrete entails the following *superior properties*:

1. Increased tensile and bending strengths;
2. Improved ductility and resistance to cracking;
3. High impact strength and toughness;
4. Spalling resistance;
5. High energy absorption capacity.

**7.12.2.4. Applications**

Due to above mentioned superior properties fibre-reinforced concrete has found special application in:

- Hydraulic structure;
- Arifield and highways pavements;
- Bridge decks;
- Tunnel linings;
- Heavy duty floors.

**7.12.3. Polymer Concrete**

**7.12.3.1. Introduction**

- Through continued research it has been established that addition of polymer in concrete brings about *marked improvement* in its *compressive strength, fatigue resistance, impact resistance, toughness and durability.*
- Polymer concrete is *highly impermeable and resistant to attack by acids, alkalies and other chemicals.*

**7.12.3.2. Types of polymer concrete**

Polymer concrete may be classified into the following *three* categories:

1. Polymer impregnated concrete (PIC)
2. Polymer cement concrete (PCC)
3. Polymer concrete (PC).

**1. *Polymer impregnated concrete (PIC)***

- PIC is a *hardened portland cement concrete that has been impregnated with a monomer* (low viscosity liquid organic material) and *subsequently polymerised in situ.*
- In this case, the cement concrete is cast and cured in the conventional manner.

- After the concrete product gets hardened and dried, *air from its voids is removed under partial vacuum* and *a low viscosity monomer* (styrene, vinyl chloride etc.) is *diffused through the pores of the concrete.* The concrete product is then finally *subjected to polymerisation by radiation or by heat treatment thereby converting the monomer filled in the voids into solid plastic.*

**Applications.** PIC products on account of their high resistance to wear and tear, impact etc. are used as/in:

— precast slabs for bridge decks;
— roads;
— marine structures;
— food processing buildings; etc.

**2. *Polymer cement concrete (PCC):***

- PCC is produced by *incorporating an emulsion of a polymer or a monomer in ordinary portland cement concrete.*
- The ingredients comprising cement, aggregates and monomer are mixed with water and *monomer in the concrete mix is polymerised after placing concrete in position.* The resultant concrete has *improved:*

— strength;
— adhesion;
— chemical resistance;
— impact resistance;
— abrasion resistance;
— increased impermeability;
— reduced absorption.

**Application.** *Marine works.*

**3. Polymer concrete (PC) :**

- In PC polymer/monomer is employed to *act as binder* in place of cement.
- The monomer and aggregate are mixed together and the *monomer is polymerised after placement of concrete in position.* It is imperative to preheat the coarse and fine aggregates while mixing momomer.

**Application.** *Irrigation works.*

### 7.12.4. Superplasticized Concrete

- In this type of concrete highly effective plasticisers (known as *superplasticisers*) are incorporated as admixtures to produce flowing concrete which is *more or less self-compacting type.*
- Addition of superplasticisers *increases workability of concrete, permits use of less quantity of water,* thereby *producing high strength concrete, without increasing cement.*
- Superplasticized concrete can be adopted in the following *three* possible situations:
  - (*i*) When it is required to produce concrete of *required strength with reduced quantity of cement.*
  - (*ii*) When it is required to produce *high strength concrete having a very low water-cement ratio.*
  - (*iii*) When it is required to produce *concrete having good flexibility without any danger of bleeding, segregation or reduction in strength.*
- The quantity of superplasticisers to be added in the concrete mix depends upon the purpose for which it is being laid. In general, superplasticiser can be added in concrete in doses upto 0.7 per cent by *weight of concrete.*

### 7.12.5. Ultra High Strength Concrete

- By suitable selection of material and by adopting special method of production it is possible to make ultra high strength concrete having compressive strength exceeding 100 MPa (strength of normal concrete being 10 to 30 MPa).

— It can be *easily produced by improved compaction* (by pressure and vibration etc.) and *adhesion of cement matrix to aggregates or by adoption of cementatious aggregates.*

— It can be easily produced by use of high quality coarse aggregates, synthetic aluminous fine aggregates and cement.

— Use of high temperature to increase the lime silica bond and use of spiral working of concrete to induce triaxial stress are other techniques being tried to develop ultra high strength concrete.

- Its use increases the load bearing capacity of the columns considerably.

**Applications.** Owing to its high compressive as well as tensile strength, ultra high strength concrete finds wide *application in prestressed concrete.*

### 7.12.6. Sulphur Impregnated Concrete

- It consists of a *mixture of sulphur, coarse aggregate and fine aggregate, there being no cement or water.*
- It is formed by *adding molten sulphur to the pre-heated aggregate in a mixture.* The hot concrete mix is poured in moulds to form *pre-cast units.*
- It needs no curing and the mould can be stripped soon after the concrete mix gets solidified.

**Properties:**

(*i*) It develops high compressive strength at an early age.

(*ii*) It can be re-used and re-moulded without any wastage.

(*iii*) It has good chemical durability and high strength.

(*iv*) It is more durable in acidic environments but is unstable in alkaline conditions.

**Applications.** Sulphur impregnated concrete is used in *industrial plants* and for *lining canals and tunnels.*

*Note:* Owing to its brittleness and corrosive action of sulphur on reinforcement, this type of concrete is *not* suitable for structural work.

### 7.12.7. Gap-graded Concrete

- Gap-graded concrete is obtained when a *gap-graded aggregate* is used in the production of the concrete. In case of gap-grading certain *undesirable sizes* of aggregates are omitted from the conventional continuous gradings (The undesirable sizes are those which prevent the efficient packing of the other sizes). Sometimes available *single-sized aggregate* only is used.
- The gap-grading is normally aimed at achieving strength from the *efficient packing of the aggregate.* A well packed aggregate will *require minimum cement paste to fill the minor voids.*
- *Gap-grading enables leaner and drier mixes* than conventional concrete of equivalent strength to be used resulting in lesser *shrinkage.* The leaner mix, however, makes vibration almost essential.
- The *creep* associated with this type of concrete is *low.*
- Due to the use of single-sized aggregate the segregation tendency is checked.
- The gap-grading is very sensitive to undesirable properties and mix obtained will be of reduced efficiency.

## 7.13. CONCRETING UNDER WATER

Concreting under water is a complicated problem. It may be required in case of deep foundations and marine works. The placing of concrete under water is usually confined to plain concrete (not reinforced) construction because of danger of loss of cement and segregation of concrete. The use of the following methods is restored to, under the circumstances when the provision of cofferdams, caissons or other such dewatering measures are difficult to practice.

**1. By tremie:**

- *Tremie* is the name given to a *steel pipe of 150 mm to 300 mm in diameter and sufficiently long to reach the bed of water shed keeping its one end above the water level.* A hopper is provided at its upper end for pouring concrete inside the pipe. The tremie is equipped with a *check valve at its lower end.*

- When the concrete is poured in the hopper of the pipe, it displaces the air and water from pipe and reaches the bed.
- The tremie, during the process of concreting, is gradually raised keeping the lower end of the pipe *submerged in the concrete mass* already poured below.

— A rich mix of concrete should always be used so that the possible loss of cement under water is compensated.

- It should be borne in mind that concrete thus laid under water should *never be compacted or consolidated.*

**2. By ship or bottom opening bucket:**

- In this method cubical hinged bottom bucket or cylindrical bucket equipped with a rolling gate or a ship is used for concreting under water.
- In order to minimize the concrete loss during the process of placing, it must be ensured that the bucket or ship is lowered to the bed of working place before the concrete is discharged.

**3. Placing in bags:**

The method consists in partially (usually about two-third) filling of cloth or gunny sacks with concrete, and tying them in such a way that they can readily be accommodated in a profile of the surface on which they are placed. The properly filled bags are lowered into water and placed carefully in a header-and-stretcher fashion as in brick masonry construction with the help of divers.

The *advantages* of this method is that, in several cases, *no framework in necessary and comparatively lean mixer may be used provided sufficient plasticity in retained.*

The *disadvantage* of this technique are:

*(i)* The accurate positioning of the bags in place can be accomplished only by the divers, the work is consequently *slow and laborious.*

*(ii)* Voids between the adjacent bags are difficult to fill, there is little bonding other than that achieved by the chemical interlock between bags.

*(iii)* The bags and labour necessary to fill and tie them are relatively expensive.

- The method is only suited for *placing the concrete in rather shallow water.*

**4. Prepacked concrete:**

The technique, also called "*grouted concrete*" consists of placing the coarse aggregate only in the forms and thoroughly compacting it to form a prepacked mass. This mass is then grouted with cement mortar of the required proportions. The aggregate should be wetted before being placed in position. The mortar that grouts the concrete displaces water and fill the voids.

- This technique is very much *suited for underwater construction and repair work of mass concrete structures, such as dams, spillway' etc.*

## 7.14. PLACING CONCRETE IN COLD WEATHER

*Any concreting operation done at a temperature below 5°C is termed* **cold weather concreting**. Most codes do not advocate concreting to be done at an atmospheric temperature below 5°C without special precautions.

Due to low temperature, the problems are mainly due to the slower development of concrete strength; the concrete in the plastic stage can be damaged if it is exposed to low temperatures which cause ice lenses to form and expansion to occur within the pore structure, and subsequent damage may occur due to *alternate freezing and thawing when the concrete has hardened.*

The effects of cold weather concreting may be summarised as follows:

*(i) Setting is delayed.* It retards the development of strength.

*(ii) Early freezing of concrete.* It may cause irrepairable loss to the properties of the concrete.

*(iii) Stresses due to temperature differential.* A large temperature differential may promote

cracking and exercise harmful effect on durability.

Hence, while concreting in cold weather, precautions need to be taken to keep the temperature of concrete above a *minimum value* both while placing the concrete in position in the form work, as well as while curing it.

*The following steps are taken for concreting in cold weather:*

- Heating up the aggregates or water or both with a view to keep the temperature of the resultant concrete less than 38°C. The most effective method to do so is to preheat the mixing water to the tune of 60°C (at the time of mixing), while the aggregate is heated to a temperature upto 15°C only.
- Another method to achieve adequate temperature for concrete in cold weather is to heat sand upto 40°C, if mixing water is taken at 60°C. Aggregates may be heated by embedding steam pipes; however, for small jobs, aggregate may be heated by steel drums embedded in heaped aggregates and filled with fire.
- All concrete surfaces, in cold weather, should be covered with dry blankets of straw, cement bags, tarpaulins etc. with a view to keep the heat in and to provide necessary protection from outside frost. In case of extreme frost conditions, it is essential to provide heated enclosures (by stones, steam, water pipes etc.).

*Note:* In *moderate or less severe conditions* concreting is done by *increasing the quantity of cement in the mix* or by using rapid hardening cement or high alumina cement.

In order to produce increased internal temperature during setting and hence to achieve early setting of concrete the use of *accelerators* like calcium chloride may also sometimes be made.

## 7.15. PLACING CONCRETE IN HOT WEATHER

*Any operation of concreting done at atmospheric temperature above 40°C or where the temperature of the concrete at the time of placement is expected to be beyond 40°C may be categorized as* **hot weather concreting.**

The climatic factors affecting concrete in hot weather or a *high ambient temperature* and *reduced relative humidity,* the effects of which may be more pronounced with the increase in wind velocity.

The *effects of hot weathers* may be summarised as follows:

*(i)* Accelerated setting.
*(ii)* Reduction in strength.
*(iii)* Increased tendency to cracking.
*(iv)* Rapid evaporation during curing.
*(v)* Difficulty in controlling the air content.

During normal concreting operation the temperature of concrete varies between 15°C to 25°C and full strength is obtained within 28 days. But when the concreting operation is carried out in extreme hot weather conditions, the rate of hydration gets accelerated and subsequently water in the concrete mix *dries out rapidly*. As a consequence, *stresses are formed* which cause shrinkage cracks leading to damaging of concrete.

In order to *minimise* the adverse effects of hot weather, the following precautions need be taken:

1. For making concrete cold water (sometimes crushed ice) may be used.
2. In order to lower the rate of hydration of cement, *retarding agents* may be added in concrete.
3. For lowering the temperature, the aggregates may be stockpiled in shade and sprinkled with cold water.
4. In order to minimise the losses due to evaporation and temperature rise, the mixed concrete should be transported and placed in position within the shortest possible time.
5. Just before the placement of the concrete, the framework and the reinforcement should be sprinkled over with cold water.
6. After placement of the concrete, it should be cured *continuously with sufficient water during first few hours* and thereafter the surface should be *kept moist* for the full duration

of curing.

## 7.16. DETERIORATION OF CONCRETE AND ITS PREVENTION

**Introduction :**

- The *durability of concrete* can be defined as *its resistance to the deteriorating influences of both external* (*e.g.* weathering, attack by natural or industrial liquids and gases etc.) *and internal agencies* (*e.g.* harmful alkali-aggregate reactions, volume changes due to noncompatible thermal and mechanical properties of aggregates and cement paste; presence of sulphides and chlorides from ingredients of concrete etc.).
- In the case of *reinforced concrete* the *ingress of moisture or air may lead to corrosion of steel, and cracking and spalling of concrete cover.*
- A durable concrete is:
  — *dense*
  — *workable* having *as low a premeability as possible* under the given situation.
- The recommendations for making durable concrete usually envisage limits for the following:
  — maximum water-cement ratio;
  — thickness of the cover;
  — type of cement;
  — The amounts of chlorides and sulphates in the concrete.
- *The gradual deterioration of concrete by chemically aggressive agents is called* ***concrete corrosion.*** Basically corrosion of concrete is a physico-chemical process and the extent of deterioration caused to it by the aggressive agents is dependent unon the properties of the constituents of concrete and the corrosive media.

**Types of deterioration:**

- The deterioration of concrete is caused not only by acids in the form of water solutions or acidic gases which form acids on dissolving in water, but by salt solutions and even by alkalies.
- A large member of other substances such as fertilizers, insecticides and certain organic compounds are harmful to concrete.
- The *principal forms of destruction* due to corrosion are as follows:
  (*i*) Decomposition of concrete
  (*ii*) Chemical reaction
  (*iii*) Crystallisation.

In addition to the principal forms of destruction cited above there are other specific influences, such as *destruction due to interaction between the hardened cement and the aggregate; attack by biological agents etc.*

**Prevention of concrete deterioration:**

- The *most effective means of reducing the deterioration of concrete is the making of a denser concrete having the least porosity.*
- A quantitative information regarding the effects of the range of parameters like water-cement ratio, cement content, curing conditions together with effects of cement admixtures and replacements on the corrosion of concrete helps determine the durability of concrete emperically.
- Durability of concrete can also be increased by *impregnating the pores of concrete with a suitable polymer.*
- The destructive processes in concrete being complex, a clear understanding of the destructive mechanism may help the selection of an appropriate technique to protect or improve the resistance of structural concrete to the aggressive agents.

## 7.17. ADMIXTURES

### Introduction

- An ***admixture*** *is a material other than water, aggregates and cement and is added to batch immediately before or during its mixing.*
- These are used to *improve or give special properties to concrete*. The use of admixture should offer an improvement not economically attainable by adjusting the proportions of cement and aggregates and *should not adversely affect any properties of the concrete.*
- They consist chiefly of those which accelerate and those which retard hydration or setting of cement, finely divided materials which improve workability, water-proofers, pigments, wetting, dispersing and air-entraining agents and pozzolanas.
- The admixture is generally added in a relatively *minute quantity*. The degree of control must be higher to ensure that overdosages are unlikely to occur. *Excess quantity of admixture may be detrimental to the properties of concrete.*

### Functions of admixtures:

Some of the important *purposes* for which the *admixtures* could be used are:

1. To accelerate the initial setting of concrete;
2. To retard the initial set;
3. To improve the strength of concrete;
4. To reduce the heat of evolution;
5. To improve the workability;
6. To inhibit the corrosion of concrete;
7. To increase the durability of concrete;
8. To increase the bond of concrete to the steel reinforcement;
9. To increase the resistance to chemical attack;
10. To control the alkali-aggregate expansion;
11. To reduce the segregation in grout mixtures;
12. To decrease the weight of concrete per cubic metre;
13. To produce coloured concrete or mortar for coloured surfaces;
14. To produce concrete of fungicidal, germicidal and insecticidal properties;
15. To increase the bond between old and new concrete surfaces.

### Classification:

Admixtures are classified as follows:

1. Air-entraining agents;
2. Retarders;
3. Accelerators;
4. Wate-proofers;
5. Pozzolanas;
6. Pigments, and
7. Workability agents.

### 1. Air-entraining agents:

The main objects of entraining air in concrete are :

(*i*) To increase the workability, and

(*ii*) To improve its resistance to weathering and especially to the action of frost.

There are *three* main ways of incorporating air or gas cells in concrete :

1. By the use of gas forming materials such as aluminium or zinc powder, hydrogen peroxide.
2. By the use of surface active agents which reduce surface tension. These are known as air-entraining agents and are subdivided into : (*i*) Natural wood resins and their soaps, *e.g.*, Vinsol resin; (*ii*) Animal or vegetable fats or oils, their fatty acids and their soaps; (*iii*) Alkali salts of sulfonated or sulphated organic compounds such as synthetic detergents, *e.g.*, Darex.

3. By the use of cement dispersing agents.

Air-entrainment should not be used unless site control is good, since the amount of air entrained may vary considerably with changes in sand grading, errors in proportioning, workability of the mix and temperatures.

**2. Retarders:**

The retarders, as the name suggests, delay the setting time of cement. Calcium sulphate in the form of gypsum is generally added during the manufacture of cement to retard the setting. Some other materials which are effectively used to retard the rate of hydration are : Ammonium chloride, ferrous and ferric chlorides, alkali bicarbonates, salts of carboxymethyl cellulose etc.

**3. Accelerators:**

The *agents that are added to the cement to make it set and acquire strength more rapidly are called* **accelerators.** The most commonly used accelerator is *calcium chloride. Sodium chloride* may also be used as an accelerating agent.

**4. Water-proofers:**

These may be obtained in *powder, paste or liquid form* and can consist of pore filling materials or water repelling materials. They can be chemically active or inactive.

Alkaline silicates notably silica of soda, aluminium and zinc sulphates and aluminium and calcium chlorides are the chief *pore filling materials*. These are all *chemically active* and hence may accelerate the setting time of the concrete thus making it more impervious at an early age.

Chalk, Fuller's earth and talc in a very finely ground form are *chemically inactive pore filling materials*. They improve workability and improve the density of concrete.

Materials in the *water repellent class* are soda and potash soaps.

**5. Pozzolanas:**

- A *pozzolana* is a silicious material which while itself having no cementitious properties will either be processed or unprocessed and in finely divided form, react in the presence of water with lime at normal temperatures to form compounds of low solubility having cementitious properties.
- It may be natural or artificial. The examples are :
  *Natural* : Clay sand shales (to be calcined to become active), diatomaceous earth and opaline cherts and shales and volcanic tuffs and pumicites etc.
  *Artificial* : Ground blast furnace slag and flyash.
- Pozzolanas can be used in combination with or for partial replacement of Portland cement.

Following *advantages* are obtained through the use of pozzolanas :

1. Heat of hydration is reduced.
2. Workability with less amount of water is improved.
3. Attack from salts and sulphates is reduced.
4. Calcium hydroxide leaching is prevented.

*Disadvantages :*

1. Strength development is slowed.
2. Drying shrinkage may be increased.
3. Durability may be impaired.

**6. Pigments:**

- Pigments are substances added to *produce coloured cements*. The chief pigments used in concrete are: *Brown, black, red, green, blue and yellow.*
- To obtain a good colouring effect, the pigment should be ground with cement in a ball mill.

**7. Workability agents:**

- Workability agents and water-reducing retarders are sometimes grouped together because many materials appear to perform all these functions.

- These are mineral powders ground at least as fine as the cement and usually much finer.
- They are generally classified with three chemical types : Lignosulfonates, organic acid, and carbohydrates.
- The excessive use of these agents may increase the shrinkage of concrete.

## 7.18. JOINTS IN CONCRETE

In cement concrete work the following two types of joints have to be provided :

1. Construction joints.
2. Functional joints (Expansion and contraction joints)

**1. Construction joints:**

- The concrete work, as far as possible, should be completed in one operation. If, however, the concreting has to be stopped before the entire portion is completed and the already laid concrete has started setting or hardening, it becomes necessary to provide construction joints.
- The spacing of construction joints should be well planned; the unplanned spacing may lead to the failure of the structure.
- The construction joints may be horizontal [Fig. 7.6] or vertical; for an inclined or curved member, the joint should be at right angle to the axis of the member.
- It is of paramount importance that position and arrangement of construction joints are divided well; before commencing concreting. The *concreting should be stopped* at positions/places in the following cases, as under :
  — *For beam* : At mid-span or over the centre of a column along a vertical plane at right angles to the length of beam.
  — *For slab* : At mid-span of the slab or directly over the centres of the beams making a vertical joint at right angles to the span.
  — *For column* : A few cm below its junction with the beam or to the bottom of hauncing, if any
- For ensuring a good bond between the old and the new work, following procedure is adopted:
  — The previously cast concrete surface is cleaned, roughened and thoroughly saturated with water and a coat of neat cement grout applied.

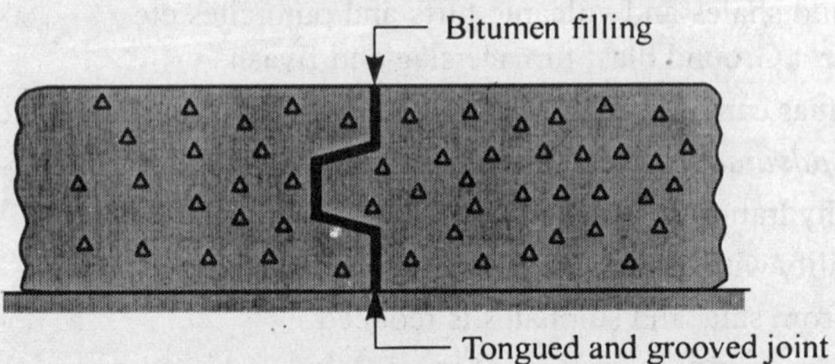


Fig. 7.6 : Longitudinal joint in a concrete road.

  — The new concrete is then deposited before the grout attains its initial set.

**2. Functional joints (Expansion and contraction joints):**

- Functional joints, also known as expansion and contraction joints, are provided to allow *variations in length due to temperature changes to take place and to preserve the appearance and original shape of the concrete structures*. These joints render the structure safe against *shrinkage, expansion, sliding and warping of concrete.*
- These joints are provided in all the concrete structures of length exceeding 12 m.
- These joints are made by forming continuous breaks in large continuous areas of structure at suitable distances apart. The spacing of joints may be 6 to 38 mm wide, depending upon the

type of structure, type and quality of the ingredients, and atmospheric conditions.

- The dowels or keys are provided in these joints to transfer the load.

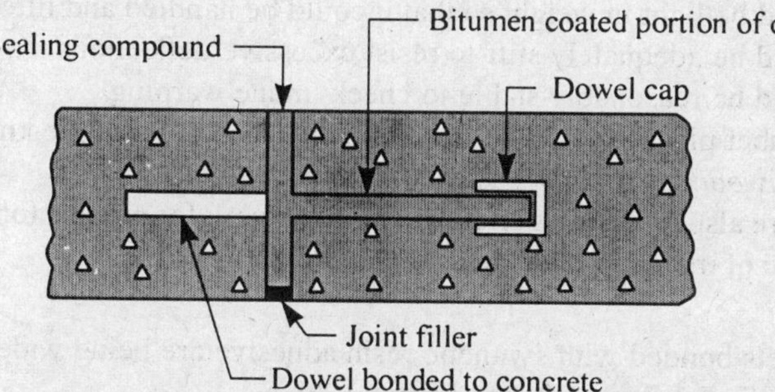


Fig. 7.7 : Transverse expansion and contraction joint.

- The joint filler used should be compressible, rigid, cellular and resilient. It should not become brittle in cold weather, and should be easy to handle.

The commonly used joint fillers are:

— Bitumen-treated felt;
— Cork bound with rubber or resin;
— Built-in strips of metal;
— Soft wood free from knots;
— Dehydrated cork etc.

## 7.19. FORMWORK

**Formwork** (also called *false work, centring* and *shuttering*) may be defined as the *timber or other materials which are used to form moulds into which concrete is poured at site.*

**Requirements of good formwork:**

Following are the *requirements of good formwork* :

1. It should be adequately strong.
2. It should be stiff.
3. It should be smooth.
4. It should provide a good shape at the concrete face.

The following points about formwork are worth noting :

- The joints made in the formwork should be simple. The joints of the formwork should be *preferably tongued and grooved so as to ensure water-tightness and minimise the possibility of the slurry oozing out.*
- The surface of formwork which is to come in contact with concrete is usually coated with crude oil or soft soap solution, to prevent the possibility of concrete getting stuck to it.
- All formwork must be so made that it can be *easily stripped after concrete sets.*
- The formwork should be so designed and constructed that they can be struck in the desired order.
- Every effort shall be made to minimise timber wastage during construction of formwork since the cost of formwork may amount to about 20 per cent of the cost on some of the jobs.
- The formwork should be carefully inspected by the Engineer-in-Chief and only after making sure that the formwork is properly made, the concreting process must be allowed.
- The formwork, during concreting, should be continuously in observation for bulges and other signs of failure.

**Timber formwork:**

A good timber formwork should have the following *qualities*:

(*i*) It should be easily available.

(*ii*) It should have reasonable price.
(*iii*) It should be easily workable (both by hand and machine).
(*iv*) It should be light in weight so that it could be handled and lifted easily.
(*v*) It should be adequately stiff to resist excessive deflections.
(*vi*) It should be reasonably stable to check undue warping.
(*viii*) The timber pieces should be free from the usual defects like knots, shaft grains etc.

- Generally *softwood varieties are used for formwork.*
- Hardwoods are also sometimes employed as formwork for moulds for precast work, wherein frequent reuse of moulds is expected.

**Plywood formwork:**

- Plywood sheets bonded with synthetic resin adhesive are being widely used as formwork.
- The size range from 1.2 m × 1.2 m to 3.0 m × 1.20 m

The plywood-formwork claims the following *advantages* :

(*i*) High impact resistance.
(*ii*) During use, it does not warp, swell or shrink.
(*iii*) Large level surfaces are available.
(*iv*) Formwork can be fixed rapidly.

**Steel formwork:**

- This type of formwork is widely used in situation where the work is large and the *moulds have to be used repeatedly.*
- In normal practice steel forms are fabricated from steel sheets strengthened with angle iron and are clipped together by keys and wedges.
- The steel forms are best suited for circular columns and flat slab construction.

The steel forms claim the following *advantages* (over timber formwork) :

1. They can be easily erected and do not need many adjustments as the units are standardised.
2. They give smooth surface, which needs very little finishing.
3. They are economical for large works, as there is no damage of any kind.
4. They are not liable to shrinkage or distortion.
5. There is no danger of formwork absorbing any moisture from concrete.
6. Unskilled labour can be used for their erection.
7. They can be used for a number of times.

## 7.20. GENERAL PRECAUTIONS IN CEMENT CONCRETE CONSTRUCTION

Following *precautions* should be taken during cement concrete construction:

- The cement used should be *fresh*.
- The aggregates should be *well graded* and *free* from dirt, clay, silt and other deleterious materials.
- Harmful chemicals and foreign materials should not be present in mixing water.
- A *rigid, impervious and watertight platform* should be used for preparing the concrete, to check any *loss* of cement or water.
- All the ingredients should be measured correctly and their complete mixing either by hand or by mixer should be ensured.
- During transporting or placing concrete, care should be taken to *avoid bleeding* and *segregation*.
- Before laying concrete, formwork or surface on which concrete is to be laid, should be *cleaned and moistened with water*.
- Before concreting operation, *shuttering* and *centering* should be *checked* and ensured thoroughly.
- Before laying the concrete, reinforcement and other components to be embedded in concrete should be thoroughly checked for their *position, size, concrete cover and correct quantity.*

- The quantity of concrete prepared each time should be used and finished within 30 minutes or initial setting time of cement.
- Laid concrete should be *compacted thoroughly* by manual tamping or mechanical vibration in layers of 150 to 450 mm. Care should be taken to avoid segregation and bleeding due to improper compaction.
- During intermittent or concreting in different spells suitable construction joints should be formed to obtain a proper bond between the previously laid concrete and freshly laid concrete by using special techniques.
- Suitable expansion and contraction joints should be provided in long concrete structures.
- Curing of concrete construction should be done for sufficient period as specified.
- Shuttering and centering should be removed after a predetermined period and in a predetermined sequence.

## 7.21. HOLLOW BLOCK PARTITIONS OF CONCRETE (OR CLAY, TERRA-COTTA)

There days the hollow blocks made from concrete (clay, terra-cota) are commonly used for the construction of partition walls. The *size of the blocks differ with the nature of material.* The thickness of this type of partition wall varies between 6 to 15 cm. These wall are constructed in similar manner as structural load bearing walls. Such walls have the following *advantages*:

*(i)* Light, rigid and strong.

*(ii)* Fire resistant.

*(iii)* Good sound insulating properties.

Fig. 7.6 : shows different types of hollow blocks for partition walls.

Fig. 7.7 : shows a view showing hollow block partition walls.

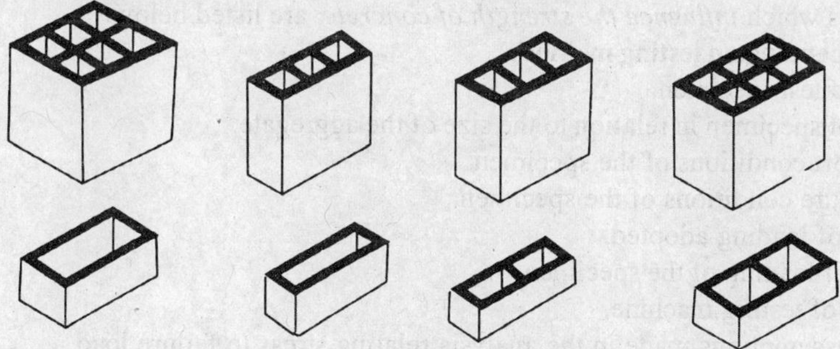

**Fig. 7.6 :** Different types of hollow blocks for partition walls.

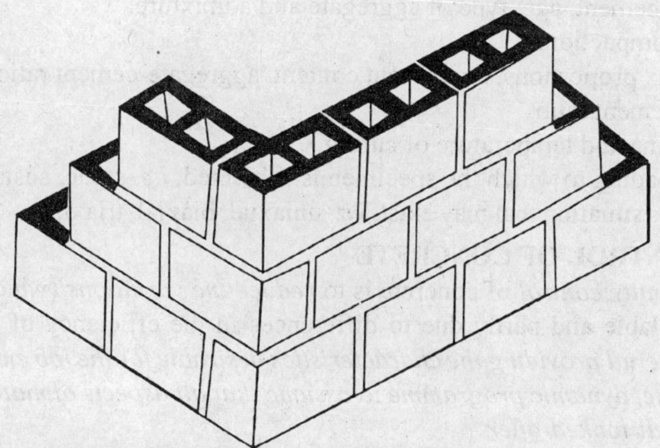

**Fig. 7.7 :** A view showing hollow block partition walls.

## 7.22. STRENGTHS OF CONCRETE

The various types of strengths of concrete are:

**1. Compressive strength.** Since the concrete is primarily meant to withstand compressive stresses, the compressive strength (of the various strengths of concrete) has received a large amount of attention.

The three types of compression test specimens used to determine the compressive strength are: Cubes (usually of 100 mm or 150 mm side), cylinders (150 mm diameter by 300 mm height) and *prisms* (100 mm × 100 mm × 500 mm size). The specimens are cast, cured and tested as per standards prescribed for such tests.

The *compressive strengths* given by different specimens for the *same concrete mix are different*. The cylinders and prisms of a ratio of height or length to the lateral dimensions of 2 may give a strength of about 75 to 85 per cent of the cube strength of normal-strength concrete.

**2. Flexural strength.** In order to estimate the load at which the concrete members may *crack* it is essential to determine the flexural strength. As it is difficult to determine the tensile strength of the concrete by conducting a direct tension test, it is computed by flexural testing. The flexural tensile strength at failure or the *modulus of rupture* is thus determined and used when necessary.

**3. Tensile strength.** Owing to the difficulties involved in conducting the direct tension test, a number of *indirect methods* have been developed to determine the tensile strength. In these tests, *in general a compressive force is applied to a concrete specimen in such a way that the specimen fails due to tensile stresses induced in the specimen.* The tensile stress at which failure occurs in the *tensile strength* of concrete.

**Splitting tests** are well known indirect tests used for determining the tensile strength of concrete, sometimes referred to as the *splitting tensile strength* of concrete.

**Factors Influencing the Strength of Concrete :**

The factors which *influence the strength of concrete* are listed below:

1. **Factors depending on testing method:**
   *(i)* Size of test specimen.
   *(ii)* Size of specimen in relation to the size of the aggregate.
   *(iii)* Support conditions of the specimen.
   *(iv)* Moisture conditions of the specimen.
   *(v)* Type of loading adopted.
   *(vi)* Rate of loading of the specimen.
   *(vii)* Type of testing machine.
   *(viii)* The assumptions made in the analysis relating stress to failure load.
2. **Factors independent of the type of test:**
   *(i)* The type of cement, age, type of aggregate and admixture.
   *(ii)* Degree of compaction.
   *(iii)* Concrete mix proportions, i.e. cement content, aggregate-cement ratio, amount of air-voids and water-cement ratio.
   *(iv)* Type of curing and temperature of curing.
   *(v)* Nature of loading to which the specimen is subjected, *i.e.* static, sustained, dynamic etc.
   *(vi)* Type of stress situation that may exist, *viz,* uniaxial, biaxial, triaxial.

## 7.23. QUALITY CONTROL OF CONCRETE

The *aim of quantity control* of concrete is to *reduce the variations* (which occur partly due to quality of plant available and partly due to differences in the efficiency of techniques used) and *produce uniform material providing the characteristics desirable for the job envisaged.* Thus **quality control** *is a corporate, dynamic programme to assume that all aspects of materials, equipment and workmanship are well looked after.*

Quality control is *conformity to the specification, no more no less.*

The general feeling that quality control means extra cost is *not correct*, the advantages due to quality control offset the extra cost. Some of the *advantages of quality control* are:

1. It results into lower costs of construction as higher stresses can be assumed.
2. It reduces the maintenance cost.
3. It helps in improved utilization of scarce resources and in extended utilisation of low-grade materials.
4. It helps to minimise failures.
5. Checks at every stage of the production of concrete and rectification of the faults at the right time expedites completion and reduces the delay.

- *The process of ensuring compliance to specifications which take into account the actual variability of concrete is termed* ***quality control****. The statistical quality control procedures are* used to ascertain the range of values that can be expected under the existing conditions. The basic parameter of statistical quality control are: *(i) Sampling: (ii) Distribution of results.*
- Quality assurance (QA) *is planned and systematic pattern of all actions necessary to provide adequate confidence that a product will confirm to established requirement.* It is a system of procedures for selecting the levels of quality required for a project or a portion there of to perform the functions intended and assuring that those levels are obtained. The quality assurance activity has to start right at the planning and design stage.

**Quality Audit (QA).** This is a system of tracking and documentation of *Quality Assurance and Quality Control programmes. Quality Audit* is the responsibility of the owner, and has to be performed at regular intervals through the tenure of the project. Quality Audit covers both the design as well as construction phases.

## HIGHLIGHTS

1. *Concrete* is a mixture of cement (or lime), sand, brick or stone ballast and water, which when placed in forms and allowed to cure, becomes hard like stone.
2. The various ingradients of plain cement concrete are: cement; sand (fine aggregate), coarse aggregate (broken stone, gravel, broken bricks etc), water.
3. The increase in volume of sand due to presence of surface moisture upto some extent is called *bulking of sand*. The extent of bulking depends on the percentage of moisture present in the sand and on its fineness.
4. Water-cement ratio is the ratio of water to cement and is expressed as the ratio of the weight or volume of water to the weight or volume of cement in concrete mixture.
5. The various methods of proportioning concrete are :
   (*i*) Arbitrary method (*ii*) Trial mixtures method
   (*iii*) Minimum voids method (*iv*) Fineness modulus method.
6. *Curing* is the process of keeping the set concrete damp for some days in order to enable the concrete gain more strength. For Portland cement, the curing period is about 7 to 14 days. The concrete is cured by the following methods:
   (*i*) By ponding; (*ii*) By using wet-sacks and sprinkling water frequently; (*iii*) By steam; (*iv*) By alternating curent; (*v*) By sealing curing compounds; and (*vi*) By chemicals.
7. *Reinforced cement concrete* (R.C.C.) is the cement concrete in which reinforcement is provided for taking tensile or excessive compressive and shear stresses. Usual proportions of ingredients in reinforced cement concrete are :
   1 part Portland cement : 1 to 2 parts clean sand : 2 to 4 parts shingle or crushed stone. R.C.C. is commonly used for the construction of the following : slabs, lintels, beams, columns and their footings, raft or mat foundations etc.
8. *Admixture* is a material other than water, aggregates and cement and is added to batch immediately before or during its mixing.

## OBJECTIVE TYPE QUESTIONS

**Choose the Correct Answer:**

1. Kennedy's method of proportioning concrete is based on the principle that
   (*a*) twenty percent of water is absorbed by cement
   (*b*) concrete having more excess cement paste, after filling the voids in the aggregate, in relation to the

total surface area of the particles is workable

(c) concrete having deficient cement paste in relation to the total surface area of the particles is easily workable

(d) water and cement cannot give workable paste at low water-cement ratios.

(e) none of the above.

2. The porosity of concrete depends largely upon
   (a) the quantity of water and mixing and compaction
   (b) the grading of aggregates
   (c) the cement content
   (d) the judicious use of sand
   (e) all of the above.
3. The durability of cement concrete is usually improved by
   (a) increasing the quantity of cement
   (b) decreasing the water-cement ratio
   (c) increasing the quantity of coarse sand
   (d) reducing the proportion of fine aggregates.
4. Shrinkage in concrete increases its
   (a) compressive strength
   (b) flexural strength
   (c) tensile strength
   (d) bond strength.
5. The shrinkage of concrete
   (a) increases with increase in the percentage of cement
   (b) depends on water-cement ratio only
   (c) is inversely proportional to water contents at the time of mixing
   (d) none of the above.
6. If a concrete gives more strength than the designed value
   (a) water cement ratio is changed
   (b) cement content is reduced
   (c) proportion of sand is changed
   (d) water percentage is increased
   (e) curing period is reduced.
7. The inert material of a cement concrete mix is
   (a) water
   (b) aggregate
   (c) cement
   (d) entire mix
   (e) none of the above.
8. After casting, normal cement concrete
   (a) shrinks on drying
   (b) shrinks when still wet
   (c) expands on drying
   (d) sometimes shrinks and sometimes expands depending on the ratio of various ingredients.
9. While designing concrete mix the point to be noted is
   (a) workability
   (b) water-cement ratio
   (c) aggregate-cement ratio
   (d) all of the above.
10. Good concrete in plastic state should have
    (a) segregation
    (b) shrinkage
    (c) bleeding
    (d) good workability
    (e) all of the above.
11. Presence of oils in water for concreting
    (a) reduces strength
    (b) improves strength
    (c) gives more slump
    (d) gives smooth surface.
12. The coefficient of thermal expansion of cement concrete is
    (a) $3 \times 10^{-4}$ per °C
    (b) $3 \times 10^{-5}$ per °C
    (c) $3 \times 10^{-6}$ per °C
    (d) $3 \times 10^{-8}$ per °C.
13. A surface vibrator for concrete compaction is preferred for all of the following *except*
    (a) raft foundations
    (b) road pavements
    (c) R.C.C. slabs
    (d) columns.
14. Formwork for concrete should be made of
    (a) soft wood
    (b) hard wood
    (c) shisham wood
    (d) teak wood.

15. The shrinkage of ordinary concrete is about
(*a*) 0.01 to 0.1 mm/metre (*b*) 0.3 to 0.6 mm/metre
(*c*) 6 to 8 mm/metre (*d*) 40 to 70 mm/metre.

16. The strength of concrete mainly depends on
(*a*) water-cement ratio (*b*) quality of fine aggregates
(*c*) quality of coarse aggregates (*d*) fineness of cement.

17. Sea water is not used for concreting work because of all the following reasons *except*
(*a*) it reduces setting time (*b*) it reduces strength
(*c*) it reduces durability of cement (*d*) it corrodes steel reinforcement.

18. In R.C.C. work it is often necessary to extend a bar by joining it with another. It may be done by
(*a*) welding the bar
(*b*) providing a lap between the two bars
(*c*) providing threads at the rod ends and putting a nut
(*d*) all of the above.

19. Construction joints in concrete work
(*a*) should not be provided at corners
(*b*) should be located at places of minimum shear force and bending moment
(*c*) in concrete walls it can be provided at floor level, sofit level of lintels and sill level of windows.
(*d*) all of the above.

20. In case of cement concrete work the smallest size of aggregate will be used in case of
(*a*) walls (*b*) pipes
(*c*) rams (*d*) heavy foundations.

21. The average value of reinforced cement concrete is taken as nearly
(*a*) 28 kN/m$^3$ (*b*) 26 kN/m$^3$
(*c*) 24 kN/m$^3$ (*d*) 20 kN/m$^3$.

22. Which of the following impurities in water for concrete is destructive?
(*a*) Alkalies (*b*) Calcium chloride
(*c*) Lead nitrate (*d*) Sugar.

23. For R.C.C. construction of water retaining structures, the grade of concrete mix used is generally
(*a*) $1:1:2$ (*b*) $1:1\frac{1}{2}:3$
(*c*) $1:1\frac{1}{2}:4$ (*d*) $1:2:4$.

24. According to ISI the full strength of concrete is achieved after
(*a*) 7 days (*b*) 14 days
(*c*) 21 days (*d*) 28 days.

25. The poisson's ratio for concrete lies between
(*a*) 0.15 to 0.25 (*b*) 0.25 to 0.30
(*c*) 0.30 to 0.35 (*d*) 0.35 to 0.40.

26. The capacity of a concrete mixture is expressed in terms of
(*a*) volume of mixture handled per batch
(*b*) weight of aggregate per batch
(*c*) total volume of mixture produced per hour
(*d*) total volume of mixture produced per day.

27. For concreting, aggregate should be in
(*a*) wet condition (*b*) bone dry condition
(*c*) surface dry condition (*d*) moist surface condition.

28. An aggregate which is generally not preferred for use in concrete is one having surface texture as
(*a*) rough (*b*) smooth
(*c*) glassy (*d*) granular.

29. Separation of water on the fresh concrete is called
(*a*) curing (*b*) hydration
(*c*) segregation (*d*) bleeding.

30. Bleeding of concrete is said to occur when
(*a*) excess water comes upon the surface of the concrete

(b) fine particles collect at some places in the structure
(c) coarse particles get separated from finer materials
(d) all of the above.

31. Difficulty in finishing of concrete surface is experienced when the slump of concrete exceeds
(a) 5 mm (b) 10 mm
(c) 25 mm (d) 40 mm.

32. Presence of algae in concrete
(a) reduces strength (b) reduces bond strength
(c) causes a large amount of air to be entrained (d) all of the above.

33. In cold weather concreting it is better to have cement in
(a) less quantity (b) average quantity
(c) required quantity (d) extra quantity.

34. ................. vibrator is generally used in concrete work.
(a) Form (b) Screen
(c) Surface (d) Needle.

35. For finding compressive strength of concrete, the minimum number of cubes required is
(a) 2 (b) 3
(c) 4 (d) 6.

36. The strength of concrete is decreased by
(a) direct impact (b) reversible stress
(c) vibration (d) all of the above.

37. The steel used in RCC work is
(a) stainless steel (b) mild steel
(c) high speed steel (d) chromium steel.

38. Disintegration of concrete can take place due to
(a) use of less durable aggregates
(b) use of unsound cement which undergoes volumetric changes, after the concrete has hardened
(c) excessive pores and voids (d) any of the above
(e) none of the above.

39. The behaviour of concrete under instantaneous loads is
(a) brittle (b) plastic
(c) elastic (d) anelastic
(e) visco-elastic.

40. In case concrete is to be transported by pumping, the slump should be between
(a) 3 cm and 5 cm (b) 5 cm and 7.5 cm
(c) 7.5 cm and 10 cm (d) none of the above.

41. Which of the following methods is used to remove oil stains from a concrete surface?
(a) A paste of oxalic acid, wheatflour and water is applied on the surface
(b) The surface is washed with soap
(c) The surface is rubbed with stone (d) None of the above.

42. The entrained air in concrete results in
(a) decreased workability (b) increased workability
(c) increased strength (d) decreased resistance to weathering.

43. The strength of concrete at 28 days as percentage of the strength at one year is
(a) 30 percent (b) 60 percent
(c) 80 percent (d) 98 percent.

44. The expansion joints are provided if the length of concrete structure exceeds
(a) 10 m (b) 30 m
(c) 45 m (d) 55 m.

45. The shrinkage of concrete can be reduced by reducing
(a) the temperature (b) the water content of mix
(c) the cement in the mix (d) the sand content of the mix.

46. Which of the following do not react with concrete?
(a) Sulphuric and hydrochloric acids (b) Alcohols

(*c*) Vegetable oils and sewage water (*d*) Any of the above.

**47.** Modulus of elasticity of concrete improves by

(*a*) higher water-cement ratio (*b*) age
(*c*) short curing periods (*d*) all of the above.

**48.** The strength of cement concrete depends on

(*a*) water-cement ratio (*b*) quality of aggregate
(*c*) quantity of aggregate (*d*) quality of cement.

**49.** Which of the following factors influences the workability of concrete?

(*a*) Particle interlock (*b*) Size of particles
(*c*) Water (*d*) Coarse and fine aggregate ratio.
(*e*) All of the above.

**50.** The lower water-cement ratio in concrete introduces

(*a*) greater water resistance
(*b*) greater density and smaller premeability
(*c*) smaller creep and shrinkage
(*d*) improved frost resistance
(*e*) all of the above.

**51.** Most of the methods of concrete mix design follow

(*a*) the assumption that concrete is always cured
(*b*) the principle that there is no air entrainment in the mix
(*c*) water-cement ratio law as a criterion of strength
(*d*) the principle that sand-cement ratio ultimately decides the strength.

**52.** The breaking up and remixing concrete that has begun to set is called

(*a*) time-setting (*b*) knocking
(*c*) strain-setting (*d*) resetting.

**53.** The light weight concrete may be prepared by

(*a*) formation of air voids in cement paste by the addition of substances causing foam
(*b*) using light aggregates
(*c*) formation of air voids in concrete by omitting sand
(*d*) any of the above or combinations there of.

**54.** As per ISI specifications concrete should be cured under a humidity of

(*a*) 20 percent (*b*) 50 percent
(*c*) 90 percent (*d*) 98 percent.

**55.** Water-proofing compounds in concrete

(*a*) fill the voids more thoroughly (*b*) drive out water from the mix quickly
(*c*) develop an impervious layer on concrete surface
(*d*) bring the sand particles closer together.

**56.** The slump test of concrete is used to measure its

(*a*) consistency (*b*) homogeneity
(*c*) tensile and compressive strength (*d*) impact value
(*e*) all of the above.

**57.** The workability of concrete can be improved by adding of the following except

(*a*) calcium chloride (*b*) copper sulphate
(*c*) flyash (*d*) hydrated lime.

**58.** Shrinkage in concrete can be reduced by

(*a*) using tight and non-absorbent formwork
(*b*) using low water- cement ratio
(*c*) using presaturated aggregates (*d*) all of the above.

**59.** Tensile strength of concrete is nearly .............. of its compressive strength.

(*a*) 2 percent (*b*) 5 percent
(*c*) 15 percent (*d*) 30 percent.

**60.** Light weight concrete is used for

(*a*) non-load bearing walls (*b*) heat resistance
(*c*) air-conditioned buildings (*d*) reducing thickness of structures.

**61.** Segregation results in

(*a*) sand streaks (*b*) porous layers

(c) honeycombing (d) surface scaling
(e) all of the above.

62. When using mechanical compactor, the slump of concrete should not exceed
(a) 20 mm (b) 50 mm
(c) 90 mm (d) 130 mm.

63. Generally the diameter of cylindrical test-specimen for concrete is kept as
(a) 5 cm (b) 10 cm
(c) 15 cm (d) 20 cm.

64. Workability of concrete is affected by which of the following?
(a) Cement content (b) Size of the aggregate
(c) Water content (d) Shape, angularity and surface texture of aggregate
(e) All of the above.

65. As per ISI specifications the temperature for curing is
(a) 5°C (b) 10°C
(c) 27°C (d) 42°C.

66. Surface vibrator is effective only when the thickness of concrete does not exceed
(a) 5 cm (b) 10 cm
(c) 15 cm (d) 20 cm.

67. Lime concrete is used in
(a) building underground (b) flooring at ground
(c) roof slab (d) columns.

68. The concrete is usually transported to the site by
(a) pumps (b) pans
(c) wheel borrows (d) any of the above.

69. Low temperature during laying of concrete
(a) decreases strength (b) increases strength
(c) depends on other factors (d) has no affect on strength.

70. The pigment generally used for imparting red colour to concrete is generally
(a) barium chromate (b) oxide of manganese
(c) red lead (d) red oxide of iron.

71. Concrete is considered as unsuitable for compaction by vibration if it is
(a) dry (b) semi-plastic
(c) plastic (d) none of the above.

72. If the slump of a concrete mix is 6 cm, its workability is considered
(a) low (b) medium
(c) high (d) very high.

73. Creep in concrete is undesirable particularly in
(a) prestressed structures (b) RCC columns
(c) continuous beams (d) all of the above.

74. Steam curing is used in
(a) columns only (b) long slabs and columns
(c) mass production of precast concrete
(d) all of the above.

75. Slump test for concrete is carried out to determine
(a) strength (b) durability
(c) water content (d) none of the above.

76. A badly mixed cement concrete results in
(a) segregation (b) bleeding
(c) cracks (d) honey-combing.

77. In order to have segregation in concrete its dropping height should not be less than
(a) 20 cm (b) 40 cm
(c) 100 cm (d) 200 cm.

78. Which defect is likely to develop in case concrete is not placed properly?
(a) Hair cracks (b) Separation of ingredients
(c) Segregation (d) All of the above.

**79.** The average value of creep coefficient for concrete is

(*a*) 1.0 (*b*) 2.0
(*c*) 3.0 (*d*) 4.0.

**80.** For the slabs and beams, the grade of concrete mix generally used is

(*a*) $1 : 1\frac{1}{2} : 4$ (*b*) 1 : 2 : 4
(*c*) 1 : 2 : 6 (*d*) 1 : 3 : 6.

## ANSWERS

| | | | | | | | | |
|---|---|---|---|---|---|---|---|---|
| **1.** (*d*) | **2.** (*e*) | **3.** (*a*) | **4.** (*d*) | **5.** (*a*) | **6.** (*a*) | **7.** (*b*) | **8.** (*a*) | **9.** (*d*) |
| **10.** (*d*) | **11.** (*a*) | **12.** (*c*) | **13.** (*d*) | **14.** (*a*) | **15.** (*b*) | **16.** (*a*) | **17.** (*a*) | **18.** (*d*) |
| **19.** (*d*) | **20.** (*b*) | **21.** (*c*) | **22.** (*c*) | **23.** (*b*) | **24.** (*d*) | **25.** (*a*) | **26.** (*a*) | **27.** (*c*) |
| **28.** (*c*) | **29.** (*d*) | **30.** (*a*) | **31.** (*d*) | **32.** (*d*) | **33.** (*d*) | **34.** (*d*) | **35.** (*b*) | **36.** (*d*) |
| **37.** (*b*) | **38.** (*d*) | **39.** (*c*) | **40.** (*b*) | **41.** (*a*) | **42.** (*b*) | **43.** (*c*) | **44.** (*c*) | **45.** (*b*) |
| **46.** (*b*) | **47.** (*b*) | **48.** (*a*) | **49.** (*e*) | **50.** (*e*) | **51.** (*c*) | **52.** (*b*) | **53.** (*d*) | **54.** (*c*) |
| **55.** (*a*) | **56.** (*a*) | **57.** (*b*) | **58.** (*d*) | **59.** (*b*) | **60.** (*a*) | **61.** (*e*) | **62.** (*b*) | **63.** (*c*) |
| **64.** (*e*) | **65.** (*c*) | **66.** (*d*) | **67.** (*b*) | **68.** (*d*) | **69.** (*a*) | **70.** (*d*) | **71.** (*c*) | **72.** (*b*) |
| **73.** (*a*) | **74.** (*c*) | **75.** (*d*) | **76.** (*d*) | **77.** (*c*) | **78.** (*c*) | **79.** (*c*) | **80.** (*b*). | |

## THEORETICAL QUESTIONS

1. What is concrete ?
2. What are the characteristics of good concrete?
3. How are concretes classified?
4. Explain briefly various ingredients of concrete.
5. Write a short note on "bulking of sand ".
6. What is water-cement ratio? How does it affect the strength of concrete ?
7. Enumerate methods of proportioning concrete.
8. Explain "fineness modulus method" of proportioning concrete.
9. What do you mean by the term "workability"?
10. What is a slump test? How is it performed?
11. Give the recommended values of slumps for the following works :
    (*i*) Vibrated concrete; (*ii*) Mass concrete;
    (*iii*) Road work: (*iv*) Ordinary beams and slabs;
    (*v*) Columns, retaining walls, and their vertical section.
12. Enumerate and discuss briefly the various steps required for the preparation of plain cement concrete
13. Discuss briefly the following :
    (*i*) Compaction of concrete, and (*ii*) Curing of concrete.
14. Give the effects of improper curing of concrete.
15. Discuss briefly any three of the following properties of concrete :
    (*i*) Workability; (*ii*) Segregation;
    (*iii*) Bleeding; (*iv*) Strength;
    (*v*) Durability and (*vi*) Impermeability.
16. Write a short note on water-proof concrete.
17. What is precast concrete? State its advantage and disadvantages.
18. What is reinforced cement concrete (R.C.C.)?
19. Give the properties, advantages and disadvantages of reinforced cement concrete.
20. What are the general requirements of R.C.C. design?
21. List the uses of reinforced cement concrete.
22. What is an admixture? How are admixtures classified?
23. Give the general precautions in cement concrete construction.
24. Explain briefly any two of the following *special concretes*.
    (*i*) Fibre-reinforced concretes (*ii*) Ferrocement
    (*iii*) Ultra high strength concrete (*iv*) Gap-graded concrete.
25. Write short notes on any two of the following:
    (*i*) Placing concrete under water
    (*ii*) Placing concrete in cold weather
    (*iii*) Deterioration of concrete and its prevention.

# 8

# Timber and Wood-based Products

*8.1. Introduction. 8.2. Characteristics of good timber. 8.3. Advantages and disadvantages of timber.* 8.4. Uses of timber. 8.5. Classification of trees. 8.6. Structure and growth of tree. 8.7. Felling of trees. 8.8. Defects in timber. 8.9. Seasoning of *timber-introduction-objects of seasoning–advantages of seasoning-methods of seasoning of timber. 8.10. Decay of timber. 8.11. Preservation of timber. 8.12. Conversion of timber. 8.13. Testing of timber. 8.14. Timber suitable for various uses. 8.15. Timber trees of India. 8.16. Wood-based products–veneers-plywood-laminboard–block board–batten board-fibre board–hard board. Highlights–Objective Type Questions–Theoretical Questions.*

## 8.1. INTRODUCTION

**Timber** is the *wood suitable for building or engineering purposes* and it is applied to trees measuring not less than 0.6 m in girth.

Important *technical terms* relating timber are given below :

- **Standing timber** : It is the timber available in a living tree.
- **Green timber :** It is the freshly felled tree which has not lost much of its moisture or the timber which still contains free water in its cells.
- **Rough timber :** It is the timber obtained after felling a tree.
- **Converted timber** : It is the timber which has been sawn into various market sizes such as beams, battens, planks etc.
- **Dressed timber :** It is the timber which been sawn, placed and worked to the exact required condition.
- **Structural timber :** It is the timber used in framing and load bearing structures.
- **Clear timber :** It is the timber free from defects and blemishes (stains).

## 8.2. CHARACTERISTICS OF GOOD TIMBER

Following are the *characteristics* of good timber:

1. It should have a *uniform colour.*
2. A freshly cut surface should give a *sweet smell.*
3. It should have *regular annular rings.*
4. It should be *sonorous when struck.*
5. It should have *straight and close fibres.*
6. It should be *heavy in weight.*

7. It should be *free from shakes, flaws, dead knots or blemishes of any kind*
8. There should be *firm adhesion of fibres and compact medullary rays.*
9. The cellular tissue of the medullary rays should be *hard and compact.*
10. When *planed*, its surface should present a *firm bright appearance with a silky lustre.*
11. Its *fibrous tissues should adhere firmly together and should not clog the teeth of the saw and the freshly-cut surface should not show wooliness.*
12. A good timber should be *durable.* It should be capable of resisting the actions of fungi insects, chemicals, physical agencies and mechanical agencies.
13. A good timber should possess the property of *elasticity*. The timber is elastic when it regains its original shape and size when the load is removed. This property of timber would be essential when it is to be used for bows, sports goods, carriage shafts etc.
14. It should be *fire-resistant.* Wood having dense texture offers great resistance to fire. It should not contain resins and other inflammable oils which accelerate the action of fire.
15. A good timber should be *hard* (*i.e.,* it should offer resistance when it is being penetrated by another body). The chemicals present in heart wood and density of wood impart hardness to the timber.
16. A good timber should be *strong* for working as structural member such as joint, beam, rafter, etc. It should be capable of taking loads slowly or suddenly.

    **Note :** The heavier and dark coloured timber is usually strong.

## 8.3. ADVANTAGES AND DISADVANTAGES OF TIMBER

**Advantages :** The advantages of timber are :

1. It is relatively easy to handle and can be planed, sawn and jointed with simple carpenter's tools.
2. It is easily available and can be quickly transported by simple means.
3. It is light in weight yet strong.
4. The floor joints in an average dwelling weighs less than rolled-steel beams of equal strength.
5. The individual units of wooden dwelling are light enough so that they may be assembled by a small crew of men without special machinery.
6. Boards may be cut rapidly by a saw and fastened firmly together with nails.
7. It is a good insulator of heat and sound.
8. When properly protected timber structures may give good service for hundred of years.
9. It stands shocks and bumps, a good deal better than iron and concrete.
10. On account of its light weight, timber is generally preferred for building works in earthquake regions.
11. Repairs, additions and alterations to timber construction are easy.
12. It is considered to be an ideal material of construction in sea water or marine works as it can resist corrosion.

**Disadvantages :** The disadvantages of timber are few, but serious. There are :

1. The *greatest disadvantage* is its ready *combustibility,* which can be *diminished but not eliminated even by expensive treatment.*

2. Frame buildings built closely together present a serious conflagration hazard.

3. Timber is destroyed by decay induced by fungi, and by insects that feed upon the timber under favourable considerations. Decay may, however, be prevented by various methods that produce an environment unfavourable for the growth of the causative organisms.

4. Timber swells and undergoes shrinkage with changing atmospheric humidity.

## 8.4. USES OF TIMBER

Timber is put to different *uses*, which are :

1. It is generally used in the form of piles, posts, beams, lintels, door-window frames and leaves, roof members, *e.g.*, rafters, purlins, trusses etc.

2. It is employed for flooring, ceiling, panelling and construction of partition walls.

3. It is used for formwork for concrete, for the timbering of trenches, centering for arch work, scaffolding, transmission poles and fencing.

4. It is used in wagon and coach building, marine installations and bridges.

5. It is also used in making furniture, agricultural implements, sports goods, musical instruments, well curbs, motor bodies carts and carriages, railway sleepers, packing cases etc.

## 8.5. CLASSIFICATION OF TREES

According to their *manner of growth*, the tree may be divided into two main classes :

1. Exogenous tree 2. Endogenous trees.

### 1. Exogenous trees :

- These trees *grow outwards* and *increase in bulk by the formation of successive annular rings on the outside under the bark.*
- In these trees, each annual ring represents a layer of wood, deposited every year.
- The timber that is used for building and engineering purposes belongs to this variety.

*Examples : Deodar, chir, sal, kail, shishum, teak, etc.*

Exogenous trees may be further divided into the following two types:

(*i*) Conifers or evergreen trees (*ii*) Deciduous or broad-leaf trees.

— **Conifers trees** : They remain evergreen and bear fruits in cone form with needle-pointed leaves (and hence the name). These trees yield **soft woods** which are generally light coloured, resinous, light in weight and weak. They show *distinct annual rings.*

— **Deciduous trees** : They shed their leaves in autumn and put on new leaves in the spring season. These trees yield **hardwoods** which are usually close-grained, strong, heavy, dark coloured, durable and non-resinous. They *do not show distinct annual rings.* The timber used for engineering purposes is *mostly derived from deciduous trees.*

### 2. Endogenous trees :

- These trees *grow inwards or endwards*
- The stems of these trees are too flexible and thus they are not much suitable for engineering works.
- They are found generally in tropical countries and have their own limited applications for temporary construction.

*Examples* : Canes, Bamboos, Palms etc.

Table 8.1 gives the comparison between soft woods and hard woods.

**Table 8.1. Comparison between Soft Woods and Hard Woods**

| S.No. | Aspects | Soft woods | Hard woods |
|---|---|---|---|
| 1. | *Trees from which obtained* | Conifers trees | Deciduous trees |
| 2. | *Annual rings* | Distinct | Indistinct |
| 3. | *Medullary rays* | Indistinct | Distinct |
| 4. | *Weight* | Light | Heavy |
| 5. | *Colour* | Light | Dark |
| 6. | *Structure* | Resinous and splits easily | Non-resinous and close-grained |

| 7. | *Strength* | Strong for direct pull and weak for resisting thrust or shear. | Equally strong for resisting tension, compression and shear. |
|---|---|---|---|
| 8. | *Fire resistance* | Poor | More |

## 8.6. STRUCTURE AND GROWTH OF TREE

### 8.6.1. Structure of Tree

Basically, a tree consists of the following *three* parts :

(*i*) Trunk; (*ii*) Crown; (*iii*) Roots

— The trunk *supports the crown* and supplies water and nutrients from the roots to the leaves through branches and from the leaves back to the roots.

— The *roots* are meant to implant the trees in the soil, to absorb moisture and the mineral substances it contains and to supply them to the trunk.

The structure of an exogenous tree is shown in Fig. 8.1. Brief description of each part is given below :

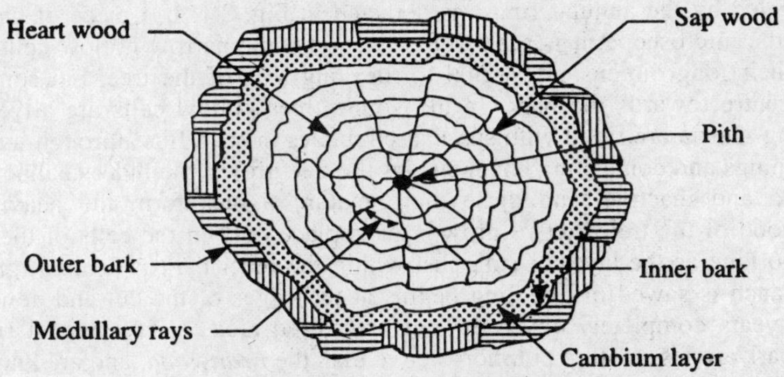


Fig. 8.1 : Cross-section of an exogenous tree.

1. **Pith :** The *pith* or *medula* is the *innermost central portion or core of the tree.* It is larger in young trees than that in matured trees. It consists entirely of cellular tissues and it nourishes the plant in its young age. It is usually about 12.5 mm in diameter but sometimes barely visible. This is the fibrous tissue representing the original sapling.

2. **Heart wood:** It consists of the *inner annual rings round the pith.* It is *compact, strong, durable and darker in colour.* This portion in reality is dead and does not take an active part in the life of a tree, except to give it rigidity. The heartwood increases in size as the tree ages. This part provides the best timber for engineering works or structures.

3. **Sapwood** : The *outer annual rings between heartwood and cambium layer is known as the "**sapwood**"* also called *alburnum*. This is usually light in colour and weight. It *indicates recent growth* and contains *sap.* It is liable to decay or attack by insects. Being the living part of a standing tree, it carries sap from the roots to leaves in a standing timber.

Sapwood being an *active layer is moist and rich in food material and is less durable.*

4. **Cambium layer** : The *thin layer below bark, not converted to sap wood as yet, is called the* **cambium layer**. This is mainly responsible for the circumferential growth of the tree. If the bark is removed for any reason, the cambium layer gets exposed and the cells cease to be active resulting in the death of tree.

5. **Inner bark or bast :** It is the *inner skin or layer covering the cambium layer.* It *protects* the cambium layer from the injury.

6. **Outer bark or cortex** : It *consists of wood fibre and is the outermost cover* or *skin of stem*. It often shows fissures and cracks. It prevents the tree from external injuries and weathering action.

7. **Annual rings** : The *annual rings* consist of *cellular tissue and woody fibre arranged in distinct concentric circles or rings round the pith generally one in every year.* These are formed due to the deposition of sap below bark. In temperate climate the outer part of the annual ring is darker in colour and denser than the inner part but it is homogeneous in tropical climate. So the *number of annual rings indicates the age of a tree in a tropical climate.* So the *number of annual rings indicates the age of a tree in a tropical climate.*

8. **Medullary rays :** These are the *thin radial fibres extending from pith to combium layer.* The function of medullary rays is to store and conduct food material from the sap to the vessels or cells in the interior of a tree radially. Only few of them are visible to the naked eye and they hold the annual rings together. These rays are sometimes broken and in some varieties of tree, they are not very important.

### 8.6.2. Growth of Tree

- A tree becomes taller only through growth at the tips of the branches. It becomes thicker through formation of cells just within the inner bark. A young tree grows taller and thicker every year, with a period of rapid growth each spring and one of slower growth in the late summer and early fall. After a tree has been cut down, the method of growth is revealed by the annular rings, as depicted in Fig. 8.1. If a piece of wood is examined under the microscope, it is seen to be made of long, narrow hollow cells, most of which have their long dimension parallel to the long axis of the tree, but some of which run from centre towards the bark. In the young trees hollow cells are alive and during the growing season are filled with sap that circulates in each fibre through an elaborate series of openings and connecting channels. As the tree grows, the older channels become partly clogged and inactive, and these more mature tissues form the heavier and stronger heartwood of the tree. Active growth takes place only in the cells of the *inner bark* (Fig 8.1). So long as the layer of cells just within the outer bark is healthy, growth continues. If a branch is sawed off, healing begins at the edges of the cut and new growth may in a few years completely cover the cut or scarred area. A few annual rings next to the inner bark are less mature but more active than the *heartwood,* and are known as *sapwood.* There is usually no great difference in strength between heartwood and sapwood, but *heartwood is more resistant to decay.*
- During the cold winter months the cells of the tree do not contain any liquid sap, but living tissues of the cell walls do contain moisture. In the early spring the sap begins to rise in the tree and fill the cells. If the gash is made through the bark at this season, some of the sap will run out. It is during the vigorous period of spring growth that the thin-walled spring cells are rapidly formed. During the drier summer months when the leaves are evaporating very large amounts of moisture and the ground is becoming drier, the amount of sap in the tree decreases. A young tree cut in the spring may have two-thirds of its weight made up of water, and only one-third of wood substance. The same tree cut in the late summer may be composed of only one-third of moisture and two-thirds of wood substances. There is almost always more moisture in a tree when cut than will be retained when the wood dries in the air.

## 8.7. FELLING OF TREES

- The process of knocking down or cutting down or causing the trees to fall to the ground is known as *felling of trees.*
- The tree should be felled when they have just matured or when they are very near to maturity. If they are felled before *they have attained maturity,* the *sap wood would be in excess and timber obtained from such tree wood not be durable and it will be in less quantity.* On the other hand, after the tree has reached maturity and allowed to stand, *it is the valuable heartwood which first begin to decay and the timber is liable to develop shakes and cracks on drying of soap.* Good timber trees generally arrive at maturity between 50 and 100 years.

- The *best time for felling trees is mid-summer or mid-winter* when the *sap is minimum in quantity as it is then at rest*. The timber is liable to decay if felled in spring or autumn when the trees contain maximum quantity of sap due to its vigorous movement.
- **Method of felling a tree :** The tree should be felled (by experienced persons) as follows:
  — First of all a cut is made at the bottom of the trunk (of the tree to be felled), on the side opposite to which it is desired to be felled, with the help of an axe and a saw. The cut is taken a little beyond its centre of gravity and a wedge is inserted into it.
  — Then on the opposite side another cut, parallel and a few centimetres above the first is made as shown in Fig. 8.2.
  — Finally the tree is tied at the top, by means of wire ropes having an angle of 60° between them.
  — The tree is then swung in the desired direction and pulled down slowly by applying force on ropes.

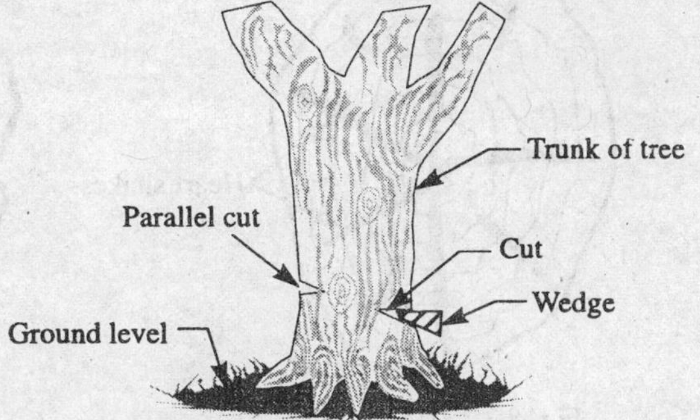


Fig. 8.2 : Felling of tree.

The following points should be taken into consideration in connection with felling of trees:

1. Before felling the trees, they should be inspected and serially numbered.
2. The direction of felling should be determined before hand before starting the cut.
3. The butt end should be rounded to facilitate the removal of fitness and ensure direction of felling. In case of hardwood good quality timber trees, this is done two to three years earlier in which case the process is known as "**Girdling**". This prevents further forming of the sap and evaporation of some of the moisture content of the tree.

## 8.8. DEFECTS IN TIMBERS

In all kinds of timber several *natural defects* occur. There are caused by the *nature of the soil upon which the tree grew and also by the changes in nature to which it was subjected while growing*. As far as possible these defects should be avoided or removed during conversion for use.

Following are the *most common defects in timber* :

| | |
|---|---|
| 1. Heartshakes | 2. Starshakes |
| 2. Cupshakes or ringshakes | 4. Radial shakes |
| 5. Rind-galls | 6. Upsets or rupture |
| 7. Twisted fibres or wandering hearts | 8. Burns or excrescence |
| 9. Wind cracks | 10. Knots |
| 11. Honeycombing | 12. End-splits |
| 13. Deadwood | 14. Druxiness |
| 15. Foxiness. | |

### 1. Heart shakes :

Refer Fig. 8.3. These are the splits or cracks widest at the centre and diminishing towards the outside circumference. This defect usually *occurs in overmatured trees* and is *probably caused due to the shrinkage of the heartwood.* Heartwood shake which follows a straight course along the timber is not so serious as compared with that which follows a twisted course.

**2. Star shakes :**

Refer Fig. 8.4. These are *radial splits or cracks widest at the circumference and get diminishing towards the centre of the tree.* Star shakes are *more serious* as they tend separate the log into a number of pieces when it is sawn. These may arise mostly from *severe frost* and *fierce heat of sun.* They are *mostly confined to sapwood.*

Fig. 8.3 : Heart shakes. Fig. 8.4 : Star shakes.

**3. Cup shakes or ring shakes :**

Refer Fig. 8.5. These are formed by the rupture of the tissues in a circular direction across the cross-section of a log, usually along annular rings. When the rupture extends only a part around, it is called a *cup shake* and when the whole way round or almost so a *ring shake.*

**4. Radial shakes :**

Refer Fig. 8.6. These are similar to star shakes but they are fine, *irregular* and *numerous.* They usually occur when tree is exposed to sun for seasoning after being felled down. They run for a short distance from bark towards the centre, then follow direction of annual ring and ultimately run towards pith.

**5. Rind galls :**

Refer Fig. 8.7. The *rind* means bark and *gall* indicates abnormal growth. Hence peculiar curved swellings formed on the body of a tree are known as the *rind galls.* These are caused generally by the growth of layers over the wounds left after branches have been improperly cut off or removed. They are rarely found in a tree and the timber in this part is *very weak and not durable.*

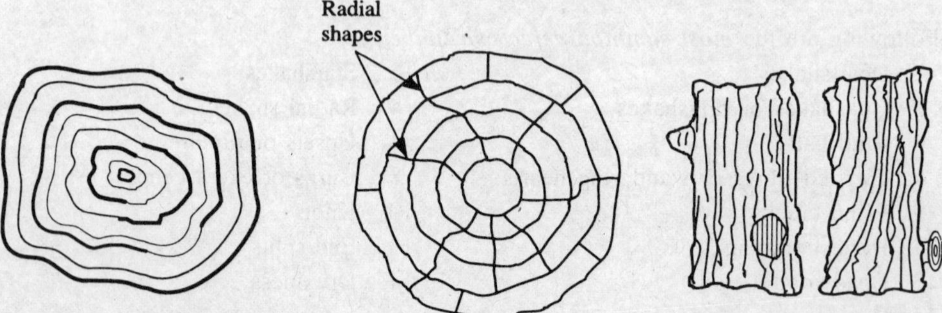


Fig. 8.5 : Cup or ring shakes. Fig. 8.6 : Radial shakes. Fig. 8.7: Rind galls.

**6. Upsets or rupture :**

Refer Fig. 8.8. This defect is due to crushing of fibres which are deformed as a result of the injury thus caused. These are due to unskilful felling and also the result of violent wind.

**7. Twisted fibres or wandering hearts :**

Refer Fig. 8.9. This defect is developed in a tree owing to its peculiar position where the prevailing winds turn the tree constantly in one direction. The fibres of such a tree are twisted longitudinally and if planks or any sections are cut from it, many of the fibres would be cut through and the strength of the pieces is reduced. The *timber with twisted fibres is unsuitable for sawing*. It can however be used *for posts and poles in an unsawn condition.*

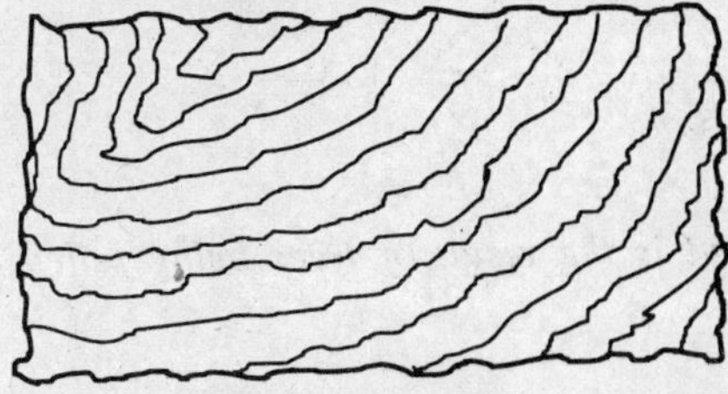

Fig. 8.8 : Upset.

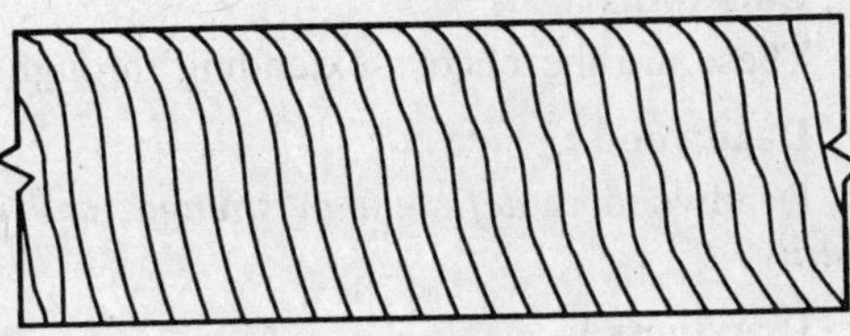

Fig. 8.9 : Twisted fibres.

**8. Burrs or excrescences :**

These are either due to unsuccessful attempts at the formation of branches or due to injuries inflicted when the tree was young, so that undeveloped, dormant buds grew and produced masses of contorted tissue. Technically, they are defects but, in practice, they produce valuable figure-wood for veneers.

**9. Wind cracks :**

When wood is exposed to atmospheric agencies, its exterior surface shrinks. Such a shrinkage results into cracks (Fig. 8.10), known as *wind cracks.*

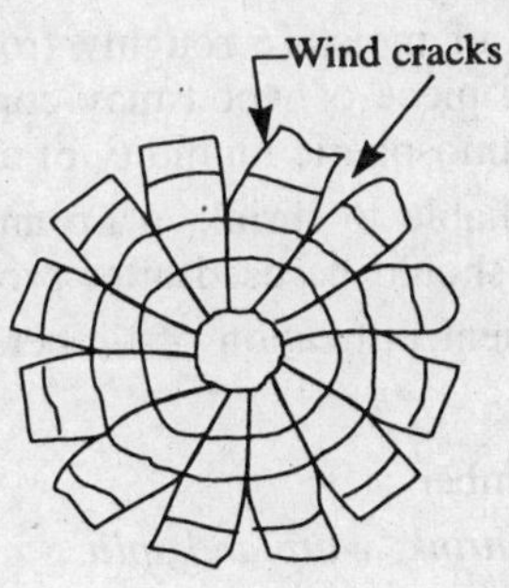


Fig. 8.10 : Wind cracks.

Fig. 8.11 : Knots.

**10. Knots :** Refer Fig. 8.11.

- *Knots* are the *roots of small branches of the tree.* Knots disturb the homogeneity of wood texture. The fibres of wood get twisted or curled in the vicinity of knots and this effects the strength of the wood. The position, size of the knot, and the degree of grain distortion around it, determine the magnitude of the reduction in strength that the knot can cause. *Knots on or near the edges of a beam at places of greatest stress have the greatest weakening effect.*
- When the formation of knot is free from decay and other defects and is firmly intact with the surrounding wood, it is called a *live knot* or *sound knot.* A knot which is not held firmly in place is called a *dead knot or loose knot.*

There are various types of knots, most common being ;

*Round knot, oval knot, spike knot, grouped knot* and *pencil knot.*

- Timber containing large dead (loose) knots or many smaller ones should be rejected as they are poorer in appearance and strength.

**11. Honeycombing :**

This defect is *due to separation of the tissues in the interior of the timber due to drying stresses*, caused by incorrect kiln seasoning as the interior wood dries earlier than the external surface.

**12. End-splits :**

These are the checks extending through two faces.

**13. Deadwood :**

Deadwood is *deficient in strength and weight and is the result of trees being felled after maturity.*

**14. Druxiness :**

This defect is indicated by white spots which are due to access of *fungi*, probably through a broken branch.

**15. Foxiness**

This *defect* is indicated by reddish or yellowish brown stains, caused by over maturity. It is also caused by *bad ventilation during storage.*

## 8.9. SEASONING OF TIMBER

### 8.9.1. Introduction

- ***Seasoning of timber** is the process of drying timber or removing moisture or sap, present in a freshly felled timber, under more or less controlled conditions.*
- Freshly felled timber contains a large quantity of moisture roughly from 100 to 200%, *based on dry weight of wood.* A well seasoned piece of wood may contain about 10 to 12% moisture and will be in equilibrium with atmospheric humidity of a particular place.
- If the timber is used without seasoning, it is liable to shrink, warp and crack, if used, and may even rot and decay. Thus all timbers should be used after proper seasoning.
- Seasoning should be the first step for the efficient utilization of timber.

### 8.9.2. Objects of Seasoning

Following are the *main objects* of seasoning of timber :

1. To *check/minimise the tendency of timber to shrink, warp and split.*
2. To *increase* strength, durability and electrical resisting power of the timber.
3. To maintain the shape and size of the components of the timber articles which are expected to remain unchanged in form.
4. To *make* timber *safe* from attack of fungi and insects.
5. To *reduce* the weight for transport purposes, handling, and thereby *reduction* in cost.
6. To make timber *fit* for receiving treatment of paints, preservatives, varnishes etc.
7. To make timber easily *workable* and to facilitate operations during conversion.
8. To make timber suitable for *gluing* (*i.e.* effectively joining two members of timber with the aid of glue).

### 8.9.3. Advantages of Seasoning

Following are the *advantages* of seasoning timber :

1. Seasoning makes the timber less liable to attack by insects and fungus.
2. It decreases the weight of timber and makes it lighter.
3. It enables timber to be easily painted, polished and preserved.
4. It avoids defects appearing in timber during its use.
5. It improves the strength properties of the timber and makes it more stable.
6. It improves working qualities of timber.

7. It provides dimensional stability and retcntion of shape to the timber used in carpentry or joinery.
8. Seasoning increases the power in timber to held metal fasteners.

### 8.9.4. Methods of Seasoning of Timber

Seasoning of timber may be carried out in two ways :

1. **Natural seasoning :** There are two methods of natural seasoning.
   (*i*) Air drying / seasoning (*ii*) Water seasoning.
2. **Artificial seasoning :** Artificial seasoning methods consist of;
   (*i*) Kiln seasoning (*ii*) Chemical seasoning (*iii*) Electrical seasoning.

### 8.9.4.1. Natural seasoning

(*i*) **Air drying/seasoning :** Refer Fig. 8.12. In this method of seasoning timber the sawn timber is stacked in a dry place about 30 cm above floor level with longitudinal and crosspieces arranged one upon another, leaving a space of a few cms between, for *free circulation of air.* To keep wood clear of the ground, a suitable foundation is provided. To *protect the stacks of wood from direct sunlight either a pitched roof covering is used or stacks are kept in shade.*

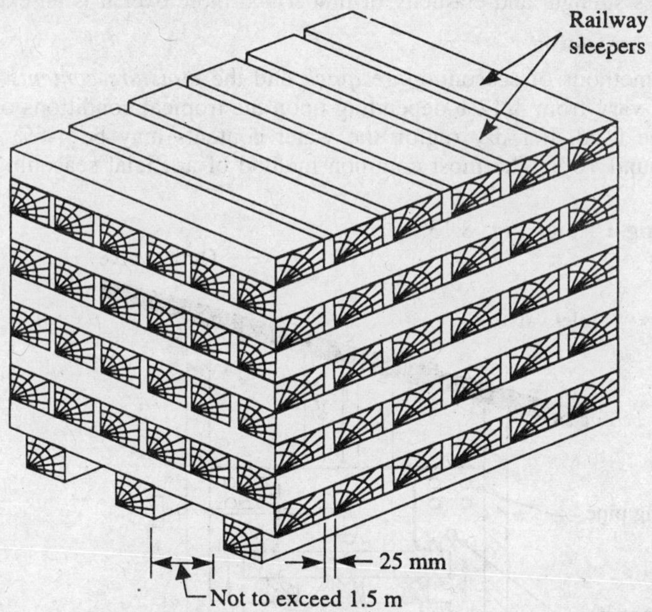


Fig. 8.12 : Air seasoning ("Close crib method" of stacking railway sleepers).

Wood seasoned by this method is generally fit for carpenter's work after two years and for pointer's work after nearly four years.

**Advantages** : (*i*) Natural or air seasoning *does not necessitate much attention.*

(*ii*) It is a *simple and cheap method* in case the supplies are small.

(*iii*) There are comparatively *less chances of damage* to the timber.

**Disadvantages :** (*i*) The process is *very slow* and extends over years.

(*ii*) For large stacks *considerable space* is required.

(*iii*) It *blocks the capital* for quite long period.

(*iv*) A *rigid control cannot be exercised.*

(*v*) Even under favourable conditions, it is *difficult to reduce the moisture content to suit the requirements of certain jobs.*

(*vi*) Timber may get *damaged* by insects and fungi during the seasoning period.

(*ii*) **Water seasoning :**

This method of seasoning timber consists in keeping logs of wood completely immersed in a running stream of water; the longer ends of the logs being kept pointing up-stream. By this process, the sap, sugar and gum etc., are leached out of the wood and replaced by water. The logs are then taken out and left to dry in an open place.

**Advantages :** Water seasoning is associated with the following advantages :

(*i*) It is a *quick process.*

(*ii*) As immersion in water *causes more rapid and regular drying, the tendency of wood to shrink or warp is reduced.*

(*iii*) As all the organic food materials present in the sap wood are washed off, the wood is rendered *less liable to be eaten away by worm* or *to decay by dry rot.*

**Disadvantages :** (*i*) The process *reduces* the elasticity and *durability of the timber.*

(*ii*) The timber is rendered brittle.

**Seasoning by boiling in water** is a very quick method and takes only few hours. It causes less shrinkage but reduces strength and elasticity of timber and more over it is an expensive process.

### 8.9.4.2. Artificial seasoning

The artificial methods of seasoning are *quick* and the *moisture contents can be kept under control*, which may vary from 4-12% depending upon the tropical conditions of the region where the product is to be used. For dry region the water contents may be 4-6% whereas in humid areas it may be around 10%. The most common method of artificial seasoning is *kiln seasoning method.*

(*i*) **Kiln Reasoning :** Refer Fig. 8.13.

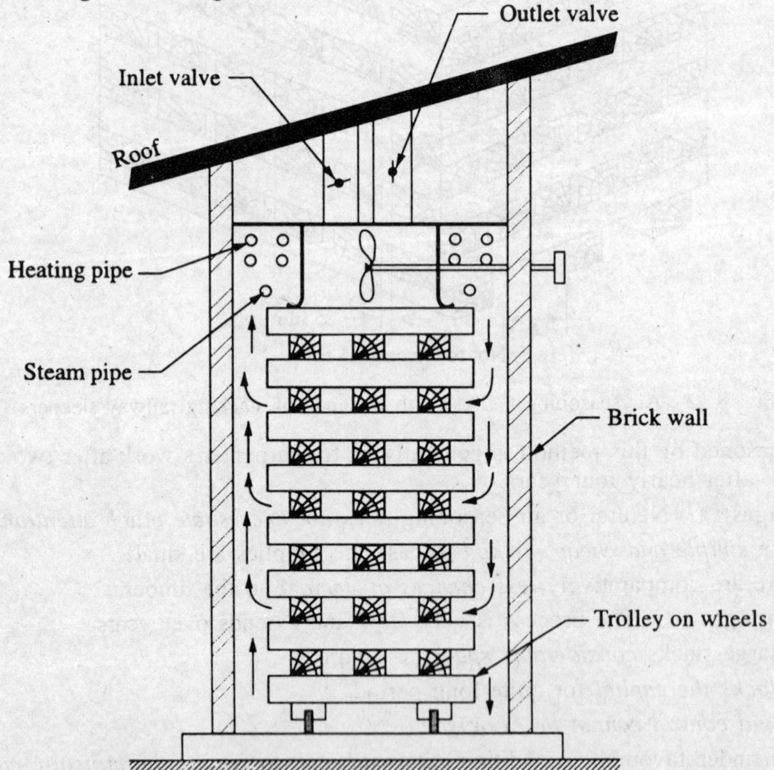


Fig. 8.13 : Kiln seasoning.

In this system the *timber is seasoned under controlled temperature and humidity conditions with proper air circulation and ventilation system.* The rise in temperature should be such that the timber retains the original strength and elastic properties. The required humidity level is maintaned to avoid warping and cracking. The drying of timber at uniform rate is well maintained by circulating air. The ventilation is provided to avoid overheating and excessive humidity.

The timber, inside the hot chamber, on the trolley is kept under controlled conditions for about a fortnight or so depending upon the initial water contents and the required moisture level. The quality of wood obtained by this method is *inferior* as compared to the one seasoned by natural methods.

The following points are worth noting :

- The wood may, if desired, be brought to the "*bone-dry*" state with almost no mositure. It is not usual to carry the drying so far because such drywood would *regain* moisture from the atmosphere. *Timber for frame work should be seasoned to about 15% moisture content and for leaves of door/window shutters, the moisture content of timber should be kept between 10 to 12%.*
- The *operation of kiln requires careful control of temperature and humidity of the air surrounding the wood at all stages of the operation.* The aim is to lower the moisture content of the wood to the required amount without causing the pieces to warp or check and without leaving any residual strains in the wood. *If the drying continues too rapidly, cracks will form. Even with careful work, when the centre of the piece does become dry and shrinks, unequal stresses are set up, with the outer portion in compression and the inner in tension. This compression, and hence increased hardness of the outer layers is named* **"casehardening"**. Casehardening may be *minimized* by expert dry-kiln practice and may be remedied by storage at a temperature higher than 70°C under controlled conditions of humidity. Under these conditions wood becomes plastic enough to permit the strains to relieve themselves.

**Advantages :** Following are the *advantages of kiln seasoning :*

1. The moisture content can be reduced as per requirement.
2. The timber can be seasoned thoroughly well and in short span of time under controlled conditions.
3. The timber seasoned by the method is less liable to shrinkage.
4. The wood can be used immediately as and when required.
5. The drying is controlled and there are practically no chances for the attack of fungi and insects.
6. The drying of different surfaces is even and uniform.

**Disadvantages :** Kiln seasoning entails the following *disadvantages* :

1. This method is costly (though the space required is less).
2. It requires skilled labour.
3. Due to the process of drying being quick a continuous attention needs to be given to check seasoning defects such as warping, internal cracks, surface cracks and end-splits.

Table 8.2 gives the comparison between *natural seasoning* and *kiln seasoning.*

**Table 8.2 : Comparison between Natural Seasoning and Kiln Seasoning**

| S.No. | Aspects | Natural seasoning | Kiln seasoning |
|---|---|---|---|
| 1. | *Nature* | Simple and economical. | Expensive and quite technical. |
| 2. | *Strength* | It gives stronger timber. | It gives a little weaker timber. |
| 3. | *Quality of timber* | More liable to the attacks of insects and fungi. | Less liable to the attacks of insects and fungi. |

| | | | |
|---|---|---|---|
| 4. | *Moisture content* | Difficult to reduce the moisture content below 15 to 18 per cent. | The moisture content can be reduced to any desired level. |
| 5. | *Speed* | Slow process | Quick process. |
| 6. | *Stacking space required* | More | Less |

(*ii*) **Chemical seasoning :**

This is also known as the *salt seasoning.*

- In this method, the timber is immersed in a solution of soluble salt.
- It is then taken out and seasoned in the ordinary way. The interior surface of timber dries in advance of exterior one and chances of formation of external cracks are reduced.

(*iii*) **Electrical seasoning :**

- This method of seasoning works on the *principle that heat is produced when poor conductors are placed in the field of high frequency.*
- The wooden plants are made to pass through an induction coil producing field of high frequency. Due to electric induction effect the moisture contents in the wood instantaneously evaporate and the wood is dried quickly.
- The main *advantage of this method is that the rise of temperature, as a result of which moisture gets evaporated, is uniform and results in superior quality of timber.*
- This method of seasoning is not popular because of *prohibitive cost, lack of control of moisture content;* sudden drying action changes the wooden fibres-resulting in defective wood.
- This method of drying is employed in *plywood manufacturing process.*

## 8.10. DECAY OF TIMBER

The life of sound and well seasoned timber in a well ventilated position or entirely immersed in water is several years. Various causes or situations which favour early decay of timber are :

(*i*) Moisture;
(*ii*) Imperfect seasoning;
(*iii*) Alternation of dry and wet states;
(*iv*) Vegetable growth;
(*v*) Attacks of insects etc.
(*vi*) Bad storage or stacking of timber;
(*vii*) Using unseasoned timber with the application of protective coat of paint or tar.
(*viii*) Using seasoned timber without applying suitable preservative on its surface.

The *decay of wood is caused by fungi colonies of plants of microscopic size* which feed upon some of the materials in the wood and form thread like filaments which spread through the wood in all directions.

In general, the same means that protect against decay by fungi also protect against destruction by animal organisms.

The following conditions must be present simultaneously in order that fungi may thrive :

(*i*) Oxygen enough to support the life processes;
(*ii*) Proper amount of moisture;
(*iii*) Proper temperature;
(*iv*) Adequate food.

If any of those above four factors is rendered too unfavourably, decay of wood cannot take place.

### 8.10.1. Diseases of Wood

The common diseases of timber are :

1. Dry rot 2. Wet rot

1. **Dry rot :**

- **Dry rot** *is the decomposition of felled timber by the action of various fungi which feed upon the wood and reduce it to a dry powdery condition.* This disease is *highly infectious* and brings a great amount of destruction to timber. It occurs where there is *not proper circulation of* air and the *timber is imperfectly seasoned.*
- Charring, tarring, painting or varnishing unseasoned timber also cause dry rot.
- In timber dry rot generally occurs in warm cellars, under unventilated wooden floors, basements, ends of timber built into walls without protection or ventilation etc.

*Detection of dry rot* : Dry rot can be easily detected by *tapping* or *scratching* at one end and placing the ear at the other end of the log. The log is sound if the tapping is distinctly heard otherwise it is a sign of internal decay. A log apparently sound from outward look may be full of dry rot inside, which can be detected by the *appearance of dry powder extracted by an auger* from the timber or by its smell.

*Prevention of dry rot* : Dry rot can be prevented by :

(*i*) Keeping the timber dry by adequate ventilation, reliable damp proof courses and the avoidance of leakage from roofs, water pipes and drains. If timber has moisture content of 20% or below no fungus can develop; fully dry timber remains immune.

(*ii*) If dry rot disease is discovered decayed wood should be removed and burnt; any suitable wood left should be treated with preservative (copper sulphate).

2. **Wet rot :**

- **Wet rot** *is the decomposition of tissue or timber caused by damp and moisture and by alternate wetting and drying.* This type of decay is *not* caused by fungial attack. Timber used in uncovered situation is very much likely to be subjected to the attack of wet rot. The timber suffering from this disease is reduced to a greyish brown powder.
- It is commonly found in a living tree. The attack takes place through wounds in the bark by the access of water. It is so caused by high percentage of moisture and general wet condition around the region of growth.

Wet rot can be *avoided by using thoroughly seasoned wood for exterior and underground work. The seasoned wood should be protected from moisture by tarring or painting.*

The best *remedy* for treating wet rot is by using a *suitable preservative.*

## 8.11. PRESERVATION OF TIMBER

**Preservation of timber** *refers to the art of treating the timber with some chemical so as to increase its life.*

The objects of preservation of timber are :

(*i*) To protect the timber structure from attack of destorying agencies such as fungi, insects, etc.

(*ii*) To lengthen the life of timber structures.

(*iii*) To make the timber structures durable.

### 8.11.1. Requirements of a Good Preservative

Following are the *requirements of a good preservative* :

1. It should be cheap and easily procurable.
2. It should be safe to the user but poisonous even in small doses to fungi and insects.
3. It should be free from obnoxious smell.
4. It should be not be washed away by water or deteriorated by heat, light etc.
5. It should be capable of allowing coats of point or decorative treatment without discolouration.
6. It should neither reduce the strength of the timber nor corrode metals in contact.
7. It should be non-inflammable.
8. It should have good power of penetrating into the fibres.

9. It should be highly penetrative.

10. It should not corrode the metals with which it comes into contact.

### 8.11.2 Types of Preservatives

As per IS: 401-1961 code preservatives shall be of the following types :

**Type 1** (*Oil type*) : Coal tar *creosote* with or without admixture of soluble oils.

**Type 2** (*Organic solvent types*) : These are the toxic chemicals in non-aqueous solvents, such as *copper napthenate, zinc napthenate, benzene hexachloride* and *DDT (Dichloro-diphenyl-trichloroethane).*

**Type 3** (*Water borne*) : These are toxic chemicals in water, such as *zinc chloride, boric acid, copper chrome arsenic composition* (*Ascue*). ***Ascue*** is a chemical used as preservative ; its composition being : *1 part Arsenic pentoxide : 3 parts hydrated copper sulphate : 4 parts sodium or potassium-dichromate.*

### 8.11.3 Methods of Preservation

The different *methods of preservation of timber* are enmuerated and discussed below :

1. Tarring.
2. Charring.
3. Painting.
4. Creosoting.
5. Ascue treatment.
6. Fire proofing.
7. Abel's process.

1. **Tarring** : It consists in applying a coat of tar or tar mixed with pitch. It is employed for rough types of works, such as *timber fences, ends of door and window frames (built into the walls) etc.*

2. **Charring :** It is done to *prevent dry rot* and the *attacks of insects* and is carried out at the lower ends of the posts to be embedded in the ground. The posts are charred to the depth about 1.5 cm over a wood fire and then quenched in water.

3. **Painting :** Painting should be carried out on the thoroughly seasoned wood otherwise the paint may confine the sap and cause the decay. Oil paint preserves the wood from moisture and prolongs its life. The well known oil paints are *solignum paints* which are available in various shades. They can *effectively bear white ants' attack.*

4. **Creosoting** : Creosote oil is the most widely used preservative treatment where dark colour and odour do not make the product too objectionable. Creosote oil is obtained by the distillation of coaltar or wood tar. Its toxic ingredients are complex phenols, of which the simplest is the phenol or carbolic acid, used as a germicide by surgeons. A secondary advantage of creosote oil is that it *imparts some water repellent properties to the wood.*

The pieces of wood to be treated are placed in cylinders or retorts which can be closed tightly. Creosote oil under a pressure of 9 bar and at a temperature of 50°C is pumped in until the desired absorption is reached. *Creosoting preserves the timber from rot and attacks by white ants.* It is usually employed for *piles, poles and railway sleepers.*

5. **Ascue treatment** : Ascue is in powder form. For use 6 parts by weight of the powder is dissolved in 100 parts by weight of water. The solution (odourless) is sprayed on the timber. Wood treated by this method can be painted, varnished, polished or waxed.

6. **Fire proofing** : Wood cannot be fire proof without destroying its structure or modifying it so greatly that the product should scarcely be called wood. *A coating of any non-combustible material* protects the surface of the wood from direct contact with flying embers as well as from radiant heat. If these materials are applied as a paint, the binder should also be non-combustible.

Another method of increasing the resistance to fire is *to impregnate the wood with a fire-retarding chemical.* The chemical should preferably be one that absorbs heat and dissociates when it is heated. *Water solutions of ammonium phosphate, ammonium sulphate, borax and boric acid either singly or in admixture are used.* All require high absorption of the chemical for their full affect, and hence are not cheap. Wood treated with zinc chloride as a preservative against decay kindles less readily than untreated wood. To *acquire much protection against fire, the amount of zinc chloride should be two or five times the minimum specified as a preservative.*

7. **Abel's process :** In this process the surface of the timber is first painted with dilute solution of sodium silicate, then with a cream like paste of slaked fat lime and finally with a concentrated solution of soda.

### 8.11.4. Types of Preservative Treatment

Timber for preservative treatment *should be seasoned to the desired moisture content and fabricated to the required size.* Cutting or boring, if any, should be done prior to the treatment :

Various types of *preservative treatments* are discussed below :

1. Surface application
2. Soaking treatment
3. Hot and cold process
4. Boucherie process
5. Pressure processes.

**1. Surface application method :**

- This is also known as *brush and spray method.*
- In this method the timber to be treated is brushed, sprayed or dipped in the preservative solution for a suitable period.
- At least two coats are applied, the second coat to be applied only after the first one has dried.
- As far as possible, the hot solution of preservative should be used.
- It is mostly used for treating materials at site but of little use over the surface of wood already in service and for retreatment of cut surfaces.
- This method is *recommended only when naturally resistant timbers are used, related from the ground* which does not have much termite trouble.

**2. Soaking treatment :**

This method is also known as *steeping.*

In this type of treatment the *timber is submerged in the preservative solution for sufficiently long time* until the required absorption is obtained. Water soluble type preservative is used to avoid excessive evaporation loss.

For a seasoned timber, a soaking period of 7 to 10 days is desirable and a penetration 3 to 6 mm is expected.

It gives good protection *even to non-resistant timbers* used isolated from the ground which has mild termite trouble.

It is *especially good for timber members containing a lot of sapwood and for light and medium density timbers.*

**3. Hot and cold process :**

- In this process timber is stacked in the tank and cold preservative (usually creosote) is then run into the tank, till the timber is completely submerged. The preservative is then heated to about 95°C and maintained at that temperature for a suitable period. After that it is allowed to cool until the required absorption of presentative is obtained. By this process, a penetration upto 5 cm is obtained.
- This treatment ensures sterilization of the timber against fungi and insects that may be present.
- This is considered to be *most efficient non-pressure treatment process.*
- This treatment is recommended for the timber containing sapwood and easily treatable heartwood.

- This treatment is even suitable for non-durable timbers and in areas where there is heavy termite trouble.

**4. Boucherie process :**

- This process is employed for treatment of sapwood of almost all green timbers (with bark on) and of bamboos (in gree condition) soon after felling. Any of the inorganic water-soluble preservatives may be used for the treatment.
- It is carried out by attaching the butt end of a pole (inclination being 45°, with butt end up) to a rubber hose connected to a reservoir (placed at a higher level) containing the preservative solution. The preservative (due to hydrostatic pressure) displaces the sap which is then forced out at the narrow end and thus the required penetration is secured.

**5. Pressure processes :**

Injecting preservative under pressure into the timber is the most effective method of treating timber with preservative. This method proves to be essential for treating non-durable timbers which are to be used at places where there is danger of attack by fungi and insects.

Various pressure processes are enumerated and described below :

(*i*) Fuel cell or Bethel process

(*ii*) Empty cell or Rueping process

**(*i*) Fuel cell or Bethel process**

This process is useful when it is desired that the *preservative solution should be retained for a fairly long duration.* Timber for *marine works, railway sleepers* etc. are treated by this method. This method is usually adopted in creosoting.

This process of treating timber is carried out as follows :

- The timber is stacked in a strong steel cylinder the door of which is tightly closed.
- An initial vacuum of 550 mm of mercury is created with the help of a vacuum pump. This expels air from the wood. This partial vacuum is maintained for about half an hour.
- Preservative is then injected into the cylinder without allowing an entry of air, thus the cylinder is filled with preservative.
- The pressure of the cylinder is then increase to 4 to 12 bar (depending upon the type of timber, the size of the sawn members etc).
- Finally final vacuum of 400 mm to 600 mm of mercury is created so that excess solution of preservative is separated from timber.

**(*ii*) Empty cell or Rueping process**

This process requires *much less preservative* to give the desired absorption and as such this process is comparatively cheaper to full-cell process. This method is specially *used for preserving timber of mixed species* and *also for timber containing sapwood and heartwood.*

In this process timber is treated as given below :

- The timber is stacked in strong steel cylinder and the door is closed tightly.
- Depending upon the sapwood content of the timber, the timber in the cylinder is then subjected to an air pressure of 1.75 to 5 bar from half an hour to one hour. This pressure is maintained whilst filling up of the cylinder with the preservatives.
- When the cylinder is completely filled, an antiseptic pressure of 5 to 12 bar is applied.
- This pressure is maintained till required absorption has taken place.
- The cylinder is drained off and a vacuum is applied to free the timber from dripping preservative.

## 8.12. CONVERSION OF TIMBER

After felling the tree, it is generally cut into standard sections to minimise the wastage of wood. The *process by which the timber is cut and sawn into suitable sections is known as the* **conversion.**

The following points regarding conversion of timber are worth noting :

1. The conversion of timber should be done in such a way that *wastage of useful timber is minimum.*
2. An allowance of 3 mm to 6 mm should be made for squaring, planing and shrinkage.
3. The saw cuts should be made tangential to the annual rings and practically parallel to the direction of medullary rays. By doing so, strong timber pieces are obtained.
4. The wooden beams should be sawn in such a fashion that they do not contain pith in their cross-section; this is achieved by sawing the timber first through pith into two halves.
5. The probable loss in conversion of timber from log to the required size of timber may be taken as follows :

   Round log to planks = 40 per cent; Round log to scantlings = 50 per cent; square log to planks = 30 per cent; square log to scantlings = 40 per cent.

### 8.12.1. Methods of Sawing Timber

Timber may generally be sawn in one of the following ways :

(*i*) Ordinary sawing (*ii*) Tangential sawing

(*iii*) Radial sawing (*iii*) Quarter sawing.

**1. Ordinary sawing :** Refer Fig. 8.14.

- It is also called *bastard, flat* or *slab sawing.*
- The cuts are made into the log of wood according to the required thickness. The saw cuts are tangential to the annual rings and right through the cross-section of timber piece.
- The log is moved backward and forward on a saw which cuts parallel slices or planks. Different thickness can be obtained by varying the width of cutting.

*Advantages* : (*i*) It is *very easy and quick* method of sawing; it is widely adopted in our country.

(*ii*) It is almost most economical method and *wastage of timber is minimum.*

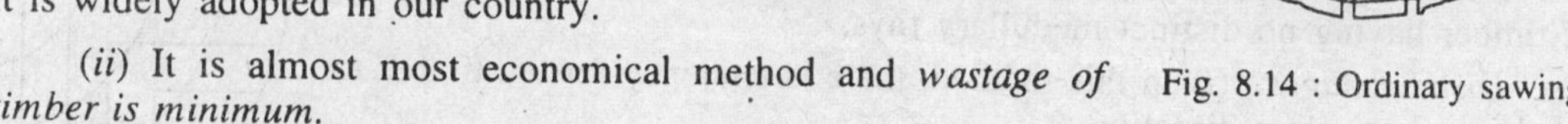

Fig. 8.14 : Ordinary sawing.

*Disadvantages :* (*i*) The planks are liable to warp and twist as a result of unequal shrinkage.

(*ii*) The surface of the planks obtained show *varying features.*

**2. Tangential sawing**

- It is also sometimes known as *plain sawing* and gives flat grained timber.
- In this method, the boards or planks are sawn out of the log of wood tangentially to the annual *rings* as shown in the Fig. 8.15.
- This method is adopted when the annual rings are very distinct and medullary rays are not clearly defined.

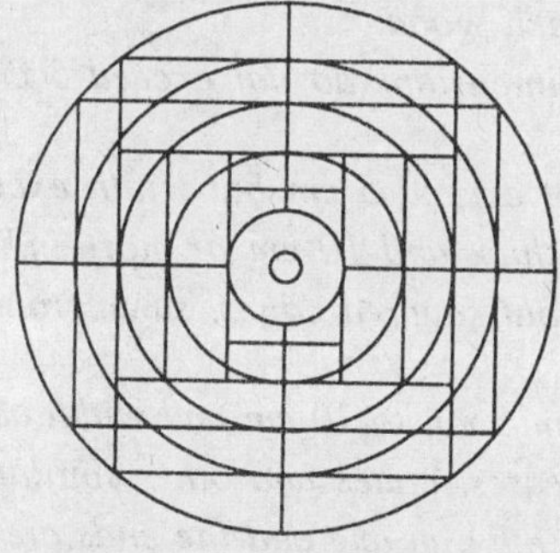

Fig. 8.15 : Tangential sawing.

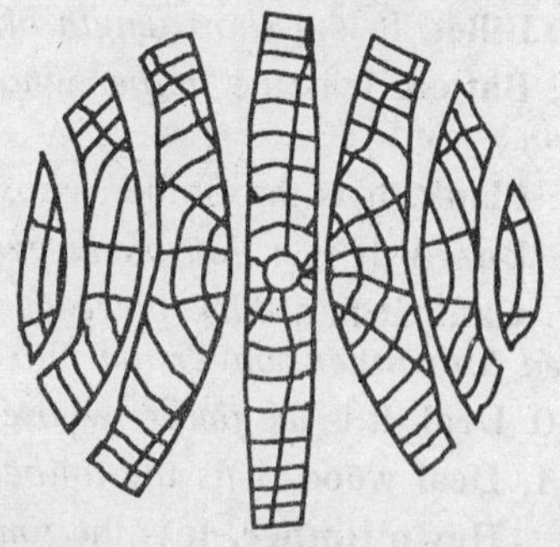

Fig. 8.16 : Shrinkage in tangential sawing.

*Advantages* : It is in economical due to *less wastage* and *involves less labour.*

*Diadvantages* : (*i*) Sections obtained by this method are weak and do not suit for heavy work as for flooring etc.

(*ii*) The planks obtained by this sawing *warp too much* (because the sections are weak as the medullary rays which impart strength to the longitudinal fibres are cut), as shown in Fig. 8.16.

(*iii*) The timber obtained by this method cannot be polished evenly.

**3. Radial sawing :** Refer Fig. 8.17.

- The sawing is done parallel to the medullary rays and perpendicular to the annular rings:
- This method is used for *conversion of hard timber.*

*Advantages* : (*i*) The cut section shrinks at a uniform rate and the *warping* is therefore *less*. The wood, therefore, can be used for high quality timber works.

(*ii*) It gives wood with decorative effect.

*Disadvantages* : (*i*) There is considerable quantity of wastage.

(*ii*) The cost of sawing proves to be high as more time and labour are required in turning wood several times.

**4. Quarter sawing :** Refer Fig. 8.18.

- In this method, the log is cut into four quarters. These quarters have their faces essentially radial.
- Boards are cut alternately from each of these faces, thus cutting the annual rings at angles not less than 45°.

*Advantages :* It may produce fine figurewood when adopted in case of timber having no distinct medullary rays.

*Disadvantages* : (*i*) In this method, there is tendency for the timber to bend in a transverse direction.

(*ii*) It requires more labour and involves more wastage.

Fig. 8.18 : Quarter sawing.

### 8.12.2. Common Market Forms

The various market forms in which the timber is cut and sold are:

1. **Log.** It is the *trunk of tree left after cutting all the branches.*
2. **Baulk.** It is the cut timber whose cross-sectional dimensions exceed 5 cm in one direction and 20 cm in the other. It is usually a roughly squared sawn log.
3. **Bole.** It is the main stem of *a tree.*
4. **Bolt.** It is a short log 1.25 m or less in length.
5. **Billet.** It is a *short length of a thin stem or branch wood.*
6. **Batten.** It is the *timber whose cross-sectional dimensions do not exceed 5 cm in either direction.*
7. **Plank.** It is the timber *whose thickness does not exceed 5 cm but width exceeds 5 cm.*
8. **Board.** It is a *thin plank generally under 5 cm thick and 10 cm or more in width.*
9. **Cant.** It is a *thick piece of timber with or without squared edges, sawn from a log and intended for further conversion into smaller sizes.*
10. **Deal.** It is the *timber whose thickness varies from 5 cm to 10 cm and width about 22 cm.*
11. **Deal wood.** It is the *timber used for packing cases, crates and other similar work.*
12. **Hewn timber.** It is the *timber converted to size by an axe and the ends are sometimes sawn.*

13. **Pole.** It is a long, *solid, straight trunk of a tree 10 to 30 cm in diameter at breast height* and tapering gradually to the top, of a diameter of about 10 cm or more.

14. **Post.** It is a *timber member used in an upright position in building fencing or other structural work.* Its usual size is between 15 cm to 50 cm side or diameter.

15. **Scantling.** It is a *piece of timber, the cross-sectional dimensions of which exceed 5 cm in both the directions but do not exceed 20 cm.*

16. **Sleeper.** It is a *piece of timber used as transverse support under rails in railway lines, usually square sawn.* Most commonly used sections are : 25 cm × 12.5 cm or 20 cm × 11.5 cm in section.

17. **Strip.** It is a *piece of timber under 5 cm thickness and less than 10 cm width.*

## 8.13. TESTING OF TIMBER

Some important tests which are performed to determine the mechanical properties of timber are discussed below :

1. Tensile strength test
2. Compressive strength test
3. Shear strength test
4. Bending strength test
5. Moisture content test

**1. Tensile test :**

- This test is carried out on a specimen which is rectangular (5 cm × 5 cm) in shape and 20 cm long.
- The load may be applied either parallel to the grains or perpendicular to the grains.

**2. Compressive strength test :**

- In this test the rectangular specimen 5 cm × 5 cm × 20 cm is used and load is applied parallel to the grains. The specimen shall be absolutely free from any defects and shall not have a slope of grain more than 1 in 20 parallel to its longitudinal edges.
- A cylindrical specimen is also used sometimes.

**3. Shear strength test :**

- Shear strength is important in the case of beams, slab, etc.
- Shear strength is determined *only parallel to the grain* (as shear strength perpendicular to the grain cannot be determined because of failure of the specimen due to other causes earlier/before the ultimate shear strength is reached).
- The size of a standard shear test specimen shall be 5 cm × 5 cm × 6.25 cm, notched in a corner so as to produce a failure on 5 cm × 5 cm surface in the radial or tangential surface as desired.

**4. Bending strength test :**

- This refers to the strength of timber as a beam. Modulus of rupture and modulus of elasticity of timber can be determined from the static bending strength.
- Static bending test is made on specimens 5 cm × 5 cm × 7.5 cm in size. The specimen shall be absolutely free from any defect and shall not have a slope of grain more than 1 in 20 parallel to its longitudinal edges.
- *Impact bending* is obtained by dropping a specified weight on to the beam from a certain height. Shock resistance properties are important in handles of hammers, athletic goods. etc.

**5. Moisture content test :**

- In this test the moisture is calculated on the weight of oven dry (or absolutely dry) wood.
- Expressing the moisture content of wood as a percentage of its oven dry weight, we have,

$$\%\text{ Moisture content} = \frac{\text{Weight of moisture in sample}}{\text{Oven dry weight of sample}} \times 100$$

This is determiend by *weighing a sample of wood and drying the same in an oven until its weight becomes constant.*

The moisture content of the wood can also be determined by *moisture meters of distillation method.*

## 8.14. TIMBER SUITABLE FOR VARIOUS USES

| | Applications/Uses | Timber |
|---|---|---|
| 1. | Sports goods and baskets. | Ash, Oak Mulberry. |
| 2. | Match industry. | Pine and Simul. |
| 3. | Bullock carts | Babul. |
| 4. | Musical instruments | Jack. |
| 5. | Railway sleepers. | Deodar, Kail, Sal, Ash. |
| 6. | Well curbs. | Mango, Jack, Simul, Jarul and Jack. |
| 7 | Doors and windows | Sal, Deodar. |
| 8. | Packing cases | Deodar (heavy packing), Mango and Fir (light packing). |
| 9. | Scaffolding. | Bamboo, Sal. |
| 10. | Agricultural implements and tool handles. | Babu, Ash, Mulberry. |
| 11. | High class joinery and furniture | Teak, shishum, walnut |

## 8.15. TIMBER TREES OF INDIA

Description of some important timber trees in India is given below:

**1. Deodar :**

It is a tall free with long pointed leaves and is generally found in hills. It is most important timber tree providing soft wood.

*Characteristics* (*i*) It yields soft wood having well defined grains.

(*ii*) It is highly durable.

(*iii*) It is light coloured wood and have distinct annual rings.

(*iv*) Its average weight is 5450 N/m$^3$ (at 12 per cent moisture content).

(*v*) It is strongly scented, oily and polishes well.

(*vi*) It is moderately strong

(*vii*) It can be easily worked.

**Uses :** It is used for making cheap and rough furniture, railway carriages, railway sleepers, packing boxes, structural work etc.

**2. Babul :**

This tree grows all over India and is available in abundance. It seldom attains a height greater than 15 m.

*Characteristics* : (*i*) Its colour is whitish red.

(*ii*) It takes up good polish.

(*iii*) Its average weight is 7000 N/m$^3$.

(*iv*) It is not much durable.

**Uses** : (*i*) It is employed for *constructional purposes, agricultural implements and tool handles.*

(*ii*) It is also used in *low cost house construction and bullock carts.*

**3. Teak :**

Teak wood is one of the most available hard woods. Its durability is due to the presence of aromatic oil which largely preserves it from the attack of white ants. It grows in Malabar (South India) and Central India.

*Characteristics* : (*i*) It is strong and durable.

(*ii*) Its colour is deep yellow to dark brown.

(*iii*) Its average weight is 6400 N/m$^3$.

(*iv*) It is not difficult to work and can be finished and polished to obtain an excellent surface.

(*v*) It can be easily recognised by virtue of its distinct smell.

(*vi*) It is resistant to moisture and certain chemicals.

(*vii*) It does not corrode iron fastenings.

(*viii*) It shrinks little.

(*ix*) It can be easily seasoned.

(*x*) It is fire-resistant.

(*xi*) It is not attacked by white ants and dry rot.

(*xii*) It is moderately hard.

**Uses :** (*i*) It is used for heavy construction of permanent character like ship-building, piling, railway sleepers etc.

(*ii*) It is also employed for high class joinery and furniture.

Its use is limited to superior work only as it is comparatively very costly.

**4. Shisham (Tali) :**

It grows widely in Punjab, Uttar Pradesh and Madhya Pradesh.

*Characteristics* : (*i*) It is broad leaved tree.

(*ii*) It is moderately durable.

(*iii*) Its average weight is 7000 N/m$^3$

(*iv*) The wood is dark brown in colour.

(*v*) It is coarse grained.

**Uses:** It is used for constructional purposes, furniture and cabinet making, agricultural implements and bridge-piles etc.

**5. Mango :**

This tree is very much esteemed for fruit and is found all over India.

*Characteristics:* (*i*) Its wood is of inferior quality.

(*ii*) It is coarse and open-grained.

(*iii*) It has deep grey colour.

(*iv*) The wood is readily attacked by white-ants and decays on exposure to wet atmosphere.

(*v*) Its average weight is 6900 N/m$^3$.

(*vi*) Easy to work but keeps its shape well.

**Uses:** It is used for *packing cases of all sorts, cheap furniture, tonga bodies, pattern making and ornamental cabinet work*.

**6. Mahogany :**

It is a native tree of Cuba in West Indies and Central America. It is now becoming popular in India also.

*Characteristics:* (*i*) It has fine wavy grains.

(*ii*) Reddish brown colour.

(*iii*) Durable in water.

(*iv*) Hard but easy to work.

(*v*) Average weight 6700 N/m$^3$.

(*vi*) Takes high polish.

**Uses :** it is used for *high class furniture, pattern making* and *ornamental cabinet work.*

**7. Walnut :**

It is available mostly in hilly areas.

*Characteristics : (i)* Light brownish colour.

*(ii)* Broad leaved tree.

*(iii)* Average weight 5750 N/m$^3$.

*(iv)* Less durable.

*(v)* Resists the attack of white ants.

*(vi)* Easy to work.

*(vii)* Tough and flexible.

**Uses :** It is employed for furniture, decorative work and cabinet work.

**8. Mulberry :**

This tree yields, strong, tough and elastic wood.

*Characteristics: (i)* Wood can be easily turned, carved and finished to a smooth surface.

*(ii)* Wood is of brownish colour.

*(iii)* Broad leaved tree.

*(iv)* Average weight is 6750 N/m$^3$.

*(v)* Seasons well.

*Uses :* It is used for making *sports goods like hockey sticks, tennis and badminton rackets etc.*

**9. Sal :**

It is available mostly in hilly areas of U.P., Bihar, Assam and Visakhapatnam.

*Characteristics : (i)* Wood is hard, close-grained, heavy and durable.

*(ii)* Not easily attacked by white ants.

*(iii)* Average weight is 8600 N/m$^3$.

*(iv)* Easily worked but does not take good polish.

*(v)* Seasons slowly.

*Uses :* it is used in *bridge contruction, ship building, piling etc.* It is however, *not* suitable for ornamental work.

**10. Chir :**

It is available in Himalayan region.

*Characteristics: (i)* Reddish brown in colour.

*(ii)* Average weight is 5700 N/m$^3$.

*(iii)* Less durable.

*(iv)* Easily worked but cracks.

*(v)* Seasons well.

*Uses :* It is employed for *inferior wooawork in low cost house building.*

**11. Kail :**

It is mostly grown in Himalayan region.

*Characteristics : (i)* Wood is close grained, heavy, moderately hard and durable.

*(ii)* Brown in colour.

*(iii)* Average weight is 5000 N/m$^3$.

*(iv)* Seasons well.

*Uses : (i)* It is used for door and window frames and panels, furniture making, beams, posts etc.

(*ii*) It is also used in heavy engineering works such as bridge construction, ship building, piling etc.

**12. Bamboo :**

It grows in large varieties in Bengal, Assam and South India. Small varieties are found in low hills also.

*Characteristics :* Strong and durable.

*Uses :* (*i*) It is used in *scaffolding, landers* and *bridge building.*

(*ii*) It is also employed for partitions, flooring, ceiling, roofing and almost all parts of house construction except for the fire place.

Other trees are *Kikar, Toon, Oak, Jarul, Simul, Spruce, Pine, Hickory Tamarind and Ash.*

## 8.16. WOOD-BASED PRODUCTS

- Wood-base products are the products manufactured from different types of wood.
- These products (such as *plywood, ballen boards, fibre boards, lamin boards* etc.) are manufactured out of *veneers, strips, fibres* of *wood* after suitable treatment.

### 8.16.1. Veneers

- *Veneers* are *thin sheets or slices of wood of superior quality, having thickness varying from 0.4 mm to 6 mm or more.*
- These are obtained with the aid of a circular rotary saw or a rotary peeling machine. These are peeled off the log by a sharp knife in a long continuous sheet by rotating the log on its longitudinal axis or otherwise as shown in Fig. 8.19.
- The veneers are sliced from logs of timber having extremely attractive and curly figures

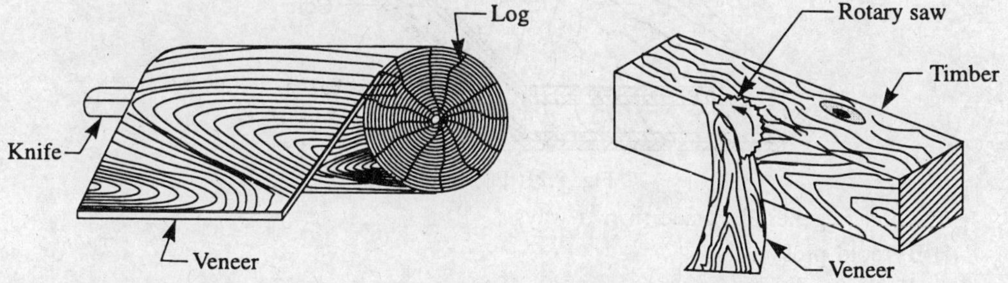


Fig. 8.19 : Veneering

such as *Teak, Toon, Mahogany, oak,* etc. The process of preparing long and continuous sheets of veneers is called **veneering.**

- The veneer is too thin to be used as a separate, but is glued or cemented to other veneers or to a heavier state of wood. Veneers may also be cemented in layers, with the grain of successive layers crossing each other so as to neutralise shrinkage and decrease the danger of cracking. Such multiple veneers are very similar to playwood, and are used in the construction of aircraft.

### 8.16.2. Plywood

*Plywood* is made by cementing together several layers of wood which may be thin veneers or thicker boards. The *advantages* of plywood are :

(*i*) Better appearance.

(*ii*) Easily workable and capable of being shaped to numerous designs.

(*iii*) Elastic material, least affected by changes in atmosphere.

(*iv*) Uniform tensile strength in all directions.

(*v*) Light in weight and greater strength.

(*vi*) Highly resistant to cracking, spalling and warping as it has great stiffness and rigidity in comparison with solid wood.

(*vii*) Available in large sizes (which are impossible with solid wood).

Three *ply* panels (Fig. 8.20) are made from three veneers; thicknesses from 6 mm to 25 mm and over. *Multiply* panels are made in any odd number of veneers 5 to 9, thickness from 6 mm to 25 mm and over. The odd number is taken so that the shrinkage stresses are symmetrical

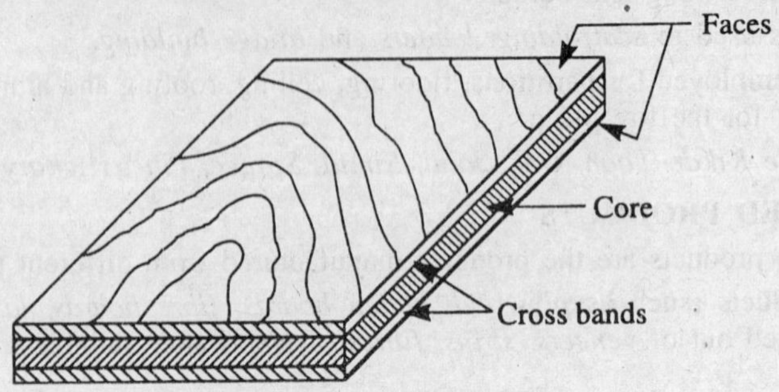


Fig. 8.20 : Three ply panel.

about the middle ply. *Five ply* panel is shown in Fig. 8.21.

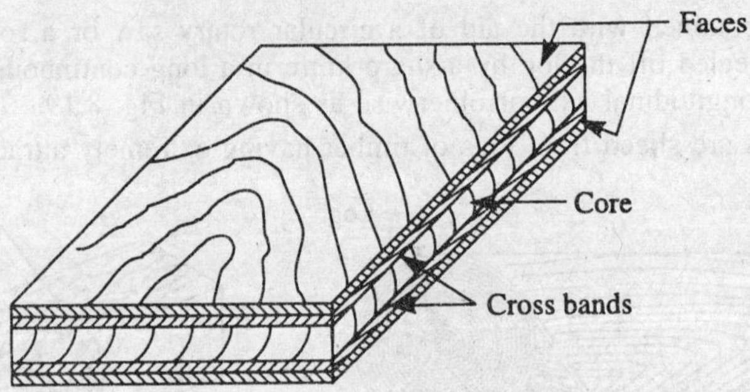


Fig. 8.21: Five ply panel.

Plywood is generally made in *two* ways :

(*i*) By cold pressing.

(*ii*) By hot pressing.

- In *cold pressing* the adhesives can set at room temperature.
- In *hot pressing* the glued veneers with the requisite number of plies are pressed in large hydraulic presses. The leaves of these presses are electrically heated.

The usual range of pressure and heat applied for the fabrication of plywood veneers is about 7 bar and 150° to 250°C respectively.

*Uses of plywood :* (*i*) It is mainly used for the construction of *furniture, partitions, ceilings, covers, doors, windows packing cases,* for decorative purposes inside and outside surfaces of building walls and shops, buses, etc.

(*ii*) It is also employed for the construction of aircraft.

### 8.16.3. Lamin Board

- A lamin board is a board having a core of strips, each not exceeding 7 mm in thickness, glued together face to face to form a slab which in term is glued between two or more veneers, with the direction of the grain of the core strips running at right angles to that of the adjacent outer veneers (Refer Fig. 8.22).
- The lamin boards are *light, strong and do not split or crack easily.*

*Uses* : They are used for *walls, ceilings, partitions* and *packing cases.*

### 8.16.4. Block Board

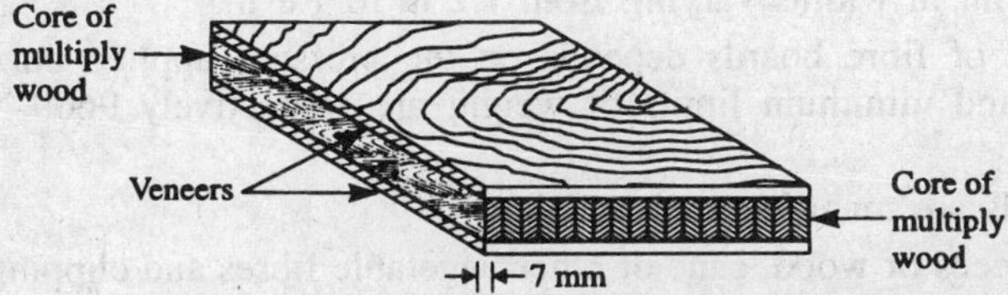


Fig. 8.22 : Lamin board.

- A *block board* is constructed in the same way as lamin board. In this case core consists of smaller timber block upto 25 mm in width. These blocks are cemented edge to edge and on each face plies upto 3 mm thickness are glued (Refer Fig. 8.23).
- The wooden strips of the core shall be cut from specified timber and seasoned to moisture

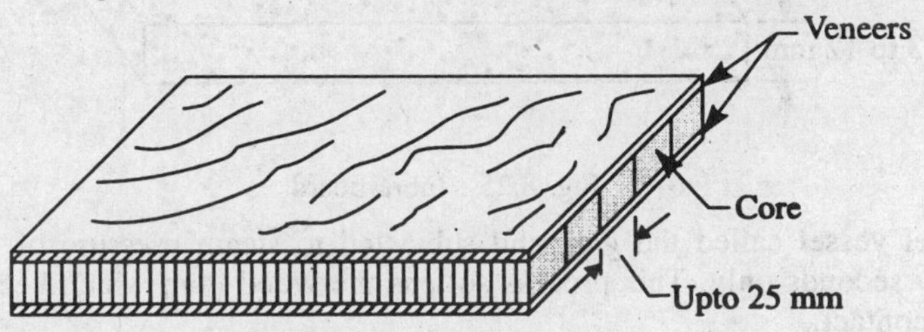


Fig. 8.23 : Block board.

content not exceeding 12 percent. The core strips should be of one species in a board and may consist of small lengths placed end-staggering lengths. The construction should be well balanced around the centre line.

**Uses :** Blockboard are extensively used for *construction of railway carriages, bus bodies, marine and river crafts and for furniture making, partitions, panelling, prefabricated houses,* etc.

### 8.16.5. Batten Board

- A batten board is a board heaving a core made up of strips of wood usually 80 mm wide, each laid separately or glued or otherwise joined to form a slat which is glued between two or more outer veneers with the direction of the grain of the core running at right angles to that of the adjucent outer veneers.
- Batten boards are light and strong.

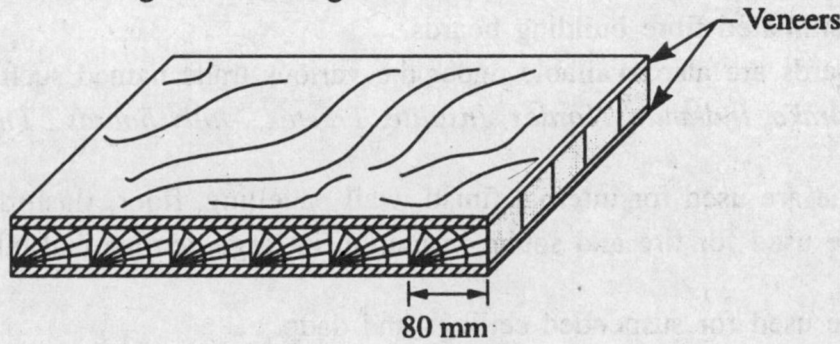


Fig. 8.24 : Batten board.

**Uses :** These boards are used for *door panels, table tops,* etc.

### 8.16.6. Fibre Boards

Refer Fig. 8.25.

- These are rigid boards and also known as *pressed wood* or *reconstructed wood.*
- They are made from a mass of felled wood or other vegetable fibres. They may be either *homogeneous* or *laminated.*

- They vary in thickness from 3 mm to 12 mm. They are available in lengths varying from 3 m to 4.5 m, in widths varying from 1.2 m to 1.8 m.
- The weight of fibre boards depends on the pressure applied during manufacture. The maximum and minimum limits of weight are respectively 9600 N/m$^3$ and 500 to 600 N/m$^3$.

The fibreboards are manufactured as follows :

— The pieces of wood, cane or other vegetable fibres and clippings are put into airtight

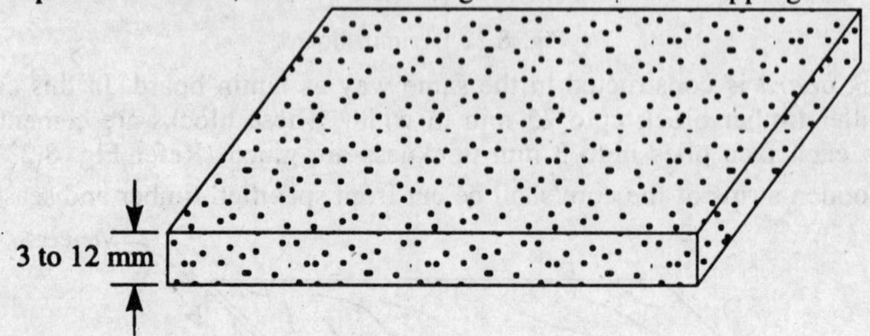


Fig. 8.25 : Fibre board.

steel vessel called the *gun*, and subjected to steam pressure of 30 to 65 bar for a few seconds only. This process softens the wood gum which keeps the wood fibres in contact.

— The valve located at the bottom of the vessel is opened and the steam is allowed to expand. The sudden release of pressure makes the wood pieces to explode and in doing so, the natural adhesive contained in the wood fibres is separated out.

— The wood fibres are then allowed to flow out and cleaned of all superfluous or extra gums.

— These cleaned fibres are spread on wire screens in the form of loose sheets or blankets of required thickness and subjected to heat and great pressure in hydraulic presses to form solid dry sheets, known as *fibre boards* (a little quantity of glue is also added, if required).

- The fibre boards, depending upon their form and composition, are classified as follows :
  (*i*) Insulating boards
  (*ii*) Light and hard boards
  (*iii*) Laminated fibre building boards.

The fibre boards are also available under the various trade named such as :

*Masonite, Euraka, Indianite, Nordex, Insulite. Treetex, 'Jolly boards', Thermofrig, 'Thermocoustic'* etc.

*Uses :* (*i*) The are used for internal finish, wall panelling, floor, flush doors, table tops.

(*ii*) These are used for fire and sound insulation in large commercial buildings and cinema houses.

(*iii*) They are used for suspended ceilings and dado.

(*iv*) They are also used in making partitions and finishing cover to furniture.

### 8.16.7. Hard Boards

- These boards are hard pressed and hence are more compact, strong and durable.
- They impart internal appearance and finish to a structure.
- They are least affected by the change in temperature and humidity of surroundings.
- They are impregnated with oil when properties of high non-absorbent nature is required.

## HIGHLIGHTS

1. *Timber* is the wood suitable for building or engnieering purposes and it is applied to trees measuring not less than 0.5 m in girth.
2. A *good timber* should have a uniform colour, smell sweet, be sonorous when struck, have straight and close fibres, be heavy in weight, and be free from flaws etc.
3. The trees may be *exogenous* or *endogenous*. The trees should be filled when they have just matured or when they are very near to maturity.
4. The most *common defects in timber* are :

   Heartshakes, starshakes, cupshakes, radial shakes, rindgalls, upsets, twisted fibres, burns, wind cracks, knots, honeycombing, end splits, deadwood, druxiness.
5. *Seasoning of timber* is the process of drying timber or removing moisture or sap, present in a freshly felled timber, under more or less controlled conditions.
6. Seasoning of timber decreases the weight of timber, improves its working qualities, enables it to be easily painted, polished and preserved and provides it dimensional stability.
7. Seasoning of timber may be carried out in the following two ways:

   (*i*) Natural season :

   (*a*) Air drying/seasoning (*b*) Water seasoning.

   (*ii*) Artificial seasoning :

   (*i*) Kiln seasoning (*ii*) Chemical seasoning (*iii*) Electrical seasoning.
8. *Decay of timber* may be caused by :

   Moisture, imperfect seasoning, alternation of dry and wet states, vegetable growth, attacks of insects etc, bad storage or stacking of timber etc.
9. *Diseases of wood* are :

   *Dry rot :* It is the decomposition of felled timber by the action of various fungi which feed upon the wood and reduce it to a dry powdery condition.

   *Wet rot :* It is the decomposition of tissue or timber caused by damp and moisture and by alternate wetting and drying.
10. The different *methods of preservation of timber* are : Tarring, charring, painting, creosoting, Ascue treatment, fire proofing and Abel's process.
11. *Veneers* are thin sheets or slices of wood of superior quality, having thickness varying from 0.4 mm to 6 mm or more.
12. *Plywood* is made by cementing together several layers of wood which may be thin veneers or thicker boards.

## OBJECTIVE TYPE QUESTIONS

### Choose the Correct Answers :

1. Creosote oil is applied on wood

   (*a*) to protect its annular layers against disintegration
   (*b*) to protect it against the action of fungi and insects
   (*b*) to remove free moisture
   (*d*) to fill pores with oil.
2. Compressive strength of wood is usually

   (*a*) equal in all directions
   (*b*) less along the grains, more across the grains
   (*c*) more along the grains, less across the grains
   (*d*) maximum at 60° to the direction of grains.
3. Cup shake is a

   (*a*) resin found in trees
   (*b*) property of broad leaved trees
   (*c*) defect in wood
   (*d*) kind of timber in north India.
4. The life of a tree can be estimated on the basis of

   (*a*) height of tree
   (*b*) diameter of trunk
   (*c*) moisture content of bark
   (*d*) number of annual rings.
5. Steeping is a method of treatment of wood in which wood is

(a) painted by brush (b) dipped in brine solution
(c) spray painted (d) dipped alternately.

6. Which of the following is the hardest wood?
(a) Teak (b) Babul
(c) Shisham (d) Chir.

7. ................ is known as natural defect of wood.
(a) Knot (b) Medullary ray
(c) Cambium layer (d) None of the above.

8. The moisture content of freshly felled tree may be about
(a) 2 percent (b) 10 percent
(c) 50 percent (d) 150 percent.

9. Which of the following wood is resistant to white ants?
(a) Sal (b) Deodar
(c) Teak (d) Chir.

10. The most valuable timber is
(a) chir (b) shisham
(c) deodar (d) sal.

11. The decay of timber caused by alternate wetting and drying is called
(a) wet rot (b) dry rot
(c) foxiness (d) none of the above.

12. Teak wood is suitable for which of the following?
(a) Sports goods (b) Furniture
(c) Railway sleepers (d) Wooden electric poles.

13. Cracks in a live tree may develop due to
(a) frost (b) drying of core
(c) storm (d) swinging by wind
(e) any of the above.

14. Which one of the following chemicals is not used as a wood preservative?
(a) Sodium chloride (b) Phenol
(c) Zinc chloride (d) Mercuric chlorid.

15. In a tree the layer between bark and wood is called
(a) pitch layer (b) medullary ray
(c) sapwood layer (d) cambium layer.

16. Age of a tree can be estimated by the
(a) diameter of the bark (b) number of rings
(c) the height of the tree (d) cambium layers.

17. Which one of the following is not an endogeneous tree?
(a) Palm tree (b) Coconut tree
(c) Cane tree (d) Teak tree.

18. Plywood is specified by
(a) volume (b) thickness
(c) weight (d) none of the above.

19. The purpose of seasoning of wood is to
(a) reduce the voids (b) remove the curves
(c) reduce the moisture content (d) change the direction of grains.

20. Tensile strength of wood
(a) is minimum in the direction parallel to the grains (b) is maximum in the direction parallel to the grains
(c) is maximum in the direction across the grains (d) is same in all the directions.

21. Uniform seasoning of wood is done by

(*a*) electric seasoning (*b*) air seasoning
(*c*) kiln seasoning (*d*) water seasoning.

**22.** The state of wood which has no capillary water but contains hygroscopic water only is known as
(*a*) capillary seizure point (*b*) fibre saturation point
(*c*) drying point (*d*) dead wood point.

**23.** As moisture content of wood is reduced, its size
(*a*) increases (*b*) reduces
(*c*) does not change (*d*) any of the above.

**24.** In plywood manufacturing, the veneers are so placed that the grains of adjacent veneers run at
(*a*) 30° (*b*) 45°
(*c*) 60° (*d*) right angles.

**25.** The effect of the attack due to dry rot is to
(*a*) reduce the timber to powder (*b*) reduce strength of wood
(c) crack the wood (*d*) shrink the timber.

**26.** In chamber drying
(*a*) hot gases are passed around wood (*b*) wood is directly put on the fire
(*c*) moisture of wood is extracted by vacuum pumps (*d*) hot and humidified air is passed around wood.

**27.** Plywood is specified by size and
(*a*) volume (*b*) grain direction
(c) number of plys (*d*) strength in longitudinal and transverse directions.

**28.** The moisture content of a well seasoned wood is in the range
(*a*) 5 to 6 percent (*b*) 8 to 10 percent
(*c*) 10 to 12 percent (*d*) 13 to 17 percent.

**29.** Which of the following is broad leaved species of wood?
(*a*) Oak (*b*) Pine
(*c*) Bamboo (*d*) Fir.

**30.** Variation in moisture content of wood also causes variation in
(*a*) colour of heart wood (*b*) colour of bark
(*c*) direction of grains (*d*) volume of wood.

**31.** Wood can be protected against fire by
(*a*) coating with gypsum (*b*) plastering
(*c*) coating with asbestos cement (*d*) any of the above.

**32.** To reduce hygroscopicity and water absorption, wood is
(*a*) painted or varnished (*b*) coated with cement
(*c*) heat treated (*d*) preserved in dry places only.

**33.** The cambium layer exists between
(*a*) heartwood and pith (*b*) bark and heartwood
(*c*) bark and bast (*d*) none of the above.

**34.** Thermal conductivity of wood depends on
(*a*) bulk density (*b*) moisture content
(*c*) porosity (*d*) orientation of fibres
(*e*) all of the above.

**35.** Average thermal conductivity of wood is around
(*a*) 0.3 W/m°C (*b*) 0.8 W/m°C
(*c*) 5.0 W/m°C (*d*) 12.0 W/m°C.

**36.** The disadvantage of plywood is
(*a*) rough edges (*b*) high cost
(*c*) irregular grains on surface (*d*) poor strength to weight ratio.

**37.** .................. is not an exogeneous tree.

(*a*) Babul (*b*) Sal
(*c*) Teak (*d*) Palm.

**38.** Volumetric shrinkage of wood, from green condition to oven dry condition, may be of the order of .................. per cent.

(*a*) 2 to 5 (*b*) 5 to 7
(*c*) 7 to 20 (*d*) 20 to 40.

**39.** Water permeability of wood depends on

(*a*) initial moisture content (*b*) species
(*c*) Width of annual rings (*d*) location of wood in the trunk
(*e*) all of the above.

**40.** By seasoning of wood

(*a*) tendency to warping reduces (*b*) weight reduces
(*c*) workability of wood improves (*d*) susceptibility to decay and insect attack reduces.

**41.** If moisture content of wood is suddenly reduced, it is likely to result in

(*a*) high compressive strength (*b*) fatigue failure
(*c*) development of cracks (*d*) high tensile strength.

**42.** Electrical resitivity of wood largely depends on

(*a*) surface conditions (*b*) direction of grains
(*c*) moisture content (*d*) all of the above.

**43.** By air drying the mositure content of wood can be reduced upto about

(*a*) 5 percent (*b*) 12 percent
(*c*) 20 percent (*d*) 30 percent.

**44.** A 3-ply plywood may have thickness of

(*a*) 3 to 6 mm (*b*) 6 to 10 mm
(*c*) 10 to 12 mm (*d*) 12 to 15 mm.

**45.** Which of the following is water soluble antiseptic for wood?

(*a*) Copper sulphate (*b*) Silica fluoride
(*c*) Sodium flouride (*d*) All of the above.

**46.** Water soluble antiseptics are used as wood preservatives for

(*a*) seasoned wood (*b*) freshly felled wood
(*c*) wood not exposed to moisture (*d*) wood exposed to moisture.

**47.** The shrinkage of wood, from green to over dry conditions, will be maximum in

(*a*) longitudinal direction (*b*) tangential direction
(*c*) radial direction (*d*) wood exposed to moisture.

**48.** The major drawback of air seasoning of wood is

(*a*) loss of strength (*b*) development of knots
(*c*) longer duration (*d*) distortion of grains.

**49.** Which of the following insects generally attack the wood?

(*a*) Beetle (*b*) Powder post
(*c*) Termites (*d*) All of the above.

**50.** Insects attacking wood immersed in sea water are known as

(*a*) marine borers (*b*) powder post
(*b*) white ants (*d*) carpenter ants.

## ANSWERS

| | | | | |
|---|---|---|---|---|
| **1.** (*b*) | **2.** (*c*) | **3.** (*c*) | **4.** (*d*) | **5.** (*b*) |
| **6.** (*c*) | **7.** (*a*) | **8.** (*c*) | **9.** (*c*) | **10.** (*b*) |
| **11.** (*a*) | **12.** (*b*) | **13.** (*e*) | **14.** (*a*) | **15.** (*d*) |
| **16.** (*b*) | **17.** (*d*) | **18.** (*b*) | **19.** (*c*) | **20.** (*b*) |

| | | | | |
|---|---|---|---|---|
| **21.** (*a*) | **22.** (*b*) | **23.** (*b*) | **24.** (*d*) | **25.** (*a*) |
| **26.** (*d*) | **27.** (*c*) | **28.** (*c*) | **29.** (*a*) | **30.** (*d*) |
| **31.** (*d*) | **32.** (*a*) | **33.** (*b*) | **34.** (*e*) | **35.** (*a*) |
| **36.** (*b*) | **37.** (*d*) | **38.** (*c*) | **39.** (*e*) | **40.** (*e*) |
| **41.** (*c*) | **42.** (*c*) | **43.** (*c*) | **44.** (*a*) | **45.** (*d*) |
| **46.** (*c*) | **47.** (*b*) | **48.** (*c*) | **49.** (*d*) | **50.** (*a*) |

## THEORETICAL QUESTIONS

**1.** What are the characteristics of good timber?

**2.** Give the croos-section of a mature tree and name the different parts.
What is the best-season for felling a tree?

**3.** Define, with sketches, where necessary, the following :
Heartwood, sapwood, medullary rays, rindgall, wet rot.

**4.** How is decay in timber caused? Describe a preservation treatment of a wooden railway sleeper.

**5.** Give the chief characteristics and engineering uses of the following timber :
(*i*) Shisham (*ii*) Deodar, (*iii*) Teak, (*iv*) Sal.

**6.** What are the various defects in timber?

**7.** What are the common uses of timber in building industry.

**8.** State the principal causes of decay of timber.

**9.** What is plywood? State its use in modern building.

**10.** What is the best time for felling a timber tree and why?

**11.** Explain the term seasoning of timber. What are the different methods adopted for the principal timbers in India for constructional purposes? Describe with sketches one of the methods.

**12.** Distinguish between 'dry rot' and 'wet-rot'. What steps are taken in executing wood work to see that these defects do not occur?

**13.** Describe briefly the following :
(*i*) Kiln seasoning of timber
(*ii*) Creosoting of timber.

**14.** What are the common defects in timber? Describe the various methods employed in preserving timber and increasing its durability.

**15.** Enumerate the principal destructive agents which generally attack timber in India and describe the methods generally adopted for preservation of such attacks. Also give the names of timber which has natural properties to resist such attacks.

**16.** What are hard-woods and soft-woods? Name a few species of each. How do you differentiate them?

**17.** What are the requirements of a good wood preservative? What are the main types?

**18.** Distinguish between plywood, battenboard and fireboard by means of sketches. Explain what is masonite.

**19.** What is plywood? Describe how it is manufactured.

**20.** State the advantages of plywood as a structural material over ordinary timber.

**21.** How can sapwood of trees be improved in strength and durability?

**22.** What is the object of seasoning timber and what changes are effected by seasoning it?

**23.** What effect has partial immersion in water on timber?

**24.** How is timber protected from (*a*) moisture and (*b*) internal decay?

**25.** Describe any method of protecting timber from white ants.

**26.** What is the object of painting woodwork and what is the effect of paint upon unseasoned timber?

**27.** What timber is best suited for (*a*) Tent poles, (*b*) Tent pegs (*c*) Cart-wheels

**28.** How is timber for engineering and building purposes classified? Classify and describe the following timber, giving the principal uses of each : Deodar, chir, teak, mango, babool, bamboo and sal.

9

# Metals and Alloys

**A. Ferrous Materials :** *9.1. Iron ores. 9.2. Pig iron. 9.3. Composition of Pig iron. 9.4. Effect of impurities on iron. 9.5. Cast iron. 9.6. Classification, composition, properties and uses of cast iron. 9.7. Wrought iron. 9.8. Aston process. 9.9. Manufacture of steel. 9.10. Composition, properties and uses of carbon steels. 9.11. Composition of cast iron, wrought iron, mild steel and hard steel. 9.12. Alloy steels.*
**B. Non-ferrous Materials :** *9.13. Aluminium. 9.14. Copper. 9.15. Lead. 9.16. Tin. 9.17. Zinc. 9.18. Magnesium. 9.19. Nickel. 9.20. Non-ferrous alloys—Copper alloys—Aluminium alloys—Magnesium alloys—Nickel alloys. 9.21. Bearing or antifriction alloys. 9.22. Selection, specifications and forms and availability of materials.—Objective Type Questions—Theoretical Questions.*

## A. FERROUS MATERIALS

### Introduction

- The *ferrous metals* are *iron base metals* which include all varieties of irons and steels. These metals go a long way in bringing prosperity to a country.
- Ordinarily, the terms, *iron, cast iron and steel* in reference to a metal in which the element iron (Fe) is the major element do not refer to a specific metal or alloy, but are loosely used to indicate a general type of iron alloy. The term *iron* should be used only when reference is made to the element iron (Fe). In speaking of the commercial forms of iron such terms as *Pig iron, grey cast iron, wrought iron* etc. may be used. Each of these terms represents some commercial form of element iron, and each form may occur in many variations of chemical composition which influence the functions within each class. Due to tremendous production tonnage of these metals, and to their many forms and varied uses, a detailed study is not a simple and easy task.
- *Steel* (master metal) is obtainable in great quantities, both in wrought and cast form. Its plasticity, whether at room temperature or at elevated temperatures, allows it to be worked either hot or cold. Its combination of strength with plasticity makes it the most important metal for use in large structures. By varying the carbon content and by suitable heat treatments, we can alter the properties from a very soft, workable steel of the type used, in pressed metal parts, wire and similar materials to a hard, strong steel suitable for use in tools and machinery where great strength and hardness are required.
- *Wrought iron* is the oldest form of iron made by man. It was originally produced by slow reduction of the metal from the ore in the forge fire. This reduction process resulted

in a very impure iron which required further refining by mechanical working, that is by hammering or shaping to the form in which it is used. Wrought iron is a metal containing high purity iron and iron silicate in physical association. *It is very low in carbon and the iron silicate or slag is distributed throughout the base metal in fibres which gives it a woody or fibrous appearance when fractured.*

- *Cast iron* is fundamentally an alloy whose chief elements are iron, silicon and carbon. Irons are available with a wide range of properties. Pig iron, grey cast iron, white cast iron, chilled cast iron and malleable iron are all referred to as cast iron, chiefly because these forms of irons are not plastic enough, even when hot to be forgeable; therefore they are always produced commercially by a process of melting and casting into shape, the commercial form of each of these metals is in castings.

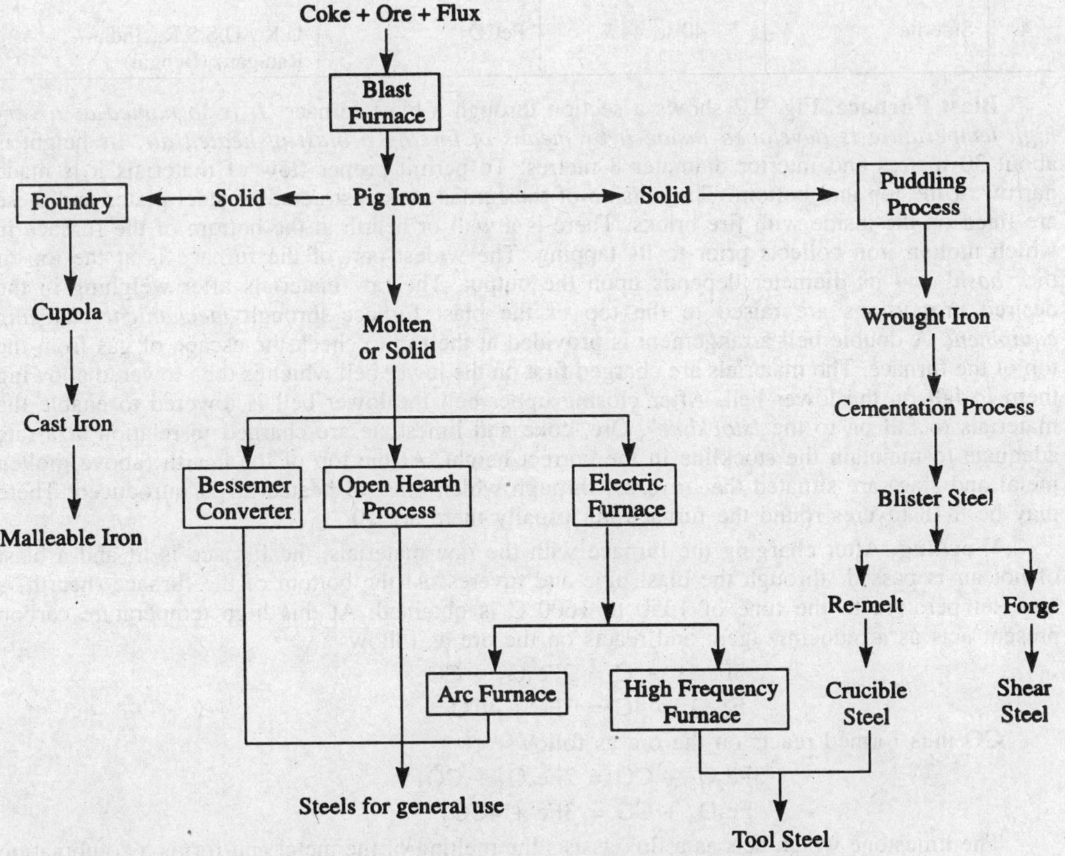


Fig. 9.1: Flow sheet for production of iron and steel.

## 9.1. IRON ORES

The principal ores which yield ferrous metals, the percentage of iron content and the countries where available are given in the Table 9.1.

## 9.2. PIG IRON

Pig iron is the basic material from which, wrought iron and steel are manufactured. It is extracted from the above mentioned ores in a *tall, continuous working furnace* called '**Blast furnace**'. The *product obtained from the blast furnace is crude and impure iron.*

**Table 9.1. Iron Ores**

| S. No. | Name of the ore | Iron content | Chemical formula | Countries where available |
|---|---|---|---|---|
| 1. | Magnetite | 72.5% | $Fe_3O_4$ | India—Salem district (Madras), Sweden, U.S.A., U.S.S.R. |
| 2. | Haematite | 65 to 70% | $Fe_2O_3$ | India (Bihar, Orissa, Andhra, Madhya Pradesh, Mysore), U.S.A. |
| 3. | Limonite | 60% | $2Fe_2O_3.3H_2O$ | France, U.K., Spain, India. |
| 4. | Siderite | 40 to 44% | $FeCO_3$ | U.K., U.S.S.R., India—Raniganj (Bengal) |

**Blast Furnace.** Fig. 9.2 shows a section through a blast furnace. *It is so named as a very high temperature is developed inside it by means of forcing a blast of heated air.* Its height is about 30 metres and interior diameter 8 metres. To permit proper flow of materials it is made narrow at the top and bottom. The outside of the furnace is constructed of steel plates and these are lined on the inside with fire bricks. There is a wall or hearth at the bottom of the furnace in which molten iron collects prior to its tapping. The widest part of the furnace is at the top of the '*bosh*' and its diameter depends upon the output. The raw materials after weighing in the desired proportions are raised to the top of the blast furnace through *mechanical charging equipment.* A double bell arrangement is provided at the top to check the escape of gas from the top of the furnace. The materials are charged first on the lower bell which is then lowered allowing them to fall on the lower bell. After closing upper bell the lower bell is lowered to enable the materials to fall on to the '*stockline*'. Ore, coke and limestone are charged in relation at a rate adequate to maintain the stockline in the correct height. At the top of the hearth (above molten metal and slag) are situated the '*tuyeres*' through which blast of heated air is introduced. There may be 8-18 tuyeres round the furnace but usually there are 10.

**Working.** After charging the furnace with the raw materials, the furnace is lit and a blast of hot air is passed through the blast pipe and tuyeres. At the bottom of the furnace (hearth) a high temperature to the tune of 1350 to 1600°C is obtained. At this high temperature carbon present acts as a reducing agent and reacts on the ore as follows :

$$3Fe_2O_3 + C = 2Fe_3O_4 + CO$$
$$Fe_3O_4 + 4C = 3Fe + 4CO$$

CO thus formed reacts on the ore as follows :

$$3Fe_2O_3 + CO = 2Fe_3O_4 + CO_2$$
$$Fe_3O_4 + CO = 3Fe + 4CO_2$$

The limestone which acts as a flux assists the melting of the metal and forms a combination with the non-metallic portions of the ore to form a metal slag. The floating slag is removed from time to time (usually after two hours) and the molten metal is trapped (generally after four hours) through the hole near the bottom. Blast furnace-slag has a composition within the following range :

| | | |
|---|---|---|
| Silica | = | 28 to 30% |
| Lime | = | 30 to 48% |
| Magnesia | = | 2 to 14% |
| Alumina | = | 12 to 20% |
| Iron oxide | = | 1 to 2% |
| Sulphur | = | 2 to 2.5% |

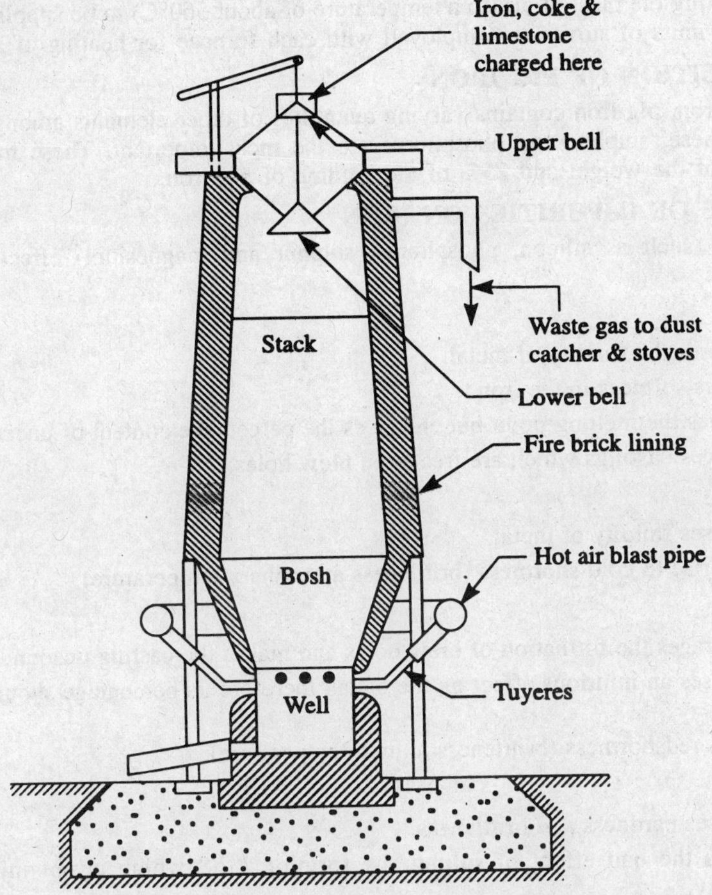


Fig. 9.2: A section through a blast furnace.

The tapped molten metal is run into channels and branches formed the sandy bed. The solidified iron after cooling is cut into pieces in same lengths. These are called *sows* and *pigs* (Fig. 9.3) because when inverted their tops have a curved, hog-black profile.

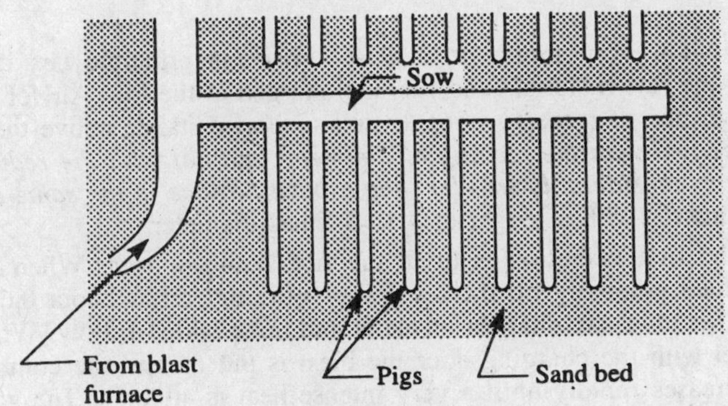


Fig. 9.3: Sows and pigs.

The gases leaving the blast furnace contain about 30% carbon monoxide (CO) which is combustible. The gases after travelling through dust catchers are burnt in a chamber, the heat of which is used in heating firebricks in a stove. The heat stored by the fire-bricks (in a stove) is utilised for heating the blast of air (to a temperature of about 560°C) to be supplied to the furnace Usually 3 to 5 units of stoves are employed with each furnace for heating of air.

## 9.3. COMPOSITION OF PIG IRON

In addition to iron, pig iron contains varying quantities of other elements amongst which carbon, silicon, manganese, sulphur and phosphorus are the most important. These may amount to as much as 10% of the weight and 25% of the volume of pig iron.

## 9.4. EFFECTS OF IMPURITIES ON IRON

The impurities (such as silicon, phosphorus, sulphur and magnesium) effect the iron in the following way :

**1. Silicon :**

(*i*) Increases the fluidity of metal.

(*ii*) Induces softness in the iron.

(*iii*) Reduces the melting point but enhances the percentage content of uncombined carbon.

(*iv*) Produces castings which are free from blow holes.

**2. Phosphorus :**

(*i*) Increases fluidity of metal.

(*ii*) Gives rise to cold shortness (brittleness at ordinary temperature).

**3. Sulphur :**

(*i*) Encourages the formation of blow holes and makes the casting unsound.

(*ii*) Exercises an injurious effect on the metal, therefore its percentage should be kept below 0.1%.

(*iii*) Causes redshortness (brittleness at high temperature).

**4. Manganese :**

(*i*) Increases hardness and brittleness.

(*ii*) Checks the bad effect of sulphur by forming MnS which is not injurious in small quantity.

## 9.5. CAST IRON

The product of the blast furnace *i.e.*, pig iron is unsuitable for castings as it contains impurities in high percentage. To render it suitable for desired purpose it is refined in the furnace known as *cupola*. The refined product is termed as *cast iron*.

### Cupola

Refer Fig. 9.4. It is very similar to a blast furnace in principle *i.e.*, it is a vertical shaft furnace, into which the raw materials and fuel are charged at the top. Air for combustion of fuel is introduced through one or more rows of tuyeres a short distance above the bottom. *Since the cupola is only concerned with the melting of the metal and not with the reduction of ores as in the blast furnace, it is considerably smaller than a blast furnace of the same output.* Its diameter varies from 1 to 2 metres with a height of 4 to 5 times diameter.

In a cupola, the first operation is to lit the fire at the bottom. When the fire is burning strongly, coke is added gradually till the level above the tuyeres is about 0.6 metres. This coke serves as a bed for the alternate charges of metal and coke which follow. When the shaft of the cupola is filled level with the charging door the blast is put on and the combustion of the coke near the tuyeres increases rapidly until a very intense heat is attained. The gases of combustion move upwards and pass on a portion of the heat to the metal and coke waiting to descend. In 5 to 10 minutes the first charge of metal starts melting and trickles down through the coke and finally collects at the bottom of the cupola. When an adequate quantity (say 1 or 2 tonnes) has accumulated the plug of clay called '*bout*' is removed from the tap hole and metal allowed to run into the ladle. The temperature of tapping metal is 1200-1400°C. After melting a number of charges as per requirements the bed coke is removed through a drop-bottom door and quenched with water so as to be available for use the next day.

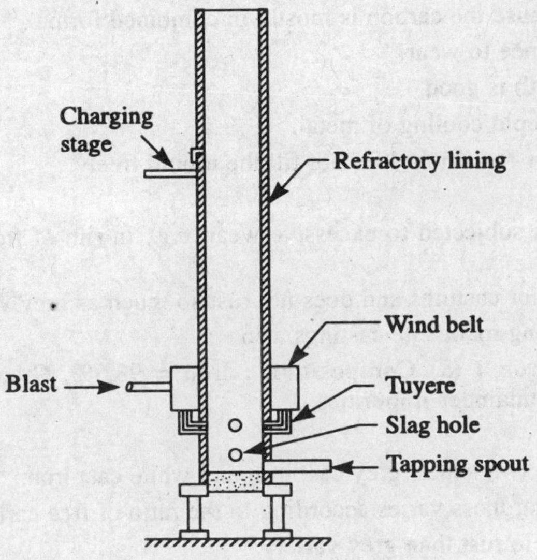


Fig. 9.4 : Cupola.

Although it is usual practice to operate a cupola with cold blast (since no reduction of ores is required) a few cupolas have recently been equipped for hot blast. It may be noted *whereas a blast furnace operates continuously, a cupola works intermittently.*

## 9.6. CLASSIFICATION, COMPOSITION, PROPERTIES AND USES OF CAST IRON

Cast iron may be classified as follows :

1. Grey cast iron
2. White cast iron
3. Mottled cast iron
4. Nodular cast iron
5. Malleable cast iron
6. Alloy cast iron.

**1. Grey cast iron :** Grey cast iron is characterised by presence of a large portion of its carbon in the form of graphite flakes. Although grey iron is often defined as steel containing graphite, its properties are far different from those of steel. *Owing to its low specific gravity, graphite occupies over three times as much space as an equal weight of metal.* From 6 to 10 per cent of the volume of a typical grey iron is occupied by graphite. Because graphite flakes contribute no more to strength than so many voids, the *tensile strength and elasticity of grey iron is considerably less than that of steel.* Nevertheless, grey iron is most widely used engineering material because of the *ease with which it will be used, its excellent machinability and its good antifriction properties, vibration-damping properties and stress relieving properties.*

- *The strength properties of grey cast iron are improved by reducing the flake size and distributing the flakes more uniformly.*
- *Because of the ease of casting grey iron in their sections and its electrical resistance properties, it is frequently used in resistance grids for control apparatus of heavy duty D.C. motors.*
- *The quantity of castings produced in grey iron far exceeds that of any other cast metal. The raw materials are low in cost and economical to melt. Its casting properties, such as fluidity and relatively low shrinkage, ensure a high yield of good castings.*

**2. White cast iron :** (*a*) **Composition :** Iron = 94%, Graphite carbon = 0.5%, Combined carbon = 3.5% and the remainder other impurities.

(*b*) **Properties :**

(*i*) Hard, brittle and cannot be machined.

(*ii*) So called because the carbon is mostly in combined form.

(*iii*) Highly resistance to wear.

(*iv*) Tensile strength is good.

(*v*) Obtained by rapid cooling of metal.

(*vi*) Due to its poor fluidity it does not fill the mould freely.

(*c*) **Uses :**

(*i*) Used for parts subjected to excessive wear, *e.g.*, in rim of freight car wheel or railway brake blocks.

(*ii*) Used for inferior castings and does not rust so much as grey variety.

(*iii*) Used for making malleable castings also.

**3. Mottled cast iron :** (*a*) **Composition :** Iron = 93.5%, Graphite = 1.75%, Combined carbon = 1.75% ; the remainder impurities.

(*b*) **Properties :**

(*i*) Mixture of the two states, grey cast iron and white cast iron.

(*ii*) Strength and hardness varies according to the ratio of free carbon to combined carbon.

(*iii*) Less tendency to rust than grey variety.

(*iv*) Fluidity is good.

(*v*) Hard and brittle.

(*c*) **Uses :** Used for *manhole covers and pipes.*

**4. Nodular cast iron :** It is also called *ductile cast iron* and is produced by ladle addition of nickel-magnesium alloy. It is cheaper to make and behaves like steel. It yields better machined surfaces. Machine parts requiring more toughness are often cast from this cast iron.

**5. Malleable cast iron :** It is obtained by giving a long time annealing to cast iron. It is soft, tough and easily machined. To get high cast irons, steel scrap 10 to 40% may be added to lower the total carbon contents. *It is used in differential and steering gear trousing, brake pedals, tractor springs, hangers and washing machine parts.*

The *procedure of obtaining malleable cast iron* is as follows :

The white cast iron casting along with silica is packed in a steel pot in heated muffle oven or continuous type thermal kiln. The temperature is maintained at 870°C for 60 hours and casting is cooled slowly in the furnace, iron carbide dissociates as $Fe_3C \rightarrow Fe + C$ and thus reduces to malleable cast iron.

Malleable cast iron is of two types : (*i*) *Black heart*, and (*ii*) *White heart*. White heart malleable cast iron can be obtained by keeping temperature and time comparatively of high values.

**6. Alloy cast iron :** Because the cast iron is supposed to be very hard, brittle, lacking in tensile and transverse strength and weak to withstand shocks it is alloyed with other metals to improve its properties. Amongst the alloying metals, *nickel is predominating alloying constituent whose addition to the extent of 0.5 to 1.5% avoids the tendency of chilling or hard spots.*

Two typical examples of alloy cast iron are : (*i*) *Acicular*, and (*ii*) *Spheroidal*. Acicular cast iron has nickle and molybdenum as the leading constituents and is *employed for crank shafts. Spheroidal cast iron* has graphite content in spheroidal form, which is converted from flaky form by alloying with a small amount of magnesium and cesium. *This change in graphite increases the tensile strength and produces a tough metal which can undergo bending and twisting.*

## 9.7. WROUGHT IRON

Wrought iron is virtually pure iron, containing a large number of minute thread of slag lying parallel to each other, thereby giving the metal a fibrous appearance when broken. *It contains practically no carbon and therefore does not harden when quenched in water* (this property distinguishes it from steels).

The manufacture of wrought iron from pig iron (laborious and expensive process) involves the following operations :

1. Refining ; 2. Puddling ; 3. Shingling ; 4. Rolling.

**1. Refining :** This operation consists in passing through the molten pig iron a strong current of air and keeping it well stirred in order that liquid mass may come in complete contact with air and get oxidised. The oxygen present in the air eliminates a portion of carbon and some other impurities. The liquid mass or iron is then cast into moulds and made brittle by sudden cooling.

**2. Puddling :** Puddling consists in melting the refined pig iron (broken into lumps) in a reverberatory furnace (Fig. 9.5). The term '*reverberatory*' is applicable to any furnace in which the charge does not come in direct contact with fire but receives heat from the roof by reflection.

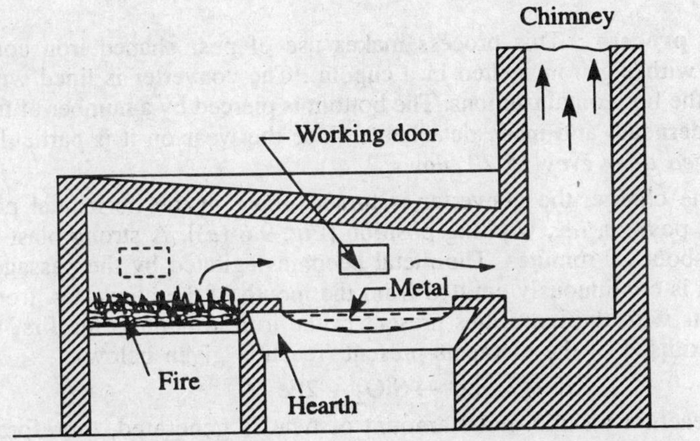


Fig. 9.5 : Reverberatory furnace.

The molten metal is mixed with oxidising substances such as haematite, oxide of iron etc., and subjected to enormous amount of heat and strong current of air. Major portion of carbon content remaining in the iron gets converted into carbonic acid gas. The silicon gets oxidised and is removed in the form of slag. As the iron is purified, it thickens up and removed from the furnace in the form of balls. These plastic balls are known as '*puddle balls*'.

**3. Shingling :** In this operation, the puddle balls are hammered to remove any particles of cinder associated with them and the iron particles welded together to form a "*bloom*".

**4. Rolling :** The blooms obtained during the previous operation are passed through grooved rollers as a result of which they get converted into puddle bars. These bars are wrought iron of lowest quality. Their quality is improved by subsequent processes of piling, reheating and rerolling.

## 9.8. ASTON PROCESS *(for manufacturing wrought iron)*

This process was invented by James Aston (U.S.A.) in 1925. In this process the molten steel, obtained from Bessemer converter (Fig. 9.6) which contains large amount of dissolved gas and at a temperature of 1500°C is poured into liquid slag maintained at 1200° C. As the steel strikes the slag it gets frozen and the dissolved gases are given off. Thus a spongy mass composed of spherical particles is produced at a temperature of about 1375° C. With this process just within a few minutes as much as 3 or 4 tonnes of metal is produced. This method produces wrought iron at a much cheaper cost compared to puddling process.

## 9.9. MANUFACTURE OF STEEL

The principal methods of manufacturing different grades of steel are : 1. *Cementation process*, 2. *Crucible process*, 3. *Bessemer (acid or basic) process*, 4. *Open hearth (acid or basic) process*, 5. *Electric process*, 6. *L.D. process*, 7. *Duplex process*, 8. *Kaldo process.*

**1. Cementation process :** The pure wrought iron bars are arranged in the furnace between the layers of powdered charcoal and maintained at higher temperature (about 1500°C) for 5 to 15 days depending upon the quality of steel required. The wrought iron gets covered with blisters due to combination of carbon and thus the steel so produced is termed as *'blister steel'. Blister steel has fissures and cavities and also its structure is not homogeneous. It may be employed for machined parts and facing hammer but less suitable for edge tools.*

**2. Crucial process :** This process of manufacturing steel consists in heating in fire clay crucible either fragments of 'blister steel' or short lengths of wrought iron bars with charcoal and running the metal into iron moulds. The steel thus produced is known as *cast steel. It is extremely hard and uniform in quality. It finds its use in making finest cutlery and hardest cutting tools.*

**3. Bessemer process :** This process makes use of pear shaped iron converter (Fig. 9.6) which is supplied with pig iron melted in a cupola . The converter is lined with fire bricks and can revolve about the horizontal trunions. The bottom is pierced by a number of tuyeres connecting with wind box underneath and made detachable since the wear on it is particular is very severe. *It is usually changed after evey 15-20 blows.*

To receive the charge, the converter is turned down in the horizontal position and again brought to vertical position *i.e.*, working position [Fig. 9.6 (*a*)]. A strong blast of air is supplied to the charge for about 20 minutes. The metal becomes agitated by the passage of air blast and a shower of sparks is continuously emitted from the mouth of the converter. Iron oxide is formed which in turn reacts with the impurities present in the iron. Silicon is the first to be removed as it is most easily oxidised of the elements present (reaction given below).

$$Si + 2\,FeO \rightarrow SiO_2 + 2Fe$$

Due to this reaction an enormous amount of heat is generated, therefore, it is important that iron should contain adequate quantity of silicon for the purpose. *Iron having less than 1.5 per cent of silicon will blow cold, while more than 2.5 per cent will generate too much heat.*

As the blow proceeds, a flame gradually appears at the mouth of the vessel and manganese starts oxidising. Some of the oxides of silicon, manganese and iron are ejected in the form of brown fumes but those remaining in the converter combine to form a slag which floats on the metal.

The last element to be eliminated is the carbon which follows the reaction :

$$C + FeO \rightarrow CO + Fe$$

This stage of process is known as *'the boil'* owing to the violent agitation caused by the evolution of carbon monoxide gas from the metal. *As the carbon is eliminated the boil gradually subsides and large white flame finally drops after sometime. The metal in the converter is now virtually pure iron in the liquid form.* The blast is shut-off and a quanity of ferro-manganese (an alloy of iron containing manganese upto 80% and carbon upto 5%, used in making addition of Mn to steel or cast iron), spiegeleisen (as alloy of iron containing 15-30% Mn and 4-5% carbon; added to steel as deoxidising agent and to raise the Mn-content of the steel) or some other suitable alloy steel as required is added and the blast resumed for a few minutes more to ensure the alloy throughly mixing with iron. The converter is then swung into discharging position shown in Fig. 9.6(*b*) and discharged into ladles from which the metal is poured into moulds where it solidifies to form ingots. *To remove the blow holes formed due to bubbles of nitrogen or oxygen, a small quantity of aluminium or ferro-silicon is also added.*

*A Bessemer converter is furnished with an acidic lining (ganister) if the pig iron is free from sulphur and phosphorus, otherwise a basic lining of dolomite is employed.* For removal of

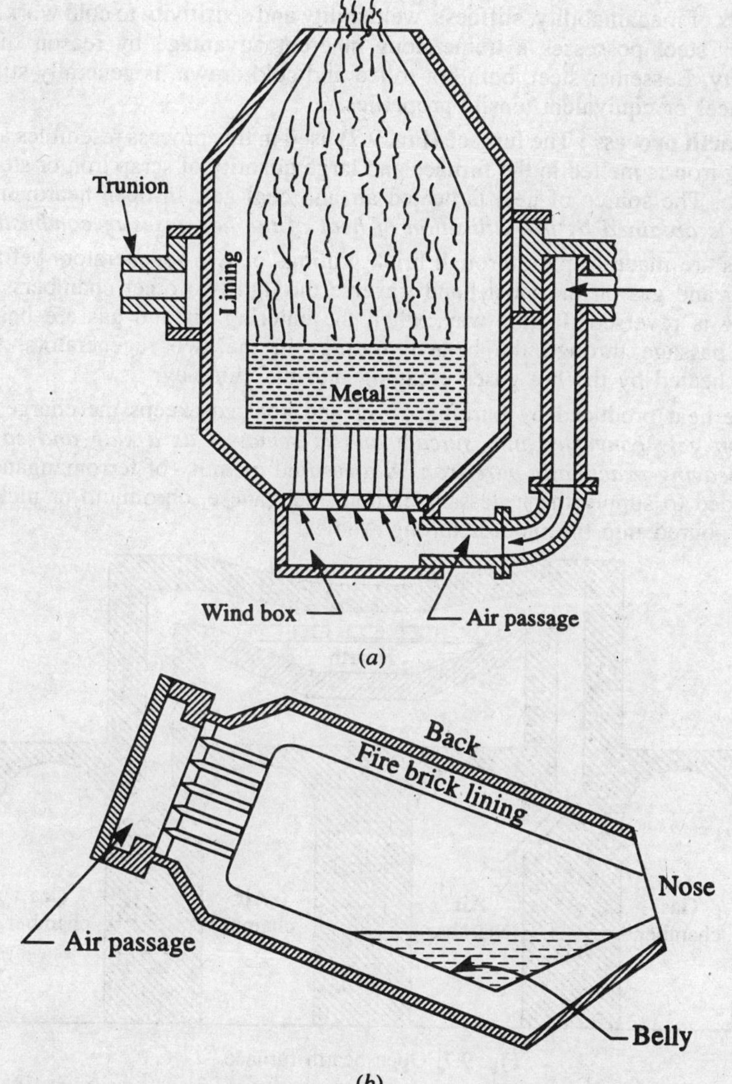


Fig. 9.6 : Bessemer converter : (a) Working position, (b) Discharging position.

sulphur, calcined lime is added to the charge and then heated to red heat. A portion of it burns out while a portion combines with lime to form slag. The basic lining is least affected by this slag containing phosphorus.

**Advantages and disadvantages of Bessemer converter :**

- The cost of building a Bessemer converter plant is comparatively less than that of an open hearth plant of equivalent capcity.
- A Bessemer converter, with its small size and ease of handling, makes it possible to produce small tonnages of many different grades of steel, actually three grades of steel in less than an hour.
- As compared to open hearth furnace, major repairs can be made in a fraction of time.
- The flexibility is all the more attractive in times of slack demand when quick deliveries of small tonnages are desirable from stand point of the manufacture.
- Probably because of the higher ferrous oxide content of acid grade, it has peculiar

properties of machinability, stiffness, weldability and sensitivity to cold work. High-sulphur Bessemer steel possesses a tremendous inherent advantage by reason of ease of machinability. Bessemer steel, both hot rolled and cold drawn, is generally stiffer than open hearth steel or equivalent tensile properties.

**4. Open hearth process :** The furnace (Fig. 9.7) used in this process resembles a reverberatory furnace. Here pig iron is melted in the furnace and large quantity of scrap iron or steel is dissolved in the liquid bath. The source of heat is heated air and coal gas. In open hearth furnace a *great economy in fuel is obtained by the utilisation of heat of the hot gases of combustion.*

These gases are made to pass through brick gratings of two regenerators before leaving the chimney. The air and gas on the other hand traverse through two other chambers. After an hour direction of flow is reversed. In this way, when the entering air and gas are being adequately heated in their passage through the heated gratings in the two regenerators, the other two regenerators get heated by the hot gases escaping into the chimney.

The intense heat produced by burning of air and coal gas keeps the charge in the molten condition. *Silicon gets converted into silicate and is removed as a slag and carbon burns to carbon dioxide leaving practically pure iron.* A measured quantity of ferromanganese or suitable alloy steel is added to supply the necessary carbon, manganese, chromium or nickel. The liquid steel, finally, is poured into the moulds through ladles.

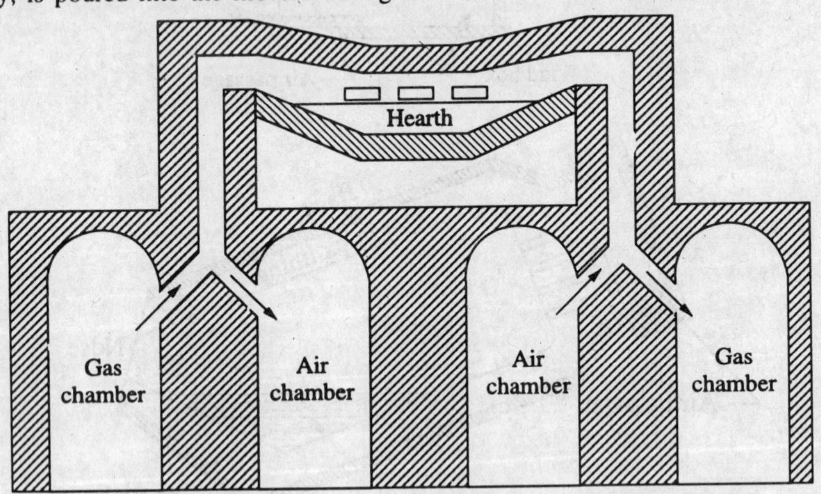


Fig. 9.7: Open hearth furnace.

The furnace *lining* may be *basic* or *acidic* depending upon the fact whether phosphorus is present or absent in the pig iron. Acid lining comprises of either silica or ganister (83% quartz, 13% clay, 4% moisture and other impurities). Basic lining is usually magnesite ($MgCO_3$) or dolomite ($CaCO_3.MgCO_3$).

The *basic open hearth* process is steadily replacing the acid open hearth due to the following reasons :

- The process is not limited to relatively low phsophorus and sulphur materials but, instead makes possible the removal of relatively large amounts of these elements and by so doing, opens up vast tonnages of ores and steel scrap which ordinary could not be available for acid steel.
- It *permits the use of all types of steel scraps;* light and bulky, heavy and dense, dirty and clean to produce high grade steel of any chemical analysis.
- It is much more flexible process permitting operator to vary the proportions of steel scrap, solid pig iron and molten metal to meet local and economic conditions.
- There is relatively, wide permissible range in the chemical composition and physical characteristics of lining materials, the slag and molten metal.

- Steel of wide range of carbon content, from as low as 0.03 to as high as 1.1 per cent can be made.

As compared to Bessemer process, the basic open hearth is a leisurely method giving the operator an ample time in which to make chemical analysis and slag and metal adjustment.

**Table 9.1: Comparison between Bessemer and Open Hearth Processes**

| S. No. | Bessemer Process | Open Hearth Process |
|---|---|---|
| 1. | It can use comparatively higher phosphatic pigs of wider ranges of composition. | It can use lower phosphatic pigs of narrower ranges of composition. |
| 2. | Scrap iron cannot be used. | Scrap iron can be used. |
| 3. | Refining and finishing require 10 to 20 minutes. | Refining and finishing are completed within 8 to 10 hours. |
| 4. | Operations depend entirely on eye judgement, so it is very difficult to produce a uniform product. | Operation is guided by laboratory analysis, so it is comparatively possible to produce a uniform product. |
| 5. | *Bessemer process produces an inferior quality steel associated with blow-holes and inclusions.* | *Open hearth steel is of much superior quality containing lesser blow-holes and inclusions.* |
| 6. | Carbon is completely eliminated first and so requires large amount of recarburizer and the product becomes less homogeneous. | Carbon is partially removed and requires smaller amount of recarburizer and the product is more homogeneous. |
| 7. | Low capital investment. | High capital investment. |
| 8. | Output low (80% of pig used). | Output high comparatively. |
| 9. | Because of rapidity of the process, the control of different operations is rather difficult and careful working is necessitated. | Operation is comparatively easy and finishing upto the desired analysis is feasible. |

**5. Electric process:** This process makes use of two types of electric furnaces :

(*a*) Arc type (*b*) High frequency type.

(*a*) ***Arc type furnace :*** Fig 9.8 shows a three phase direct arc type furnace. It consists of a circular steel casting lined inside with refractory material. The roof is removable and a spare is usually kept for rapid replacement. The roof is provided with three holes through which pass the electrodes. The electrodes may be of *graphite or amorphous carbon,* the former materials has double the conductivity and will carry 2.5 times the current ; hence, the graphite electrodes are usually about 2/3rd diameter of amorphous carbon electrodes. Though graphite electrodes are costly yet the above advantages are sufficient to account for their choice for most arc furnaces.

To maintain a desired length of arc, the electrodes are raised and lowered individually by electric motors operated by automatic regulators. The voltage between steel and electrodes may be 40-145 volts; the *longer the arc, higher the voltage required and the less the input of heat to the furnace.* Electric power is supplied in bulk in the form of three phase alternating current at 6.6 or 10 kV. A transformer set up close to the furnace reduces the voltage down to that required for the arcs and its primary winding having tappings to allow for adjustments to the arc voltage. As the power supply is a three phase circuit, three electrodes are arranged in an *equilateral triangle* over the metal. Owing to low voltage required by the arc, the current must be very high to obtain the desired output.

The hearth of an electric furnace may have acid and basic lining depending upon the process adopted. *Basic process is used for making steel ingots and some casting while the acid process mostly for making steel castings.*

The usual size of this furnace is between 5 to 10 tonnes, though 50 and 100 tonnes furnaces have been produced. This type of furnace is *used* for making alloy steels such as stainless, high speed steel etc.

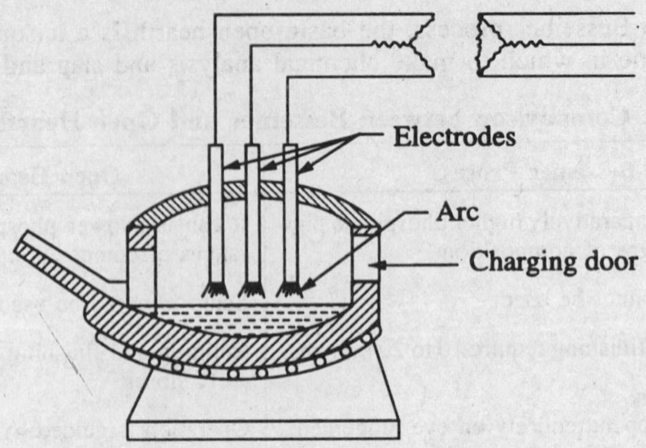


Fig. 9.8 : Three-phase direct arc type furnace.

The *advantage* of this furnace is that *purer product is obtained and composition can be exactly controlled during refining process. This is the reason that direct arc furnace even being costlier in initial as well as operating cost is preferred.* Though this furnace is employed for melting and refining but *due to higher cost its use is restricted to refining than melting.* It operates at a power factor of about 0.8 lagging.

(*b*) **High frequency electric furnace :** This type of furnace *works on the principle that when a piece of metal is held in a coil of wire carrying alternating current, eddy currents flow in steel due to the alternating magnetic field being produced. Due to these currents steel is heated up.* When the charge gets melted, the eddy currents flowing through molten metal agitate or stir it and mix the constituents throughly. Fig. 9.9 shows a high frequency induction furnace which works upto *10,000 cycles/second* and is most suitable when the charge is of the *same composition as the casting.* In this furnace *tubular shaped conductors* are used to (*i*) *produce inductive effect* and (*ii*) *withstand high temperature at high frequency.* The wire solenoid is not used as it cannot stand as high a temperature as 1800°C which may be produced in this furnace.

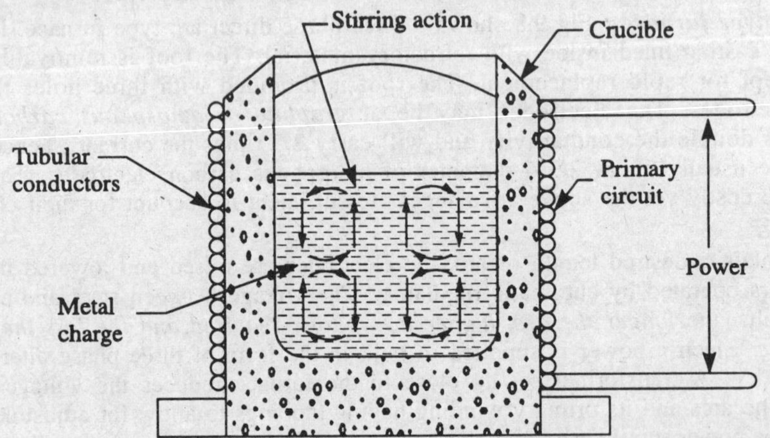


Fig. 9.9: High frequency induction furnace.

It is usual practice to operate the high frequency furnace purely as a melting furnace and only rarely are refining operations attempted. For this reason it is possible to avoid oxidising conditions during melting and steels containing high percentages of easily oxidised elements like chromium, manganese and vanadium can be melted with practically no loss of these elements.

The perfection cannot be attained in the arc furnace, so that the high-frequency furnace reigns supreme for melting such materials as stainless steel, high speed steel, magnet steel and other high alloy steels.

Even when operated intermittently, the high-frequency furnace can prove economical owing to very small weight-refractory material which has to be heated up with steel. *A high-frequency furnace costs about three times as much to install as does an arc furnace of the same capacity.*

**Advantages of the electric furnaces :**

1. The generation of heat within an electric furnace is independent of the furnace atmosphere, and this may be maintained *oxidising, neutral or reducing, at will.*
2. The *temperature is more under control* than in an open hearth furnace and very high temperature can be attained by means of the arc.
3. The close control over steel-making conditions possible in the electric furnace permits high-grade steel to be made from low grade scrap, often without using pig iron. *Electric steels are mostly of superior qualities not normally made by an open hearth process.*
4. The gas, fumes and impurities which exist in fuel bed furnaces are absent.

**6. L.D. process :** This process originated in 1953 at Linz steel works in Austria. In India it is being used in the Hindustan steel plant at Rourkela (Orissa). *It combines the high productivity of the acid Bessemer process and the superior quality of the basic open hearth steel.*

The furnace (Fig. 9.10) is a vessel similar to that of the Bessemer converter except that the capacity is between 30 to 40 tonnes (double that of Bessemer) and there is no detachable bottom with its row of tuyere bricks. The vessel has a basic lining.

In this process a jet of pure oxygen (99%) is blown at a presure of 7.0 to 10.5 bar and at a speed even greater than the sound to the molten bath of hot metal. The temperature thus produced is about 2550°C due to which the impurities like carbon, nitrogen, phosphorus and sulphur etc., are burnt. For a converter of capacity 50 tonnes, the blow lasts for about half an hour while the complete operation consumes about one hour.

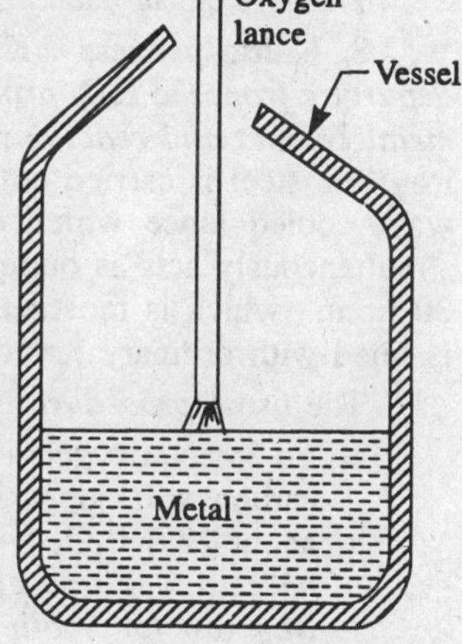


Fig. 9.10

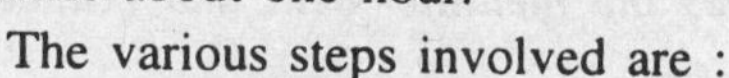

The various steps involved are :

- The converter still hot from previous blow is charged with 15% steel scrap and 85% molten pig iron from the mixer.
- Oxygen is blown through water cooled nozzel at a pressure of 7 to 10.5 bar and lime added.
- The blow lasts for about half an hour, the dropping of flame indicates that the molten mass has been refined.
- The separation of low carbon steel is effected by pouring out steel into a ladle and the liquid slag thickened with burnt lime is left mostly in the converter.

L.D. process is the *only process where sulphur can be effectively reduced.* The *reasons* for this are as follows :

- Large volume of slag which is highly rich in lime is formed and is kept fluid due to intense heat. Sulphur is eliminated as gas.
- The residual manganese in bath is higher and MnS gets oxidised giving out $SO_2$ gas.

**7. Duplex process :** This process is designed to combine the advantages of the Bessemer and open hearth process while avoiding their disadvantages. For duplexing *two methods* are used :

— In one method, hot metal from the blast furnace or mixture is partially blown in a basic converter and then transferred to a basic open-hearth furnace, where refining is completed in 5-6 hours.

— In another method, five acid Bessemer converters are used to eliminate silicon, manganese and carbon and the soft steel so made is passed on to a 200 tonnes basic tilting furnace for removal of sulphur and phosphorus. A sixth converter is used to supply partially blow metal for recarburizing the steel in the tilting furnace, which is tapped every $4\frac{1}{2}$ hours, leaving 50 tonnes of steel in the furnace.

Duplex process has the following *characteristics* :

- The life of open hearth is increased because of the absence of silica oxide which has detrimental effect on the basic lining.
- Pig iron associated with silica and phosphorus contents is converted into a good quality of steel. Silica undergoes oxidation in acid Bessemer converter while phosphorus in the basic open hearth process.
- Since the time of open hearth heat is reduced, labour, maintenance and overhead costs are less.

This system is used at Tata Iron and Steel Works, Jamshedpur (Bihar).

**8. Kaldo process** : Kaldo process was introduced by Professor Killing. *The fundamental departure from the L.D. process is the rotation of the bath during oxygen injection where by slag metal contact and reactions are affected.* In the Kaldo rotary oxygen process conversion of pig iron into steel is carried out into a rotating converter into which oxygen is introduced through a water-cooled lance which enters the converter through a central opening at one end, which simultaneously acts as outlet for the exhaust gases. The speed of the rotation can be varied upto 30 r.p.m., which is most suitable speed for main stage of the process. The furnace or converter is lined with ordinary tarred dolomite and has a *life of 50 heats.*

The usual *procedure* is as follows :

- To inject oxygen until most of the phosphorus has been removed at which stage the carbon will approximate 1.5%, when the first slag is removed. The first slag contains about 22% $P_2O_5$ with only 3% Fe. Owing to heat evolved during oxygen injection it is *necessary to add either iron ore or scrap as a cooler. The usual practice is to employ iron ore for cooling purposes, since this increases the metallic yield and reduces the oxygen.*
- After the *first slag* has been removed blowing is continued until the carbon has been reduced to about 1.0% by which time the phosphorus will be less than 0.1%. At this stage *second slag* is removed. The second slag contains about 17 to 20% $P_2O_5$ with about 6% iron. These two phosphoric slags, which average 16 to 32% $P_2O_5$ can therefore be sold for fertilizing purposes.
- After removing the second slag, a final slag is collected with *lime additions* and blowing is continued until the metal has been refined to the desired specification. The final slag remains in the vessel as a base for subsequent heat. The metallic yield is about 92% and the average blowing time is 35 to 40 minutes for 30 tonnes yield.

Kaldo process entails the following **advantages** :

1. Gives high rates of productivity at a low capital cost.
2. Particularly suitable for manufacturing high grade steel from high phosphorus pig iron and the surplus heat produced in the process permits the employment of an appreciable quantity of iron ore as a cooling agent ; thereby increasing the molten yield.
3. The satisfactory by-product value of high phosphoric acid slag produced is another economic advantage.

The main *disadvantage* of this process is the *number of moving parts and their weight which probably exercise the adverse effect on the maintenance cost.*

## 9.10. COMPOSITION, PROPERTIES AND USES OF CARBON STEELS

*Carbon steels* are classified as follows :

(*i*) Low carbon steels or mild steels

(*ii*) Medium carbon steels

(*iii*) High carbon steels.

**Table 9.2. Composition, Properties and Uses of Carbon steels**

| Types of Steel | Carbon (%age) | Tensile strength ($MN/m^2$) | Uses |
|---|---|---|---|
| | Below 0.1 | 320 to 400 | Galvanished plates, tin plates and wires. |
| **Mild Steels** (good for welding) | 0.1 to 0.18 | 400 to 450 | Boiler plates, ships plates, case-hardening for gear wheels, cams. |
| | 0.15 to 0.25 | 400 to 500 | General engineering purposes, crank axles, shafting. |
| | 0.25 to 0.35 | 500 to 600 | Hydraulic cylinders, rams, turbine motor shafts, spindles. |
| **Medium carbon steels** | 0.35 to 0.45 | 550 to 700 | Rails, turbine discs, connecting rods, railway and tram axles. |
| | 0.45 to 0.55 | 700 to 850 | Rifle barrels, gun parts, shells, gear wheels. |
| | 0.55 to 0.65 | 800 to 950 | Die blocks, gears, wheels tyres and mandrels. |
| | 0.65 to 0.75 | 950 to 1100 | Hammers, crusher rolls, general tools. |
| **High carbon steels** | 0.75 to 0.85 | 950 to 1100 | Hand chisels, scissors and ball mill parts. |
| | 0.85 to 1.5 | 950 to 1100 | Drills, taps, ball races, dies and wood tools. |
| | 1.0 to 1.3 | 850 to 1050 | Razors, drills, tools and wire dies. |

## 9.11. COMPARISON OF CAST IRON, WROUGHT IRON, MILD STEEL AND HARD STEEL

**Table 9.3. Comparison of Cast iron, Wrought iron, Mild steel and Hard steel.**

| S. No. | Cast iron | Wrought iron | Mild steel (Low carbon steel) | Hard steel (High carbon steel) |
|---|---|---|---|---|
| 1. | It contains 2 to 5% carbon. | Purest form of iron containing 0 to 0.25% carbon. | Contains 0.08 to 0.35% carbon | Contains 0.55 to 1.3% carbon. |
| 2. | It has a crystalline, coarse granular structure. | Fibrous structure of bluish colour. | Bright fibrous structure. | Fine granular structure. |
| 3. | Hard and brittle | Tough and more elastic than cast iron. | Tough and more elastic than wrought iron. | Tough and more elastic than mild steel. |

| | | | | |
|---|---|---|---|---|
| 4. | Can be hardened by heating and sudden cooling but cannot be tempered. | Can neither be hardened nor tempered. | Can be hardened and tempered but not easily. | Can be hardened and tempered readily. |
| 5. | Can neither be forged nor welded. | Can be easily forged and welded. | Can be readily forged and welded. | Cannot be easily forged and welded. |
| 6. | Cannot be magnetised. | Can be temporarily magnetised. | Can be permanently magnetised. | Can be permanently magnetised. |
| 7. | Melting point = 1200°C. | Melting point = 1530°C. | Melting point = 1400°C. | Melting point = 1300°C. |
| 8. | Neither malleable nor ductile. | Very malleable and ductile. | Malleable and ductile. | Brittle and less ductile than mild steel. |
| 9. | Does not rust easily. | Rusts more rapidly than cast iron. | Rusts readily. | Rusts rapidly. |
| 10. | Cannot absorb shocks. | Can stand sudden and excessive shocks. | Absorbs shocks. | Absorbs shocks. |
| 11. | Tensile strength fair and compressive strength good. | Tensile strength better and compressive strength less than cast iron. | Tensile strength is better than cast iron and wrought iron while compressive strength better than wrought iron but less than cast iron. | Both tensile and compressive strength better than cast iron, wrought iron and mild steel. |
| 12. | Becomes soft in salty water. | Stands salty water better than cast iron. | Not much affected by salty water. | Not much affected by salty water. |
| | *Uses* : (*i*) It is used for making bed plates, columns, rail chairs, brackets and machine parts not subjected to heavy shocks or tension. | (*i*) It is used for making rolled iron joints, angle iron etc. | (*i*) Used for all kinds of structural work in bridges and buildings, for making channels, griders, rails, angle iron etc. | (*i*) Used for dies, cutlery and edge tools. |
| | (*ii*) As it does not rust easily so it is used for making sewers, drain pipes, water pipes etc. | (*ii*) Since it can withstand shock it may be used for crane hooks, chains, railway couplings etc. | (*ii*) Also used for rivets, bolts, wire, tapes and for making sheets. | (*ii*) Also used in prestress concrete. |
| | (*iii*) As it is poor in tension, therefore least suitable for structural purposes. | (*iii*) Also used for small size water pipes fittings, corrugated sheets, cores of electro-magnets etc. | | |

## 9.12. ALLOY STEELS

When certain special properties are desired some elements such as nickel, chromium, manganese, vanadium, tungsten etc. are added to the carbon steels. The steels thus obtained are called **alloy steels**.

The first investigation on the effect of alloying elements in steel was made from 1875 to 1890. But the use of alloyed steel found little application until 1901, when reduced cost of alloys made their use practicable.

*The alloying elements are added to accomplish one or more of the following :*

- *To impart a fine grain size to steel.*
- *To improve casehardening properties.*
- *To improve elasticity.*
- *To improve corrosion and fatigue resistance.*
- *To improve hardness, toughness and tensile strength.*
- *To improve machinability etc.*

**The Effects of Alloying Elements :**

| Metal | Remarks |
|---|---|
| *Nickel* | • Increases toughness. |
| | • Improves response to heat treatment especially in large sections. |
| | • In large amounts, provides special electrical and magnetic properties. |
| | • Improves forming properties of stainless steel. |
| *Chromium* | • Provides stainless property in steel. |
| | • Used widely in tool steels and in electric plates. |
| *Manganese* | • Improves properties and response to heat treatment. |
| | • Provides control of structure. |
| | • Used in high speed tool steels. |
| *Tungsten* | • Retention of hardness and toughness at high temperature. |
| | • Used in tools, dies, valves, magnets etc. |
| *Silicon* | • Provides high electrical resistance and magnetic permeability used in electrical machinery. |
| *Copper* | • In small amounts improves atmospheric corrosion resistance. |

Some of the important alloys are discussed below :

**Silicon steel**

- Silicon improves considerably the *electrical properties of steel.*
- Silicon imparts *fatigue, strength and resistance to steel.*
- Steel containing silicon is more ductile than a plain carbon steel.
- *Steel containing 3.5 per cent silicon has very low magnetic hysteresis.*

**Uses :** 1. Steel with Mn = 1%, Si = 2%, C = 0.4 to 0.6% has very high elastic limit and is used for springs.

2. With Cr = 5 to 7%, Si = 2 to 4%, C = 0.4 to 0.5%, steel retains its hardness and resistance to oxidation even at red heat. Such steel is used for internal combustion engines.
3. Silicon steel containing 2.5 to 4% silicon and low percentage of carbon and manganese is employed for making *laminations of electrical machines*.
4. 13% silicon content steel has a *very high corrosion resistance* and so is used in chemical industries.
5. Steel containing 0.6 to 1% silicon and 0.75 to 0.95% manganese is used for structural purposes.

**High Speed Tool Steels :**

- High speed tool steels are widely used for cutting of metals where *hardness must be retained at elevated temperatures.*
- These steels are obtained by alloying tungsten, chromium, vanadium, cobalt and molybdenum with steel. This alloying produces metals which remain hard at temperatures at which normal steels become quite soft. *A common analysis is 18% tungsten, 4% chromium and 1% vanadium with a carbon content of 0.6 to 0.7%.* This alloy is termed as 18 : 4 : 1 while an increase of vanadium to 0.2% produces 18 : 4 : 2 steels. Cobalt is often added to 18 : 4 : 1 and 18 : 4 : 2 steel as it improves the red hardness and cutting ability; from 5 to 10% is used.
- Another class of high speed steel contains a lower percentage of tungsten, this being compensated by the addition of molybdenum; steel of this type is being increasingly used today.
- In addition to heat resistance, high speed steels have the desirable properties of high hardness, high compressive strength and outstanding wear resistance. They are close competitors to carbides for metal cutting-tool material such as drills, reamers, milling cutters etc.

**Heat Resistance Steels :**

- Steels which must be *resistant to creep at high temperatures* must contain *molybdenum. Silicon and chromium impart resistance to oxidation and scaling.*
- Steels which are satisfactory upto about 700°C operating temperature are : C = 0.15% max., Si = 0.5 to 20%, Mn = 0.5% max., Ni = none, Cr = 1.0 to 6%, Mo = 0.5%.

**Uses :** These are used in the as-rolled or as-forged condition, particularly for the values of internal combustion engines. For higher temperatures upto 1000°C, steels containing upto 22% nickel and 26% chromium are used.

**Spring Steels :**

**1. Carbon-manganese spring steels**

- C = 0.45 to 0.65%
  Si = 0.1 to 0.35%
  Mn = 0.5 to 1.0%
- These steels are *quenched and tempered to* give a Brinell, hardness of about 350.

**Uses :** They are widely used are for *laminated springs* for railway and general purposes.

**2. Hyper-eutectoid spring steels :**

- C = 0.9 to 1.2%
  Si = 0.30% max.
  Mn = 0.45 to 0.70%
- These steels are oil quenched and tempered at a low temperature.

**Uses:** They are used for *volute* and *helical springs.*

**3. Silicon-manganese spring steels :**

- C = 0.33 to 0.6%
  Si = 1.5 to 2%
  Mn = 0.6 to 1.0%
- These steels are hardened and tempered to give a Brinell hardness of about 450.

**Uses :** Employed for the manufacture of *railway and road springs* generally.

**Stainless Steel**

The stainless steels are classified in two groups :

1. Plain chromium and high chromium low nickel steel.

2. Chromium nickel steel.

**1. Plain chromium and high chromium low nickel steel :**

- Out of the group the former has C = 0.8% and Cr = 12 to 20%, whilst the latter has C = 0.1 to 0.2%; Cr = 12 to 20% and Ni = 2%.
- These steels can be *heat-treated.*
- *Used for dies, valves and cutlery.*

**2. Chromium nickel steel:**

- These steels are non-magnetic and cannot be hardened.
- They have varieties due to the varying contents of chromium and nickel respectively *e.g.*, 18 : 8, 12 : 2 and 18 : 9 (called *stay bride*). Small quantity of copper, tungsten and molybdenum is added to these steels.
- They have *high resistance to corrosion and may be cold or hot worked, pressed welded, brazed or soldered.*
- The percentage of carbon in these steels is kept upto 0.5.
- These steels are *poor conductor of heat and electricity.*

**Chrome Vanadium Steels :**

- These steels usually contain 0.8 to 1.1% Cr, 0.25 to 0.33 C and less than 0.25% V. Even such a small amount of V has marked effect on the properties of steel. It helps in producing cleaner steel because of the deoxidising action. Vanadium is distributed both between the ferrite and the carbide involving both. Cr intensifies the effect of V more than that of Ni.
- Chrome-vanadium steels are used where *great strength, toughness and resistance to fatigue are required such as axles and shafts of aeroplanes, automobiles and locomotives.*

**Nickel Steel :**

- The percentage of nickel varies from 2 to 40.
- Nickel from 3 to 5% *raises elastic limit and improves toughness.*
- Steel containing 20% nickel has *very high tensile strength.*
- If nickel is present upto 27% it makes the steel *non-magnetic and non-corrodible.*
- Nickel containing 3.5% nickel and 0.15% to 0.45% carbon has good ductility, high elastic ratio and resistance to fatigue. It is used for *long span bridges.*
- *Invar* (Ni = 36%) and *super-invar* (Ni = 31%) are the popular materials for least co-efficient of expansion and are used for *measuring instruments, surveyor tapes and clock pendulums.*
- *Nickel tends to retard the grain growth* in steel, resulting in a wide range of heating, without damage to the steel, or it may be maintained above the critical range for long periods of time without great damage.
- The addition of nickel to annealed carbon steel apparently strengthens the ferrite, with the result that the *endurance ratio is raised.* The full effect of nickel on the endurance ratio, however, is obtained only by quenching and tempering.

**Vanadium Steel :**

- Addition of vanadium even in small proportion to an ordinary low carbon steel considerably raises its *elastic limit and improves the fatigue resistance property.*
- When vanadium is added upto 0.25%; the elastic limit of the steel is raised by 50%, becomes exceedingly tough and can *resist high alternating stresses and severe shocks.*
- Vanadium makes the steel *strong and more ductile.*
- Steel with C = 0.15% to 1.1%, Cr = 0.5 to 1.5% and V = 0.15 to 0.3% *has high tensile strength, elastic limit, endurance limit and ductility.*
- Improves response to heat treatment.

- Provides control of structure.

**Uses :** 1. It is widely used for making tools.

2. It may be used for shafts, spring, gears and drop forged parts.

**Manganese Steel :**

- Manganese from 1 to 1.5% makes the steel strong and tough to withstand severe duty but manganese content from 1.5 to 5% renders the steel hard and brittle.
- When manganese is from 11 to 14%, carbon 0.8 to 1.5%, steel becomes very hard; tough, non-magnetic and possesses considerably high tensile strength.
- More manganese *reduces strength and ductility.*
- Presence of manganese even upto 2% reduces the formation of sulphide and deoxidation of molten metal.
- Manganese steels show *high percentage of elongation.*
- It may be forged easily but difficult to machine and thus it is usually ground.
- The austenitic manganese steel is practically non-magnetic having a permeability of less than 2.0, so that it has special applications for parts that have to combine non-magnetism with high strength and resistance to wear.
- Heat-treated cast manganese steel in bar form is *so ductile that it can be bent double when cold, without fracture;* similarly castings of this metal will deform when fracturing. This deformation, however, is not continuous owing to the fact that cold working raises the yield point in the stressed areas.
- Specific gravity = 7.92, melting point = 1343° C.
- Manganese steel *can be forged* but special care is necessary to avoid degrading the steel. After forging the steel should be heat-treated by raising it to temperature of 1010°C and quenching in water.
- Manganese steel is *weldable*, a nickel manganese welding rod being employed for repairing fractures, an alternative is an 18/8 stainless steel rod. Care is necessary in welding to avoid overheating, as this tends to cause degrading of the steel.

**Uses :** 1. The properties of *work-hardening* renders this steel eminently suitable for such parts as the *jaws of stone and ore crushers, tramway and railway points and crossing, etc.*

2. Manganese steel has been much used for *helmets* and *shields*; it is also a good material for "*burglar proof*" *safes and vaults* as it is too hard to be cut with ordinary explosives.
3. Manganese steel in the hot worked condition is much used for *rails*.
4. Manganese steel in the form of bars is now widely used for screening coke. Ordinary steels wear away rapidly under the impact of the coke, but with manganese steel it has a life about a *hundred times as long*.
5. Owing to its *non-magnetic property* it has important use in the *cover plates of lifting magnets* for handling heavy iron steel articles, and also for ship structure situated near the magnetic compass, since it has no effect on the latter.
6. Other applications include *agricultural implements* such as shovels; these are made from rolled manganese steel plates.

**Tungsten Steel :**

- Tungsten when added to steel improves its *magnetic properties and hardenability.*
- Addition of tungsten to an extent of 6% high carbon steel enables it to retain the magnetic properties to high degree and produce field more intense than an ordinary steel.
- 8% tungsten attributes so much hardness in steel that it can *scratch glass even.*
- Steel containing 1.5% is too hard to be machined by ordinary methods.
- High percentage of tungsten makes the steel to *retain its hardness even at high temperatures.*
- Tungsten used in certain known percentage, *imparts air hardening properties to the steel.*

**Uses :** It is used for making permanent magnets and high speed cutting tools.

**Magnet Steels :**

Carbon, chromium, tungsten and cobalt steels are used for permanent magnets. The superiority, as well as price, increases in the order given above. These stees are used in the hardened condition. The carbon steels contain from 0.80 to 1.20% carbon and are, in fact, carbon tool steels. The chromium magnet steels contain from 0.70 to 1.00% carbon and 2% to 3% chromium. The tungsten steels contain about 0.70% carbon and 5% tungsten. The best and most expensive cobalt steels contain 35% cobalt, together with several percent of both chromium and tungsten. An important permanent magnet alloy (Alnico) contains approximately 60% iron, 20% nickel, 8% cobalt and 12% aluminium. This alloy cannot be forged and is used as a casting hardened by precipitation heat treatments.

## B. NON-FERROUS MATERIALS

### 9.13. ALUMINIUM

Although aluminium is very abundant in the earth it is never found free, nor can it, by present methods be obtained commercially from clay in which it exists in such great quantities. There is but not commercial ore called *'bauxite'*. *Bauxite is hydrated aluminium oxide* generally believed to consist of mixture of monohydrate $Al_2 O_3 . H_2O$ and trihydrate $Al_2 O_3 .3H_2O$. The chief impurities are oxide, silica, clay and titanium oxide. It is found in districts of Bihar, Madras and Madhya Pradesh.

**Manufacture :** Bauxite is first purified and then dissolved in fused cryolite (double fluoride of aluminium and sodium). The aluminium is then separated from this solution by electrolysis at about 910° C.

**Physical and Mechanical Properties :**

(*i*) Pure aluminium has silvery colour and lustre, while the commercial grades show a characteristic bluish tinge.

(*ii*) The high purity has a much a greater resistance to corrosion than the ordinary steel.

(*iii*) It is ductile and malleable.

(*iv*) Its specific gravity is 2.7.

(*v*) In proportion to its weight it is quite strong.

(*vi*) Melting point = 658° C, Boiling point = 2057° C.

(*vii*) Its electrical resistivity is 2.669 micro ohms/$cm^3$ at 20° C.

(*viii*) Its tensile strength varies from 95 to 157 $MN/m^2$.

(*ix*) It is a good conductor of heat and electricity.

(*x*) It forms useful alloys with iron, copper, zinc and other metals.

(*xi*) It is unaffected by ordinary atmospheric influences but is corroded in sea water. It is soluble in solutions of caustic alkalies and in hydrochloric acid. *When there is any excess of silicon present in the metal it does not withstand atmospheric actions.*

(*xii*) Aluminium is electron positive to most other metals, *e.g.*, iron, chromium, zinc, copper, nickel, tin, lead etc. Care is necessary, therefore, to prevent it from coming into metallic contact with other metals under conditions where moisture is present in order to avoid electrolytic action *i.e.*, corrosion.

**Uses.** 1. Because of its softness and difficulty of making sound castings, *little pure aluminium is used in the cast form. The largest quantity is employed after it has been mechanically worked in some manner,* as by rolling, wire drawing, drop forging or extruding. After being rolled into sheets, it may be stamped into a variety of shapes.

2. It is employed, often alloyed with small amounts of other metals, in the manufacture of *furniture, rail-road and trolley cars, automobile bodies and pistons, electric cables and bus bars, rivets, kitchen utensils and collapsible tubes for pastes.*

3. In a finely divided flake form, aluminium is employed as a *pigment in paint*. Aluminium paint is used as a priming coat for wood, as a protective coat for metals.
4. Aluminium is used in *deoxidizing molten iron and steel,* especially in the top of the ingot mould. In a similar manner it is used to prepare the metals from their oxides by heating a mixture of powdered aluminium and the oxide of the metal to be reduced. The mixture is known as *Thermit.*
5. *Aluminium foil is used as silver paper for packing chocolates* etc.

**Annealing :**

Aluminium sheet which has been hardened by cold working such as hammering or rolling can be hardened by heating it to about 350° C and afterwards cooling in air or water. The period of heating need only be for a few minutes. As a rough guide to correct temperature of heating for annealing, the surface may be rubbed, during the heating process, with a dry match stick from time to time, the heating being stopped when wood begins to chart.

The annealing of aluminium wire for electrical purposes require more care, since the aim is to obtain a high electrical conductivity by obtaining the correct structure. For high conductivity a long exposure at a lower temperature, namely 250° to 300°C appears to give the best results. In passing, it may be remarked that the *aluminium alloys require higher annealing temperature, from about 350° to 400°C.*

## 9.14. COPPER

Copper is one of the comparatively few metals that are found in the metallic state. It occurs in some minerals such as copper glance ($Cu_2S$), malachite ($CuCO_3.CuO_2H_2$) and azurite ($2CuCO_3.CuO_2H_2$). Copper ores are found in Burma, Sikkim and Singhbhoom (Bihar).

**Manufacture :**

- Copper is extracted from its ores by several different methods that chosen depending upon the character of the ore and local conditions.
- Refining of the metal is usually considered to begin when the copper is in the *blister stage*, the surfaces of the cast material being irregular and blistered due to the generation of gases during cooling. This copper is 99% pure and is *further refined in the furnace by oxidation process which removes sulphur and other impurities.* The excess of oxygen is removed from the metal by operation known as *poling*. Green wooden poles or tree trunks are thrust under the surface of the molten metal, which is covered with charcoal, coke or similar material rich in carbon. Although this process may seem crude it is still almost invariably used, owing to its cheapness and efficiency. Poling is discontinued when sample casting indicates that the oxygen content has been reduced to 0.08-0.025 percent. The copper is then known as *tough pitch copper.*
- *Deoxidized copper is needed for intricate castings, welding and certain other processes. Special reducing agents such as phosphorus, silicon, lithium, magnesium, beryllium or calcium are added to the molten metal to eliminate the oxygen just sufficieint of the deoxidizer being used to prevent any residue in the metal.*
- *Arsenic,* in amounts upto approximately 5 percent is added to improve the strength and toughness of the metal, and most copper products, other than electrical gear, are manufactured from arsenical copper.

**Physical and Mechanical Properties :**

(*i*) Copper is a *reddish-brown* metal.

(*ii*) Although pure copper is *one of the best conductors of heat and electricity,* its electrical conductivity is highly sensitive to the presence of impurities.

(*iii*) If copper is heated to red heat and cooled slowly it becomes brittle; but if cooled rapidly it becomes soft, malleable and ductile. The brittleness is due to the coarsely crystalline structure that developed during slow cooling.

(*iv*) Copper *can be welded at red heat.*

(*v*) Like aluminium, *pure copper does not cast well.* When molten it absorbs gases, such as carbon monoxide, hydrogen and sulphur dioxide which separate out on cooling and cause blow holes.

(*vi*) Melting point = 1084° C, Boiling point = 2595° C.

(*vii*) Specific gravity = 8.9, Electrical resistivity = 1.682 micro ohms per cm.

(*viii*) It is *highly resistant to corrosion by liquids.*

(*ix*) It tensile strength varies from 300 to 470 MN/$m^2$.

(*x*) It forms important alloys like *bronze and gun metal.*

(*xi*) It is strongly attacked by nitric acid but only very slowly by dilute hydrochloric and sulphuric acids in the absence of air; ammonical solutions also attack copper.

For copper that is to be worked, lead is sometimes added in order that metal may be worked more easily, but if more than about 0.5 per cent is employed, it causes the copper to be brittle. In copper alloys, for casting, as much as 10 to 20 per cent may be added to cheapen the product but this lead largely separates out in globular masses on cooling. At the melting point of lead, copper is practically insoluble in liquid lead and less than 0.05 per cent of the lead enters into the solid solution in the copper.

**Uses :** 1. It is largely used in *making electric cables and wires and electric machinery and appliances.*

2. Used in *electroplating, electrotyping* and *for soldering iron bits.*

3. Used as a *damp proof material* and *for making alloys.*

4. It is used for *sheeting, roofing, spouts, boilers, condensers and other purposes where corrosion resistance with fair strength and flexibility is essential.*

**Annealing Temperature :**

The anaeling temperature of copper varies between 200°C and 600°C according to the impurities present and condition of the metal, *i.e.*, upon the amount of cold work it has been subjected. Work-hardened copper of high purity may be partly softened at temperature as low as 120°C, but the *minimum softening temperature is usually at least* 200°C. For most commercial copper a temperature of 500°C is employed but for heavy sections the furnace temperature may be raised to 600°C. After raising to the anealing temperature it may be cooled in any convenient manner, namely, in air or by water quenching; the later method facilitates the removal of dirt and scale.

## 9.15. LEAD

The chief lead ore is lead sulphide, PbS, called *galena* or *galenite*. Its distribution is very wide. There are practically no important deposits of lead in which galena is not present to some extent. Silver sulphide is scarcely ever absent from galena; the amount is small, however, generally being less than 0.3 per cent.

**Manufacture :**

- The ores are roasted in the reverberatory furnace when load oxide and sulphate are formed.
- To the roasted ore, silica, coke, metallic iron and lime are added and then smelted in a blast furnace. Lead oxide and sulphate react with iron to form ferrous oxide and sulphate, thereby releasing metallic lead. Ferrous oxide combines with silica to form *slag.*
- The heavy metal obtained from the bottom of the furnace is further oxidised in Bessemer's converter to remove most of the impurities.

**Physical and Mechanical Properties :**

(*i*) Lead is *bluish-grey* metal with a high metallic lustre when freshly cut.

(*ii*) It is the *softest and heaviest* of all the common metals.

(*iii*) It can *readily be scratched* with fingernail when pure.

(*iv*) It is *least tenacious* of the common metals.

(*v*) It is *very malleable* and may be readily formed into foil.

(*vi*) When hot, it may be extruded by hydraulic press into tubes, rods and wires.

(*vii*) Melting point = 327.4°C, Boiling point = 1744°C.

(*viii*) Its specific gravity is 11.35 and elctrical resistivity 20.65 micro ohms.

(*ix*) It marks paper.

**Uses.** 1. It is employed for *chemical laboratory* and plant drains.

2. It finds extensive applications as *sheaths* for electric cables, both overhead and underground.

3. In the form of sheet lead it serves excellently for *roofs, gutters* and *cornices.*

4. Large quantities of the metal are consumed in the *manufacture of white lead* and *other paint pigments.*

5. Lead enters into greater variety of alloys such as *brasses and bronzes, low melting point alloys for sprinkler heads, safety plug in boilers, fire door releases, and fuses.*

## 9.16. TIN

The chief source of tin is tinstone (or cassiterite) $SnO_2$. Large deposits of tinstone occur in Tairy (Burma) and small quantities in Hazaribagh (Bihar).

**Manufacture :**

- The ore is crushed, calcined, washed and then smelted in a furnace using anthracite coal and sand, to obtain crude tin.
- The crude tin is then refined in a reverberatory furnace to get commercially pure tin.
- Chemically pure tin is made by electrolytic deposition from commercial tin.

**Physical and Mechanical Properties :**

(*i*) It is a *brightly shining white metal.*

(*ii*) It is *soft and malleable* and can be hammered into *thin foils.*

(*iii*) Melting point = 232°C, Boiling point = 765°C.

(*iv*) Its specific gravity is 7.3 and electrical resistivity 7.59 micro ohms.

**Uses :** 1. Used alone as coating on other metals and alloys owing to its resistance to corrosion.

2. It is employed especially in *low-melting point alloys* as a complete or partial substitute for bismuth. Cadmium is used chiefly in alloys.

3. It is used as *moisture proof packing.*

## 9.17. ZINC

The common ores of zinc are zincite (ZnO), calamine ($ZnCO_3$) and zinc blende (ZnS). These ores are available in Burma.

**Manufacture :** The oxide is heated in an electric furnace where zinc is liberated as vapour. The vapour is then cooled in condensers to get metallic zinc.

**Physical and Mechanical Properties :**

(*i*) It is bluish-grey metal.

(*ii*) It becomes brittle at 200° C and can be powdered at this temperature.

(*iii*) Its specific gravity is 6.2 and melting point 481°C.

**Uses :** 1. It is used in *sheet form as a corrosion resistance surface* or *lining* or as a *protective coating* on iron and steel in the form of a galvanized or sprayed surface.

2. It is also used in electric cells and in making brass and other alloys.

## 9.18. MAGNESIUM

It occurs in nature in the following minerals :

Magnesite ($MgCO_3$), dolomite ($CaCO_3$, $MgCO_3$), kieserite ($MgSO_4.H_2O$) and carnallite ($MgCl_2.KCl.6H_2O$). In India magnesite occurs in Madras and Mysore and dolomite in many states.

**Manufacture :** Metallic magnesium is prepared commercially chiefly by two methods: the *chloride* and *oxide processes.*

- In the '*chloride process*' a melted mixture of the chlorides of sodium, potassium, and magnesium is electrolysed. The presence of alkali chlorides is necessary to prevent the decomposition of magnesium chloride during heating. The electrolysis is carried out in an air tight iron vessel, the walls of which act as the cathode, with a cetrally located graphite anode, which is protected by a porcelain hood from which the chlorine escapes. As the magnesium is liberated it rises to the surface and is ladled out. The metal is refined by remelting in iron pots and fluxing it with sodium magnesium chloride to remove oxide inclusions. If especially pure magnesium is desired, it is prepared by distillation.
- '*Oxide process*' consists in dissolving the oxide in bath containing the fused fluorides of magnesium, barium and sodium and separating the metal by electrolysis. The process, however, is no longer used.

**Physical and Mechanical Properties :**

(*i*) It is the *lightest metal* used as engineering material.

(*ii*) The *tensile strength of cast metal is the same as that of ordinary cast aluminium*, but when reduction of weight is important magnesium is preferable because of its low density of 1.74 which is a little less than one sevenths of that of aluminium.

(*iii*) The tensile strength of rolled and annealed magnesium is about the same as that of good quality grey cast iron.

(*iv*) It is *harder than aluminium.*

(*v*) Melting pont = 650°C, Boiling point = 1107°C.

(*vi*) Its electrical resistivity is 4.46 micro ohms.

(*vii*) Its brinell hardness is 33 (500 kgf load, 10 mm ball).

(*viii*) In casting magnesium and its alloys, special care is necessary because of its tendency to oxidise readily.

(*ix*) Because of its lightness, it tends to trap gases during casting and form gas flames.

(*x*) It is *readily machined* and under the buffing wheel it takes a high polish.

(*xi*) It hardens very rapidly with cold working, but it is suitably plastic and workable at 350°C.

(*xii*) It is readily attracked *by weak acids* and also by saline solution, it is not affected, however, by most alkaline solutions.

(*xiii*) It *ignites easily at above its ignition temperature.*

(*xiv*) It *can be welded,* but because it combines with oxygen more readily than any other common metal, a *special technique is required.*

(*xv*) It may *be soldered but the joints must afterwards be protected to prevent them from corroding.*

**Uses.** 1. It is used in the form of sheets, wires, rods, tubes etc.

2. The ribbon is chiefly used *for degasification of radio tubes.*

3. In the powdered form, mixed with an oxidising agent such as barium nitrate or potassium chlorate, magnesium is employed in the manufacture of flash light powder and in military pyrotechnics for production of rockets, signals and flares.

## 9.19. NICKEL

The most important known deposits of the world are those of the Sudhbury district of Ontario, Canada where nickel bearing iron sulphides containing not over 3 percent of the nickel are found. About 90 per cent of the world's production comes from this source.

**Physical and Mechanical Properties :**

(*i*) It is *silvery white metal* capable of taking a *high polish.*

(*ii*) It is *almost as hard as steel.*

(*iii*) When it contains a small amount of carbon, it is quite *malleable*. In fact, it may be satisfactory rolled with as little as 0.005 per cent of cabron.

(*iv*) It is somewhat *less ductile* than soft steel, but small amounts of magnesium improves the ductility considerably.

(*v*) Nickel is *resistant to the attack of most acids, but it dissolves readily in nitric acid.*

(*vi*) It is very permanent upon exposure to atmosphere.

(*vii*) It is ferromagnetic below 360°C.

(*viii*) Melting point = 1452°C, Boiling point = 2732° C.

(*ix*) Its electrical resistivity is 10.9 micro ohms per cm cube.

(*x*) Specific gravity is 0.85.

(*xi*) The annealing temperature lies between 750° and 850°C.

(*xii*) The forging temperature is 1100° to 1250°C.

**Uses :** 1. Used as *an alloying metal* in some types of steel and cast iron, as also in alloys like monel metal, permalloy, cupro-nickel and nickel-silver.

2. It is extensively used to *plate iron and brass as a protective coating.*

## 9.20. NON-FERROUS ALLOYS

An alloy may be defined *as coherent metallic mass produced by the intimate association of two or more metals.* In some cases also the introduction of non-metals such as carbon, silicon, phosphorus, boron, sulphur and even nitrogen or oxygen may produce marked effects. It is important that thousands of combinations resulting in useful alloys are possible.

*Important non-ferrous alloys* are discussed below.

### 9.20.1. Copper Alloys

The copper alloys can be divided into two principal classes : (*i*) Brasses and (*ii*) Bronzes. The *'brasses' are primarily alloys of copper and zinc and 'bronzes' are of copper and tin,* but there are numerous modifications of them produced by the addition of smaller amounts of other elements.

In addition there are cupro-nickels, containing principally copper and nickel, and nickel silvers (German silvers) in which zinc, of brass is partially replaced by nickel.

**9.20.1.1. Brasses :** With range of composition from 5 to 45 percent of zinc, the brasses are among the most useful alloys. *They possess excellent mechanical properties, and are corrosion resistant and are readily machinable.*

The classification of these alloys is given in the Table 9.4.

**Table 9.4. Classification of Brasses**

| Name | Composition | | Colour | Typical uses |
|---|---|---|---|---|
| | Cu | Sn | | |
| Muntze metal | 59 | 41 | Reddish | Architectural work, welding rod, condenser tubes, valve stems. |
| Cartridge or spinning brass | 70 | 30 | Typical brass colour | Cartridges, tubes, spinning, drawing. |
| Brazing brass | 75 | 25 | Typical brass colour | Drawing, spinning, springs, particularly suitable for brazing. |

| | | | | |
|---|---|---|---|---|
| **Red brass** | 85 | 15 | Red | Hardware, radiator cores, Plumbing pipes, condenser tubes, flexible hose. |
| **High brass** | 66 | 34 | Typical brass colour | Stamping, blanking, drawing, spinning, springs, rivets, chains. |
| **Low brass** | 80 | 20 | Red gold | Drawing, forming, flexible hose. |

**Modified Brasses :**

The mechanical properties or corrosion resistance of the brasses is in some cases greatly improved by the introduction of relatively small quantities of one or more other metals. Some of the more common of these are given below :

**Manganese Brass.** The alloy referred to under this head is generally spoken of as *Manganese bronze*, but it is in reality a brass since the main constituents are copper and zinc. The composition varies considerably, but the following percentages may be regarded as typical; copper = 60%; zinc = 38 to 42%; tin = 0.5 to 1.5%; iron = 0.5 to 1.5%; manganese 0 to 0.5% and sometimes a little aluminium. The manganese is generally added in the form of ferromanganese; this accounts for the presence of iron, which in itself has important effects.

**Properties :**

(*i*) It is exceedingly tough.

(*ii*) It has high resistance even in sea water.

(*iii*) It is very active in reducing the oxides of other metals, an action that is highly useful.

(*iv*) In addition to its action as deoxidiser, manganese hardens and strengthens the alloy.

(*v*) In cast form, the tensile strength of the manganese brass lies in the neighbourhood of 500 $MN/m^2$ and this is somewhat improved by working, as by rolling and forging.

**Uses :** It is used for *pump rods, hydraulic rams, valves and cylinders, tubes, propellers, nuts, bolts etc.*

**Iron Brass .** One of the most widely used alloys of this type is known as *delta metal,* its composition is : Cu = 60%, Zn = 37%, Fe = 3% (also some quantity of Ni or Mn).

**Properties :**

(*i*) It is hard, strong and tough.

(*ii*) It is easily cast.

(*iii*) Its tensile strength is about two fifths greater than brass of similar composition with the iron omitted.

(*iv*) It resists corrosion.

**Uses :** It is used for mild steel if corrosion is to be resisted.

**Tin Brass.** Tin is one of the most useful metals added to brass. It increases the hardness and tensile strength but amount employed should not exceed about 2%, because with higher amounts the ductility begins to decrease. Its composition is : Cu = 60%, Zn = 39%, Zn = 1%. It is also sometime called *naval brass* since it is used in naval construction.

**Properties :**

(*i*) It has *high strength* and *hardness.*

(*ii*) It has *excellent corrosion resistance property.*

(*iii*) *Its tenacity is high and has good working qualities.*

**Lead Brass.** In brass that is intended for filing or turning, 1 to 2% of lead is employed to prevent fouling of tools and to cause the turnings to break more readily; this prevents logging of automatic machines. *Lead increases the softness of the brass.* It is sometimes added to the brass that is to be worked, but care should be taken to add but a small amount, because it reduces the ductility and strength. Only about 3% will alloy with the brass; if more than this is *added*, it has a tendency to liquate. On this account '*leaded*' brasses should be chilled quickly when cast.

**Aluminium Brass.** The amount of aluminium added to brass does not usually exceed 3%. It raises the tensile strength but decreases the ductility. The product has a deep golden colour and resists corrosion better than ordinary brass. It casts well and may be forged or rolled.

**9.20.1.2. Bronzes :** These are essentially alloys of *copper and tin*. The effect of composition on physical perperties is given below :

- The tensile strength of bronze increases gradually with the amount of tin, reaching a maximum with about 20% of tin, but as the tin increases bevond this amount the tensile strength very rapidly decreases.
- Bronze is *most ductile when it contains about 5% of tin,* with this amount it may be rolled satisfactory at red heat. Bronze is used chiefly for casting.
- As the amount of tin increases above 5%, the ductility gradually lessens and *practically disappears with about 20% of tin;* since ductility is co-ordinate with toughness, these alloys are *very brittle*. They are also *very hard*.
- The most useful of bronzes are those that contain from 8 to 11% of tin, since the maximum combined strength and toughness are secured with about these amounts. Bronze containing tin within these limits was *formerly,* known as gun metal, since because of this strength it was used for making guns, but now *steel has entirely replaced it for this purpose.* At the present time, the term gun metal is very loosely used and cannot be said to have definite significance.
- As cast the alloy containing 9% of the tin has a tensile strength of about 215 $MN/m^2$.
- The bronze containing 4 to 8% tin is called '*coinage bronze*' and used for making coins and metals.
- The copper-tin series of alloys containing 15 to 25% of tin is known as '*bell metal*'. Such *alloys are very hard and brittle, but are sonorous and are employed, therefore, in making bells.*
- By alloying 2 parts copper with 1 of tin, a very hard, brittle white alloy is produced called '*spectrum metal*'. When highly polished it serves excellently for mirrors and reflectors.

**Modified Bronze**

The zinc bronzes usually contain; Cu = 88%, Sn = 8 to 10% and Zn = 4 to 2%.

The presence of the zinc in the bronzes has the following effects :

- It increases the fluidity of the melted metal and in this way tends to prevent gas flaws in the casting.
- It also increases the strength and ductility.
- The specific gravity when cast in sand is about 8.58; the melting point is about 980°C.

**Phosphor Bronze**

- It is certain, however, that whatever good qualities the bronzes might have, they are *greatly increased by the addition of phosphorus*. The tensile strength, the elasticity and the resistance to fatigue are enormously increased; so much so that it may also be considered as an entirely new alloy.
- Phosphorus is best added to bronze in the form of phosphorus copper, a hard brittle substance of white fracture, containing about 8% phosphorus. Upon being added to molten bronze nearly all of the phosphorus oxidizes and escapes, so that most phosphorus bronzes contain no more than a few tenths of 1%. The crystalline structure of two bronzes, one with and the other without phosphorus, as shown by microscope, seems to be the same in both cases. Phosphor bronze is then essentially a deoxidized bronze.
- Although phosphorus produces very beneficial results through its activity as deoxidising

agent, if *there is much of it left in the alloy, it may be decided detriment.*

- Phosphor bronze of *proper composition can be forged, drawn, cold rolled and cast.* It seems to resist corrosion better than ordinary bronze, especially by sea water, so that it is much used for propeller blades.
- On account of its toughness, elasticity and strength it may serve as a substitute for steel as in the manufacture of *corrosion-resistant mine cables, ship sheathing, valve parts, springs etc.*

**Silicon Bronze**

- These are copper alloys having 4% of silicon and upto 1% of manganese with small amounts of tin, zinc, iron and aluminium.
- These are obviously not bronzes since tin is very small.
- They are available in the market under various trade names.
- These always *have strength of soft steel and at the same time are corrosion resistant.*
- They are used in screws, belts, tubings, pumps etc

**Aluminium Bronze**

Its composition is : Cu = 88%, Al = 8%; Fe = 3%; Sn = 0.5%. Thus, it is primarily a copper-aluminium alloy.

**Properties :**

(*i*) This alloy has a *high strength.*

(*ii*) It has *marked resistance to corrosion.*

(*iii*) Its melting point is about 1040°C.

(*iv*) Some what difficult to cast due to oxidation.

(*v*) Castings of metal are sound and not liable to segregation, but the aluminium used must be free from impurity, a small quantity of which is found to produce a very marked deterioration.

(*vi*) The commercial aluminium bronzes are practically non-magnetic, but certain complex bronzes containing appreciable proportions of iron are magnetic.

**Uses :** 1. Aluminium bronze is at present most satisfactory copper alloy for die casting, since it causes less corrosion of dies than brass.

2. It is used in making hot stampings although it is not easy to work in this respect as brass or manganese.

3. It is also used in massive work, bearing metal in locomotive.

**Note.** A *disadvantage* of aluminium bronze is that it usually develops surface films of high electrical resistance when used for current carrying purposes; these films tend to give rise to high electrical resistance and render jointing a matter of difficulty.

**Beryllium Copper**

This alloy is also sometimes called *beryllium bronzes.* It contain 2 to 3% of beryllium.

**Properties :**

(*i*) It has a *high yield point and high fatigue limit.*

(*ii*) It has *excellent cold and hot resistance.*

(*iii*) These *alloys can also be heat treated.*

(*iv*) It has brinell hardness 340 (when heated to 800°C and then hardened by subsequent precipitation treatment).

**Uses :** 1. It is used for springs, heavy duty electrical switches, cams and bushings.

2. Being non-sparking, it is utilized for making chisels and hammers under conditions where spark might cause an explosion.

**Copper-Nickel Alloys**

- These are copper base alloys.
- Addition of nickel to copper improves mechanical properties and resistance to corrosion.
- '*Manganin*' a resistance wire with a very low temperature co-efficient of resistance, consists of 80% of copper, 5 per cent of nickel and 15% of manganese.

*Nickel silvers* or *German silvers* are ternary alloys of copper containing from 5 to 45% of zinc and from 5 to 30% of nickel. They have *fair strength, good plasticity* and low thermal conductivity. They are used extensively as the base metal on which silver is plated for table ware and for high grade plumbing and hardware. They are used for *electrical wares* and *numerous fittings.*

### 9.20.2. Aluminium Alloys

- The principal elements which are alloyed with pure aluminium to improve its tensile strength and hardness are copper, silicon, manganese, zinc, magnesium and nickel. One element may be used alone but often two, three or four additions are made to the base metal to produce a metal having specific physical properties.
- *Copper* (for instance) is the *main hardening element* added while the addition of a small percentage of magnesium to an aluminium copper alloy still further improves the hardness and strength after heat treatment. Strength may also be improved by addition of small quantities of manganese and nickel to a copper-aluminium alloy.
- *Silicon*, is next in importance to copper as a main alloying element, since in combination with magnesium it forms a hard compound known as magnesium silicide which is largely responsible for the hardness obtained on the treatment.
- *Manganese* acts as a strengthening agent and prevents the formation of a course crystalline structure during heat treatment.

Some of the important aluminium alloys are discussed below :

#### 9.20.2.1. Duralumin

**Composition :** Al = 94%, Cu = 4%, Mg, Mn, Si, Fe 0.5% each.

**Properties :**

(*i*) It can be *cast, forged* and *stamped.*

(*ii*) It has *high tensile strength.*

(*iii*) It possesses *high electrical conductance.*

(*iv*) It hardends spontaneously when exposed to room temperature.

(*v*) The alloy is soft enough for a workable period after it has been quenched.

(*vi*) The temperature employed for the solution heat treatment of the alloy is the lowest that is applicable to any commercial light alloy.

(*vii*) Specific gravity = 2.8.

(*viii*) Melting point = 650°C.

(*ix*) Brinell hardness; Annealed = 60, age hardened = 100.

**Uses :** 1. It is widely used for *sheets, tubes, forgings, rivets, nuts, bolts and similar parts.*

2. Used in making cables.

3. It is also extensively used for *air planes* and *other machines where weight is a deciding factor.*

4. It is also employed in *surgical and orthopaedic work* and for non-magnetic and other instrument parts.

**Heat Treatment :**

- Annealing is carried out at 360-400°C and the metal is cooled in air. It is then ductile and can be cold worked. Normalising is carried out as a temperature of 490°C plus or minus 10 degrees. After normalising, the metal is quenched in clean cold water.

- A salt bath is used for heat treatment, generally consisting of 50% silver nitrate, with 50% potassium nitrate. For built up fittings a muffle furnace is sometimes used.

**9.20.2.2. Y-Alloy**

**Composition :** Al = 92.5%; Cu = 4%; Ni = 2%; Mg = 1.5%.

**Properties :**

(*i*) Its strength at 200° C is better than aluminium.

(*ii*) It retains its high strength and hardness at high temperature.

(*iii*) It can be easily cast and hot worked.

**Uses :** 1. It is extensively used for such components as *piston cylinder heads* and *crank cases of internal combustion engines.*

2. It is also used for *die casting, pump rods and in sparking chisel in place of steel.*

**Heat Treatment and Age Hardening :**

- The heat treatment range of Y-alloy is 500°C – 520°C; this is higher than for aluminium.
- It has been found that *muffle furnace* is best medium for heat treating Y-alloy. The period of treatment at 500-520° C required to bring about solution of age-hardening constituents depends on the fineness of grain and distribution of constituents. For chilled cast bars, 2.5 cm diameter, about 6 hours are required. For coarser grain material a longer period becomes necessary, whilst for large castings 24 hours or more may be required. In all cases the period of treatment at 500° – 520° C (solution treatment) must be regulated in relation to the micro-structure. Treatment at the temperature 500°C to 520°C is followed by *quenching in boiling water.* After quenching, *age hardening* at room temperature, substantially completes in 5 days.
- Age hardening can be accelerated by retaining the alloy immersed in the water at boiling temperature after quenching. Under these conditions age hardening is substantially completed in from 1/2 hour for the wrought to 2 hours for the cast alloy.
- By subjecting normally heat-treated material, cast or wrought, to temperatures between 150°C to 250°C the tensile strength and hardness are substantially increased but the ductility is diminished. Above 250°C permanent softening results.

**Note.** Y-alloy is annealed in the wrought state by heating it to 350° C to 400° C and allowing it to cool in the air.

**9.20.2.3. Hindalium**

- Hindalium is an alloy of *aluminium, magnesium, manganese, chromium* and *silicon* etc. and is the trade name of the aluminium alloy produced by Hindustan Aluminium Corporation Ltd., Renukoot Distt. Mirzapur (U.P.).
- It is manufactured as a rolled product (16 gauge) mainly for *anodized utensil manufacture.* During processing special care is taken to maintain the necessary mechanical and surface characteristics.

*Hindalium utensils* possess the following **advantages :**

1. *Strong and hard.*
2. *Cannot be easily scratched.*
3. *Can take fine finish.*
4. *Do not absorb much heat and thus save fuel while cooking.*
5. *Can be easily cleaned.*
6. *Do not react with the food acids.*
7. *Low cost* (about one-third of stainless steel).

**9.20.2.4. Magnelium**

- It is an alloy of *aluminium, magnesium, copper, nickel* and *tin* etc.

Typical composition of magnelium is:

Al = 85 to 95%
Cu = 0 to 25%
Mg = 1 to 5.5%
Ni = 0 to 1.2%
Sn = 0 to 3%
Fe = 0 to 0.9%
Mn = 0 to 0.03%
Si = 0.2 to 0.6%

**Properties**

(*i*) Light weight.

(*ii*) Tensile strength — annealed state : 200 MN/m$^2$; cold worked state : 280 MN/m$^2$

(*iii*) Elongation — annealed state : 30%; cold worked state : 7%

(*iv*) Alloy is brittle.

(*v*) Castability poor.

(*vi*) Machinability good.

(*vii*) Can be welded.

**Uses.** Mostly used in the aircraft and automobile industries. Some of the parts made from magnelium are:

*Gearbox housings, vehicle door handles, luggage racks, coffee-grinder parts and ornamental fixtures.*

## 9.20. MAGNESIUM ALLOYS

**Composition :** Magnesium alloys contain 3 to 10% aluminium, 1 to 3.5% zinc and 0.4% manganese.

**Dow metal** (an alloy of magnesium) contains 9.1% magnesium and 9% aluminium.

**Electron metal** (a trade name of magnesium) base alloy contains 4% zinc and small percentages of copper, iron and silicon.

**Properties :**

(*i*) The tensile strength of magnesium alloys is not high, averaging between 90 and 180 MN/m$^2$. This can be improved to 220-260 MN/m$^2$ when the alloys is in sheet form, and to 290-350 MN/m$^2$ when forged or extruded.

(*ii*) Resistance to shock is increased by heat treatment after casting-heating to temperature between 330°C and 415°C, followed by cooling in air.

(*iii*) Owing to the fierceness with which magnesium burns, a certain fire risk is bound to exist when machining the alloy.

(*iv*) On exposure to the atmosphere, magnesium alloys develop a dark oxide film which resists corrosion. To afford increased protection, and to provide a suitable base for paint or varnish, it is usual to subject the parts to chrome pickling process.

(*v*) Their weight is only two thirds that of aluminium and a quarter of that of steel.

(*vi*) Magnesium alloys (Dow metal) can be forged, welded and drawn in wires.

**Note**. Alloys of aluminium and magnesium are called LIGHT ALLOYS.

### 9.20.4. Nickel Alloys

The important nickel alloys are: *Iconol, Monel metal,* Nichrome Brightray alloys, Manganese nickel.

**Iconol**

**Composition :** Ni = 75%, Cr = 15%, Fe = 9%

**Properties :**

(*i*) It can be cast, forged, rolled and cold drawn.

(*ii*) It can be forged at 1000-1300° C but is brittle between 650°C and 950°C.

(*iii*) It has high corrosion resistance at ordinary and high temperatures.

(*iv*) Its tensile strength is 500 MN/$m^2$ and brinell hardness 160.

(*v*) Melting point = 1395° C; specific gravity = 8.55.

(*vi*) Brinell hardness is about 160.

(*vii*) It can be soft soldered as the ordinary tin smith's solder using a neutral zinc chloride solution as a flux.

(*viii*) It can be welded by the oxyacetylene and metallic arc methods without difficulty.

**Uses :** 1. It is used for making *springs which have to stand high temperature and are exposed to corrosion.*

2. It is also used for *exhaust manifold* (manifold is the main pipe which carries the explosive mixture from the carburettor to the cylinders of an internal combustion engine) of air craft engines.

**Monel Metal**

**Composition :** Monel metal can said to be *two third nickel, one third copper with small percentage of the elements iron, silicon, manganese and carbon.* A number of different grades are: "K" Monel metal contains 66% nickel, 29% copper and 2.75% aluminium. "A" Monel metal contains 67% nickel and has no aluminium. This is a free maching alloy of slightly lower strength than the "K" type. Several grades contain manganese upto about 2%.

**Properties :**

(*i*) Monel metal is superior to brass or bronze in resisting corrosion and in retaining its strength at high temperatures.

(*ii*) It is magnetic at ordinary temperatures when rolled, becoming non-magnetic on heating to 100-150°C and regaining its magnetism on cooling.

(*iii*) Its mechanical properties are improved by cold working.

(*iv*) Addition of elements like aluminium and beryllium makes the alloy amenable to heat treatment.

(*v*) It can be welded but requires a very special technique owing to its peculiar flow properties. It must be hammered after the deposition of each bead, to disperse the internal stresses set up in cooling.

(*vi*) It does not corrode under atmospheric conditions, even in air conditions of industrial centres.

(*vii*) It is only slightly attacked by acetic, citric and tartaric acids.

**Uses :** 1. Owing *to its excellent corrosion resistance properties, it is widely used for parts of water-pumps, propellers, domestic water storage tanks and parts subjected to high temperatures, such as internal combustion engines valve seatings particularly in light alloy cylinder-heads.*

2. It is also used for making *turbine blades and chemical food handling plant.*

**Nichrome**

**Composition :** Ni = 60%, Cr = 15% and Fe = 20%.

**Properties :**

(*i*) It is practically non-corrosive.

(*ii*) It can withstand high temperatures without oxidation

(*iii*) Alloyed with cast iron, it increases resistance of the latter to corrosion and heat, as well as to wear.

**Uses :** It is used in making electrical resistance wire for electric furnaces and heating elements.

**Brightray Alloy**

These are nickel-chromium alloys. They are highly heat resistant. Brightray contains Ni = 80%, Cr = 20%. It can be used for temperature upto 1150°C. Brightray H contains a small percentage of aluminium and can be used upto 1250°C.

## 9.21. BEARING OR ANTIFRICTION ALLOYS

A bearing alloy should have the following *characteristics:*

1. Good wearing quality.
2. Low co-efficient of friction.
3. High thermal conductivity.
4. High melting point.
5. Good casting qualities.
6. Ability to withstand continuous bearing pressure and impact.
7. Ability to work satisfactory at the rubbing speed at which it is required to run.
8. Low shrinkage after casting.
9. Desired plasticity under the load it is called upon to bear.
10. Economy in cost.
11. Non-corrosive property.

**Bearing Bronzes**

The description of typical bearing bronzes is given below :

**Hard bearing bronze**

**Composition :** Cu = 85%, Sn = 15%

**Properties :**

(*i*) Its tensile strength is 220 MN/m$^2$.

(*ii*) It has brinell hardness 100.

(*iii*) Elongation in 5 cm length is 4%.

**Use :** It is used for *heavy compressive loads, locomotive slide valves, bearings for turntables* etc.

**Phosphor bronze**

**Compostion :** Cu = 89%; Sn = 11%.

**Properties :**

(*i*) Tensile strength is 280 MN/m$^2$.

(*ii*) It has brinell hardness 100.

(*iii*) Elongation in 5 cm length is 4%.

**Uses :** It is widely employed for *heavy loading.*

**Admiralty gun metal**

**Composition :** Cu = 88%, Sn = 10%, Pb = 2%.

**Properties :**

(*i*) Tensile strength is 270 MN/m$^2$.

(*ii*) Brinell hardness is 65.

(*iii*) Elongation in 5 cm length is 20%.

**Uses :** It is used for general casting purposes, specially to resist marine corrosion, suitable for bearings when lubrication is good.

**Load bronze**

**Composition :** Cu = 80%, Sn = 10%, Pb = 10%.

**Properties :**

(*i*) Tensile strength is 230 MN/m$^2$.

(*ii*) Brinell hardness is 65.

(*iii*) Elongation in 5 cm length is 15%.

**Uses :** It has good *antifriction properties and is used where lubrication is doubtful.*

**White metals**

These metals are copper-lead alloys and are generally *employed for heavy duty bearings.*

**Babbit metal**

- It is a *white metal bearing* alloy. The term now covers a range of alloys of similar characteristics but varying composition.
- The exact composition of Babbit metal is not known, although it contained the following constituents in approximately the following proportions :

| | | |
|---|---|---|
| Tin | = | 89.2% |
| Copper | = | 3.8% |
| Antimony | = | 7.0% |

- A metal of this composition possesses *excellent antifriction properties*, but is expensive *owing to high tin content.*
- A cheaper Babbit which is suitable for bearings subjected to moderate pressure is made up as follows: Tin = 59.5% minimum; Copper = 2.25 to 3.75%; antimony = 9.5 to 11.5%; lead = 26% maximum; iron = 0.08% maximum; Bismuth = 0.08 % maximum.
- Babbit metal possesses a number of *advantages as a bearing metal* which are :

1. Soft enough to allow high-spots to be smoothed down in use.
2. It possesses sufficient mechanical strength for normal uses.
3. It is easy to cast and to scrap to a good lift.
4. Should a bearing overheat, the Babbit lining will melt and prevent the bearing from seizing up.

- The performance of the metal is dependent upon the way in which it is manufactured, whether from ingots, melted and poured into pockets in the bearing concerned, or cast in position by centrifugal action derived from a special machine designed to minimise blow-hole formation.

**Cadmium silver bearing alloy**

It contains ; Cadmium = 97.6%; copper = 1.9%; and slilver = 0.5%. These bearings are used for loads upto 14.5 MN/m$^2$.

**Beryllium copper bearing**

- It contains 97.5% copper and 2.25% beryllium.
- It has been used as a hard bearing metal in place of phosphor bronze, its p 2 wear resistance being five times that of phosphor bronze.
- Beryllium copper possesses a film forming and self lubricating property which makes it more suitable as a bearing metal.

**Graphite bearing metals**

- A type of bearing metal sometimes known as *spongy bearing bronze* has resulted from investigations into the effect of graphite upon certain alloys. Those metals contain *graphite uniformly distributed* throughout their composition, and when used for bearings they are therefore self lubricating.

- *Such bearings are applicable to non-accessible slow running machinery, both marine and stationary.*

**Note**. Porous bearings are produced by powder metallurgy. Powders of copper, tin and graphite *are sintered, and then passed through a sizing die* to the correct dimensions. The metal thus produced is of a porous nature capable of holding upto one third of its volume of oil. A pressure on the bearing surface, or a temperature rise, will cause the oil to exude, so that lubrication is ensured continuously and automatically where it is needed.

## 9.22. SELECTION, SPECIFICATIONS AND FORMS AND AVAILABILITY OF MATERIALS

### 9.22.1. Considerations for Selection of Materials for Different Components

Selection of right or proper materials for manufacturing/producing different components is not an easy job. Several factors need to be considered in order to make a judicious selection of materials of specific applications. Selection of materials amply depends upon the following factors.

— Stresses to which a component may be subjected.

— High temperatures or corrosion actions it may have to withstand.

— Flexibility or rigidity requirements.

— Adaptability for manufacturing processes.

— Cost

— Appearance and aesthetic appeal.

— Availability as an when required etc.

Particularly, for metals and alloys, the following points in respect of '*Selection for use*' are worth noting.

**(*i*) Strength, shaping, cost :**

Strength, ease of shaping and relatively low cost are of great importance for *major structural purposes*. For these purposes, *steel* is ideally suited. Steel is used for the structural members of buildings, for rails, locomotives and ships. For automobile parts, and wherever greater strength and toughness are required, more expensive special steels are used. As a general rule, if strength alone is the main consideration, non-ferrous alloys would not be used. However, where the requirement is for strength combined with resistance to rusting, aluminium - bronze or monel metal may be used. There are purposes for which strength is not so important as ease of machining. In making screws, for instance, another kind of steel, or perhaps a brass, may be used. Where the finished form is to be produced by casting, and great strength is not required, cast iron or cast brass are employed.

**(*ii*) Light metals :**

For making aeroplane parts, and in other applications where strength must be combined with light weight, metals such as aluminium or magnesium and their alloys are used.

**(*iii*) Softness, ease in bending :**

For uses requiring softness and ease in bending, as in cable sheathing or plumbers' pipe and where certain chemical properties are needed, lead and its alloys may be employed.

**(*iv*) Electrical conductivity :**

Copper and aluminium are used for conducting electricity since they offer less resistance to the passage of the current. Silver offers even less resistance but is too expensive for commercial use. Copper offers less resistance than aluminium for the same size wire, but aluminium, due to its lighter weight, offers less resistance per unit of weight. Despite the fact, from a cost stand-point the advantage is still with copper. In the passage of current through a conductor, resistance results in the giving off of heat; the greater the resistance, the greater the heat for the passage of a given current. Hence for electrical heating, metals with high electrical resistance are needed. These are found in alloys; among the best are alloys of nickel and chromium.

(*v*) **Resistance to heat :**

The resistance to flow of heat is of the same order as the resistance to electricity; hence ordinarily we use copper for boiler tubes, heating coils and soldering irons. For cooking utensils, aluminium is used because of the combinations of high heat conduction with resistance to attack by foodstuffs. In magnetic properties, iron and its alloys stand alone.

(*vi*) **Susceptibility for corrosion :**

Metals vary greatly in their susceptibility to atmospheric and chemical corrosion. The rusting of iron is the commonest example. A list of the elements in order of their susceptibility to attack would start with potassium, which actually catches fire in contact with water, and would end with platinum and gold, which are unattacked by almost all chemical reagents.

Iron or steel when used in exposed positions, are protected by paint or by a thin coat of zinc (galvanized sheet) or tin. If *greater resistance to corrosion is required, copper or its alloys, brass, bronze or nickel silver are used. Where real permanence is required, the precious metals must be chosen.*

The development of stainless steels (steels containing varying percentages of carbon, chromium and nickel) has made it possible to use this material without protective coating against corrosion.

### 9.22.2. I.S.I. Methods of Specifications of Metals and their Alloys

For specifying metals and alloys the Indian Standards Institution (I.S.I.) has published certain standards. While designating metals, alloys and other materials their properties are given ample consideration. A code number is sufficient for specifying a material. A '*Handbook of I.S.I. publications*' assimilates a full list of I.S.I. publications and for particular materials particular I.S.I. pamphlets may be consulted.

### 9.22.2.1. Extracts from I.S.I Codes for Designation of Some Metals and Alloys

**Copper and copper alloys (IS : 2378 - 1963)**

- Copper and copper alloys may be designated by a group of symbols indicating the important characteristics, in the following order :
- (*a*) Condition, if necessary (*b*) Chemical composition (*c*) Special requirements.
- The Symbol 'Cu' is to be followed for copper. Similarly, internationally accepted symbols for other elements are to be used.
- Designation for different grades of copper will be as follows :

| *Grade* | *Designation* |
|---|---|
| Electrolytic high conducting copper | CuEHC |
| Cathode copper | CuCATH |
| Arsenical, tough pitch copper | CuATP |
| Phosphorised, arsenical copper | CuDDA |
| Phosphorised, high residual phosphorus, copper | CuDHP |

- For copper alloys, the chemical symbol of copper is to be followed by the chemical symbol of the next most significant element, after which two other significant elements are to be stated in the order of decreasing percentage or, when equal, in alphabetical order. The nominal or average percentage of each alloying element is to be indicated by an index number and is immediately to follow the respective chemical symbol. Only the minimum number of symbols necessary to identify the alloy are to be used.
- When two or more significant alloying elements have the same alloy index in an alloy, the chemical symbols are to be grouped together, followed by their alloy index.
- When minimum contents of some minor alloying elements are specified for certain specific purposes, they are, if necessary, to be considered as significant alloying elements and indicated at the end of the designation, by their chemical symbols only.
- Where rigid limits for impurities are specified for workability, to differentiate between

two similar alloys it is to be indicated by the symbol 'K'.

- The following symbols are to be used to indicate surface finish :
  - (*i*) J-Bright rolled or drawn
  - (*ii*) J7-Bright pickled
  - (*iii*) J8-Bright annealed
  - (*iv*) M-Machined
- The following symbols are used for the treatment :
  - (*i*) a–Annealed
  - (*ii*) 1/4 H–Quarter hard
  - (*iii*) 1/2 H–Half hard
  - (*iv*) 3/4 H–Three fourths hard
  - (*v*) H–Full hard
  - (*vi*) h–Hot rolled
  - (*vii*) d–Cold worked (rolled or drawn)
  - (*viii*) e–Extruded
  - (*ix*) f–Forged

**Aluminium and aluminium alloys**

Elements are identified by the group number assigned to each as follows

| Element | Group Number |
|---|---|
| Unalloyed Aluminium (Wrought and cast form) | 1 |
| Copper | 2 |
| Manganese | 3 |
| Silicon | 4 |
| Magnesium | 5 |
| Magnesium Silicide ($Mg_2Si$) | 6 |
| Zinc | 7 |
| Other elements (such as nickel, titanium, chromium, lead, bismuth etc.) | 8 |
| Unassigned | 9 |

- The new alloy designation system adopted for aluminium and its alloys is based on *Five-Digit System* for wrought aluminium alloys, aluminium (unalloyed) and a *Four*-Digit System for casting alloys castings (unalloyed) and primary ingots.

**Five digit System**

**(*a*) Wrought aluminium alloys**

- *First Digit* identifies the *major alloying element.*
- *Second Digit* indicates *rounded off mean value of* the percentage of the major alloying element except for Group 4 containing Silicon, when the digit refers to the mean percentage halved and rounded off; and for Group 6 containing Magnesium silicide, the digit refers to five times the mean magnesium percentage rounded off.
- *Third, Fourth* and *Fifth Digits* identify the major alloying elements in the descending order of their percentage and in the case of same alloy percentage in the serial order, except for Group 6 containing the intermetallic compound $Mg_2Si$, when the third digit refers to either magnesium or silicon, which is in excess of that required for magnesium silicide (that is 1 : 7, which is ratio of Mg to Si). For the balanced composition, the third digit is to be zero. In case of high purity aluminium base alloy, the fifth digit is to be 1.

**Example :** *Alloying elements and their mean percentages : Cu 1.5, Si 1.0, Mg 0.85 Designation : 22450.*

(*b*) **Wrought Aluminium (unalloyed)**

- *First-Digit* is always shown as 1, representing unalloyed aluminium in the wrought condition.
- *Second-Digit* is always shown as 9, the unit digit of 99 minimum purity percentage.
- *Third and Fourth digits* indicate the decimal purity percentage.
- *Fifth Digit* indicates variants, if any.

**Example :** *Aluminium Purity, percent* : All 99.5 Designation : 19500

**Four-digit System**

(*a*) Casting Alloy :

- *First Digit* indicates the major alloying element.
- *Second Digit* indicates the mean value percentage of the major alloying element halved and rounded off.
- *Third and Fourth Digits* identify the minor alloying elements in the order of their decreasing percentage or in the case of same alloy percentage in serial order.

**Example :** *Alloying elements : Si 11.5, Ni 3.0 Designation : 4680*

(*b*) Primary ingots or castings (unalloyed)

- *First Digit* is always shown as 1, representing unalloyed aluminium.
- *Second Digit* is always shown as 9, the unit digit of 99, minimum purity percentage.
- *Third Digit* indicates the decimal purity percentage.
- *Fourth digit* indicates variants if any.

**Example :** *Aluminium purity, percent : Al 99.6 (*EC grade*) Designation : 1961.*

**Steel**

A steel may be designated by a group of symbols indicating the important characteristics in the following order : (*a*) Tensile strength; (*b*) Carbon content ; (*c*) Alloy content ; (*d*) Sulphur and phosphorus limits ; (*e*) Weldability ; (*f*) Surface finish ; (*g*) Surface condition ; (*h*) Steel quality and (*i*) Treatment.

- For *steels* specially those which are standardised on the basis of their tensile strength, without detailed chemical composition, the symbol 'SE' shall be followed by the minimum ultimate tensile strength.

**Example :**

Tensile strength : 32 to 44 kg/mm$^2$ Symbol : St 32

- For *plain carbon steel*, letter 'C' shall be followed by the average carbon content in the hundredths of a percent, *e.g.*,

**Example :**

Carbon : 0.1 to 0.18%
Manganese : 0.40 to 0.80%
} Designation : C 14

- The *alloy index* shall consist of chemical symbols of significant elements arranged in descending order of percentage contents. The nominal or average percentage of each alloying element shall be indicated by an *index number following its chemical symbol.*
- *Symbol for sulphur and phosphorus limits - the letter 'K' shall be used to indicate steel with special limits for maximum 'S' and 'P'.*

**Example :**

Tensile strength : 34 to 36 kg/mm$^2$ Designation : St 34 K

(Properties : Special limits for maximum Sulphur and Phosphorus)

*Weldability symbols.* Guaranteed weldability of steel shall be indicated by the following symbols.

'w' Fusion weldable
$w_p$ Pressure weldable
$w_r$ Weldable by resistance
$w_s$ Weldable by spot welding

**Example :**

Tensile strength : 55 to 65 kg/mm$^2$ Designation : St 55 'w'
(Properties : Fusion weldable)

— *Steel Quality Symbols*

A — Non ageing
E — Stabilised against stress corrosion or intergranular attack.
L — Control cooled to ensure freedom from flakes.
D — Fully killed (deoxidized)
D$^2$ — Semi-killed
R — Rimming quality
G — Grain size controlled
H — Hardenability controlled
I — Inclusion controlled
M — Structural homogenity guaranteed by macro-etch test.

— *Treatment Symbols*

a — Annealed or softened
c — Case carburised
d — Hard drawn, cold reduced
h — Hot rolled
n — Normalised
o — Spheroised
p — Patented
q — Hardened and tempered
s — Stress relieved
t — Temperated

**— Uses of Indian Standard Steel (IS : 1871 – 1965 Appendix A)**

| | |
|---|---|
| St 30 | Structural steel sheets for plain drawn or enamelled parts, tubes for oil well casing, steam, water and air passage, cycle, motorcycle and automobile tubes, rivet bars and wire. |
| St 32, St 34 | Steels of car structures, screw stock. and other general engineering purposes. |
| St 37 | Structural steel for chemical pressure vessels, and other general engineering purposes. |
| St 39 | Structural steel for ships, chemical pressure vessels, air receivers, fasteners, concrete reinforcement, and other general engineering purposes. |
| St 42 | Structural steel bridges and building construction, railway rolling stock, screw spikes, oil well casing tube piles and other general engineering purposes. |
| St 44 | Structural steel for railway stock, pressure vessels, fasteners, colliery arches, valve fittings for compressed gas cylinders and other general engineering purposes. |

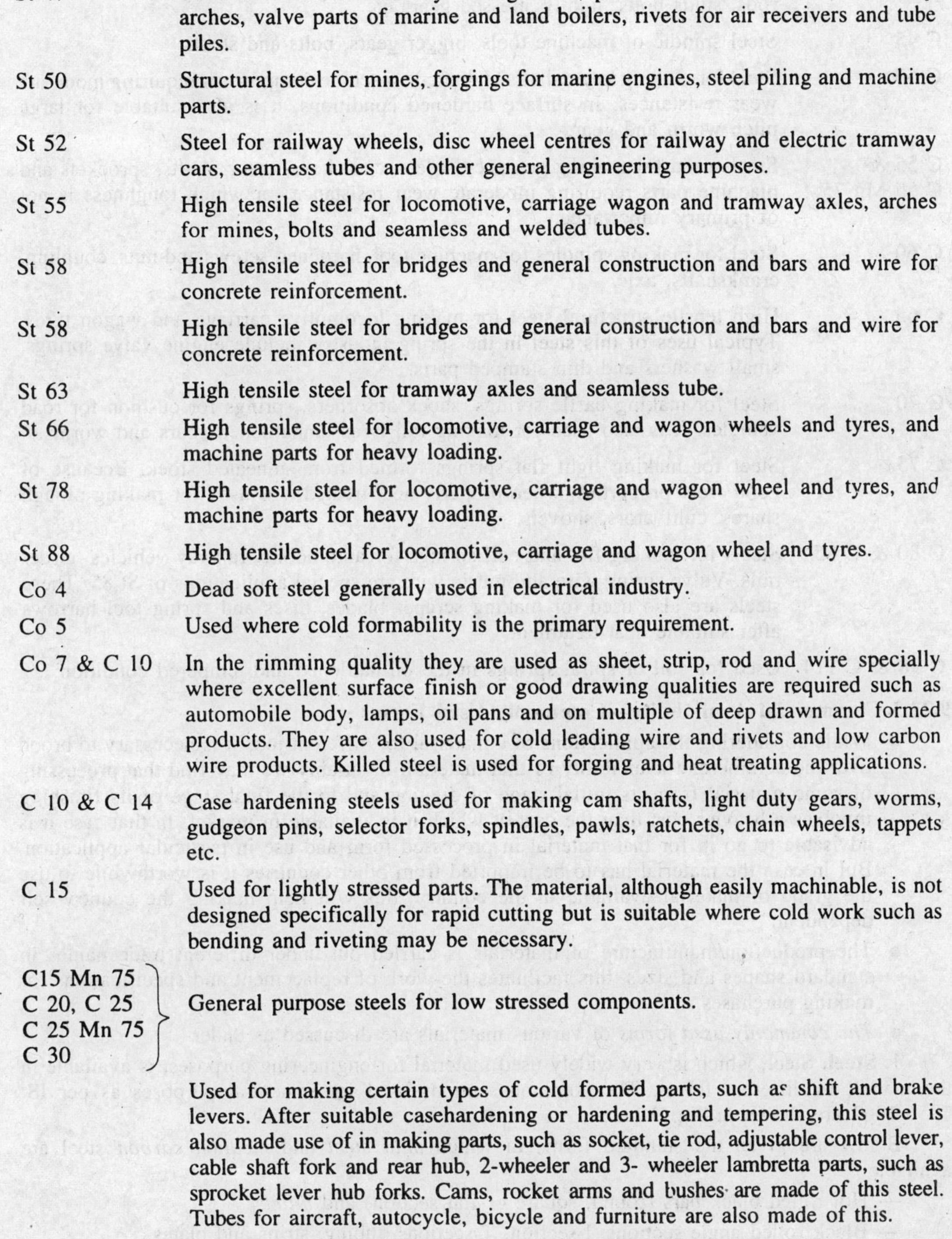

| Grade | Uses |
|---|---|
| St 47 | Structural steel for railway rolling stock, pressure vessels, fasteners, colliery arches, valve parts of marine and land boilers, rivets for air receivers and tube piles. |
| St 50 | Structural steel for mines, forgings for marine engines, steel piling and machine parts. |
| St 52 | Steel for railway wheels, disc wheel centres for railway and electric tramway cars, seamless tubes and other general engineering purposes. |
| St 55 | High tensile steel for locomotive, carriage wagon and tramway axles, arches for mines, bolts and seamless and welded tubes. |
| St 58 | High tensile steel for bridges and general construction and bars and wire for concrete reinforcement. |
| St 58 | High tensile steel for bridges and general construction and bars and wire for concrete reinforcement. |
| St 63 | High tensile steel for tramway axles and seamless tube. |
| St 66 | High tensile steel for locomotive, carriage and wagon wheels and tyres, and machine parts for heavy loading. |
| St 78 | High tensile steel for locomotive, carriage and wagon wheel and tyres, and machine parts for heavy loading. |
| St 88 | High tensile steel for locomotive, carriage and wagon wheel and tyres. |
| Co 4 | Dead soft steel generally used in electrical industry. |
| Co 5 | Used where cold formability is the primary requirement. |
| Co 7 & C 10 | In the rimming quality they are used as sheet, strip, rod and wire specially where excellent surface finish or good drawing qualities are required such as automobile body, lamps, oil pans and on multiple of deep drawn and formed products. They are also used for cold leading wire and rivets and low carbon wire products. Killed steel is used for forging and heat treating applications. |
| C 10 & C 14 | Case hardening steels used for making cam shafts, light duty gears, worms, gudgeon pins, selector forks, spindles, pawls, ratchets, chain wheels, tappets etc. |
| C 15 | Used for lightly stressed parts. The material, although easily machinable, is not designed specifically for rapid cutting but is suitable where cold work such as bending and riveting may be necessary. |
| C15 Mn 75<br>C 20, C 25<br>C 25 Mn 75<br>C 30 | General purpose steels for low stressed components. |
| | Used for making certain types of cold formed parts, such as shift and brake levers. After suitable casehardening or hardening and tempering, this steel is also made use of in making parts, such as socket, tie rod, adjustable control lever, cable shaft fork and rear hub, 2-wheeler and 3- wheeler lambretta parts, such as sprocket lever hub forks. Cams, rocket arms and bushes are made of this steel. Tubes for aircraft, autocycle, bicycle and furniture are also made of this. |
| C 35 | Steel for low stressed parts, automobile tubes and fasteners. |
| C 35 Mn 75 | Steel for making low stress parts in machine structure, cycle and motor cycle tubes, fish plates for rails and fasteners. |

| | |
|---|---|
| C 40 | Steel for crankshaft, shafts, spindle, automobile axles, beams, push rod, connecting rods, studs bolts, lightly stressed gears etc. |
| C 45 | Steel spindle of machine tools, bigger gears, bolts and shafts. |
| C 50 | Steel for making keys, shafts, cylinders, machine components requiring moderate wear resistances. In surface hardened conditions, it is also suitable for large pitch worm and gear. |
| C 55 & C 55 Mn 75 | Steels used for making gears, cylinders, cars, keys, crank shafts, sprockets and machine parts requiring moderate wear resistance for which toughness is not of primary importance. |
| C 60 | Steel for making spindles for machine tool, hardened screws and nuts, coupling, crankshafts, axle. |
| C 65 | High tensile structural steel for making locomotive carriage and wagon tyres. Typical uses of this steel in the spring industry include engine valve springs, small washers and thin stamped parts. |
| C 70 | Steel for making baffle springs, shock absorbers, springs for cushion for road vehicles. It is also used for making rail tyres, unhardened gears and worms. |
| C 75 | Steel for making light flat springs formed from annealed stock. Because of good wear properties, when properly heat treated it is used for making plough shares, cultivators, shovels. |
| C 80 & C 85 | Steels for making flat and coil springs for automobile, railway vehicles, girder rails. Valve spring wire and music wire are special applications of St 85. These steels are also used for making scraper blades, discs and spring tool harrows after suitable heat treatment. |
| C 98 & C 114 | Used for coil or spiral springs in the oil-hardened and tempered condition. |

### 9.22.3. Commercial Availability—Commonly Used Forms

- While considering the applications of a material for different jobs it is necessary to brood over the commercial availability of that material. Invariably we may find that processing of some material from its initial stage of digging and to the final stage of its finishing involves a heavier cost than the cost at which it is available in market. In that case it is advisable to go in for that material in processed form and use in particular application. But in case the material has to be imported from other countries it is worthwhile to use the grade of material available in the country; this will help making the country self dependent.
- The production/manufacture of materials is carried out under different trade names in standard shapes and sizes, this facilitates the work of replacement and specifications for making purchases.
- *The commonly used forms* of various materials are discussed as under :

1. **Steel.** Steel, which is very widely used material for engineering purposes, is available in the market in different forms. The common sections used for structural purposes as per IS: 1956-1962 are shown in Fig. 9.11.

Below are given the standard forms in which *mild steel* and *medium carbon* steel are available.

— Hot rolled black bars (round, square or flat section) and *billets.*

— Black rolled angle sections, I-section, T-sections, tubing, strips and plates.

— Bright rolled flats and strips.

— Bright drawn tubing and bars of round, square, rectangular and hexagonal sections.

— Galvanised iron (G-I) corrugated sheets.

— Miscellaneous bright and black forged articles *e.g.*, bolts, nails, pipe fittings washers etc.

— Special sections.

— *High carbon steel* rolled black bars (round, square, flat, octagonal etc. sections).

— *Spring steel wire.*

Some important standard shapes (in which the rolled steel sections are available in the market) are discussed below :

(*i*) **Bars :**

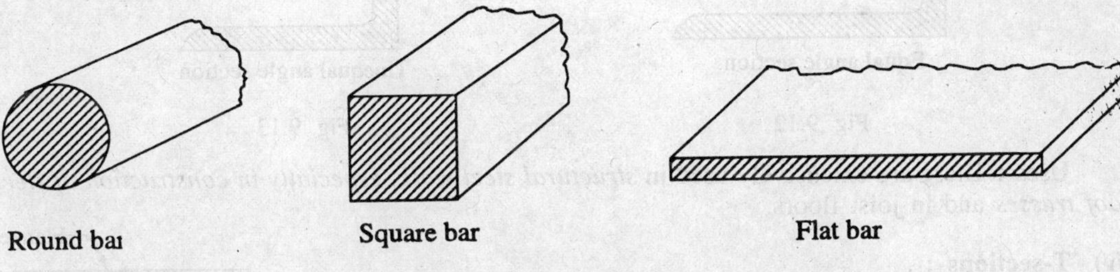


Fig. 9.11

(*a*) *Round bars :*

- These are available in circular cross-section with diameters varying from 5 mm to 250 mm.
- The bars below 4 mm diameter are known as *wires* and the bars below 12 mm diameter are called *rods.*
- The commonly used cross-sections are 5 mm to 25 mm with corresponding weights per metre length as 1.50 N and 38 N respectively.
- As per Indian standards, they are designated as ISRO.

*Uses* : They are widely used as *reinforcement in concrete structures, steel grill work in windows* etc.

(*b*) *Square bars :*

- They are available in square cross-section with sides varying from 5 mm to 250 mm. The commonly used cross-sections have sides varying from 5 mm to 25 mm with corresponding weights per metre length as 2N and 49 N respectively.
- As per Indian standard, they are designated as ISSQ.
- *Uses :* They are widely used in the construction of *steel grill work, gates, windows* etc.

(*c*) *Flat bars :*

- They are available in varying widths and thicknesses.
- They are usually 25 mm × 5 mm to 300 mm × 25 mm in cross-section and available upto 6.5 m in length.

Uses : They are widely used in the construction of *steel grill work for windows and gates.*

(*ii*) **Plates :**

They are available in different sizes with thickness varying from 5 mm to 50 mm. The corresponding weights per square metre are 392 N and 3925 N respectively.

*Uses :* They are employed as *tension members of steel roof* trusses, for *extension of beams* and to *form build-up sections of steel.*

(*iii*) **Angle sections :**

The angle sections may be of equal or unequal legs as shown in Fig. 9.12 and Fig. 9.13.

- *The equal angle sections* are available in sizes varying from 20 mm × 20 mm × 3 mm to 200 mm × 200 mm × 25 mm. The corresponding weights per metre length are 9 N and 736 N respectively.

- *The unequal angle sections* are available in sizes varying from 30 mm × 20 mm × 3 mm to 200 mm × 150 mm × 18 mm; corresponding weights per metre being 11 N and 469 N respectively.

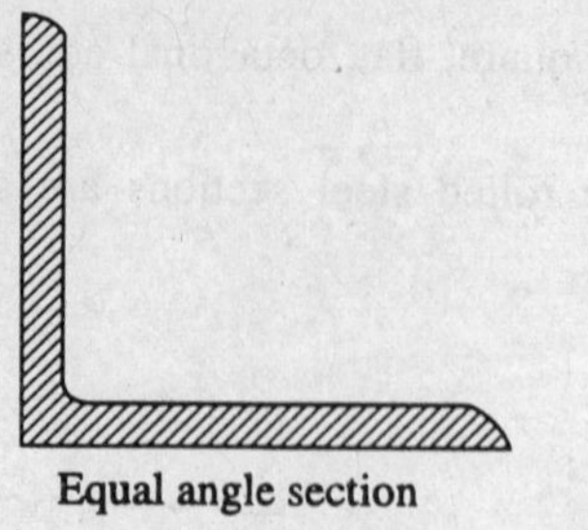
Equal angle section

Fig. 9.12

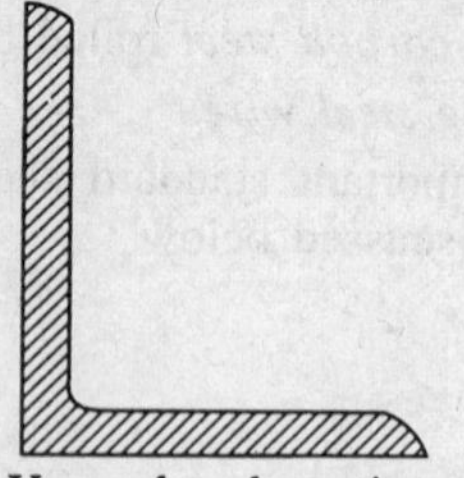
Unequal angle section

Fig. 9.13

**Uses :** These are extensively used in *structural steel work, especially in construction of steel roof trusses* and in joist floors.

(*iv*) **T-sections :**

- The shape of this section is like that of T and consists of flange and web as shown in the Fig. It is designated by overall dimensions and thickness.
- These sections are available in sizes varying from 20 mm × 20 mm × 3 mm to 150 mm × 150 mm × 10 mm; corresponding weights per mitre length being 9 N and 228 N respectively.

*Uses :* These are widely used as *structural members of steel roof trusses* and also in *build-up sections.*

T-section

Fig. 9.14

(*v*) **Channel sections :**

- The channel section consist of a web with two equal flanges as shown in Fig. 9.15. It is designated by the height of web and width of flange.
- These sections are available in sizes varying from 100 mm × 45 mm × 400 mm × 100 mm corresponding weights per metre length being 58 N and 494 N respectively.
- The channel sections, according to Bureau of Indian Standards have been classified as junior channel, light channel and medium channel and accordingly are designated as ISJC, ISLC and ISMC respectively.

*Uses* : These are widely used as structural members of the framed steel structures such as *purlins, built-up columns* etc.

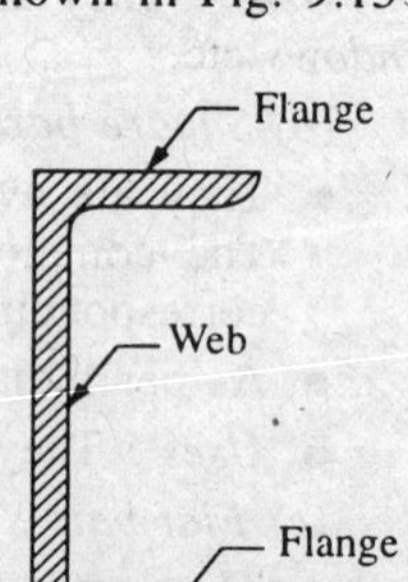

Channel section

Fig. 9.15

(*vi*) **I-sections :**

- These are commonly known as *rolled steel joints* (RSJ) or *beams.* An I-section consists of two flanges connected by a web as shown in Fig. 9.16. It is designated by overall depth, width of flange and weight per metre length.
- These are available in sizes ranging from 75 mm × 50 mm at 61 N to 600 mm × 210 mm at 995 N. The wide flange beams are available in sizes varying from 150 mm × 100 mm at 170 to 600 mm × 250 mm at 1451 N.

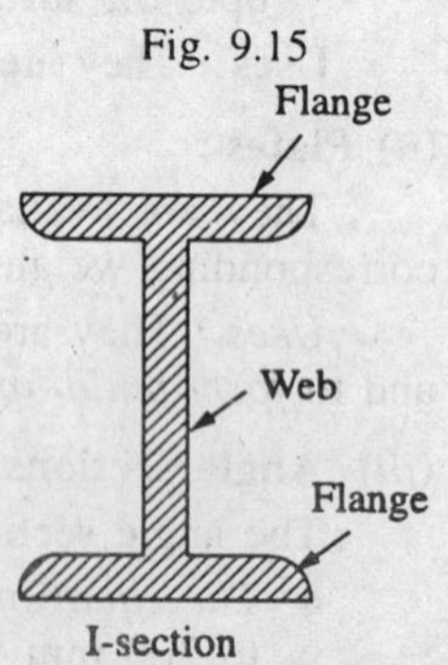

I-section

Fig. 9.16

The beams suitable for columns are available in H-section which vary in sizes from 150 mm × 150 mm at 271 N to 450 mm × 250 mm at 925 N.

- According to Bureau of Indian Standards the I-sections are classified into junior beams, light beams, medium beams, wide-flange beams and heavy beams and these are accordingly designated as ISJB, ISLB, ISMB, ISWB and ISHB respectively.

**Uses :** These are widely used as *floor beams, door lintels, stair treads,* columns etc. These are economical because the concentration of material is only on the flanges, where bending stresses are maximum.

(*vii*) **Corrugated sheets :**

- These sheets (See Fig 9.17) are formed by passing steel sheets through the grooves; these grooves bend and press steel sheets and corrugations are formed on the sheets.
- The corrugated sheets are usually galvanised with a layer of *zinc* and are commonly known as *galvanised iron sheets or GI-sheets.*
- *Uses :* These are widely used as a *roof covering materials.*

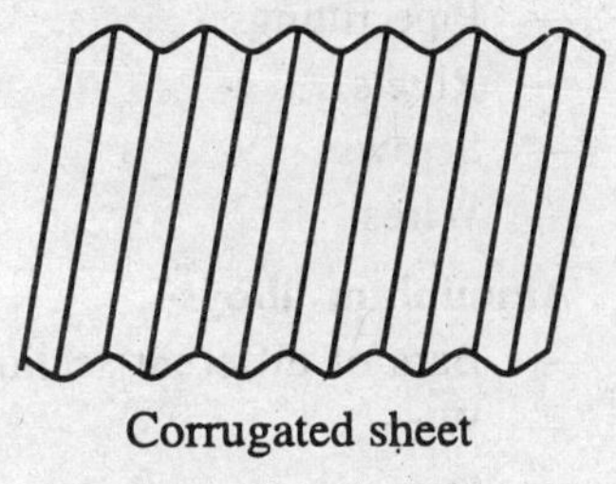

**Corrugated sheet**

Fig. 9.17

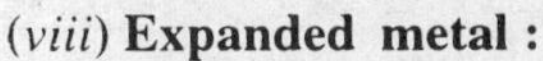

(*viii*) **Expanded metal :**

- This form of steel is available in various shapes and sizes.
- Fig. 9.18 shows a plain expanded metal.
- It is prepared from sheets of mild steel which are machine cut and drawn out or expanded. A diamond mesh appearance is thus formed throughout the whole area of the sheet.

*Uses :* (*i*) Expanded metal is widely employed for reinforcing concrete in *foundations, roads, floors, bridges, partitions etc.*

(*ii*) The sheets of expanded metal in various sizes and shapes are also used as a *temporary fencing around building and other constructions.*

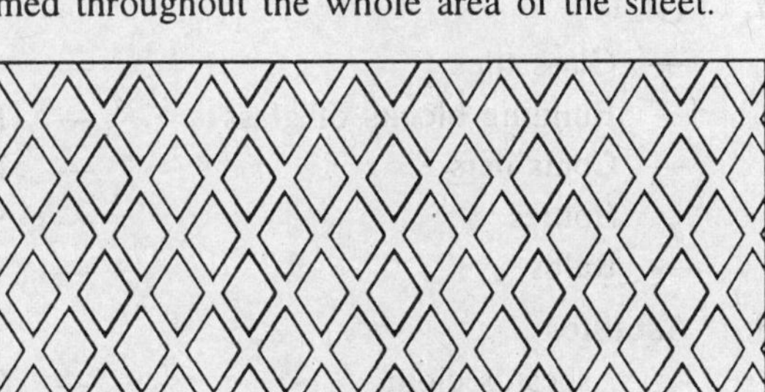

Plain expanded metal

Fig. 9.18

**2. Cast iron**

Cast iron is available in the form of

— Pipes

— Pipe fittings

— Cast plates

— Bars

— Castings of particular shapes etc.

**3. Copper Alloys**

— Drawn and rolled bars of different sections.

— Wire

— Strip

— Sheet

— Ingots

— Short cast sticks

— Special castings (made to order)

— Other forms (in case of brass) e.g. powder, brazing rods, screws, rivets, washers, pipe fittings, forgings and pressings of special shapes.

**4. Aluminium**

- — Strips
- — Tubes
- — Sheets
- — Pipe fittings
- — Rivets
- — Screws
- — Wires.
- — Foils
- — Ingots
- — Powders
- — Rods of various sections etc.

**5. Aluminium alloys**

- — Bars of different sections
- — Wire
- — Strip
- — Sheet
- — Tubing
- — Rivets
- — Washers
- — Castings
- — Pressing
- — Ingots etc

**6. Plastics**

- — Sheets
- — Rods
- — Tubes
- — Powder
- — Laminated sheets
- — Thread (like Nylon)
- — Special castings.

**7. Glass**

- — Glass tiles
- — Building blocks of glass
- — Containers
- — Bottles
- — Bulbs
- — Cut glass vessels
- — Decoration pieces
- — Phials
- — Scientific vessels etc.
- — Tableware

**8. Asbestos**

- — Asbestos cloth
- — Board
- — Felt etc.
- — Sheet
- — Yarn

**9. Cork**

- — Sheet
- — Stoppers
- — Pads
- — Tiles etc.

## OBJECTIVE TYPE QUESTIONS

**A. Choose the Correct Answer :**

**1.** Vanadium steel is often used for

(*a*) bearings (*b*) electromagnets

(*c*) machine beds (*c*) axles and springs.

**2.** Nickel is used in

(*a*) pressure sensitive elements (*b*) electrodes of thermionic valves

(*c*) bulb filaments (*d*) automatic voltage regulators.

**3.** ............ is the property of steel due to which it can withstand the blow of hammer.

(*a*) Hardness (*b*) Toughness

(*c*) Malleability (*d*) Plasticity.

4. High purity copper is obtained by
(*a*) casting (*b*) electric refining
(*c*) induction heating (*d*) rolling.

5. Which one of the following alloys has magnesium as the principal constituent?
(*a*) Y-alloy (*b*) Babbit metal
(*c*) Dow metal (*d*) Bell metal.

6. German silver is an alloy consisting of
(*a*) copper, aluminium and nickel (*b*) copper and lead
(*c*) copper, zinc and nickel (*d*) chromium and copper.

7. The main constituent of duralumin is
(*a*) aluminium (*b*) manganese
(*c*) copper (*d*) magnesium.

8. For melting one tonne of cast iron in a cupola the volume of air required is
(*a*) 50 $m^3$ (*b*) 150 $m^2$
(*c*) 700 $m^3$ (*d*) 7000 $m^3$.

9. The steel used for rails under heavy traffic and on sharp curves is
(*a*) manganese steel (*b*) chrome steel
(*c*) cast steel (*d*) mild steel.

10. Gun metal is an alloy of
(*a*) copper, tin and zinc (*b*) aluminium, lead and tin
(*c*) lead, copper and tin (*d*) aluminium, lead and nickel.

11. .............. is the process by which wrought iron is manufactured from pig iron.
(*a*) Annealing (*b*) Shingling
(*c*) Puddling (*d*) Roasting.

12. Application of tin is in
(*a*) low current fuses (*b*) bulb filaments
(*c*) transducers (*d*) conductors for starters.

13. Which of the following is not valid for blister steel?
(*a*) It cannot be forged (*b*) It cannot be welded
(*c*) It is produced by cementation process (*d*) It is full of fissures and cavities.

14. General silver is an alloy of
(*a*) copper and tungsten (*b*) copper and nickel
(*c*) silver and tin (*d*) tungsten and silver.

15. ............. has the lowest point.
(*a*) Tin (*b*) Lead
(*c*) Copper (*d*) Cast iron.

16. Muntz metal is an alloy of
(*a*) zinc and lead (*b*) copper and zinc
(*c*) tin and zinc (*d*) lead and white metal.

17. Iron ore rich in Fe content is
(*a*) Siderite (*b*) Magnesite
(*c*) Haematite (*d*) Limonite.

18. In a puddling furnace pig iron is remelted to produce
(*a*) cast steel (*b*) wrought iron
(*c*) cast iron (*d*) chilled iron.

19. Product of blast furnance is known as
(*a*) pig iron (*b*) cast iron
(*c*) wrought iron (*d*) malleable iron.

20. Corrosion resistance of stainless steel is due to
(*a*) chromium (*b*) vanadium

(c) carbon (d) sulphur.

21. The mild steel to be used as a reinforcing material should be free from
(a) loose rust (b) oil and grease
(c) loose mill scale (d) paint
(e) all of the above.

22. Cast iron is used for
(a) column and struts (b) wire nails
(c) members subjected to tension (d) trusses.

23. Brass is an alloy of
(a) lead and tin (b) copper and zinc
(c) tin and silver (d) nickel and zinc.

24. Galvanizing is done with a thin layer of
(a) copper (b) zinc
(c) lead (d) cadmium.

25. The percentage of chromium in stainless steel used for cutlery is usually in the range
(a) 0.5% 1.1% (b) 1.2% to 2.7%
(c) 10 to 20% (d) 20 to 30%.

26. The specific gravity of cast iron is usually
(a) 4.1 (b) 5.8
(c) 6.5 (d) 7.25.

27. Which of the following is *not* the characteristic property of aluminium?
(a) Least affinity for oxygen (b) High thermal conductivity
(c) Lightness (d) Good electrical conductivity.

28. Which variety of copper has the best conductivity?
(a) Induction hardened copper (b) Pure annealed copper
(c) Hard drawn copper (d) None of the above.

29. Which variety of copper has the best mechanical strength?
(a) Hard drawn copper (b) Cast copper
(c) Soft paper (d) Annealed copper.

30. Which material will have highest limiting strength?
(a) Aluminium (b) Cost iron
(c) Mild steel (d) Wrought iron.

## ANSWERS

| | | | | |
|---|---|---|---|---|
| 1. (d) | 2. (b) | 3. (b) | 4. (b) | 5. (c) |
| 6. (c) | 7. (a) | 8. (c) | 9. (a) | 10. (a) |
| 11. (c) | 12. (a) | 13. (b) | 14. (b) | 15. (a) |
| 16. (b) | 17. (b) | 18. (b) | 19. (a) | 20. (a) |
| 21. (e) | 22. (a) | 23. (b) | 24. (b) | 25. (c) |
| 26. (d) | 27. (a) | 28. (b) | 29. (a) | 30. (c) |

## A. Fill in the Blanks :

1. ................ iron is the oldest form of iron made by man.
2. ................ iron is fundamentally an alloy whose chief elements are iron, silicon and carbon.
3. ................ iron is the basic material from which wrought iron and steel are manufactured.
4. ................ furnace is so named as a very high temperature is developed inside it by means of forcing a blast of heated air.
5. The gases leaving the blast furnace contain about ................ percent CO.
6. ................ induces softness in the iron.
7. ................ is a vertical shaft furnace used to produce cast iron.

8. ................ cast iron is characterised by presence of a largest portion of its carbon in the form of graphite flakes.
9. ................ cast iron is hard, brittle and cannot be machined.
10. ................ iron is virtually pure iron.
11. ................ steel has fissures and cavities and also its structure is not homogeneous.
12. The cost of building a Bessemer converter plant is comparatively ................ than that of an open hearth plant of eqivalent capacity.
13. ................ process produces an inferior quality steel associated with blow holes and inclusion.
14. In ................ furnace a great economy in fuel is obtained by the utilisation of heat of the hot gases of combustion.
15. A high-frequency furnace costs about three times as much to install as does an ................ furnace of the same capacity.
16. ................ process is the only process where sulphur can be effectively reduced.
17. Cast iron contains ................ to ................ percent cabon.
18. ................ iron can neither be hardened nor tempered.
19. ................ iron cannot absorb shocks.
20. Melting point of cast iron is ................ °C
21. ................ is hydrated aluminium oxide.
22. Specific gravity of aluminium is ................
23. ................ foil is used as silver paper for packing chocolates etc.
24. ................ is a reddish brown metal.
25. ................ is largely used in making electric cables and wires and electric machinery and appliances.
26. ................ is a bluish-grey metal with a high metallic lustre when freshly cut.
27. ................ is least tenacious of the common metal.
28. The chief source of tin is ................
29. ................ is brightly shining white metal.
30. The common ores of zinc are ................, ................ and ................
31. ................ is a bluish grey metal.
32. Specific gravity of zinc is ................ .
33. ................ is the lightest metal used as engineering material.
34. ................ is a silvery white metal capable of taking a high polish.
35. ................ are primarily alloys of copper and zinc.
36. ................ are primarily alloys of copper and tin.
37. ................ is one of the most useful metals added in brass.
38. Two important aluminium alloys are: ................ and ................
39. Dow metal contains : magnesium = ................ % and aluminium = ................ %.
40. ................ is practically non-corrosive.

**ANSWERS**

| | | | | |
|---|---|---|---|---|
| 1. Wrought | 2. Cast | 3. Pig | 4. Blast | 5. 30 |
| 6. Silicon | 7. Cupola | 8. Grey | 9. White | 10. Wrought |
| 11. Blister | 12. Less | 13. Bessemer | 14. Open Hearth | 15. Arc |
| 16. L.D. | 17. 2, 5 | 18. Wrought | 19. Cast | 20. 1200 |
| 21. Bauxite | 22. 2.7 | 23. Aluminium | 24. Copper | 25. Copper |
| 26. Lead | 27. Lead | 28. Tinstone | 29. Tin | |
| 30. Zincite, calamine and zinc blende | | 31. Zinc | 32. 6.2 | 33. Magnesium |
| 34. Nickel | 35. Brasses | 36. Bronzes | 37. Tin | |
| 38. Duralumin, Y-alloy | | 39. 91,9 | 40. Nichrome. | |

**B. Answer the following in 'Yes' or 'No' :**

1. Cast iron is an alloy.

2. Magnetite is an ore of aluminium.
3. Limonite is an iron ore.
4. A Bessemer converter is used to produce pig iron.
5. A blast furnace produces steel.
6. Silicon (in iron) increases fluidity of metal.
7. Phosphorus gives rise to cold shortness.
8. Sulphur causes red shortness.
9. Manganese reduces hardness and brittleness.
10. Grey cast iron contains carbon in free form (graphite flakes).
11. White cast iron is highly resistant to wear.
12. Mottled cast iron is used for man hole covers and pipes.
13. Nodular cast iron is also called ductile cast iron.
14. Malleable cast iron is obtained by giving a long time annealing to cast iron.
15. Wrought iron contains a large amount of carbon.
16. L.D. process is used to produce cast iron.
17. Wrought iron is produced by open hearth process.
18. Duplex process of manufacturing steel combines the advantages of the Bessemer and open hearth processes.
19. Chromium provides stainless property in steel.
20. Tungsten steel is used for making permanent magnets.
21. Bauxite is commercial ore of aluminium.
22. Pure aluminium has silvery colour and lustre.
23. Malachite is an ore of lead.
24. Copper is a reddish brown metal.
25. Melting point of copper is 700°C.
26. Lead is least tenacious of all metals.
27. Specific gravity of lead is 11.35.
28. The chief source of tin is cassiterite.
29. Melting point of tin is 232°C.
30. Calamine is an ore of zinc.
31. Magnesium is the lightest metal used as engineering material.
32. Nickel is almost as hard as steel.
33. Brasses are primarily alloys of copper and zinc.
34. Bronzes are primarily alloys of copper and tin.
35. Aluminium bronze has marked resistance to corrosion.
36. Specific gravity of duralumin (an alloy of aluminium) is 2.8.
37. Hindalium cannot be easily scratched.
38. Dow metal is an alloy of aluminium.
39. Iconol is an alloy of nickel.
40. Nichrome is practically non-corrosive.

## ANSWERS

| | | | | |
|---|---|---|---|---|
| 1. Yes | 2. No. | 3. Yes | 4. No | 5. No. |
| 6. Yes | 7. Yes | 8. Yes | 9. Yes | 10. Yes |
| 11. Yes | 12. Yes | 13. Yes | 14. Yes | 15. No |
| 16. No | 17. No | 18. Yes | 19. Yes | 20. Yes |
| 21. Yes | 22. Yes | 23. No | 24. Yes | 25. No |
| 26. Yes | 27. Yes | 28. Yes | 29. Yes | 30. Yes |
| 31. Yes | 32. Yes | 33. Yes | 34. Yes | 35. Yes |
| 36. Yes | 37. Yes | 38. No | 39. Yes | 40. Yes |

## THEORETICAL QUESTIONS

1. Enumerate various ores which yield ferrous metals.
2. Draw the flow sheet for production of iron and steel.
3. Describe briefly with a neat diagram the construction and working of a 'blast furnace'.
4. What is the composition of 'Pig iron'? How is 'Pig iron' produced?
5. State the effect of following impurities on iron:
   Silicon, Phosphorus, Sulphur and Manganese.
6. How is cast iron produced?
7. Explain the construction and working of a 'cupola'.
8. Give the composition, properties and uses of any two of the following :
   1. Grey cast iron, 2. Mottled cast iron, 3. Nodular cast iron 4. Malleable cast iron.
9. How is wrought iron manufactured from pig iron? Explain the process step-wise.
10. Explain the 'Aston Process' for manufacturing wrought iron.
11. Enumerate various methods of manufacturing steel and explain any one of them in detail.
12. Explain any two of the following methods of manufacturing steel :
    1. Cementation process; 2. Bessemer process; 3. Open hearth process; 4. Duplex process; 5. Kaldo process.
13. State advantages and disadvantages of 'Bessemer process' over Open hearth process'.
14. Explain the various steps involved in L.D. process of manufacturing steel.
15. State composition, properties and uses of carbon steels.
16. State the comparison of cast iron, wrought iron, mild steel and hard steel.
17. What do you accomplish by adding alloying elements to steel?
18. State the effects of adding following alloying elements to steel:
    Nickel, Chromium, Manganese, Vanadium, Tungsten, Silicon and Copper.
19. Given the composition, properties and uses of the following alloy steels : High speed steel, Silicon steel, Stainless steel, Vanadium steel and Tungsten steel.
20. Enumerate the physical and mechanical properties of aluminium.
21. State the importance of commercially pure aluminium as a general engineering material.
22. (*a*) How is copper manufactured?
    (*b*) List the physical and mechanical properties of copper.
    (*c*) State the fields of application of copper.
23. Explain the process of manufacture of lead from its chief lead ore (lead sulphide).
24. Describe how is 'zinc' manufactured? What are its physical and mechanical properties?
25. Which is the lightest metal used as engineering material? State its physical and mechanical properties and uses.
26. Enumerate physical and mechanical properties and uses of nickel.
27. What alloys of copper are chiefly used in engineering constructions? State their approximate compositions, properties and fields of use.
28. Explain the characteristics and uses of low brasses.
29. (*a*) Differentiate between brass and bronze.
    (*b*) Why tin bronzes are suitable for use as bearings?
    (*c*) Discuss the effect on corrosion resistance of copper by increasing additions of zinc, tin and nickel.
30. Can copper be hardened by alloying? Discuss in brief.
31. What is bronze? How does it machine?
32. What are the compositions and characteristics of the copper tin bronzes? Why are these alloys used for springs?
33. (*a*) Why are aluminium alloys so important in modern engineering practice?
    (*b*) What is Duralumin? Describe its properties and uses.
34. What are the effects of the following alloying elements on aluminium?
    (*a*) Copper

(*b*) Manganese
(*c*) Zinc
(*d*) Silicon.

**35.** Write a short note on 'Magnesium alloys'.

**36.** Describe briefly any two of the following nickel alloys:
Iconol, Monel metal, Nichrome and Brightray alloys.

**37.** What are the requirements of an alloy to be used as a bearing metal?

**38.** Why white metal and certain copper alloys are used extensively as bearing metals? Give a typical composition for a bearing metal.

**39.** What is '*babbit*'?

10

# Paints, Varnishes, Distempers and Anti-termite Treatment

*10.1. Introduction. 10.2. Paints–definition of paint–functions of paints–constituents of an oil paint and their functions–preparation of an oil paint–characteristics of a good paint–types of paints–the painter's kit – painting of new wood work–painting of old woodwork–painting of metallic surfaces–painting of plastered surfaces–general precautions in the process of painting–failure of paint–defects in painting. 10.3. Varnishes – definition of varnish – functions of varnish–constituents/ingredients of varnish-characteristics of a good varnish – types of varnishes – process of varnishing–furniture polish–lacquer and lacquering. 10.4. Distempers–properties of distempers–types of distempers–distempering process. 10.5. Wall paper–whitewashing and colourwashing–wall paper–white washing–colour washing. 10.6 Anti-termite treatement – Highlights–Objective Type Questions–Theoretical Questions.*

## 10.1 INTRODUCTION

- The **paints** are (coatings of fluid materials) applied over the surfaces of timber and metals as *protective coatings*. Metals will corrode, wood will warp and decay, concrete and masonry may retain corrosive dust and finally disintegrate unless they are protected by some sort of coatings. In addition, coatings of paint also serve the purpose of *decoration*, sanitation and *improved illuminaiton.*
- The **varnishes** are transparent or nearly transparent solutions of resinous materials ano they are *applied on unpainted furniture and other woodwork to decorate the surfaces and protect them against harmful effects of atmosphere.* Painted surfaces are varnished to *enhance the appearance of the paint and to increase its durability.*
- The distempers are comparatively cheap decorative paints for walls and ceilings. They are less durable than oil or emulsion paints. They are applied on brickwork, cement and lime plastered surfaces and building boards such as plaster board and insulating boards.

## 10.2 PAINTS

### 10.2.1. Definition of Paint

- Paint, as usually defined, *is a dispersion of pigments in a drying oil, with addition of driers and thinners, the former to accelerate film formation and hardening, the latter to allow ease and uniformity in brushing and spraying.* To reduce the defect of cracking in films formed by paints and also to get smooth film from paint *plasticizers* are also added.

- *Paint* is *fluid paste prepared by dissolving a base into a vehicle (carrier) alongwith a colouring pigment.*
- *Paint* is usually a coloured liquid mixture with which suitable surfaces can be coated by a brush or other means where it solidifies as an impermeable film. The film bonds strongly to the surface and is sufficiently durable both in physical and mechanical properties as well as in decorating qualities.

### 10.2.2. Functions of Paints

The functions of applying coats of paints to the surfaces of timber, iron or steel and plaster work in the engineering field are as follows :

1. To protect the surfaces from the weathering effects of the atmosphere and actions by other liquids, fumes and gases.
2. The provide pleasing, colourful and decorative appearance to the surfaces.
3. To prevent decay of wooden members (by rot).
4. To prevent corrosion of metallic surfaces.
5. To provide a smooth surface for easy cleaning.

### 10.2.3. Constituents of an Oil Paint and their Functions

Following are the *constiutents/ingredients* of an oil paint :

1. A base.
2. A vehicle.
3. A pigment.
4. A thinner.
5. Drier.
6. Inert filler.
7. Antiskinning agents.
8. Plasticizers.

1. **Base :** The *base* is a *principal contituent* of a paint and possesses the *binding properties.* The commonly used *bases* are : *White lead, zinc white, red lead, iron oxide, titanium white and lithophone.*

The base performs the following *functions :*

(*i*) It gives body to the point.

(*ii*) It reinforces the film of paint after it has dried *i.e.,* makes the film of the paint harder and more resistant to abrasion.

(*iii*) It minimises shrinking cracks usually formed in drying.

(*iv*) It prevents penetration of the paint to the underlying surface.

Table 10.1 gives the list of bases which are commonly used for paints.

**Table 10.1 : Bases for Paints**

| S. No. | Base | Properties | Uses/Suitablity |
|---|---|---|---|
| 1. | *White lead* | – It is carbonate of lead and forms the base of lead paints<br>– It is *dense, permanent* and *water proof.*<br>– It has a good covering power. | – Most suitable for wood surfaces.<br>– Not suitable for iron surfaces as it does not afford protection against rusting. |
| 2. | *Zinc white* | – It is an oxide of zinc, and forms the base of all zinc paints. | – It is suitable for painting woodwork. |

| | | | |
|---|---|---|---|
| | | – It is *smooth, transparent* and *non-poisonous.*<br>– It has less bulk and hence it is costly.<br>– It has a less covering power.<br>– It is not discoloured when exposed to the sulphur vapours.<br>– It is less durable and difficult to work. | – Not commonly used. |
| 3. | *Red lead* | – It is an oxide of lead and forms the base of lead paints<br>– It solidifies in a short time with linsed oil and hence it is used as a drier also. | – It is quite suitable for painting iron surfaces.<br>– It is suitable for providing primary coat to wooden surfaces. |
| 4. | *Iron oxide* | – It is an oxide of iron and forms the base of all iron paints.<br>– The tint of paint varies from yellowish brown to black<br>– It mixes easily with vehicle.<br>– It is cheap and durable | – It is generally used for prining coat of the iron surfaces. |
| 5. | *Titanium white* | – It is hydrated titanium dioxide mixed with barium sulphate in various proportions.<br>– This material possesses intense opacity.<br>– It is non-poisonous and provides and thin transparent film | – It is used for receiving the coat of an enamel. |
| 6. | *Lithophone* | – It is a mixture of zinc sulphide and barytes.<br>– It is similar in appearance to zinc white.<br>– It is cheap.<br>– It can be easily applied on the surfaces.<br>– It changes colour, when exposed to daylight. | – It is generally used for *interior* work of *inferior* nature. |

2. **Vehicle :** *Vehicle or carrier* is the liquid which carries *solid materials of base and helps them to spread evenly on the surface to be painted.* It is also known as the *drying oil.* It acts a binder to the base and pigment and allows it to be applied. It should dry out on exposure to atmosphere.

The various vehicles are : *Linseed oil, soyabean oil, dehydrated castor oil, fish oil, casein, latex emulsions, varnishes* etc.

The vehicle peforms the following **functions:**

(*i*) It holds the pigments on the surface.

(*ii*) It forms the protective film by evaporation or by oxidation and polymerisation of the unsaturated constituents of the drying oil.

(*iii*) It acts as binder for the base and the pigment.

(*iv*) It provides glow to the painted surface.

Table 10.2. gives the list of vehicles which are commonly employed in preparing a paint.

**Table 10.2. Vehicles for Paints**

| S. No. | Vehicle | Properties | Uses/Suitability |
|---|---|---|---|
| 1. | *Linseed oil* | – It is extracted from flax seeds.<br>– It is clear, transparent and practically colourless.<br>– It is pale cream in colour and sweet to taste.<br>– The acid present in it reacts with the oxygen of the atmosphere and forms a thin and hard film of the paint on the surface being painted. | – It is the most common material used as vehicle of paints in various grades. |
| | (*a*) *Raw linseed oil* | – Thin and pale. | – Suitable for interior work of delicate nature. |
| | | – Requires more time for drying | – Used for exterior surfaces. |
| | (*b*) *Boiled linseed oil* | – Thicker and dark in colour than raw oil. | – Used for exterior surfaces. |
| | | – Dries quickly | – Cannot be used for interior delicate work. |
| | (*c*) *Pale boiled linseed oil* | – Similar to boiled linseed oil but does not possess a dark colour. | – More suitable for painting plastered surfaces. |
| | (*d*) *Double boiled linseed oil* | – Dries very quickly. | – Suitable for external work. |
| | | – Requires a thinning agent like turpentine. | – Suitable for external work. |
| | (*e*) *Stand oil* | – Prepared by heat treatment to raw linseed oil till it becomes thick like honey. | – Suitable for interior work. |

| | | | |
|---|---|---|---|
| | | – Dries slowly and provides a durable clear and shining finish. | |
| 2. | *Poppy oil* | – Prepared from poppy seeds.<br>– Dries slowly<br>– Its colours last long. | – Suitable for making paints of very delicate colours. |
| 3. | *Nut oil* | Extracted from ordinary mal-nuts.<br>Nearly odourless.<br>Dries rapidly.<br>Does not provide a durable finish.<br>Cheap. | Suitable for ordinary work. |

3. **Pigment :** A colouring pigment or stainer is *a finely divided solid material which provides required shade and colour to the paint.* The best pigments do not change their colour when exposed to heat.

Following are the **functions** of a colouring pigment in an oil paint:

(*i*) It provides required shade and colour to the paint.

(*ii*) It reduces the intensity of development of cracks due to drying of the vehicle.

*Pigment volume concentration number (P.V.C.N.)*· The P.V.C.N. is worked out by using the following equation :

$$\text{P.V.C.N.} = \frac{V'}{V' + V''}$$

where $V'$ = volume of pigment in the paint; and

$V''$ = volume of non-volatile vehicle or carrier in the paint.

The durability and gloss of the paint *decrease* with the *increase* in P.V.C.N.

P.V.C.N. for different paints are : Paints for prime coat on metal = 25 to 40; for exterior surfaces of house = 28 to 40; for prime coat on wood = 35 to 40; semi-glass paint = 35 to 45; faint paint = 50 to 75.

The commonly used pigments are given below :

*White* : White lead, zinc oxide, lithophone, titanium oxide.
*Blue* : Iron blue, cobalt blue, sublimed blue.
*Green* : Chrome green, hydrated chromium oxide.
*Yellow* : Chrome yellow, zinc yellow, cadmium yellow.
*Orange* : Mixture of chrome yellow and the red pigments of basic lead chromates.
*Brown* : Umber, copper oxide.
*Red* : Red lead, cadmium red, iron oxides. mercuric oxide.
*Black* : Lamp black, carbon black, drop black, graphite, antimony sulphide.

4. **Thinner :** A *thinner, solvent* or *diluent* is a *volatile substance that is added to the paint to make its application easy and smooth.*

The various **types of thinners** used in paint industry are :

(*i*) *Aliphatic hydrocarbons* such as *mineral spirits, napthas* and other *petroleum fractions.*

(*ii*) *Aromatic hydrocarbons such as toluol, xylol. methylated naphthalene.*

(*iii*) *Turpentine and dipentine.*

*Thinners* perform the following **functions :**

(*i*) They dissolve the film forming material.

(*ii*) They suspend pigments.

(*iii*) They thin the concentrated paints for proper handling.

(*iv*) They evaporate quickly and thus facilitate the drying of paint.

5. **Drier :** A drier is a substance which acts as catalyst and quickens the drying of the vehicle used in the paint.

Driers used are heavy metallic soaps such as *napthenates, resinates, octates, linoleats* etc. of Co, Mn, Pb and Zn. Quantity used is *very little.*

6. **Inert filler** : It is an adulterant usually mixed in an oil paint (*i*) to modify its weight, (*ii*) to improve its durability and (*iii*) to effect economy in the cost of base used.

The commonly used fillers are : Silica, charcoal, barytes, gypsum etc.,

7. **Antiskinning agent** : Antiskinning agents are used to *prevent gelling and skinning of the paint.* These are generally polyhdroxy phenols.

8. **Plasticizer** : Plasticizers are used to *give elasticity to the film and they minimize or prevent cracking.* Some oils are used as plasticizers.

### 10.2.4. Preparation of an Oil Paint

The preparation of an oil paint involves the following **steps** :

1. The base (usually white lead) is first thoroughly ground in the vehicle (linseed oil) so as to form a thick paste.
2. The thick paste is then *thinned down* by adding a suitable quantity of the thinner/solvent to *impart necessary workability to the paint.*
3. The clouring pigment well ground and mixed or with drier is then added in the base already prepared (as mentioned above) and thorough mixing is done to get the required paint. Grinding in roller mills is an important operation in the mixing of paints.

While preparing an oil paint following *precautions* should be observed:

(*i*) All the ingredients should be properly mixed so as to avoid spots on the work to be painted.

(*ii*) Drier should not be added to a paint until it is about to be used.

(*iii*) Excess of drier should not be used in the paint as it injures the paint.

(*iv*) The use of more than one kind of drier in the same paint should be avoided.

(*v*) In order to avoid the discolouring of litharge (oxide of lead), sulphate of managanese is frequently used with zinc paints.

**Note :** Some of the important trade marks for ready made paints are :

'Nerolac paints', 'Blunded paints', ICI paints etc. These paints are usually too thick to be applied in a thin uniform film, therefore, these are thinned by adding a suitable quantity of oil while using them.

### 10.2.5. Characteristics of a Good Paint

Following are the *characteristics* of a good oil paint.

1. It should be *able to resist the atmospheric conditions* to which it will be put.
2. It should have *high hiding power* and the *required colour.*
3. It should *dry fairly quickly*.
4. It should form a *tough, elastic and durable film when dry*.
5. It should *resist corrosion* when applied on a metal.
6. It *should not crack on drying.*
7. The film produced by the paint should be *glossy* and *washable.*
8. Paint should have *necessary consistency* for a particular purpose for which the paint is to be used.

9. It should be *fluid enough to be spread evenly* in a thin coat with a brush.
10. It *should not show brush marks after drying.*
11. It should be free form defects of *chalking, cracking* and *blistering.*
12. It should have lasting pleasing appearance.

**10.2.6. Types of Paints**

The different types of paints are :

1. Oil Paint
2. Aluminium paint
3. Asbestos paint
4. Bituminous paint
5. Cellulose paint
6. Cement paint
7. Colloidal paint
8. Emulsion paint
9. Enamel paint
10. Graphite paint
11. Luminous paint
12. Silicate paint
13. Anti-corrosive paint
14. Plastic paint
15. Synthetic rubber paint.

**1. Oil paint :**

- This is an ordinary paint, cheap and easy to apply.
- It is generally applied in three coats namely : Primes, undercoats and finishing coats..
- It possesses good opacity and low gloss.

Oil paint should not be applied during humid and damp weather.

**2. Aluminium paint :**

An aluminium paint consists of *aluminium powder suspended in spirit varnish or oil varnish.*

Following are the **advantages** of an aluminium paint :

(*i*) Heat resisting.
(*ii*) Visible in darkness.
(*iii*) Does not corrode in sea water or in acidic atmosphere.
(*iv*) High electrical resistance.
(*v*) Impervious to moisture.
(*vi*) Provides goods appearance to the surface.
(*vii*) Possesses a high covering capacity.

**Uses.** The aluminium paint is widely used for *painting radiators, hot water pipes, gas tanks, marine piers, oil storage tanks* etc.

**3. Asbestos paint :**

- This a peculiar type of paint.
- It is applied on the surfaces which are exposed to the acidic gases and steam.

**4. Bituminous paint :**

- It is prepared by dissoving asphalt or vegetable bitumen or mineral pitches in any type of oil or petroleum.
- It presents a black appearance.
- Employed for painting ironwork under water.

**5. Cellulose paint :**

- It is prepared from *nitro-cotton, celluloid sheets, photographic films* etc.
- It hardens by evaporation of thining agent.
- It is not affected by contact with hot water; the surface can stand extreme degrees of cold and hot.
- It presents a flexible, hard and smooth surface.

**6. Cement paint :**

- A cement paint consists of *cement and hydrated lime mixed alongwith a colouring pigment.*
- *Composition* of cement paint : *White cement* = 60 to 75%; *hydrated lime* = 10 to 20%; *hygroscopic salts* (calcium or sodium chloride) = 4 to 5%; *calcium or aluminium stearate* = 1%; white pigment (titanium dioxide or zinc sulphide) = 2 to 5%; *colouring pigments* = upto 8%.
- An inert filler (sand) is often added to make the paint *economical.*
- Cement paints are available in powder form and require mixing only in water before use.
- *Preparation of the paint* : In order to prepare this type of paint, the cement is first intimately mixed with white and colouring pigments. The other ingredients are added by thoroughly stirring in a suitable container.
- The cement paint is applied in *two coats.*
  — Before the *first coat* is applied, the surfaces are wetted to even and control suction and to assist the hardening process of the cement paint. Liquid water should not be present on the surface when the paint is applied.
  — The *second coat* is applied not less than 24 hours after the first coat. It is essential to keep the surface wet after the final coat for about two days by frequent splashing of water to get the best performance.

**Advantages of cement paints :**

Following are the *advantages* of cement paints :

(*i*) Can be applied with a brush or sprayer.
(*ii*) Impart pleasing appearance to the structure.
(*iii*) Can be applied over new and damp walls which cannot be painted over with oil paints until they are sufficiently dried.
(*iv*) They can also be washed.
(*v*) They do not flake or peel off and are unaffected by climatic conditions.
(*vi*) They provide a smooth mat finish without any brush mark.
(*vii*) They prove to be economical as compared to the oil paints.
(*viii*) They are suitable for painting fresh plaster having high alkalinity.

**Uses :** Cement paints are now being commonly used as a lasting finish on outer surfaces of walls and ceiling in residential as well in public buildings etc.

**7. Colloidal paint :**

- In this type of paint no inert material is mixed.
- It requires more time to settle and in the process of settlement it penetrates through the surface.

**Uses :** Colloidal paints may be used for interior as well as exterior walls.

**8. Emulsion paint :**

- It consists of polyvinyl acetate and synthetic resin as binding materials.
- It is easy to apply and dries quickly (in about $1\frac{1}{2}$ to 2 hours)
- The paint possesses excellent alkali resistance.
- The colour of the paint is retained for a long period.
- The surface of paint is tough and it can be cleaned by washing with water.

**Uses:** These paints are mostly employed for painting steel and woodwork which is to be built in masonry.

**9. Enamel paint :**

- Enamel point consists of white lead or zinc white ground in a small quantity of oil and mixed with petroleum spirit and resinous matter.
- **Properties :**

(*i*) It dries slowly and forms a hard and durable surface.

(*ii*) It is not affected by cold and hot water.

(*iii*) It is little costlier but economical in the long run.

(*iv*) It is resistant to steam, fumes, acids and alkalies.

(*v*) It is durable and forms a tough, elastic and glossy film.

**Uses :** Employed for painting door and window frames and protecting timber and other wood-based products. These paints are used for both external and internal works.

**10. Graphite paint :**

- It presents a black colour.
- It is applied on iron surfaces which come in contact with ammonia, chlorine, sulphur gases etc. It is also used in underground railways and mines.

**11. Luminous paint :**

- It contains calcium sulphide with varnish.
- The surface on which this paint is applied shines like radium dials of watches after the light source has been cut-off.

**Uses.** The liminous paint is applied on the surfaces which are free from corrosion or any other lead paint.

**12. Silicate paint :**

- It is a prepared by mixing calcined and finely ground silica with resinous substances
- The drier used with the paint should be of a special silicate drier type.
- *Properties :*

(*i*) It can stand extreme heat.

(*ii*) It adheres firmly to brickwork also.

(*iii*) It is not affected by alkalies.

(*iv*) The paint has no chemical action on metals.

(*v*) The paint when dried forms a hard durable surface.

**Uses :** The paint can be directly applied on brick, plaster or concrete surface; these surfaces should be made wet before the paint is applied. The surface should not be painted with silicate paint in hot weather.

**13. Anti-corrosive paint :**

- It consists of oil and a strong drier.
- A pigment (such as chromium oxide or lead or red lead or zinc chrome) after mixing with some quantity of sand is added to the paint.

**Properties :**

(*i*) It has a black appearance.

(*ii*) It is durable.

(*iii*) It is cheap.

**14. Plastic paint :**

- The paint contains necessary variety of plastics.
- It is available in the market under different trade names.
- The plastic paint can be applied either by brush painting or spray painting.

**Properties :**

(*i*) The paint possesses pleasing apperance and is attractive.

(*ii*) These paints are the finest types of paints.

(*iii*) They provide durable film on drying.

**Uses :** There paints are widely used for auditoriums, show-rooms, cinema halls etc.

*Following points are worth noting regarding the use of plastic emulsion paints :*

(*i*) As these paints cannot resist the attack of enemies of paints (*e.g.*, salts, dust, gases carried by air, fog, rain etc.,), these should be widely used for interior jobs.

(*ii*) The surface to be painted should be cleaned of all dust particles and rubbed with sand paper if necessary, if required levelling putty may be applied.

(*iii*) These paints are water based and hence they are not suitable for metallic surfaces. These paints are not water repellent and may develop some fungus growth in unfavourable circumstances.

(*iv*) The thickness of coat should neither be too thin nor too thick.

(*v*) The painted surfaces need to be washed with wet cloth lightly at least once in a month, failing which the paint may loose its good appearance due to adherence of the dust particles.

**Note :**

- The typical composition of one litre of plastic emulsion paint is as follows : Binders = 2N; Pigments = 5N; Other solids = 1N; Water = 6N. When the paint dries, the water evaporates and a film of binders, pigments and other solids is left behind.
- One litre of plastic emulsion paint covers 15 $m^2$ of wall surface per coat. For interior jobs, two coats, each having a thickness of about 0.04 mm are sufficient.

**15. Synthetic rubber paint :**

This paint is prepared from resins :

**Properties :**

(*i*) Dries quickly.

(*ii*) Easy to apply on the surface.

(*iii*) Offers good resistance to water.

(*iv*) Little affected by sunlight and weather.

(*v*) Moderate in cost.

(*vi*) Covers a sizeable area.

(*vii*) Possesses excellent chemical resisting property.

(*viii*) Can be applied on the surface which may not be completly dry (*e.g.*, fresh concerete).

### 10.2.7 The Painter's Kit

The painter's kit comprises the tools with which the job of painting is done and includes the following:

**Brushes :**

- The common sizes of the brushes available in the market are 150 mm, 100 mm, 75 mm, 50 mm, 25 mm and 15 mm.
- Cutting-in-tool is a special brush, useful for the difficult job of painting window frames and straight lines.
- In addition there are other brushes like the *crevice brush, dusting brush* etc.

**Roller and tray :** These perform the job quickly and in a superior way than brushes, but they tend to use more paint.

**Scrapers knives and containers :**

- *Stripping knife* and *filling knife* (or putty knife) are employed for stripping off the old paint and putty filling to prepare a smooth surface.
- Shane hooks, of triangular and combination types, are used for moulding and difficult corners.

- The *paint kettle* is required for convenience in pouring out the desired quantity of paint for big jobs.

### 10.2.8 Painting of New Woodwork

'*Painting* is the *process of applying suitable paint with the help of painter's flat brush or by spraying.*

'Painting' involves the following *steps:*

**1. Preparation of surface :**

- Prior to the application of paint to any woodwork it should be seasoned thoroughly.
- The surface to be painted should be *carefully cleaned and rendered smooth.*

**2. Knotting :**

It consist of *covering up all the knots with substance through which the resin cannot come out.* The common methods of knotting are :

(*i*) *Ordinary or size knotting* : It comprises of two coats. The first coat is a hot coat of red lead ground with a strong glue size in water. After this coat is dried a coat of red lead ground in boiled linseed oil and thinned with turpentine is applied.

(*ii*) *Lime knotting* : In this method the knots are *covered with hot lime* for 24 hours and then ordinary size knotting is applied.

(*iii*) *Patent knotting* : In this case, two coats of a varnish made by dissoving shellac in methylated spirit are applied.

**3. Priming coat :**

- It is also called primary or first coat of painting and is applied after the knotting is finished.
- It entails the following *advantages:*

(*i*) It provides protection to the surface from the corrosive action of the atmosphere.

(*ii*) Due to this coat some of the paint sticks to the surface over which the subsequent coats can be *smoothly* applied.

It forms a *hard and opaque covering, filling the pores of the wood.*

**4. Stopping :**

- It is carried out after the priming coat is applied.
- It consists in rubbing down the primed surface with either pumic stone or glass paper and filling any holes or cracks with glazier's putty.

**5. Application of succeeding coats of paints:**

- After the surface is prepared by knotting, priming and stopping the woodwork is then painted by applying at least two (preferably three) coats of paints.
- Prior to the application of any succeeding coat, the previous one is dried and is gently rubbed down with glass paper and rendered smooth.
- Each coat is applied in longitudinal direction and care is taken that brush marks are not left on the painted surface.
- The final coat is called the *finishing coat.*

### 10.2.9. Painting of Old Woodwork

Repainting of old woodwork consists in washing the surface throughly with soap and water, rubbing down smooth with pumic stone and stopping with putty all the holes and cracks. Two or three coats of oil paint are then applied carefully so that no brush mark is visible on the surface. No succeding coat is applied until the previous one gets dried.

### 10.2.10. Painting of Metallic Surfaces

**A. New ferrous and galvanized metal surfaces :**

- *Chip, scrape* and *wire brush* to free from rust and mill scale and dust off.

- *Degrease* with white spirit and after pretreatment with a specified solution, apply an appropriate metal primer.
- Rub down lightly and apply two coats of gloss paint.

**B. Old metallic (ferrous and galvanized) surfaces :**

- Where the original paint is firmly adherent and the surface is free from rust; dry brush thoroughly and scrape to remove dirt, dust etc., and degrease.
- Areas damaged and showing localized rusting should be cleaned back to bright metal, degreased and patch primed.
- Apply, then, two coats of gloss paint. Flat oil paint may be used for interiors.

Painting or repainting work on metallic surfaces should be carried out under *dry conditions* and the *metal should be free of condensed moisture.*

### 10.2.11. Painting of Plastered Surfaces

- Painting on a freshly plastered surface should be avoided for at least six months.
- The surface should be white or colour washed in the first instance otherwise the alkalies present in the plaster will destroy, bleach and discolour the paint and also prevent it from drying.
- On drying of the new plastered surface, it should be treated with dilute sulphuric or hydrochloric acid and then washed with water.
- The surface is then given a coat of sizing (glue mixed with water). It will fill up the cracks and reduce the suction of the paint.
- Before applying the paint, the surface may also be given a coat of linseed oil.
- After the absorption of the oil by the surface, filling of cracks if any, is done by preparing a paste of *chalk powder, plaster of paris* and the paint to be used. the entire surface is then smoothened by rubbing it with sand paper.
- The surface thus prepared is given following *four coats* of the paint.
  - — The *first two coats* of the paint consist of white lead and boiled linseed oil.
  - — The third coat may be of white lead tinted to approach the desired colour and mixed with raw or boiled linseed oil.
  - — The *last-coat* may consist of large proportion of turpentine oil with a little varnish.

**Note :** Ready made or nearly mixed paints are also available in the market which can be applied directly on the newly plastered surfaces.

### 10.2.12 General Precautions in the Process of Painting

While painting any surface, the following *precautions* should be observed :

1. The surface to be painted should be perfectly clean and dry.
2. Painting work should be carried out in dry, warm weather after a dry spell.
3. Each coat of paint should be of moderate thickness for quick drying.
4. The undercoats and first coats must invariably be allowed to dry properly before applying the final coat.
5. Always rub down lightly between coats.
6. For exposed works, *boiled linseed oil* should be used.
7. The work should be planned in such a way that the specific part (door, ceiling), is completed in the same day.

### 10.2.13 Failure of Paint

A paint may undergo the following types of failure:

| | |
|---|---|
| (*i*) Chalking | (*ii*) Erosion |
| (*iii*) Flaking | (*iv*) Checking. |

**Chalking** is the gradual powdering of paint film. It is caused by continuous destructive oxidation of oil after the original drying of paint.

**Erosion** is very quick chalking.

**Flaking** is caused due to poor adherence of paint film to surface because of presence of grease or dirt on the surface or entering of water behind the paint.

**Checking** is very fine type of surface cracking.

### 10.2.14. Defects in Painting

In the painting work, the following *defects* are usually found :

1. *Fading* : Fading is the gradual loss of colour, and is primiraly due to the sunlight effect on the pigment (of paint).
2. *Blistering* : It is caused due to trapping of water vapour behind the painted surface. It may occur due to imperfect seasoning of timber, use of excess oil in final coat etc.
3. *Sagging* : This defect occurs when a vertical or inclined surface is too thickly painted.
4. *Flaking* : Flaking means *loosing of small portion of the painted surface;* it is caused when adhesion is poor.
5. *Running* : This defect relates to the *paint running back and leaving small areas of surface uncovered.* It is caused when the surface being painted is *too smooth.*
6. *Flashing :* It means glossy patches on the painted surface. This defect is caused due to poor workmanship, weather actions or cheap paint.
7. *Bloom* : It relates to the formation of dull patches on the finished polished surface. This defect is caused due to bad ventilation or defective paint.
8. *Grinning :* This defect relates to the *clear visibility of the background* when the final coat of paint does not have sufficient opacity.
9. *Wrinkling :* This defect becomes prominent when a horizontal surface is too thickly painted.

## 10.3 VARNISHES

### 10.3.1 Definition of Varnish

- A **varnish** is a *colloidal dispersion or solution of synthetic or natural resin in oil or in thinner or in oil and thinner both.*
- A *transparent or semi-transparent solution of resinous substance either in linseed oil, turpentine or alcohol* is *called* **varnish.**

When varnish is applied on the surface, it gives a transparent film. The film dries up by evaporation, oxidation and polymerisation of portions of its constituents. If only thinner is used for dissolving the resin, then the film dries up by evaporation only. However, if oil is also present, the film dries up due to oxidation and polymerisation of drying oil and evaporation of thinner.

Varnishes are modified form of paints and differ from paints in the following *two* aspects:

(*i*) Varnishes have *no pigments.*

(*ii*) In varnishes *a part or whole of the oil is substituted by resin.*

### 10.3.2 Functions of Varnish

Following are the *functions* of varnish :

1. To protect the painted surface from the effects of weathering agencies.
2. To brighten the appearance of the grain in wood.
3. To provide brilliancy to the painted surface.

### 10.3.3. Constituents/Ingredients of a Varnish

Following are the *ingredients* of a varnish :

1. Resin or resinous substance.
2. Solvent

3. Drier.

**1. Resin or resinous substance :**

- A resin or resinous substance is a natural or synthetic organic substance which is soluble in some organic solvent. It acts as *a base and provides body to the varnish.* The quality of varnish greatly depends up upon the quality of its resin.
- The commonly used resins are :

(*i*) *Copal* : A hard substance and is available in variety of forms.

(*ii*) *Lac or shellac* : It is obtained by exudation of some types of insects in India.

(*iii*) *Rosin* : It is obtained from pine trees.

Other resins are mastic, gum, amber etc.

**2. Solvent :**

- It acts as vehicle of the varnish.
- It helps in spreading the resin over the surface to be varnished.
- It must be suitable to dissolve the resin used in varnish.

The solvents for different resins are as follows :

| Resins | Solvents |
|---|---|
| Copal, amber | Boiled linseed oil |
| Lac or shellac | Methylated spirit |
| Mastic, Rosin, Gum dammer | Turpentine |
| Other cheap varieties of resins | Wood naptha. |

**3. Drier :**

- It accelearates the process of drying.
- It should be added in small proportions.
- The commonly used driers are : *Litharge, white copper* and *lead accetate.*

### 10.3.4 Characteristics of a Good Varnish

A good varnish should have the following characteristics :

1. It should not show cracks on drying.
2. It should appear uniform and pleasing.
3. Its colour should not fade.
4. It should dry rapidly and give permanent glossy film.
5. It should be tough, hard and durable.
6. It should provide a brilliant and decorative covering on wooden surfaces.

### 10.3.5. Types of Varnishes

Depending upon the solvents used, the varnishes are *classified* as follows :

1. Oil varnishes
2. Spirit varnishes
3. Turpentine varnishes
4. Water varnishes

**1. Oil varnishes :**

- In an oil varnish the *linseed oil is used as solvent.*
- These varnishes consist of resins like *copal* and *amber,* dissolved in boiled *linseed* oil.
- Oil varnishe dry slowly, but they form hard and durable surface. These are the hardest and most suitable varnishes.

**Uses :** They are specially used for the *exposed works* which require frequent cleaning. Used on coaches and fittings in houses.

**2. Spirit varnishes:**

- They dry quickly.
- They are not durable and are easily affected by weathering actions.
- The French polish is a variety of this class of varnish and the desired colouring tinge can be obtained by addition of suitable colouring pigment.

Uses: These varnishes are mostly employed for furniture and on unpainted wood surface.

**3. Turpentine varnishes:**

- These varnishes consists of soft resins like *common resin, mastic, gum gammer* etc. dissolved in the best turpentine oil.
- They are of light colour.
- They dry quickly.
- They are cheaper and more flexible than oil varnishes.
- They are not so durable and tough as oil varnishes.

**Uses:** These varnishes are used on painted surface.

**4. Water varnishes:**

These consist of lac dissolved in hot water mixed with so much ammonia, borax, potash or soda as will just dissolve the lac in hot water.

**Uses:** These varnishes are used for *pictures, varnishing maps* etc. Also used for *delicate internal work and as a covering for wall paper.*

**10.3.6. Process of Varnishing**

The varnish is applied on the woodwork in the following way:

1. **Preparation of the surface:** The surface of the woodwork is thoroughly rubbed down by means of sand paper or pumice stone and made smooth and clean.
2. **Knotting:** The process of knotting is then carried out as in case of painting.
3. **Stopping :** The surface, after knotting is stopped by applying a coat of hot weak glue size or two coats of boiled linseed oil. After the surface becomes dry, it should be rubbed down with sand paper. Stopping is done to fill up the pores existing in the surface.
4. **Application of coats of varnish:** The varnish is then applied on the surface in thin coats. The subsequent coat of varnish is applied only when the previous coat has been thoroughly dried. The varnishing should be done with fine haired varnishing brushes (not ordinary paint brushes).

**10.3.7. Furniture Polish**

*Following is the composition of an excellent furniture polish:* Linseed oil = 16 litres; Methylated spirit = 2 litres; Vinegar = 1 litre; Turpentine = 1 litre; Copal varnish = 1 litre; Hydrochloric acid = 0.75 litre.

The furniture polish is prepared by heating linseed oil first, and then mixing up all the ingredients. It is commonly used for varnishing the furniture.

**Preparation of wax polish:**

- Wax polish consists of *bee wax, linseed oil, turpentine and varnish.* Varnish, turpentine, linseed oil and bee wax are respectively taken in the ratio of 1 : 2 : 3 : 4 by weight.

First of all bee wax and linseed oil are mixed together and heated on mild fire. After some time, bee wax gets dissolved in the oil and mixture is allowed to cool down slightly. Now turpentine and varnish are also added to the mixture of bee-wax and stirred well. *Wax polish is ready for use.*

- The polish is applied in the similar manner as French polish. Wax polish applied on the prepared cleaned surface with the help of soft pad of cotton cloth, and kept continuously rubbed for about 30 minutes. Second coat is applied after the *first coat has dried.* Similarly several coats of wax polish are applied till desired surface finish is obtained.

**Oiling of timber-work**

In order to make timber more durable and good looking, they are, sometimes, *oiled.*

*The oiling mixture is prepared as follows:*

Three parts of double-boiled linseed oil are mixed with one part of bee wax and the mixture is heated on a slow fire till wax is melted. The mixture is then cooled and turpentine oil is further added, and stirred well. The oil mixture is ready for application. The mixture is freely applied on the wooden surface with the help of brush *till surface becomes saturated.*

Sometimes, a mixture of country sweet oil, vinegar, and turpentine in equal proportion can be used for oiling the timber-work.

**10.3.8. Lacquer and Lacquering**

- *Lacquers are coatings consisting of simple solution of preformed binder (polymer) in volatile solvents.* The term is generally restricted to cellulose and acrylic lacquers.
- Lacquering is done on many polished and unpolished articles.
- The *objects of lacquering* are:

*(i)* To prevent the polished surface from tarnishing;

(ii) To give an unpolished surface a protective coat and this prevents oxidation;

(iii) To improve the general appearance and make the article more pleasing to the eye and hence more saleable.

**Constituents of lacquer:**

Following are the *constituents* of lacquer:

1. **Film forming basic materials:**

***Examples***: Nitrocellulose, Cellulose acetate, phenol formaldehyde etc.

- These materials impart *water proofness, hardness and durability and gloss.*

2. **Pigments:**

***Examples***: As in paints.

- They impart resistance and pleasing colour.

3. **Solvents:**

***Examples***: Esters, Ketones, Ethers, Alcohols.

- They dissolve the film forming substances and suspend the pigments.

4. **Diluents:**

***Examples***: Solvent naptha, Petroleum naptha.

- They *reduce viscosity and cost.*

5. **Plasticizers:**

***Example***: Caster oil

- They reduce film brittleness.

**Type of lacquers:** The lacquers may be divided into the following *four* groups:

1. Cold-gum lacquers
2. Hot-gum lacquers
3. Cellulose lacquers
4. Synthetic lacquers.

**1. Cold-gum lacquers:**

- The lacquers can be directly applied and can be easily worked with a brush without heating.
- They provide a glossy finish on painted or unpainted surface.
- Some of the gum lacquers are air-drying, others are of the stoving type.

**2. Hot-gum lacquers:**

- They require the work to be heated first.
- As their application needs great skill (obtained by long training) they are little used today.
- They give a very hard, glossy surface.
- The stoving temperature is 95°C.

**3. Cellulose lacquers:**

- They are air-drying, and give a hard film which does not soften when exposed to tropical heat.

- Some cellulose lacquers give a colourless protection without altering the finished appearance of the component. Others are built to give various degree of gloss, while pigments are added to some to give an opaque, coloured film.

**4. Synthetic Lacquers:**

- They require a stoving after application; during stoving the nature of film changes and becomes extremely hard, durable and resistant.
- Where an article is to withstand exceptional conditions, such as exposure to chemical fumes, corrosive atmosphere, or fluctuating temperatures, a synthetic lacquer is required.

**Methods of applying lacquer:**

Following are the *methods of applying lacquer:*

1. **Brush lacquering:** Here the lacquer to an article is applied by brush.

2. **Dip Lacquering:** It consists of dipping the articles into a bowl containing the lacquer and then withdrawing them. This may be done by holding the work in hand, wiring the part together, or placing them in an aluminium basket before immersion.

3. **Spraying:** It consist in carrying the lacquer on to the article with the help of compressed air. It entails the following *advantages:*

(*i*) It is about three times as quick as brush lacquering.
(*ii*) It requires less training and skill.
(*iii*) An even coat is easily obtained.

**Uses of lacquer:**

Following are the *uses of lacquer;*

1. For preparing artificial leather, lacquer is coated over the cotton fabric and then it is embossed to stimulate the grain of the leather.

2. For giving a finish to automobile bodies, a coating of lacquer is applied as it can be applied easily and rapidly. It resists abrasion, cracking and chalking.

3. It is used as an interior decoration such as painting of woodwork and furniture etc. It has the advantage of getting dried up very quickly.

### 10.3.9. Enamels

When a varnish finish is to be exposed to sunlight it is coloured by pigments and is called a "brushing enamel". If this coating is on metal and is properly formulated, it may be subjected to higher temperatures and is then termed "*baking enamel*".

*Lacquer enamels:* Like varnishes and enamels lacquer enamels are merely clear lacquers plus pigment, the object being to cover as much area as possible with a firm, coloured, permanent film at the lowest cost. The cheaper varieties used on logs and other articles of short life are made using less cellulose ester of high viscosity, cheaper solvents, coarser pigments and sometimes scrap film. The pigment is incorporated by grinding it in some oil, usually caster oil, then grinding it with the lacquer.

## 10.4. DISTEMPERS

- **Distempers** are *comparatively cheap decorative paints for walls and ceilings.*
- **Distempering** is *a process of applying wash like whitewash or colour wash on the surface.*
- Finished surfaces obtained by distempering are far superior than those obtained by white or colourwash.
- Distempering the surfaces is done by *distempers.*

Distempers are available in the market under different trade names.

**Composition of distempers:**

Any distemper consists of the following:

(*i*) Base
(*ii*) Carrier
(*iii*) Colour agent
(*iv*) Size.

**Whiting** (powdered chalk) is used as *'base'* and **water** as *'carrier'*. **Colouring agents** or pigments are *added only if specific shade is to be obtained.* **Glue** is mostly used as *size.*

– Distemper may also be termed as a *water paint* having *whiting as base and water as the carrier.*

Distempers are available in powder form or plate form. They are to be *mixed with hot water before use.* There are a variety of *oil bound distemper* also in which the drying oil is so treated that is mixes with water readily. Oil bound distempers are available in paste form in different shades in sealed tins. *Before use, water is mixed with the distemper to thin it to the required consistency.* After *application on the surface, the oil content of the distemper hardens and forms a durable coating.* Distempers are applied on the surface with the help of distemper brushes which are about 10 cm wide.

• **Distempers** are available in *powder form and also in paste form. Powdered distempers are known as **'dry distempers'*** whereas paste form distempers are known as ***"oil bound distempers"***. Oil bound distempers are *superior to dry distempers.* Distempered surfaces with oil-bound distemper, are *washable.*

### 10.4.1. Properties of Distempers

The *properties of distempers* are listed below:

1. They are generally light in colour and provide a good reflective coating.
2. They are classed as *water paints* and are *easy to apply.*
3. They are less durable than oil or emulsion paints.
4. There exhibit poor workability.
5. The coatings are usually thick and are brittle than other types of water paints.
6. They prove to be unsatisfactory in damp location (*e.g.* bathroon, kitchen etc.)
7. Distempers generally shrink on drying. Hence if surface to receive distemper is weak it may lead to cracking and flaking of the distempers.
8. Distemper film is generally porous in nature and it allows water vapour to pass through it. Hence, it allows new walls to dry out without damaging the distemper film.
9. They can be applied on cement plastered surface, lime plastered surface, brickwork, insulating boards etc.

### 10.4.2. Type of Distempers

Broadly there are *two* types of distempers:

1. Non-washable (also often called *soft* or *dry*) distemper.
2. Washable distemper.

**1. Non-washable distemper:**

• It consists of *whiting with glue and pigment in suitable proportions.*

• It is marketed in powder form but is sometimes made at site by mixing other constituents into a mixture of whiting and pigment.

• Glue is water soluble hence this types of distemper has no *water resistance.* If glue is replaced by cesein and borax, a less water sensitive coating is obtained which develops sufficient water resistance in due course of time to become washable and recoatable.

• It gives off a bad odour when the glue used is of poor quality. The odour may persist for sometimes or reappear in humid surroundings.

**2. Washable distemper:**

It contains a *water resistant binder, usually an emulsion of a drying oil or varnish.* Such distempers are also called *oil bound distempers.*

• They are marketed in paste form to be thinned with water just prior to the application. A few days after application, they acquire a moderate degree of washability and can be cleaned by sponging. By careful adjustment of the oil varnish emulsion in the mixture, the durability of oil bound distemper can be enhanced. *The distemper dries and hardness first by the evaporation of water and then by oxidation of the oil in the emulsion.*

- Emulsions of high polymeric materials such as polyvinyl acetate (PVA) and certain grades bitumen have been used in making good washable distempers. They are *quick drying, easy to apply and odourless.*
- A washable distemper has a characteristic but less offensive odour at the time of the application (while PVA and bitumen bound distempers are practically odourless).

### 10.4.3. Distempering Process

The distempering process involves the followings *three stages*:

1. Preparation of surface.
2. Application of prime coat on the surface to be distempered.
3. Application of distemper.

**1. Preparation of surface:**

The surface for distempering should be prepared as follows:

- Surface to be distempered should be *absolutely dry.* Dampness in the surface spoils the distemper coat. New plastered surface should be left for at least 2-3 months so that it may dry out completely.
- If the surface to be distempered is having *efflorescence* at some places, the spots *should be wiped out* with the help of dry cloth before distemper coating is applied over it.
- Holes, patches, cracks etc. on the surface should be *filled with lime putty or gypsum* and allowed to get hard before distemper coat is applied.
- If the surface already having distemper is to be redistempered, the old *distemper coating should be removed by washing the surface with water.* The surface should be allowed to *dry* fully before new distemper is applied over it.
- The surface to be distempered should be *thoroughly rubbed with sand paper and cleaned.*

**2. Application of prime coat:**

After preparation of the surface for distempering, a priming coat is applied to ensure good bond between distemper coat and the surface.

- The priming coat may be composed of *whiting in water, or in milk.* But usually the manufacturers recommend the use of material for priming coat for their product.
- The priming coat should be allowed to dry completely before distemper coat is applied over it.
- Sometimes, cement wash is also given on the surface to act as prime coat.

**3. Application of distemper:**

- After applying the prime coat, *'first coat'* of distemper is applied on the surface. The first coat should be of a *light tint and applied with great care.*
- *'Second coat'* is applied after the *first coat has dried and become hard.*

Distempering coats are applied with the help of *broad shift distemper brushes.* Nowadays distempers can be applied with help of *spray pistols.*

– Sometimes if application of prime coat is not felt necessary, a coating of warm glue should be applied prior to the application of distemper.

**Distempering the exterior surfaces:**

Distempers to be used on cement concrete, stucco, and brick surfaces are manufactured with *weather resistance ingredients* during the process of manufacturing of the distempers. *Snocem, silvacrete, colourcrete* are nothing but trade names of different types of distempers. They are very *good for external exposed situations.*

The following points about '*distempering*' *are worth noting*:

- In case of *new construction,* generally little preparation of the surface is needed. All loose materials and efflorescence should be removed. For a soft distemper, a thin coat of lime wash is sometimes applied first over new walls, followed a few days later by a coat of the distemper. A second coat may also be applied if necessary. It is important to use distemper containing the same proportion of glue for both coats. A better method of decoration is to give a first coat of soft distemper followed by a second coat thinned with water to brushable consistency.

- On *new surfaces,* two coats of a washable distember are recommended. The first is well thinned with water to obtain maximum penetration and second coat is applied as a thicker film. For redecoration only one coat of distemper is recommended.
- *Spray painting of distemper is always superior to brushing.* Spraying provides a uniformly smoother, thicker and more durable film.
- If water gains entry into the structure, the appearance of distempered walls and ceiling may be *seriously marred by water stains, segregation of pigment and migration of salt.* These stains are very obstinate and reappear even after a fresh coat of distemper is applied. A coat of thinned PVA Emulsion pigment with titanium dioxide equal in weight to PVA, prior to the application of distemper helps to seal the stains.

## 10.5. WALL PAPER, WHITEWASHING AND COLOURWASHING

### 10.5.1. Wall Paper

- The wall papers are made exclusively from paper or combined with other materials. They may be unprimed, primed, metal coated, embossed etc.
- They are available in one colour or multi-colour patterns (printed on the face surface).
- Some types of wall papers are washable and may be painted with oil paints. The metal-coated wall papers and the wall papers with sound absorbing properties are also available.
- The wall papers are *widely employed for developing decorative finishing of interior walls.*

The following ***procedure*** is *adopted to fix up the wall papers*:

- The surface on which paper is to be fixed is dried, levelled and made free from chalk or lime solution.
- Using lime-gypsum solution, all cracks and slits are then filled up.
- By applying suitable paste the surface is pasted with wapping paper or newspapers.
- The wall paper (which is available in rolls) is then hung very carefully on the surface and the continuity of the pattern or design of wall paper is maintained.

### 10.5.2. Whitewashing

The procedure of white-washing the surface includes the following *steps*:

**1. Preparation of white-wash:**

- White-wash is prepared from fat lime (white stone lime) or shell lime. The unslaked lime is dissolved in a tub with adequate quantity of water, and it is *thoroughly stirred* with a wooden pole, *until it attains the consistency of thin cream.* Normally about 5 litres of water should be added to each kg of lime for making the cream. This mixture is allowed to stand in the tub for a period of 24 hours and then *strained through a clean coarse cloth. Clean gum dissolved in hot-water is then added in the preparation of 4 kg of gum per $m^3$ of thin cream.*
- In the thin cream obtained above, is then added Indigo (Neel) in the proportion of 3 kg of Indigo per $m^3$ of thin cream to *prevent glare and to obtain pleasing effect.*
- So that the coat of white-wash may stick well to the surface, alum or common salt taken in the same preparation as gum, is also sometimes added.

**2. Preparation of surface:**

- Prior to the application of white-wash on **new walls,** the *wall surface is thoroughly cleaned, brushed and rendered free from mortar droppings and other matter.*
- When the surface to be coated is *sufficiently smooth,* it should be rubbed with sand paper to

ensure proper adhesion of white-wash.

- When an **old surface** is to be re-whitewashed, all loose pieces and scales should be scrapped off. In case of *heavy scaling, the entire surface should be scrapped by means of steel wire brushes or steel scrapers.* The old loose white-wash is removed by rubbing with sandpaper and *holes in the wall are filled with lime putty.* The plaster repairs to the wall should be done well in advance so that they are dry before whitewashing. The *greasy spots should be given a coat of a mixture of rice water and sand so that the wash may stick to the surface.*
  — In kitchens and other places where the walls have been *discoloured by smoke or due to some other reasons,* the surface should be given a *coat of a mixture of wood ash in water before applying white-wash.*

**3. Application of white-wash:**

For getting a smooth and uniform finish *three coats* are considered necessary for *new work* and *two coats* for *old work*. Moonj brushes are usually used for white-washing.

- Each coat consist of a stroke of the brush applied from the *top downwards,* another stroke is applied vertically from the *bottom upwards* over the first stroke, and similarly one stroke is applied *horizontally from the right and another from the left.*
- Each coat should be allowed to dry before next one is applied.
- Before the start of work, as a measure of protection, all doors, windows, floors, articles, furniture etc. should be protected from being splashed upon by white-wash.
- A satisfactory work provides an *opaque smooth surface with uniform white colour* and it does not readily come off on the hand when rubbed.

**Uses of whitewash :**

As the lime is toxic for germs and reflects light (and thus it *increases the brightness of the surface*), the whitewashing is employed for the following :

— Interior wall surfaces :
— Ceilings of the houses;
— Exterior walls surfaces also.

**Points to be kept in view whole 'Whitewashing':**

Whitewashing, the *following point should be kept in view.*

1. In order to ensure uniformity in glare and appearance of the finished surface whitewash for full day's work should be *prepared at a stretch.*
2. While whitewashing, solution should be continuously stirred.
3. Whitewash solution should be strained or screened through coarse cloth so as to remove any undissolved particles from the solution.
4. For ensuring uniform distribution of the whitewash moonj brush (to be used for whitewashing) should be cut properly from time to time.
5. Dip the brush in the whitewash solution by small length, whole of the broomed portion should not be dipped.
6. In order to prevent splashing of whitewash from the dipped brush, it (brush) should not be directly carried for whitewashing but should be slightly swept against the bucket containing whitewash solution.
7. Successive coats of whitewash should be applied only after the coat already applied has fully dried.
8. Only experienced person should be engaged for the work of whitewashing.
9. Before the application of whitewash over greasy spots, loose spots, hole spots, etc, should be properly treated.

### 10.5.3. Colourwashing

**Preparation.** Colourwash is prepared by adding the colouring pigments to the screened whitewash. It may be ensured that the presence of lime does not affect the colouring pigments. Usually yellow earth is employed for colourwashing.

**Application of colour-wash:**

- The *colour-wash is applied in the same manner as white wash.* The mixture is kept constantly stirred with a bamboo stick while it is being used. Colour-wash, sufficient for the day's work only, should be prepared at a time.
- Prior to the application of colour-wash on *new surfaces,* a coat of white-wash is applied which acts as a *priming coat.* In case of *old prepared surface,* a coat of colour-wash is applied over the patched and repair work before start of colour-washing. Two or more coats of colour-wash are then applied on the entire surface till it presents a uniform tint of shade.
- A satisfactory work does not give out powder when the finished surface is rubbed with fingers.
- The colourwashing process provides *cleanliness* and *pleasant appearance* to the treated surfaces.

**Uses of colourwash.** Generally, the walls are colourwashed and ceilings are whitewashed.

- In regard to *colourwash the following facts should be remembered:*

1. Colourwash should be prepared at a stretch which would be sufficient for full day's work.
2. Before the application of the colour wash on new surfaces, a coat of whitewash should be applied. This whitewash coat acts as a primary coat.
3. In case of old surface, a coat of colourwash should be applied over the patches of repair works before two or more coats of colourwash are applied on the entire surface.
4. Colourwash should be kept continuously stirred during use.
5. Finishing surface should exhibit uniform colour and lustre.
6. A good colourwash does not give out powder when rubber with the fingers.

- **Colouring pigments for plaster and colourwash:**

Following pigments are used for the colourwash to obtain the desired shade:

**Colour pigments for plaster and colourwash**

| *Name of colour* | *Pigment* |
|---|---|
| Rose | Red oxide of iron |
| Green | Chromium oxide |
| Pink | Red oxide of iron in small quantity |
| Cream | Yellow oxide of iron in small quality |
| Brown | Brown oxide of iron |
| Yellow | Yellow earth |

## 10.6. ANTI-TERMITE TREATMENT

### 10.6.1. Introduction:

The buildings get damaged due to the following types of *termites:*

**1. Dry wood termites:**

- These termites reside in wood and prove nuisance in coastal areas.
- They can *survive without soil connection.*
- They are, however, not as common as subterranean termites and they cause less damage to the buildings.

**2. Subterranean termites:**

- These *cannot survive without soil connection* since necessary amount of moisture required is supplied by the soil on which they survive.

- These termites are mainly responsible for causing damage to the buildings and its contents. They enter the building through foundations or from ground adjacent to buildings and advance upward through/floors destroying everything that comes within their reach.

### 10.6.2. Types of Anti-termite Treatment

The termites breed and spread so fast that they can damage a building in 4 to 5 years. Keeping in view the huge damage caused by the termites it is desirable to take suitable measures to prevent the ingress of termites into the buildings.

**Measures / Precautions for "Termite Control" (Pre-construction treatment):**

The following *measures/precautions should be taken to check the detrimental effects of termites*:

- In areas having subterranean termites, site should be cleared by removing dead wood, old tree stumps, roots, etc. within seven metres of the foundation of the buildings.
- Any termite mound, within a radius of 50 m or so should be opened up and nest (from which the attack of termites originate) be destroyed by using the chemical solutions (quantity of insecticides to be used depends on the mound size). For a mound volume of about 0.25 cum, 1 litre of one of the following chemical solutions should be used :
  (*i*) Chlordane ... 1.00 percent
  (*ii*) Heptachlor ... 0.50 percent
  (*iii*) Aldrin ... 0.25 percent.
- A good protection against white ants is provided by *damp proof courses of rich and dense concrete* (1:2:4 or richer) and cement concrete floors. The following methods also offer protection :
  (*i*) *0.25 kg of copper sulphate in 10 litres of water in the mortar* (the solution also protects wood, takes two days of steeping per 25 mm of wood thickness).
  (*ii*) *Yellow arsenic in mortars* in the following proportions : Concrete = 6.5 kg, masonry and plaster = 0.8 kg, per 10 cum.
  (*iii*) A sodium arsenate solution spread over a layer of about 75 mm graded washed sand or sand and cinders.

(**Note :** *Arsenic being poison should not be used on surface works*).

- Generally, termites do not penetrate masonry or concrete of first class construction in which there are no voids. Masonry with lime mortar of mix leaner than 1:3 should not be in contact with soils where a concrete floor has vertical joints between the floor and plinth masonry, these joints should be filled with heavy grade coal tar pitch to minimize the infiltrate of termites through joints.
- All materials such as wood, boards etc. which are liable to be attacked by termites should be treated with preservatives as per IS:401. Care should be taken that untreated wood or wood not naturally resistant to termites is not placed in contact with or close to the soil.
- Proper drainage/apron should be provided to check the entry of moisture to the foundation.
- All sub-floors should be of dense cement concrete (1:3:6) and should be continuous through the walls.
- Lime or cement mortar should be used for all masonry work in the sub-structure; there should be no cracks and dry joints. Cracks/gaps, if any, should be filled with tar or cement mortar.
- The chemicals being toxic should not be allowed to contaminate with food and drinking water. During their use, hand gloves, gum boots and respirators should be used.

**Post-construction treatment:**

The post-construction treatment is applied to the existing building which have already been attacked by the termites. The type of treatment to be given to eradicate termites from the existing building largely *depends upon the extent of attack and the magnitude of cellulosic and other materials available in the building.*

The following *operations* are involved in *eradicating the termites from an existing building* :

(*i*) Inspection;
(*ii*) Soil treatment for foundations;
(*iii*) Soil treatment under floor;
(*iv*) Treatment of voids in masonry;
(*v*) Treatment of wood work.

**(*i*) Inspection:**

- Firstly the following portion of the building in contacts with or adjacent to the earth should be *inspected:*
  - — Basements;
  - — Ground floor;
  - — Steps leading from the ground;
  - — Walls and columns;
  - — Areas having damp or humid conditions (*e.g.*, lavatories, bathrooms, leaking pipes drains etc.);
  - — The places where wood work is embedded in the floor or wall;
  - — The ceilings;
  - — Wood panelling; Battens for wiring conduits and switch boards etc.
- The following should be inspected carefully in case of *multi-storeyed buildings:*
  - — Lift wells;
  - — Casings covering electrical wiring, telephone cable, water supply and soil pipes etc.
- The following actions need to be taken:
  - — Remove the wind walled shelter tubes or the termite runways wherever detected.
  - — Inject the oil or kerosene base chemical emulsion over the attached areas of woodwork and masonry, wherever possible.
  - — In order to ensure elimination of all direct contacts between soil and the effected portion of the structures, structural additions may be made, wherever necessary.

**(*ii*) Soil treatment for foundations:**

The *soil treatment for foundations involves the following procedure* :

- Preparation of 500 mm deep trenches along the external peripheral wall of the buildings with the help of a showel.
- Making of 12 mm to 18 mm dia. *holes* at 150 mm centres in the trenches, close to the wall face, with the help of *iron rod.* The holes should preferably extend upto the top of footing of foundations to a depth of at least 500 mm whichever is lesser.
- Filling of holes with chemical emulsion in water.
  - — The total quantity of chemical to be used in this treatment should be @ 7.5 litres per sq.m. of the vertical surface of the masonry in foundation.
  - — The chemical treatment, in case of *R.C.C. frame structure,* shall be applied to the soil in contact with column sides and flinth beams along external periphery of the building for a depth of 500 mm below the level of ground.

**(*iii*) Soil treatment under floor:**

The termites from soil enter the floors through the cracks. These cracks usually occur at the junction of the floor and walls, expansion joints in floor and at construction joints in a concrete floor. The defective workmanship and the use of unsound material may also cause cracks in the floors.

For the eradication of the termites underneath the floors, the following *procedure* is adopted:

- Drilling of 12 mm dia. holes at 300 mm centre to centre all along the cracks in the floors in different areas.
- Injecting chemical emulsion into the holes till the soil below gets fully saturated. The maximum

quantity of chemical emulsions, usually, not to exceed one litre per hole.

- Sealing of the holes after treatment.

**(*iv*) Treatment of voids in masonry:**

In order to check/prevent the entry of termites through the voids the following *procedure* is adopted:

- Drilling of 12 mm dia. holes at 300 mm centre to centre at downward angle of about 45° from both sides of walls at plenth level.
- Pumping of chemical emulsion into the holes until masonry gets fully saturated with the emulsion.
- Sealing of the holes after treatment.

**(*v*) Treatment of wood work:**

The following procedure is followed for treatment of woodwork which has been badly damaged by the termites:

- Replacement of woodwork badly damaged by termites by new timber which is *adequately brushed or dipped in oil* or *kerosene based chemical emulsion.*
- The *infested wood work* for door and window frames etc. should be given protective treatment by *drilling 6 mm dia. holes at 150 mm centre to centre at a downward angle of 45° to cover the entire framework and thereafter pumping oil based chemical emulsion into the holes.*
- In order to prevent possible attack of termites on the unaffected wood work, it would be sprayed over with chemical emulsion.

## HIGHLIGHTS

1. **Paint** is the fluid paste prepared by dissolving a base into a vehicle (carrier) alongwith a colouring pigment.

   The constituents/ingredients of a paint are : *Base, Vehicle, pigment, thinner, drier, inert filler, antiskinning agents and plasticizers.*

   Painting of new woodwork involves the following steps: (*i*) Preparation of the surface; (*ii*) Knotting; (*iii*) Printing coat; (*iv*) Stopping; (*v*) Application of succeeding coats of paints.

   A paint may undergo the following types of failure : (*i*) Chalking, (*ii*) Erosion, (*iii*) Flaking, (*iv*) Checking.

2. **Varnish** is a colloidal dispersion or solution of synthetic or natural resin in oil or in thinner or in oil and thinner both.

   The constituents / ingredients of varnish are : *Resin or resinous substance, solvent and drier.*

   Various types of varnishes are : (*i*) Oil varnishes (*ii*) Spirit varnishes, (*iii*) Turpentine varnishes, (*iv*) Water varnishes.

   The process of varnishing a woodwork surface involves the following steps: (*i*) Preparation of the surface; (*ii*) Knotting; (*iii*) Stopping; (*iv*) Application of coats of varnish.

   Lacquers are coatings consisting of simple solution of preformed binder (polymer) in volatile solvents.

   The constituents of lacquers are : (*i*) Film forming basic materials, (*ii*) Pigments, (*iii*) Solvents, (*iv*) Diluents, (*v*) Plasticizers.

   Types of lacquers are : (*i*) Cold-gum lacquers, (*ii*) Hot-gum lacquers, (*iii*) Cellulose lacquers, (*iv*) Synthetic lacquers.

3. **Distempers** are comparatively cheap decorative paints for walls and ceilings. A distemper is composed of base, cement, colouring pigments and size. Distempers may be of washable and non-washable types.

4. Wall papers are widely employed for developing decorative finishing of interior walls.

   A satisfactory work of whitewashing provides an opaque smooth surface with uniform white colour and it does not come off on the hand when rubbed.

   *Colourwashing* process provides cleanliness and pleasant appearance to the treated surface.

## OBJECTIVE TYPE QUESTIONS

**Choose the Correct Answer :**

1. Which of the following paints is highly resistant to fire ?
   (*a*) Aluminium paint (*b*) Asbestos paint (*c*) Cement paint (*d*) Enamel paint
   (*e*) None of the above.
2. ................ has high reflectivity.
   (*a*) Enamel paint (*b*) Cellulose paint (*c*) Bronze paint (*d*) None of the above.
3. French polish is a kind of .................. varnish.
   (*a*) oil (*b*) spirit (*c*) water (*d*) none of the above.
4. ............. is the defect in painting due to sliding of one layer of paint over another.
   (*a*) Alligatoring (*b*) Pecling (*c*) Blistering (*d*) None of the above.
5. Snocem is ............ paint
   (*a*) a cement (*b*) an oil (*c*) a cellulose (*d*) a synthetic enamel.
6. The volatile diluent added to a paint is known as ..............
   (*a*) pigment (*b*) thinner (*c*) vehicle (*d*) drier.
7. The paints used for automobiles are
   (*a*) emulsion paints (*b*) oil paints (*c*) water paints (*d*) distempers
   (*e*) none of the above.
8. Duco paint is ............... paint.
   (*a*) oil (*b*) water (*c*) cellulase (*d*) bituminous.
9. Which of the following is a function of a paint ?
   (*a*) To prevent corrosion of the metal (*b*) To give a clean, colourful and pleasing surface
   (*c*) To improve the life of the painted surface (*d*) To protect surface from weathering effect
   (*e*) All of the above.
10. A varnish essentially contains which of the following ?
    (*a*) Drier (*b*) Resin (*c*) Solvent (*d*) All of the above
    (*e*) None of the above.
11. Bitumen paints are used for providing .................. surface.
    (*a*) rough (*b*) smooth (*c*) protective (*d*) shining.
12. ............ varnish is most durable.
    (*a*) Water (*b*) Oil (*c*) Turpentine (*d*) Spirit.
13. Painting work is usually specified by
    (*a*) area of the painted surface (*b*) weight of the paint applied
    (*c*) volume of the painted surface (*d*) labour used in painting
    (*e*) thickness of the layer of paint applied.
14. Snocem is ................
    (*a*) coloured cement (*b*) powdered lime
    (*c*) mixture of chalk powder and lime (*d*) none of the above.
15. Thinner used for oil paints is
    (*a*) turpentine (*b*) water (*c*) carbon tetrachloride (*d*) any of the above.
16. Linseed oil is rapidly soluble in
    (*a*) naptha (*b*) alcohol (*c*) turpentine (*d*) all of the above.
17. Which of following oils can be used as a vehicle for paints?
    (*a*) Poppy oil (*b*) Nut oil (*c*) Tung oil (*d*) Any of the above.
18. Linseed oil is subject to adulteration by the addition of
    (*a*) mineral oil (*b*) resin (*c*) cotton seed (*d*) all of the above.
19. The small areas on painted surface enclosed by hairline cracks are known as
    (*a*) crazing (*b*) blistering (*c*) chalking (*d*) crocodiling.

20. A laquer is ...............
    (*a*) oil paint (*b*) dry paint (*c*) spirit varnish (*d*) distemper
    (*e*) none of the above.
21. Linseed oil in paint is used as
    (*a*) base (*b*) carrier (*c*) thinner (*d*) pigment
    (*e*) any of the above.
22. The base material for distemper is
    (*a*) chalk (*b*) lime (*c*) clay (*d*) lime putty.

## ANSWERS

**1.** (*b*) **2.** (*c*) **3.** (*b*) **4.** (*a*) **5.** (*a*) **6.** (*b*) **7.** (*b*) **8.** (*c*) **9.** (*e*) **10.** (*d*) **11.** (*c*) **12.** (*b*) **13.** (*a*) **14.** (*a*) **15.** (*a*) **16.** (*d*) **17.** (*d*) **18.** (*d*) **19.** (*a*) **20.** (*a*) **21.** (*b*) **22.** (*a*).

## THEORETICAL QUESTIONS

1. What are the qualities expected of a good paint? Explain why more than one coat is recommended for a painting work.
2. Describe briefly the meaning of the following terms:
   (*i*) Paints; (*ii*) Varnishes; (*iii*) Enamels.
   Discuss their uses, merits and demerits.
3. What are the different constituents of an oil paint? Describe briefly how each one of the constituent acts in imparting the desired qualities to the finished material.
4. Write short notes on.
   (*a*) Lacquer and lacquering (*b*) Enamels (*c*) Varnishes.
5. What are the ingredients of paints? Explain the same with an example of each.
6. What is the object of painting or varnishing a surface and where will you prefer a paint to varnish and vice versa ?
7. Distinguish between oil paint and French polish, explain the composition of each.
8. What is enamelling? How is it carried out on simple products ?
9. Distinguish between enamels, lacquers, varnishes and paints.
10. Mention the types of failure commonly met with in paint work and state their causes.
11. Describe the method of removing oil paint from woodwork and repainting it.
12. How would you proceed to paint (*i*) a new wood panelled door (*ii*) a new steel roof truss of a workshop?
13. What preparatory work should be done before repainting an old painted surface?
14. What is varnish and why is it used ? Classify different kinds of varnish and give brief description of each.
15. What are the qualities of a good varnish?
16. Point out the difference between colourwash, distemper and oil paint, indicating the influences of the vehicles, pigments and fillers on their properties.
17. What are the causes of failure in oil paints and how can they be avoided ?
18. Mention the different types of varnishes and describe the process of varnishing on woodwork.
19. Enumerate the various steps involved in the preparation of a paint.
20. Describe fully the painting of new woodwork.
21. What is distemper? How is it prepared? Give the uses of distempers.
22. Mention the usual defects which are found in the painting work.
23. Enumerate the properties of distemper.
24. Write short notes on the following :
    1. Distemper 2. Enamel paint
    3. Lacquer 4. Synthetic rubber paint
    5. Whitewashing 6. Colourwashing
    7. Wall paper 8. French polish.

**11**

# Asphalt, Bitumen and Tar

*11.1. Asphalt – Definition – Properties of asphalt – Uses of asphalt – Constituents of asphalt – Types of asphalt – Forms of asphalt. 11.2. Bitumen – Definition – Properties of bitumen – Uses of bitumen – Forms of bitumen. 11.3 Tar – Definition and properties – Uses of tar – Types of tar. 11.4. Pitch. 11.5. Comparison between asphalt, bitumen and tar. Highlights – Objective Type Questions – Theoretical Questions.*

## 11.1 ASPHALT

### 11.1.1 Definition

- *Asphalt* may be defined as *natural or mechanical mixture in which bitumen is associated with inert mineral matter.*
- It is a *solid* or *semi-solid* sticky product, formed by *partial evaporation or distillation of certain petroleum oils.* It is soluble in varying degrees in *carbon disulphide.*

### 11.1.2 Properties of Asphalt

Following are the *properties* of asphalt :

1. It is sticky or adhesive and binds strongly as cement. It adheres perfectly to wood, stone, concrete and metal.
2. It is usually solid or semi-solid in state.
3. It is black brownish in colour.
4. It is water-proof.
5. It is durable and retains its properties for several years.
6. It is elastic.
7. It becomes plastic and workable when heated.
8. It possesses binding properties when softened by heat.
9. It is not seriously affected by adverse weather.
10. It is a good conductor of heat, sound and electricity.
11. It is ductile and can be stretched without breaking.
12. It is soluble in varying degrees in carbon disulphide.

### 11.1.3 Uses of Asphalt

Following are the *uses* of asphalt :

1. It is used as *damp-proof course* (D.P.C.) for preventing dampness in walls, floors and roofs of buildings.

2. It is used as a water repellent layer over *flat-roofs, arches* and *basements.*
3. It is used for lining walls of tanks, swimming pools and in other such places.
4. It is used for smooth and non-absorbent flows of bathrooms, laterines etc.
5. It is used for preparing paints and roofing felts.
6. In the form of asphalt mortar and asphalt concrete it is extensively used for paving roads in different ways such as, surface dressing, grouted macadam, coated macadam etc.
7. It is used for flooring purposes.
8. It is used for preparing useful electric insulating materials when mixed with wood tar pitch, rubber, resin and other suitable compounds.

### 11.1.4 Constituents of Asphalt

Besides the inert mineral matter, asphalt contains the following *constituents* :

(*i*) *Asphaltenes*: These are soluble in resins and provide hardness to asphalt.

(*ii*) *Resins*: They enhance the binding properties of asphalt.

(*iii*) *Oil:* It imparts viscosity to asphalt.

### 11.1.5 Types of Asphalt

The asphalt is classified into the following two categories:

1. Natural asphalt
2. Residual or petroleum asphalt

#### 11.1.5.1 Natural Asphalt

This variety of asphalt is obtained from nature. Depending upon the source of occurrence, natural asphalt may be boadly divided in two classes:

(*i*) Lake asphalt; (*ii*) Rock asphalt.

(*i*) *Lake asphalt:* It is obtained from lakes at Trinidad and Bermudez (South America) at depths varying from 3 to 60 metres. It is a composite mineral containing about 40 to 70% of pure bitumen. It is refined by boiling in a tank. The water evaporates and impurities collected at the top are removed. It is widely *used for road and pavement construction.*

(*ii*) *Rock asphalt:* It is natural asphalt impregnated in limestone rocks found in some parts of Switzerland, France, Germany etc. It contains about 4 to 20% pure bitumen by volume, the rest consists of calcarious materials. It is used *in crushed and heated form for making road pavements.* It can also be used for roofing sheets, paving tiles etc.

#### 11.1.5.2 Residual or Petroleum Asphalt

- It is also known as the *artificial asphalt.*
- It is obtained by the fractional distillation of crude petroleum oils with an asphaltic base. The residual product which settles at the bottom contains asphalt.

### 11.1.6 Forms of Asphalt

Usual forms of asphalt are :

1. Cut-back asphalt
2. Mastic asphalt
3. Asphaltic cement
4. Asphaltic emulsion.

**1. Cut-back asphalt:**

- It is a liquid asphalt and is prepared by dissolving asphalt in a volatile solvent.
- It can be applied at normal temperature in cold condition.

*Uses:* Used for *preparing bituminous paints*, for *repairing roofs* etc.

**2. Mastic asphalt:**

- It is produced by heating natural asphalt with sand and mineral fillers; the resulting product is a voidless impermeable mass.

- It may either be in solid or semi-solid state.

*Uses:* It is used as a material for *damp-proofing* and *water-proofing*.

3. **Asphaltic cement:**

- It is prepared by *blowing air through melted asphalt at high temperature.* The asphalt is oxidised and the resulting product is plastic in nature.
- It is highly resistant to varying climatic conditions.

*Uses:* It is used for *roofing, flooring, water-proofing* and as *filler for expansion joints in concrete.*

4. **Asphaltic emulsion:**

- It is prepared by mixing asphalt with 50 to 60% water in presence of 1% of emulsifying agent. On the evaporation of water, the emulsion breaks and it forms a water-proofing layer.
- It can be applied in cold condition.

## 11.2 BITUMEN

### 11.2.1 Definition

- It may be defined as *a solid or semi-solid, black, sticky, ductile substance, obtained as an important byproduct from the distillation of crude petroleum.*

It may also be defined as *that portion of petroleum asphalt and tar products, which will dissolve in carbon disulphide.*

Bitumen *implies a group of hydrocarbons often mixed with some organic matter.*

- *Bitumen is known as petroleum in the fluid state, mineral tar in the semifluid state and asphalt in the solid state.*
- In natural state bitumen is found in combination with finely graded mineral matter in the form of natural asphalt.

The two terms asphalt and bitumen are used synonymously. The former is *natural product* whereas the latter is the *refined product.*

### 11.2.2 Properties of Bitumen

Following are the *properties of bitumen*:

1. It is characteristically solid or semi-solid, black and sticky.
2. It melts or softens on application of heat.
3. Its specific gravity is 1.09.
4. It is completely soluble in carbon disulphide.
5. It is a binder in all types of asphalt.
6. It possesses great chemical stability, but is affected by oil.
7. It has low permittivity and high insulation resistance.

### 11.2.3 Uses of Bitumen

Following are the *uses of bitumen:*

1. It is used as a damp-proof course in walls and underbasements and in lining tanks, swimming pools, urinals etc.
2. Since it forms good expansion joint therefore, it is used for filling up the joints in leaky roofs.
3. It is being expensively used as a road making material.
4. It is used for making heat insulating materials for buildings.
5. It is also employed in the manufacture of roofing felt, impermeable paints, and cold moulded bituminous plastics. Bituminous paints are made by incorporating red, green or brown pigments into asphalt. These paints are suitable for decorative and water proof flooring.

6. It is also used for making bituminous filling compounds for *cable boxes, sealing accumulators* etc.

### 11.2.4 Forms of Bitumen

Following are the usual *forms of bitumen*:

1. Cut-back bitumen
2. Plastic bitumen
3. Blown bitumen
4. Straight run bitumen
5. Bitumen emulsion.

**1. Cut-back bitumen:**

- It is obtained by fluxing asphaltic bitumen in the presence of some suitable liquid distillates of coal tar or petroleum.
- It can be applied cold as a bitumen paint.

**2. Plastic bitumen:**

- It comprises bitumen thinner and a suitable inert filler (40-45%).
- It is used for stopping leakages and for filling cracks in masonry structures, etc.

**3. Blown bitumen:**

- It is obtained by *passing air under pressure at high temperature*.
- It can be used as heat insulating material, as roofing and damp-proofing felts. It is also used in the manufacture of pipe asphalts and joint fillers.

**4. Straight run bitumen:**

It is the bitumen which is being distilled to a definite viscosity or penetration without further treatment.

**5. Bitumen emulsion:**

Bitumen emulsion is a liquid product containing bitumen in a very finely divided state to a great extent in an acqueous medium; the bitumen is suspended in the acqueous medium with the help of some suitable stabilising agent.

## 11.3 TAR

### 11.3.1. Definition and Properties

- **Tar** *is a black solid mass obtained during the destructive distillation of coal, peat, wood, or other organic material.*
- According to American Society of Testing Materials, tar can be defined as *black to dark-brown bituminous condensate which yields substantial quantities of pitch when practically evaporated or fractionally distilled and which is produced by destructive distillation of organic material such as coal, oil, lignite, peat and wood. It is the byproduct of the manufacture of bituminous coke and is soluble in carbon disulphide.*
- Tar contains 75 to 95% of bituminous contents.
- It contains higher percentage of carbon.
- The *properties of tar depend upon the type of raw material used for manufacture, method of distillation and proportion of residual matter incorporated.*
- It hardens much quicker than asphalt.
- It is more adhesive than asphalt
- It possesses toxicity to a high degree.

### 11.3.2 Uses of Tar

Following are the *uses of tar:*

1. It is used for roofing and road making.
2. It is used for making bituminous paints and water-proofing compounds.

3. Coal tar is used as a preservative for timber.
4. It is used as a saturant for felts.
5. It is used for painting of laterine walls.

### 11.3.3 Types of Tar

According to the raw material used in the manufacture, tar is classified as follows:

(*i*) Coal tar (*ii*) Mineral tar (*iii*) Wood tar.

(*i*) **Coal tar:** It is produced by *destructive distillation of coal or as a by-product in the manufacture of coal gas*. It is used as a *preservator for timber and in making tar-macadam roads.*

(*ii*) **Mineral tar:** It is produced by *distillation of bituminous shales*. It contains less volatile matter, however, its composition is same as of other kinds of tar.

(*iii*) **Wood tar:** It is produced by *distillation of pure and resinous trees*. It contains creosote oil and hence possesses strong preservative property.

## 11.4. PITCH

- It can be defined as *a residue obtained in the distillation of organic materials.*
- It is dark in colour.
- It is brittle, easily softened by heat and crushed under pressure.

*Uses:* It is chiefly used in roads and pavements, and as an inferior substitute for natural asphalt.

## 11.5. COMPARISON BETWEEN ASPHALT, BITUMEN AND TAR

The comparison between asphalt, bitumen and tar is given in table 11.1.

**Table 11.1. Comparison between Asphalt, Bitumen and Tar**

| S. No. | Properties | Asphalt | Bitumen | Tar |
|---|---|---|---|---|
| 1. | *State* | Usually solid or semi-sold | Usually solid | Viscous liquid |
| 2. | *Colour* | Black brownish | Deep black with tinge of redness | Jet black |
| 3. | *Toxicity* | Low degree | Low degree | High degree |
| 4. | *Adhesiveness* | Less adhesive | More adhasive | Most adhesive |
| 5. | *Percentage of carbon* | Less percentage | Moderate percentage | High percentage |
| 6. | *Resistance to water and acid* | More | More | Less |
| 7. | *Setting time* | Less | Less | More |
| 8. | *Behaviour when spread* | Loses volatile matter slowly and thus hardens slowly | Loses volatile matter slowly and thus hardens slowly. | Loses volatile matter at a faster rate and thus hardens quickly. |
| 9. | *Effect on heating* | Non-inflammable but burns with a smoke flame at a temperature of about 250°C, becomes plastic and workable when heated. | Melts when heated. | Becomes more fluid when heated. |

## HIGHLIGHTS

1. **Asphalt** may be defined as natural or mechanical mixture in which bitumen is associated with inert mineral matter.

   Besides the inert material, asphalt contains asphaltenes, resins and oil.

   The asphalt is classified into two categories: 1. Natural asphalt. 2. Residual or petroleum asphalt.

   Usual forms of asphalt are: (*i*) Cut-back asphalt; (*ii*) Mastic asphalt; (*iii*) Asphaltic cement; (*iv*) Asphaltic emulsion.

2. **Bitumen** may defined as a solid or semi-solid, black, sticky, ductile substance, obtained as an important byproduct from the distillation of crude petroleum.

   Usual forms of bitumen are: (*i*) Cut-back bitumen; (*ii*) Plastic bitumen; (*iii*) Blown bitumen; (*iv*) Straight run bitumen; (*v*) Bitumen emulsion.

   *Uses of bitumen/ashalt:* Used as damp-proof course for preventing dampness in walls, floors and roofs of buildings; used for lining walls of tanks, swimming pools and in other such places. In the form of asphalt mortar and asphalt concrete, it is extensively used for paving roads in different ways such as, surface dressing, grouted macadam, coated macadam etc; used in the making of water-proof paints and anti-corrosive coatings.

3. **Tar** is a black solid mass obtained during the destructive distillation of coal, feat, wood, or other organic material. It contains 75 to 95% of bituminous contents. It is used for roofing, road making, as a preservative for timber etc.

4. **Pitch** can be defined as a residue obtained in the distillation of organic materials. It is chiefly used in roads and pavements, and as an inferior substitute for natural asphalt.

## OBJECTIVE TYPE QUESTIONS

### Choose the Correct Answer:

1. When air is blown through melted asphalt at a hight temperature, the product obtained is
   (*a*) asphaltic cement (*b*) colas
   (*c*) mastic asphalt (*d*) cut-back asphalt.
2. Asphalt is a
   (*a*) mixture of bitumen and inert material (*b*) impure bitumen
   (*c*) pure bitumen (*d*) any of the above.
3. Bitumen felt is made from bitumen and
   (*a*) steel sheets (*b*) cloth
   (*c*) rubber (*d*) none of the above.
4. Which of the following is the property of asphalt?
   (*a*) It is black brownish in colour (*b*) It is elastic
   (*c*) It is water-proof (*d*) It is durable.
   (*e*) All of the above.
5. Which of the following is the constituent of asphalt?
   (*a*) Resins (*b*) Oil
   (*c*) Asphaltenes (*d*) All of the above.
6. ................ is a liquid asphalt and is prepared by dissolving asphalt in a volatile solvent.
   (*a*) Asphaltic cement (*b*) Cut-back asphalt
   (*c*) Mastic asphalt (*d*) Asphaltic emulsion.
7. The specific gravity of bitumen is
   (*a*) 0.8 (*b*) 0.9
   (*c*) 1.09 (*d*) 1.4.
8. ................ bitumen is obtained by fluxing asphaltic bitumen in the presence of some suitable liquid distillates of coal tar or petroleum.
   (*a*) Plastic (*b*) Cut-back
   (*c*) Blown (*d*) Straight run.
9. ................ is a black solid mass obtained during the destructive distillation of coal, peat, wood, or other organic material.

(*a*) Asphalt (*b*) Bitumen

(*c*) Tar (*d*) None of the above.

**10.** ................ **tar is produced by distillation of bituminous shales.**

(*a*) Coal (*b*) Wood

(*c*) Mineral (*d*) None of the above.

## ANSWERS

| | | | | |
|---|---|---|---|---|
| **1.** (*a*) | **2.** (*a*) | **3.** (*d*) | **4.** (*e*) | **5.** (*d*) |
| **6.** (*b*) | **7.** (*c*) | **8.** (*b*) | **9.** (*c*) | **10.** (*c*) |

## THEORETICAL QUESTIONS

1. Define 'asphalt'.
2. State the properties of asphalt.
3. What are the uses of asphalt.
4. Classify asphalt. Give the properties and uses of different types of asphalt.
5. What is bitumen?
6. Give the properties and uses of bitumen.
7. Enumerate and describe the various forms of bitumen.
8. What is tar?
9. Give the properties and uses of tar.
10. What is pitch?
11. Give the comparison between asphalt, bitumen, and tar.

# 12

# Asbestos, Adhesives and Abrasives

> *12.1. Asbestos – Properties of asbestos – Uses of asbestos – Commercial forms of asbestos. 12.2. Adhesives – Advantages and disadvantages – Types of adhesives. 12.3. Abrasives – Types of abrasives – Highlights – Objective Type Questions – Theoretical Questions.*

## 12.1 ASBESTOS

### 12.1.1 General Aspects

- Asbestos is an *acid-proof* and *fire-proof natural fibrous mineral substance* of different colours. It is a *silicate of calcium and magnesium* ($CaSiO_3.3MgSiO_3$). It also contains *small amounts of iron oxide and alumina*.
- The original source of asbestos was the mineral '*actinolite*', but the variety of '*serpentine*' known as '*chrysolite*' now furnishes most of the commercial asbestos. Its fibres are long and silky and the tensite strength is from 550 to 1370 N/mm$^2$. The colour is white, amber, grey or greenish. The melting point is 1520°C.

The *natural asbestos* can be divided into the following *two groups*:

| | | | |
|---|---|---|---|
| (*i*) | **Acid-resistance** | ............. | comprises *amosite* asbestos, *crocidolite* asbestos. *actinolite* asbestos, *tremolite* asbestos and *anthophylite* asbestos. |
| (*ii*) | **Non-acid-resistance** | ............. | comprises only chrysolite asbestos (of great industrial importance) |

**Availability:** In India, asbestos is found in Mewar (Rajputana), Garhwal (U.P.), Bhandhra (M.P.) and Hassan (Mysore state) etc. Deposits of chrysolite also occur in Andhra Pradesh. Canada and Rhodesia (Africa) are producing world's largest quantities of asbestos.

### 12.1.2 Properties of Asbestos

Following are the *properties of asbestos*:

1. Its colour is white, grey, greenish or brown.
2. It is fibrous in structure.
3. It is smooth like glass and silk.
4. It is acid-proof and fire-proof.
5. It can be cut into pieces.
6. It is flexible.

7. It is an excellent insulator for heat and electricity.
8. Its melting point varies from 1200°C to 1550°C.
9. It is incorrodible and also vermin-proof.
10. The holes can be drilled and screws can be fitted on its surface.
11. It is capable of being interwoven with webbing to produce flexible fire proof material.
12. Its specific gravity is 3.1.
13. Its molecules are strongly bound together only in one direction and that is why it possesses very high tensile strength along the fibres.
14. It can easily mix with binding materials like *bitumen* and *cement*; such a mix can be easily worked into various forms and shapes to manufacture different articles.

**12.1.3 Uses of Asbestos**

Following are the *uses of asbestos:*

1. The asbestos cement products-*sheets* and *pipes*, are prepared by mixing asbestos fibres with cement. The sheets are used as *roofing material* and the pipes are used to *convey rainwater, seepage water* etc.
2. The asbestos felt (prepared by coating asbestos fibres with bitumen) is used as damp-proof layer.
3. Being an excellent heat insulator, it is widely employed in the form for *insulating furnaces, steam and exhaust pipes, boilers* etc.
4. It is often combined with magnesium carbonate, hair, or with wool felt, and used for *lagging steam pipes*, an outer casing being employed for holding it in position.
5. Asbestos, when powdered or in the form of fibres, is sometimes kneaded into a dough with water and used for stopping up holes and cracks in small muffles, furnace bearing metal, shells etc.
6. It is also frequently employed for making moulded shapes to withstand high temperatures.
7. Asbestos is supplied commercially in the form of twine, rope, wool or flake, thin paper mill board, jointings and moulded shapes etc. *It is used with a thin copper shell for internal combustion engine exhaust pipe and silencer gaskets and sparking plug washers.*
8. Asbestos, in the form of *fibres*, is used for cable insulation, and in the *sheet form* it is used in lining and partition in switches and fuse boxes.
9. Fire proof suits of asbestos cloth have been used successfully for man engaged in fighting fierce fire.
10. Asbestos paper is used as a cover for electric wires, in electric motors and coils exposed to high temperature.
11. It is used to form asbestos paints.
12. Asbestos having small length fibres is used for sound insulation in different conditions (*e.g.* weight boards or blocks).

**12.1.4 Commercial Forms of Asbestos**

Some of the commercial forms of asbestos are described below:

**1. Asbestos boards and sheets :**

It is a construction or insulating material in sheets made of *asbestos fibre and portland cement moulded under hydraulic pressure.*

- *Asbestos boards:* The ordinary *board* is in natural mottled grey colour, but pigmented boards are also marketed. The boards are dense and rigid, but can be worked easily with carpenter's tool. The specific gravity of asbestos board is about 2.0, and it can withstand temperature upto about 540°C.
- Asbestos cement building boards are categorized into two classes:
  (*i*) *Class A*: Thickness = 6.5 mm, width = 1.2 m

(*ii*) *Class B*: Thickness = 5.0 mm, width = 1.2 m
The length ranges are 1.2 m, 1.8 m and 2.4 m.

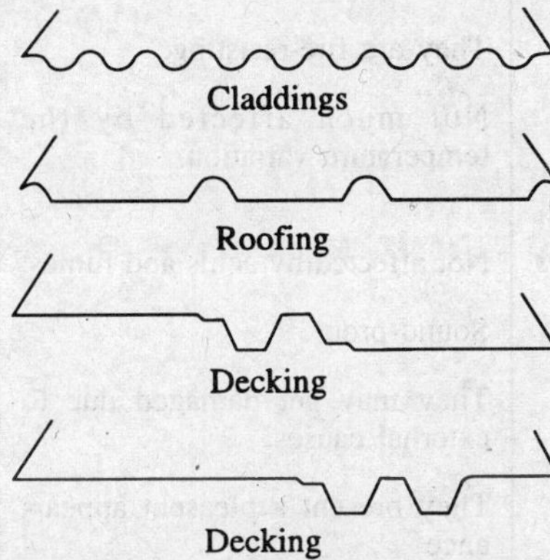


Fig. 12.1. Corrugated asbestos-cement boards/sheets.

**Asbestos sheets:**

- The *flat asbestos sheets* are classified as *compressed* and *uncompressed* according to minimum unit bending stress of 22 N/mm$^2$ and 16 N/mm$^2$ and density exceeding 1600 kg/m$^3$ and 1200 kg/m$^3$ respectively.
- The *corrugated asbestos* used for roofing and sliding industrial buildings, has a thickness of 4.76 to 6.35 mm and the usual corrugation has a pitch of 10 cm; the weight being 145 N/m$^2$.

**Uses of asbestos boards and sheets:**

(*i*) Wall-lining, panelling, partitions side-cladding and false-ceilings.

(*ii*) Toilet partitions, back up for wardrobes, cupboards and kitchen-pantry shelves.

(*iii*) Blackboards, hoardings, signboards and kiosks.

(*iv*) Window panes and door panels.

Asbestos boards claim the following *advantages* over other conventional material like plywood:

1. *Economical*; do not require any protective treatment and maintenance.
2. Completely *incombustible* and hence act as an effective fire barrier.
3. They *can take on paint*, *any shade*. They *consume less paint* and retain paint effectively.
4. Easily *workable* (can be easily cut, drilled, nailed and screwed with ordinary wood working tools).
5. *Highly durable* (do not warp in the presence of moisture, not affected by saline conditions and vermin and moth).

**Comparison Between A.C. sheets and G.I. sheets**

| S.No. | Aspects | Asbestos cement (A.C.) sheets | Galvanised iron (G.I) sheets |
|---|---|---|---|
| 1. | *Materials* | They are prepared from a mixture of cement and asbestos fibres. | They are prepared from wrought iron plates and then galvanised. |
| 2. | *Weight* | Heavy in weight. | Light in weight. |
| 3. | *Initial cost* | Cheap in initial cost. | High in initial cost. |
| 4. | *Corrosion* | Do not corrode. | Easily corroded by atmospheric actions. |

| | | | |
|---|---|---|---|
| 5. | *Maintenance cost* | Practically nil, if not damaged. | Maintenance cost is high as they require periodic painting. |
| 6. | *Fire-resistance* | They are fire-resisting. | They are not fire-resisting. |
| 7. | *Effect of temperature* | Not much affected by the temperature variation. | Much affected by temperature variation, they become hot in summer and cold in winter. |
| 8. | *Effect of acids and fumes* | Not affected by acids and fumes. | Affected by acids and fumes. |
| 9. | *Sound insulation* | Sound-proof. | Not sound-proof. |
| 10. | *Handling and fixing* | They may get damaged due to external causes. | There is less possibility of getting damaged due to external causes. |
| 11. | *Appearance* | They present a pleasent appearance | They do not present a pleasent appearance even if a neat finish is made. |

**2. Asbestos cement pipes :**

- These pipes are made of asbestos cement which is a mixture of asbestos fibres (15-20%) and cement (80-85%) and the required quantity of water. These can be prepared to any shape and length. Various forms of rain-water and drainage asbestos pipes are available in the market.
- They can be easily cut and require no painting to maintain them.
- *Asbestos-cement bends* and *junctions*, *taper pipes, shoes* etc. can also be manufactured of the required shape and size.

*Advantages of asbestos cement pipes:*

1. Adaptability
2. Easy installation
3. Light, yet strong enough to withstand normal handling.

**Uses:**

(*i*) Rain water pipes
(*ii*) Soil waste and ventilating pipes
(*iii*) Valley gutters
(*iv*) Boundary wall (leaves) gutters
(*v*) Half round gutters
(*vi*) Ogee gutters and their respective fittings.

**3. Asbestos paper:**

It is used as a cover for electric wires in electric motors and coils exposed to high temperature

**4. Asbestos paint:**

- It is prepared by *adding finely ground asbestos to an oil paint or a distemper.*
- It is *fire, heat, sound* and *vermin-proof.*
- It also prevents corrosion of metals.
- The patent form of asbestos paint is "*Limpet*" (prepared by U.K.)

**5. Asbestos fibres:**

- Asbestos is also available in the market in the form of *fibres.*
- The fibres have adequate *strength* and *flexibility.*

These fibres are used in making conveyor belts, boiler insulation, brake lining etc.

**6. Asbestos felt:**

- It is made of *asbestos fibres impregnated with mineral bitumen.*
- Asbestos felt is water-proof.

**7. Asbestos fabric:**

It is the *woven stuff from asbestos fibres* and is used as *cable insulation and for winding of motors*.

## 12.2 ADHESIVES

A *substance used to join two or more parts so as to form a single unit* is called an **adhesive**.

### 12.2.1 Advantages and Disadvantages of using Adhesives

**Advantages:**

The application of adhesive has the following *advantages* over the conventional methods of bolting, riveting and welding:

1. Adequate strength is produced.
2. It is possible to have a wide variety of combinations in joining.
3. A massive effect is created.
4. Corrosion may be prevented between different metals joined by adhesive.
5. The surfaces of plastics, wood, metal and glass can be bonded by adhesive.
6. The adhesive application process is economical, easy and speedy.
7. By the use of adhesives, the joints impermeable for water and gas can be made.

**Disadvantages:**

1. Unstable at high temperatures.
2. An adhesive requires sometime to attain the desired strength.
3. Specific adhesive is required to be used for specific substance(s).

### 12.2.2 Types of Adhesives

An adhesive is indicated by a general term, the 'glue'. Following are the various types of adhesives:

**1. Animals glue:**

- It is generally obtained from *hides and bones of animals*.

The glue is *prepared as follows*:

— The bones are crushed to a fine powder.

— The bone powder is then treated with organic solvents like benzene, or carbon tetrachloride to *remove the grease* present in it.

— The degreased bone powder is treated with hydrochloric acid for 2 to 10 days. The acid leaches out $(Ca_3PO_4)_2$ from the bone powder

— After extraction of $Ca_3(PO_4)_2$, the residue is washed and nentralized of any remaining acid and finally placed in pressure tanks for the extraction of glue.

- It is available in the form of cakes, granules, jelly, pearls and sheets.
- It develops strong and tough joints and is easy to apply.
- It is affected by damp and moist conditions.

**Uses:** It is used in the manufacture of *plywood, laminated timbers* etc.

**2. Starch glue:**

- Starch glue or starch paste may be made by *mixing wheat* or *rice flour with hot water.* Acid or alkali modifiers are used to make starch paste *thick and tacky*.
- These glues are *cheaper and easier to handle than animal glues.*
- They have poor water resistance but bond quickly to paper and textiles.

**Uses:** (*i*) These glues are extensively used as labelling adhesives and find wide use in *textile industry* as coating material for fine cottons during weaving.

(*ii*) Since these glues spread and dry easily, they are used in *automatic package machines.*

(*iii*) Also used in the manufacture of low-strength and low-water resistant plywood.

**3. Dextrine:**

- This form of glue is prepared by treating starch with acid by heating it to about 200°C. Dextrine used may be white or yellow.
- It is expensive as compared to starch adhesive.
- These glues possess *fast setting* and *tacky properties.*

**Uses:** Widely used for *fast machine packing, for pasting on wall-paper* etc.

**4. Gum arabic:**

- It contains mixed mineral salts of *arabic acids*; is obtained from acacia tree.
- It forms the most useful natural resin adhesive.
- It is soluble in water.
- When used in high concentration it is considered superior to even dextrines.

**Uses:** It is used for *joining paper and wood* and in *high-speed packaging and labelling machines.*

**5. Mucilage:**

- It is an aqueous solution in water of the adhesive prepared from starches, dextrines or gums.
- These adhesives spread easily and dry rapidly but they do give high tensile strength to bonding film.

**Uses:** These are used with paper like *stamps, envelopes* etc.

**6. Albumin glues:**

- These glues are made from protein derived from blood serum, or egg white.
- They are *excellent water-resistant* and *durable glues.*

**Uses:** They have good adhesive properties for paper, textiles, metals etc. and are largely used in *food packaging, leather dressing* and for *wood working.*

**7. Casein glue :**

- It is obtained by coagulating skimmed milk in vats and grinding the product to a fine powder.
- It can stand prolonged exposure to the harmful natural elements.
- It serves as a good adhesive for surfaces which do not make close contacts.
- It exhibits high strength even when applied in thick layers.

**Uses**: These glues have been used since long to *form strong, water-proof wooden joints and to make durable plywood.*

**8. Sodium silicate glues:**

- It is made *by fusion of the soda ash and sand in a furnace.* The fused mass is cooled and dissolved in water. On loss of water, *water-glass* is formed, which possesses adhesive properties.
- It is water and fungus resistant.

**Uses:** (*i*) It is used in the manufacture of card-board and paper boxes.

(*ii*) Silicate cement compounded with inert fillers like powdered asbestos, barium sulphate etc. are used for *ceramic applications.*

**9. Natural resin adhesives:**

These are prepared from natural resins (*e.g.* Resin, shellac and asphalt) are used for labelling building paper.

**10. Rubber-based adhesives:**

- These are prepared by *dissolving rubber in benzene.*

- Rubber adhesives have excellent property of forming non-rigid bonds. This enables the adhesive to flex and stretch under stress and recover to its original position as soon as the load is removed.
- They have *high bond strength and durability*.
- *Reclaimed natural rubber adhesive* has the advantage of low *cost as* automobile tyres form the largest source of reclaimed rubber for making of adhesive.

**Uses:** (*i*) Commonly used in *shoe industry*.

(*ii*) In the *automobile industry*, they are used to bond felt, cloth and rubber to metal.

(*iii*) They also form common house-hold cements.

**11. Nitro-cellulose glues:**

- Cellulose nitrate or pyroxylin when mixed with other resins form the basis of common household cements. The solution of pyroxylin in alcohol and ether is known as *Cellodin*.
- These glues are highly flammable and as such these must be handed with great care.
- They form good barrier against moisture and hold up well in outdoor use.
- They discolour when exposed to sunlight.

**Uses:** They are used to cement *glass, metals, leather clothes and ceramics.*

**12. Synthetic resin glues:**

- A synthetic resin adhesive has a composition consisting substantially of phenolic or aminoplastic resins including a hardening agent, fortifier, filler or expander.

— A phenolic synthetic resin is obtained from the reaction of phenol with an aldehyde (phenol formaldehyde).

— An aminoplastic resin is obtained from the reaction of urea, thio-urea, melamine, etc. with formaldehyde (urea-or melamine-formaldehyde).

- All synthetic glues are fire-proof, strong and water-proof.
- The setting time of these glues can be regulated by varying the type or kind and quality of the hardner.
- They resist the attack by fungi.
- They possess non-staining qualities.

## 12.3 ABRASIVES

*Abrasives are substances or objects that are employed to wear off or rub down objects, in order to give them desired size, shape or finish.* Abrasive particles held together by a bonding material comprise the cutting agents in grinding wheels.

Abresives are the hardest materials. These are used for the engineering purposes in the following forms:

(*i*) As abrasive stones (*ii*) As abrasive papers or cloths

(*iii*) As grinding pastes with oil (*iv*) In grinding wheels

(*v*) With some suitable bonding materials.

**Types of abrasives.** The abrasives may be:

1. Natural abrasives 2. Artificial abrasives.

**1. Natural abrasives:**

These abrasives are further classified as:

(*i*) *Hard abrasives;* (*ii*) *Siliceous abrasives;*

(*iii*) *Soft abrasives.*

(*i*) **Hard abrasives:**

(*a*) **Diamond.** It consists of crystallized carbon (chemically) and is the *best known abrasive*. It is the *hardest of all materials in scratching power.* It is employed for *drill points for drilling* *rocks, for dies in wire drawing, for polishing gems, for cutting, turning and dressing grinding wheels and for lapping.* Diamond dust is also being used as an abrasive.

(*b*) **Corundum**. Corundum is a natural aluminium oxide; the remainder consists of impurities. The three principal varieties of corrundum are: (*i*) the transparent gem stone (ruby, sapphire, topaz); (*ii*) the common or opaque (*a*) block corundum (*b*) crystal corundum associated with many rocks and loose soil (*c*) sand corundum, (*iii*) emery.

The abrasive corundum *is the opaque variety and its scratch hardness is next to diamond. Corundum is mostly used for the manufacture of wheels used in the metal industry.*

(*c*) **Emery**. It is also a *natural aluminium oxide*. It contains 55 to 65% alumina, the remainder consists of iron oxide and other impurities. The hardness of emery depends on the proportion of corundum which it contains. The coarseness of grain also is largely determined by the size of the corundum crystals.

Its utility rests in its having a combination of desirable properties, high scratch hardness, reasonable toughness, and gradual breakdown of crystal under pressure.

(*d*) **Garnet.** It consists of a group of similar silicate minerals varying in tint and having a range of hardness from 6.5 to 7.5 in Moh's scale of hardness. Garnet is used in the intermediate 'fining' step between sanding and rough polishing of plate glass. Small amounts are bonded into wheels with silicate or shellac binder. Vitrified binder cannot be employed because of the low melting point of the garnet.

(*ii*) **Siliceous abrasives :**

(*a*) **Quartz.** It is crystallized silica and is a common mineral. It is brittle and has a hardness of 7 on Moh's scale. It is *mostly used for making sand papers, scouring powder and soaps.*

(*b*) **Flint.** It is modification of silica and is as hard as quartz. It is sometimes used for making sharpening stones and mill stones.

(*c*) **Quartzite.** It is a metamorphosed sandstone which is nearly pure silica. It is harder and more denser than sandstone and the grains are not easily distinguishable.

(*d*) **Sand stone and sand.** Sandstone is used for making grind stones, pulp stones, sharpening stones, etc. The stone should be of uniform hardness throughout, the grains should be even and sharp; and it should be free from clay or other substances that reduce its strength. Sand of round grains is used in large quantities for sand blasting.

Diatomite and Tripoli are other varieties of siliceous abrasives.

(*iii*) **Soft abrasives.**

These are non-siliceous abrasives. Though they are softer than the materials to be abraded, they smooth and polish metal and other surfaces very effectively Rotton stone, vienna lime and metallic oxides are some of the important examples of soft abrasives.

**2. Artificial abrasives :**

Artificial or manufactured abrasives include chiefly: (*i*) *Silicone carbide*, and (*ii*) *Aluminium oxide*.

(*i*) **Silicone carbide.** This abrasive is manufactured from 56 parts of silica sand, 34 parts of powdered coke, 2 parts of salt and 12 parts of saw dust in a long, rectangular electric furnace of the resistance type built-up of loose brickwork. Sand furnishes carbon, sawdust makes the charge porous, salt helps to fuse it, and gases may escape through the open joints in the brickwork.

There are two types of silicon carbide abrasives: *greet grit* which contains at least 97 per cent silicon carbide, *black grit* which contains at least 95 per cent silicon carbide. This form is harder but weaker than the latter.

Silicon carbide follows the diamond in order of hardness, but it is not as tough as aluminium oxide. The crystals are very brittle but as they wear away they continue to present extremely sharp edges, so that their cutting power is not reduced. It is unreactive chemically and can withstand the action of high temperatures without damage. It is mostly *used for making grinding wheels.* These wheels are employed for grinding glass, granite, chilled iron, cemented carbides, aluminium, rubber and leather.

(*ii*) **Aluminium oxide.** It is manufactured by heating mineral bauxite, a hydrated aluminium oxide clay containing silica, iron oxide, titanium oxide, etc. mixed with ground coke and iron in arc type electric furnace. *It is not as hard as silicon carbide but it is tougher and more durable and superior for rugged work. It is used in grinding wheels and in coatings for abrasive paper and cloth.*

Boron carbide and metallic abrasives are other examples of artificial abrasives.

## HIGHLIGHTS

1. **Asbestos** is an acid-proof and fire-proof natural mineral substance of different colours. It is silicate of calcium and magnesium and contains small amounts of iron oxide and alumina.

   It is a construction or insulating material available in sheets and boards made of asbestos fibre and portland cement.

   The important advantages of Asbestos cement pipes are its adaptability, ease in installation, and lightness properties.
2. An **adhesive** is a substance used to join two or more parts so as to form a single unit.

   *Types of adhesives* : Animal glue, Vegetable glue, Natural resins glue, Synthetic resin glues, Rubber glues and Special glues.
3. **Abrasives** are the substances or objects that are employed to wear off or rub down objects, in order to give them desired size, shape or finish.

   *Types of abrasives* : 1. Natural abrasives (Hard abrasives, siliceous abrasives and soft abrasives). 2. Artificial abrasives (silicon carbide and aluminium oxide).

## OBJECTIVE TYPE QUESTIONS

**Fill in the Blanks/Say Yes or No:**

1. ................. is a silicate of calcium and magnesium.
2. Asbestos is an acid-proof and fire-proof natural fibrous mineral substance of different colours.
3. The melting point of asbestos is 1000°C.
4. Asbestos is poor insulator for heat and electricity.
5. Asbestos is incorrodible and also vermin-proof.
6. Specific gravity of asbestos is 3.1.
7. A substance used to join two or more parts so as to form a single is called an .................
8. ................. glue is generally obtained from hides and bones of animals.
9. Animal glue is affected by damp and moist conditions.
10. All synthetic glues are fire-proof, strong and water-proof.
11. Rubber glues are prepared by dissolving rubber in .................
12. The araldite is used to join the light metals.
13. ................. are substances or objects that are employed to wear off or rub down objects, in order to give them desired size, shape or finish.
14. ................. is the best known abrasive.
15. Corundum is a natural aluminium oxide.

## ANSWERS

| | | | | |
|---|---|---|---|---|
| **1.** Asbestos | **2.** Yes | **3.** No | **4.** No | **5.** Yes |
| **6.** Yes | **7.** adhesive | **8.** Animal | **9.** Yes | **10.** Yes |
| **11.** benzene | **12.** Yes | **13.** Abrasives | **14.** Diamond | **15.** Yes |

## THEORETICAL QUESTIONS

1. What is asbestos?
2. Enumerate properties of asbestos.
3. What are the uses of asbestos?

4. Describe briefly the commercial forms of asbestos.
5. Write short notes on:
   Asbestos sheets, boards and cement pipes.
6. What is an adhesive?
7. What are advantages and disadvantages of adhesives over the conventional methods of bolting, riveting and welding?
8. Describe briefly various types of adhesives.
9. Write short notes on:
   Animal glue, synthetic resins glues and special glues.
10. What are abrasives?
11. Enumerate various types of abrasives.
12. Explain briefly any two of the following abrasives :
    Diamond, Garnet, Silicon carbide, Aluminium oxide.

# 13

# Plastics and Fibres

*13.1. Definition of plastic. 13.2. Constituents of plastics. 13.3. Classification of plastics –thermoplastic materials – thermosetting materials. 13.4. Properties of plastics. 13.5. Shortcomings of plastics. 13.5. Uses of plastics. 13.7. Trade names and typical applications of some important plastics. 13.8. Commercial forms of plastics – plastic sheets – plastic tiles – plastic laminates – plastic panels – plastic pipes. 13.9. Manufacturing processes – casting – moulding – laminating – blowing – calendering – machining and cementing. 13.10. Fibre glass reinforced plastic. 13.11. Polymer/plastic coatings. 13.12. Deformation of plastics. 13.13. Testing of plastics. 13.14. Machining of plastics. 13.15. Advanced polymeric materials. 13.16. Behaviour of polymers. 13.17. Fibres. Highlights – Objective Type Questions – Theoretical Questions.*

## 13.1 DEFINITION OF PLASTIC

- A *plastic* can be broadly defined as *any non-metallic material that can be moulded to desired shape.*
- The most common definition for plastics is that they are *natural or synthetic resins, or their compounds, which can be moulded, extruded, cast or used as films or coatings.*

— Most plastics are of organic nature composed of *hydrogen, oxygen, carbon* and *nitrogen.*

— The synthetic plastic development dates from 1900 when Dr. Beekland announced the production of phenol formaldehyde. Since then several new plastics have been developed.

— In order to give desired properties to the finished plastic articles, certain moulding compounds like *fillers, plasticizers, solvents, pigments* etc. are added.

— The basic raw materials used in the manufacture of plastics are generally obtained from the following natural substances:

| | | |
|---|---|---|
| (*i*) Coal | (*ii*) Petroleum | (*iii*) Limestone |
| (*iv*) Salt | (*v*) Sulphur | (*vi*) Air |
| (*vii*) Water | (*viii*) Cellulose from cotton and wood. | |

— In the field of *building construction*, plastics have been found very useful for a variety of purposes *e.g. fittings, fixtures, structural component parts*, etc.

## 13.2. CONSTITUENTS OF PLASTICS

As earlier stated that the basic raw materials for producing plastics are obtained from coal, petroleum, air and water. A moulding composition for plastics is prepared from the following raw material groups:

**1. Binder.** Binders may be either *resins* (synthetic or natural) or *cellulose derivatives* (polymeric compounds), chemically both kinds of materials may be described as *substances made up of compounds of very large molecular weight.*

**2. Fillers**. The *materials added to the plastics to improve their mechanical properties and to make them economical* are called **'fillers'.** These are powder, fibrous and laminated fillers.

(*i*) *Powder fillers:* Quartz powder, chalk, talcum and organic substances (ground saw dust).

- These fillers improve hardness, durability, heat resistance and acid resistance of plastics and reduce cost.

(*ii*) *Fibrous fillers:* Asbestos, wood wool, saw dust and glass fibres.

- They increase strength, reduce brittleness and enhance thermal resistance and impact strength of plastics.

(*iii*) *Laminated fillers:* Paper, cotton and fibre glass cloths, asbestos, card board and wood veneer.

- They increase mechanical strength of plastics. Asbestos and cardboards also increase heat and acid resistance properties.

**3. Plasticizers.** The *chemicals added to plastics to make them soft; to improve their toughness at finished stage and to make them flexible* are called **plasticizers.** A plasticizer should be *chemically inert, non-volatile and non-toxic.*

Common plasticizers are: *Vegetable oils, aluminium stearates and dibutyl phthalate and camphor.*

**4. Colouring matter.** This is usually in the form of pigment and dyes and often added to monomers and gives the required colour to the plastics. The colouring matter should be durable and adequately fast to light.

Commonly used dyes are: *Organic* (AZO dyes, anthraquinone vat dyes) and *mineral pigments* (ochre, chromium oxide and ultramarine).

**5. Lubricants.** Common lubricants are: Mosallic soaps and stearates.

- They facilitate moulding operation by increasing the flow of plastic mix in the die and also prevent sticking of plastic to moulds.

**6. Catalysts.** These compounds are added to accelerate the chemical reaction during the process of polymerisation of plastics. These compounds also act as hardeners and accelerators.

**Polymers and Polymerisation**

The **polymers** are composed of a large number of *repeating units* (small molecules) called **monomers.** A polymer is, therefore, made up of thousands of monomers joined together to form a large molecule of colloidal dimension, called *macromolecule*. The *unique feature of a polymer is that each molecule is either a long chain or network of repeating units all covalently bounded together*. In some cases, molecules are held together by secondary bonds.

The *process of linking together of monomers is called* ***polymerisation.*** The need to start with the process of polymerisation lies on the necessity of breaking the double bond (C=C) of the monomers. This requires considerable energy. Polymerization mechanisms may be of the following two types:

1. Addition polymerisation
2. Condensation polymerisation.

**1. Addition polymerisation:**

This polymerisation process is of simplest form. When a large number of simple molecules are chemically added together to increase the average molecule size without wastage, process of addition polymerisation takes place. Such a polymerisation takes place by three steps namely. (*i*) *Initiation* (*ii*) *Chain propagation* and (*iii*) *Termination.*

**Example.** Addition polymerisation of ethylene (Fig. 13.1).

Once polymerisation process is initiated, it does not continue indefinitely since it is impossible to link all the monomers in a plastic one long continuous chain. The polymerisation is terminated by a collision between the active ends of two chains or by addition of a terminator, such as free radicals from the catalyst.

**Copolymerization.** It is another type of addition polymerisation.

*Copolymerization* is the addition polymerisation of two or more chemically different monomers. Many monomers will not polymerise with themselves but will copolymerise with other compounds.

Copolymerisation has been applied extensively, in the *artificial rubbers.*

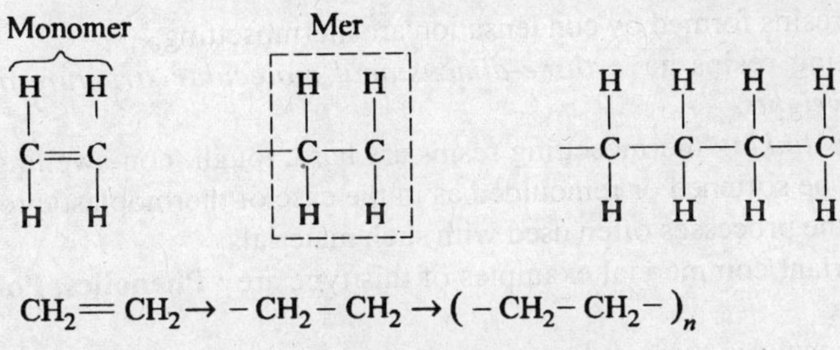


$$CH_2 = CH_2 \rightarrow -CH_2 - CH_2 \rightarrow (-CH_2 - CH_2 -)_n$$

Fig. 13.1. Addition polymerisation.

**2. Condensation polymerisation:**

*Condensation polymerisation is defined as the process of linking together of unlike monomers accompanied by splitting off a small molecule. This process usually requires a catalyst.*

In comparison to *addition* reaction in which as simple molecular summation occurs, *condensation* reactions result in *splitting out of* simple monopolymerisable molecules, *e.g.* water which are considered to be *by-products* of the process. Thus when a phenol and formaldehyde monomers are polymerised, water is released, and the resulting product is polymerised phenol formaldehyde, more commonly known as *Bakelite.*

**Comparison between Addition and Condensation polymerisation**

| S.No. | Addition polymerization | Condensation polymerisation |
|---|---|---|
| 1. | It requires molecules which are unsaturated. | It requires two unlike molecules. |
| 2. | It does not yield a by-product. | It yields a by-product. |
| 3. | Reaction is very fast and may take $10^{-2}$ to $10^{-6}$ sec. | Reaction normally takes hours and days to complete. |
| 4. | It is kinetic chain reaction. | It involves inter-molecular reaction. |

## 13.3 CLASSIFICATION OF PLASTICS

Most commonly, plastics, are classified into,

1. Thermoplastic, and
2. Thermosetting.

**Thermoplastic materials** are those which *soften on the application of heat, with or without pressure and require cooling to be set to a shape.*

They can be *heated and cooled any number of times,* only they *should not be heated above* their *decomposition temperatures.*

- They are main long chain straight or slightly branched molecules and the chains are held close to each other by secondary weak forces of type Vander Waal's forces. During heating, as the temperature increases the secondary forces are reduced and the sliding of these long chain molecules can easily occur one over the other at a reduced stress level.
- They are highly plastic and are easy for moulding or shaping.

- They have low melting temperatures and are not so strong as the thermosetting plastics.
- Since they can be repeatedly used, they have a resale value.
- Some commercial thermoplastics are : **Polythene, Polyvinyl chloride (PVC), Polystrene, Polytetrafluroethylene (PTEE) etc.**

**Thermosetting materials** are those plastics *which require heat and pressure to mould them into shape.*

- They cannot be resoftened once they have set and hardened.
- They are ideal for moulding into components which require rigidity, strength and some resistance to heat.
- In general, resins formed by condensation are thermosetting.
- Thermosetting resins have *three-dimensional molecular structure and have very high molecular weights.*
- Due to *cross-linking* thermosetting resins are hard, tough, non-swelling and brittle. Hence they cannot be softened or remoulded as in the case of thermoplastic resins. Moulding and casting are the processes often used with such materials.
- Some important commercial examples of this type are : **Phenolics, Polyesters, Expoxies, Silicones etc.**

The *difference between thermoplastic and thermosetting materials* may be explained *in terms of molecular structure.* The thermoplastics are essentially long chain macromolecules with a limited number of cross links. When heated and compressed, the chains glide over each other and fluid materials take the shape of any mould in which they are placed. The thermosetting plastics are characterised by strong cross links between the chains; once these are formed by heat and pressure, the plastics set to a rigid infusible solid.

**Difference between Thermosetting Thermoplastic materials**

| | *Thermosetting materials* | *Thermoplastic materials* |
|---|---|---|
| 1. | They have three dimensional network of primary covalent bonds with cross-linking between chains. | They are linear polymers without cross-linking and branching. |
| 2. | They are more stronger and harder than thermoplastic resins. | They are comparatively softer and less strong. |
| 3. | Once hardened and set they do not soften with the application of heat. | They can be repeatedly softened by heat and hardened by cooling. |
| 4. | Objects made by thermosetting resins can be used at comparatively higher temperature without damage. | Objects made by thermoplastic resins cannot be used at comparatively higher temperatures as they will tend to soften under heat. |
| 5. | They are usually supplied in a monomeric or partially polymerized form in which they are either liquid or partially thermoplastic solids. | They are usually supplied as granular materials. |
| 6. | It is difficult to fill an intricate mould with such plastics. | They can fill the complicated mould quite easily. |
| 7. | They cannot be recycled. | The scrap of these plastics can be recycled again and thus they are *economical.* |
| | **Uses:** *Telephone receivers, electric plugs radio & TV cabinets, camera bodies, automobile parts, circuit breaker switch panels etc.* | **Uses :** *Toys, combs, toilet goods, photographic films, insulating tapes, hoses, electric insulation, etc.* |

### 13.3.1. Thermoplastic Materials

Important thermoplastic materials are described here under :

**1. Polyethylene or polythene $(C_2H_4)_n$**

This material originated in England was produced commercially in the United States in 1943.

Polyethylenes are obtainable *as various liquids, gums and tough flexible solids suitable for moulding.*

**Properties :**

(*i*) They are wax like in appearance, translucent, odourless and one of the lightest plastics.
(*ii*) Flexible over a wide temperature range.
(*iii*) High resistivity and dielectric strength.
(*iv*) Chemically resistant.
(*v*) Do not absorb moisture.
(*vi*) Dielectric losses and dielectric constant are low.
(*vii*) They are relatively low in cost.

**Uses :**

(*i*) High voltage (upto 30 kV) applications.
(*ii*) Coaxial cables.
(*iii*) Packaging.
(*iv*) Moisture proofing.
(*v*) Coating ice-cube trays.
(*vi*) As insulation in submarine cables and radar lines.
(*vii*) Lining for lagoons to avoid seepage of polluted water into the underground.
(*viii*) Fan and blower casing (gas filled polythene)
(*ix*) Pipes and tanks for water storage.

Two types of polyethylene are manufactured depending upon the condition of polymerisation.

(*a*) High density polyethylene (HDPE)
(*b*) Low density polyethylene (LDPE)

**(*a*) High density polyethylene (HDPE) :**

- This is obtained by *low pressure polymerisation process.*
- It has specific gravity 0.96 and softening temperature 120-130°C.

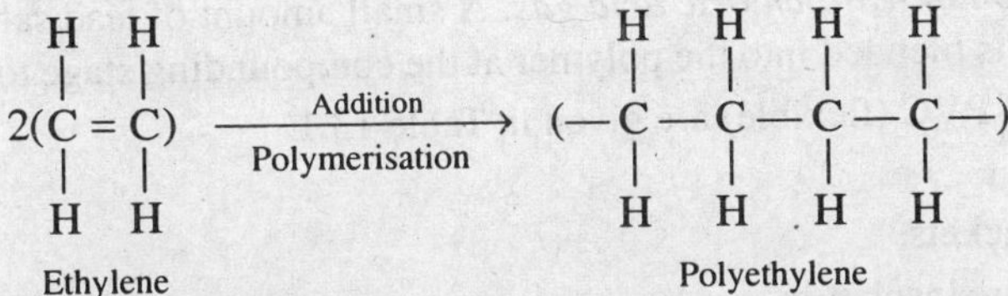


Typical properties of HDPE are given in Table 13.1.

**(*b*) Low density polyethylene (LDPE) :**

- This is obtained by high pressure process.
- It has specific gravity 0.90 and softening temperature 86°C.
- Polythene is prone to degradation and embrittlement by sunlight due to cross linking but performance can be improved by addition of black pigments such as carbon black, which absorbs rays and prevents damage.
- Underground polythene is not affected by salts in the soil and has excellent resistance to salts and bacteria. However, it is attacked by strong acids. It is a good electrical insulator.
- *Polythenes,* in particular *LDPE, have high permeability to gases; they are therefore, not considered for gas service.*

**2. Polyvinylchloride (PVC):**

- This is the cheapest and most widely used plastic.
- The vinylchlorides are formed from hydrochloric acid, limestone and natural gas or coal. The forms of vinylchloride are almost unlimited.
- It is manufactured by *addition polymerisation of vinylchloride.* Polymerisation is carried out in the presence of catalyst.
- It is different from polyethylene. In PVC one in four hydrogen atoms is replaced by a chlorine atom, hence it has *greater rigidity.* The reaction is indicated below :

$$_{n}\underset{\substack{|\\ Cl}}{CH_2 = CH} \xrightarrow[\text{Polymerisation}]{\text{Addition}} (\underset{\substack{|\\ Cl}}{—CH_2 — CH—})_n$$

- PVC can be manufactured in expanded or cellular form. It is available in two forms namely *flexible and rigid.* It can be easily extruded and moulded into the desired form. The joints are obtained by solvent welding.

**Properties :**

(*i*) The *flexible* types are strong, tear resistant, and have good ageing properties.

(*ii*) The *rigid* types have good dimensional stability, and are water resistant. They are resistant to acids and alkalies.

(*iii*) It is attacked by aromatics and soluble in ketones and esters.

(*iv*) Its resistance to impact invariably deteriorates with time.

(*v*) It becomes soft beyond 80°C.

(*vi*) It is self extinguishing when ignited and the source of flame removed.

(*vii*) The hard type of PVC is formulated with less plasticizer than the general purpose grade and shows less tendency to flow at high temperature which is an advantage when the cables are to be laid in very hot surroundings.

(*viii*) Although its electrical properties are not so good as those of rubber, it offers more resistance to oxygen, ozone and sunlight.

(*ix*) PVC has a *tendency to decompose when it is heated or exposed to sunlight with the liberation of hydrochloric acid gas.* A small amount of lead salt, lead silicate or lead stearite is blended into the polymer at the compounding stage to prevent the same.

The properties of PVC (flexible) are given in Table 13.1.

**Uses:**

(*i*) Cable jackets.

(*ii*) Lead-wire insulation.

(*iii*) Rubber substitute.

(*iv*) Fabric coating.

(*v*) Rain water goods.

(*vi*) Corrugated roofing.

(*vii*) Flooring and ceiling panels.

(*viii*) Gramophone records.

(*ix*) Cladding.

(*x*) Pipes and fittings for water service (unsuitable for hot water service because of low softening point).

Table 13.1. Properties of Thermoplastic Materials

| S. No. | *Properties* | *HDPE* | *Flexible PVC* | *Poly. propylene* | *Polystyrene* | *PTFE* | *PMMA* |
|---|---|---|---|---|---|---|---|
| 1. | *Density* ($kg/m^3$) | $0.95\times10^2$ | $11.6\times10^2$ | $9.0\times10^2$ | $10.4\times10^2$ | $21\times10^2$ | $11\times10^2$ |
| 2. | *Softening point* (°C) | 90-100° | = < 79° | 100-120° | 82-103° | 260° | 71-91°C |
| 3. | *Thermal conductivity* ($Wm^{-1} K^{-1}$) | 0.42-0.55 | 0.13-0.17 | 0.88 | 0.09-0.21 | 0.250 | 0. 17-0.25 |
| 4. | *Thermal expansion* ($K^{-1}$) | $120\times10^{-6}$ | $70\text{-}250 \times10^{-6}$ | $120\times10^{-6}$ | $60\text{-}80 \times10^{-6}$ | $100\times10^{-6}$ | $50\text{-}90 \times10^{-6}$ |
| 5. | *Specific heat capacity* ($Jkg^{-1} m^{-1}$) | 2100-2310 | 1260-2100 | 1932 | 1340-1466 | 1050 | 1470 |
| 6. | *Tensile strength* ($MNm^{-2}$) | 20-30 | 10-25 | 33-35 | 35-62 | 15-35 | 75 |
| 7. | *Compressive strength* ($MNm^{-2}$) | 20-25 | 7-12 | 35 | 90-110 | 10-15 | 80-130 |
| 8. | *Coefficient of friction* | 0.25 | 0.4 | — | — | 0.04 | — |
| 9. | *Young's modulus* ($MNm^{-2}$) | 550-1050 | 3500-4800 | 900-1400 | 2410-4130 | 3500-4620 | 2700-3500 |

**3. Polypropylene (PP):**

- It is manufactured by addition polymerisation of propylene.
- It is stronger and more rigid material than polyethylene owing to the presence of $CH_3$ group attached to the linear molecular chains.

$$2\left(\begin{array}{c}H\\|\\C\\|\\H\end{array} = \begin{array}{c}H\\|\\C\\|\\CH_3\end{array}\right) \xrightarrow[\text{Polymerisation}]{\text{Addition}} \begin{array}{c}H\\|\\C\\|\\H\end{array} - \begin{array}{c}H\\|\\C\\|\\CH_3\end{array} - \begin{array}{c}H\\|\\C\\|\\H\end{array} - \begin{array}{c}H\\|\\C\\|\\CH_3\end{array} -$$

Propylene Polypropylene

**Properties:**

(*i*) Its resistance to chemicals is better than that of polyethylene.

(*ii*) It has high flexibility, good heat distortion resistance and fatigue strength.

(*iii*) It is attacked by strong acids.

(*iv*) They are brittle at low temperature, translucent and flammable.

(*v*) They have poor resistance to ultra violet rays and weathering.

Other properties are given in Table 13.1.

**Uses:**

(*i*) Vacuum flasks

(*ii*) Hair dryers

(*iii*) Filaments and fibres

(*iv*) Refrigerator parts

(*v*) Electrical insulators

(*vi*) Films and sheets for packing

(*vii*) Flash light casings

(*viii*) Automative parts

(*ix*) Vacuum cleaner bodies (*x*) Washing machine parts
(*xi*) Pipes, tanks, spray nozzles (*xii*) Radio and television cabinets.

**4. Teflon or polytetrafluoroethylene (PTFE) :**

- The mer of PTFE is represented as $CF_2 = CF_2$.
- PTFE is manufactured by addition polymerisation technique. Polymerisation is carried out under organic peroxide or oxygen as catalyst.

$$_n CF_2 = CF_2 \xrightarrow[\text{Polymerisation}]{\text{Addition}} (-CF_2 - CF_2 -)_n$$

Tetrafluoroethylene Polytetrafluoroethylene

- It can be fabricated into desired shapes by hot processing and sintering the polymer powder.

**Properties :**

(*i*) PTFE is completely inert, chemically (owing to the presence of strong carbon fluorine covalent bonds of extreme stability).
(*ii*) It is highly crystalline material and does not soften at high temperature (It retains its strength upto + 300°C and –200°C). It loses its crystallinity on heating above 325°C.
(*iii*) It is non-flammable.
(*iv*) It is transluscent/opaque in nature.
(*v*) It is a good electrical insulator.
(*vi*) It has very low coefficient of friction and it does not stick to other materials.
(*vii*) It has excellent resistance to most chemicals and solvents.
(*viii*) PTFE with 30% glass fibre filling has a much *increased resistance to creep* (and is therefore, used for gaskets and valve seats where good creep resistance is required).

Other properties are given in Table 13.1.

**Uses :**

(*i*) Non-stick coatings (*ii*) Chemical pipes
(*iii*) Bearing bushes (*iv*) Piston rings
(*v*) Anti-corrosive seals (*vi*) High temperature electronic parts
(*vii*) Mouldings in aircrafts (*viii*) Seals
(*ix*) Pipes (*x*) Gaskets
(*xi*) Laboratory equipments (*xii*) Lining of chutes.

**5. Polystyrene :**

- A high strength plastic is obtained by addition polymerisation in presence of $SnCl_4$ catalyst.

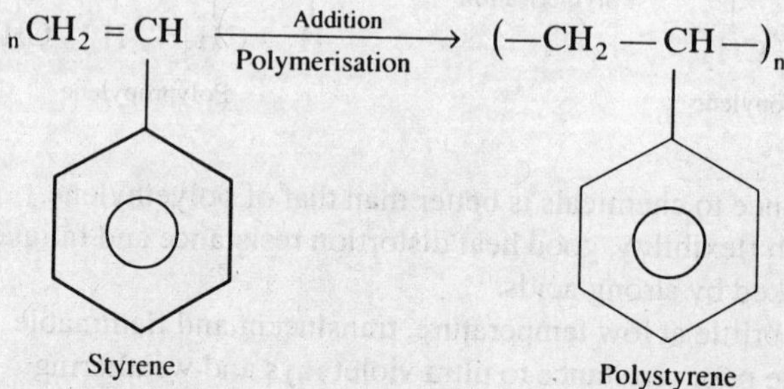


- It can be toughened by mixing another polymeric material acrylonitrile butadiene to give ABS plastics.

**Properties :**

(*i*) It has good electrical properties and negligible water absorption.
(*ii*) Hard and brittle.
(*iii*) Low dielectric constant.

(*iv*) High resistivity.
(*v*) It is transparent in nature.
(*vi*) Freedom from taste and colour and physiologically harmless nature makes polystyrene suitable for use in contact with food stuffs.
(*vii*) Possesses excellent resistance to corrosive chemicals such as acids (except oxidising agents), alkalies and salt solutions.

**Uses :**

(*i*) Lenses (*ii*) Refrigerator-cabinet compounds
(*iii*) Radar components (*iv*) Electrical insulation
(*v*) Instrument panels (*vi*) Food containers
(*vii*) Light fittings (*viii*) Automobile dash blanks
(*ix*) Office items like desk calendars, pen stands, trays etc.
(*x*) Housewares like water jugs, mugs, tumblers, plates, trays etc.
(*xi*) Sheets and films for packaging.

**6. Acrylics (Polymethyl methacrylate PMMA) :**

- They are the polymer of methyl methacrylate having trade name "***perspex***".
- PMMA is manufactured by addition polymerisation of methyl methacrylate in presence of a peroxide catalyst.

$$N\,CH_2 = \underset{\substack{|\\ COOCH}}{\overset{\substack{CH_2\\ |}}{C}} \xrightarrow[\text{Polymerisation}]{\text{Addition}} (-CH_2 - \underset{\substack{|\\ COOCH}}{\overset{\substack{CH\\ |}}{C}})_n$$

Methyl methacrylate Polymethyl methacrylate (PMMA)

**Properties :**

(*i*) It is hard, rigid and transparent; it possesses high impact strength.
(*ii*) It is attacked by acetone, petrol etc., but it is resistant to household chemicals.
(*iii*) It is highly resistant to weathering. However, it is attacked by concentrated acids, but not by alkalies.
(*iv*) It has low abrasion resistance and low softening temperature.
(*v*) It is a good electrical insulator.

**Uses :**

(*i*) Sanitaryware (*ii*) Sheeting
(*iii*) Sinks, baths (*iv*) Display signs
(*v*) Lenses (*vi*) Hospital equipment
(*vii*) Roof lights and domes (*viii*) Aircraft light fixtures
(*ix*) Light weight garments (*x*) Manufacture of fibres.

**7. ABS (Acrylonitric butadiene styrene) :**

**Properties :**

(*i*) ABS materials are tough, hard and rigid.
(*ii*) They possess hard surface, dimensional stability, chemical resistance and heat resistance.
(*iii*) They have low water absorption and resistance to creep or cold flow.
(*iv*) They are unaffected by water, inorganic salts, alkalies and many acids.
(*v*) They are soluble in ketones, aldehydes, esters and some chlorinated hydrocarbons.

**Uses :**

(*i*) Extruded sheets.
(*ii*) Extruded pipes.

(*iii*) Safety helmets.
(*iv*) Refrigerator lines.
(*v*) Protective and decorative housings for telephones, radios, business machines, portable appliances.
(*vi*) Electrical parts.

**8. Silicones :**

The basic ingredients consist of silica and coke which are mixed and placed in an electric furnace. The silica is reduced to silicon, which is one of the starting materials. The other reactant is methyl chloride, prepared by chlorinating methane gas. The two reactants are mixed in a reaction at a high temperature and moderate pressure, using copper as a catalyst. The desired products are then hydrolysed and condensed to form silicon polymers.

- They are a chemical hybrid, a cross between organic and inorganic materials. They have both the stability of inorganic products and the versatility of organic.
- As a moulding compound they have good qualities and the ability to withstand high heat (upto 260°C).
- They are mechanically weak and have poor chemical resistance.
- They are water repellent.

**Uses :** They are used as high heat resistant insulations.

**9. Polyvinylidene chloride:**

**Properties :**

(*i*) Acid resistant.
(*ii*) Adequately tough.
(*iii*) Non-flammable.
(*iv*) Solvent resistant.

**Uses :**

(*i*) Seat covers
(*ii*) Belts and gaskets
(*iii*) Woman fabrics
(*iv*) Acid-resistant tubing.

**10. Polyamides :**

- The general name of linear polyamides is *nylon.* It is produced by series of condensation reaction between an amine and organic acids. **Nylon** is *produced from the condensation polymerisation reaction between hexaethylene diamine and adipic acid.* In this process both the reactions, *i.e.*, hexamethylene diamine and adipic acid contain six carbon atoms, thus the nylon so produced is called 6.6. For all nylons similar nomenclature is used.
- **Nylon-6** is produced by self condensation caprolactam which is an amino acid with six carbon atoms. The condensation reaction takes place between the acid group at one side and amino group at other side of the same molecule.
- **Nylon-11** is produced by self condensation of 11 amino undecanove acid with a linear monomer containing eleven carbon atoms.
- **Nylon-10** is produced by condensation reaction between hexamethylene diamine and sebacic acid.

**Properties of nylon :**

(*i*) All nylons possess good abrasion resistance.
(*ii*) They are very tough and strong but flexible too.
(*iii*) They possess high impact strength.
(*iv*) They absorb water which causes reduction in strength and impact properties.
(*v*) They are resistant to most of the solvents and chemicals.
(*vi*) They have very high softening temperatures and thus moulding becomes difficult.

**Uses :**

(*i*) The main use of nylon is in the *manufacture of fibre and yarn.*

(*ii*) The nylons are also used for *moulding gears, valves, bearings, containers and electrical equipment.*

**11. Bitumen :**

- It is the oldest known plastic and was employed about 3000 B.C.
- The bitumen plastics of to-day are made from biuminious products by mixing with a suitable filter, such as slate dust or asbestos in powder form.
- They have good *electrical insulating properties.*

**Uses :** They have been used for battery-cell plugs, stoppers, electrical-mouldings of the cheaper class etc.

### 13.3.2. Thermosetting Materials

Here follows the description of important thermosetting materials.

**1. Phenol formaldehyde (PF) – "Bakelite":**

- PF is made by condensation polymerisation of phenol and formaldehyde in an acid or alkaline medium.
- Asbestos, glass fibres, pigments and other additives can be added to improve its properties.

**erties :**

(*i*) The colour range is limited to black or brown (however, it discolours on ageing by sunlight).

(*ii*) It is strong, rigid and dimensionally stable.

(*iii*) It is resistant to heat, most chemicals and solvents.

**Uses :**

(*i*) Lavatory seats
(*ii*) Thermal insulation as cellular/foamed
(*iii*) Paints and adhesives
(*iv*) Vacuum cleaner parts
(*v*) Electrical parts
(*vi*) Impregnants for paper and for mica
(*vii*) Handles, knobs for domestic appliances.

**2. Amine formaldehyde (Urea and melamine formaldehyde) :**

- These plastics are obtained by *condensation of urea or melamine with formaldehyde.*
- These are highly *cross-link polymers.*
- They can be compounded with fillers, pigments and other additives to form moulding materials of different colours.

**Properties :** Their general properties are similar to formaldehyde.

(*i*) M*elamine* has *more resistance to chemicals*, heat and moisture. However, these plastics are attacked by strong acids.

(*ii*) It is scratch free and more expensive.

(*iii*) It has better electrical properties.

(*iv*) It is slightly affected by the sunlight.

Other properties are given in Table 13.2.

**Uses :**

(*i*) Glues for plywood
(*ii*) Cellulose and foamed products
(*iii*) Paints
(*iv*) Surface coating
(*v*) Adhesive
(*vi*) Plugs, switches, buttons etc.

**Table 13.2. Properties of Thermosetting materials**

| S. No. | *Properties* | *PF* | *UF* | *MF* | *Polyesters* | *Epoxies* | *Polyurethanes* |
|---|---|---|---|---|---|---|---|
| 1. | *Density* (kg/m$^3$) | $12.5\times10^2$ | $15\times10^2$ | $15\times10^2$ | $11\times10^2$ | $12\times10^2$ | $40.8\times10^2$ |
| 2. | *Heat distortion temperature* °C | 112-125 | 75 | 100 | — | upto 60 | — |
| 3. | *Thermal conductivity* (Wm$^{-1}$ K$^{-1}$) | 0.125-0.25 | 0.29-0.42 | 0.29-0.42 | 0.17-0.19 | 0.17-0.21 | 0.035 |
| 4. | *Thermal expansion* (K$^{-1}$) | 25-60 $\times10^{-6}$ | 35-45 $\times10^{-6}$ | 35-45 $\times10^{-6}$ | 100-50 $\times10^{-6}$ | 50-90 $\times10^{-6}$ | 50-70 $\times10^{-6}$ |
| 5. | *Specific heat capacity* (Jkg$^{-1}$ K$^{-1}$) | 1550-1760 | 1670 | 1670 | 1260 | 1250-1670 | — |
| 6. | *Tensile strength* (MNm$^{-2}$) | 35-55 | 48-76 | 48-88 | 31-70 | 70-280 | — |
| 7. | *Compressive strength* (MNm$^{-2}$) | 70-210 | 70-110 | 70-110 | 90-240 | — | 0.02-0.08 |
| 8. | *Young's modulus* (MNm$^{-2}$) | 5200-7000 | 7000-13500 | 7000-10500 | 2800-7000 | — | — |

**3. Polysters (unsaturated):**

- These plastics are manufactured by *condensation polymerisation of dicarboxylic acid* (mallic acid) and *dihydric alcohol* (*e.g.*, glycol) followed by curing with cross-linking agent (styrene).
- A wide variety of products can be made by varying monomers and curing agents.

**Properties:**

(*i*) They have good resistance to heat and most chemicals except strong acids and alkalies.

(*ii*) They are affected by sunlight unless stabilized.

Other properties are given in Table 13.2.

**Uses:**

(*i*) Safety helmets (*ii*) Automobile body components

(*iii*) Cladding panels (*iv*) Paints

(*v*) Binder for glass fibres (*vi*) Fibre glass boats

(*vii*) Jointing and repair work.

**4. Epoxy resins (epoxies) :**

- These are obtained by condensation polymerisation of epichlorohydrin and polyhydroxy compound (*e.g.*, bisphenol).
- As adhesive these materials have shown extremely high bond strengths without the need for pressure curing.
- They are transparent, light amber colour and have very little shrinkage.
- As coating materials, they have shown superior toughness, elasticity and chemical resistance.
- To increase hardness and strength glass fibres or carbon fibres are compounded with expoxies. (Other properties are given in Table 13.2).

- While handling epoxies, following precautions are necessary (because styrene can cause irritation and peroxide catalyst can cause skin irritation).
  — No use of inflammable solvents for cleaning;
  — No smoking, no naked flame;
  — Proper ventilation of room;
  — Use of protective equipments like goggles, gloves etc.

**Uses:**

(*i*) They are used as an insulating material in cable-end boxes, cable joint boxes, instrument transformers etc.
(*ii*) Epoxies are used as/for adhesive (Araldite) and jointing and repair work.
(*iii*) They have also found considerable use as casting materials.

## 13.4 PROPERTIES OF PLASTICS

The plastics have a variety of significant properties. The properties common to most plastics are given below:

1. Light weight.
2. Low thermal conductivity.
3. A wide range of colours.
4. Resistance to deterioration by moisture.
5. Low electrical conductivity.
6. Fairly good resistance offered to attack by inorganic acids, bases and salts.
7. Plastics in many cases are non-inflammable, self-extinguishing, and burn very slowly.
8. They are transparent, transluscent or opaque.
9. Good dimensional stability.
10. Good sound absorption properties.
11. Bad conductor of heat.
12. Good tensile strength.
13. Can be easily fixed in position.
14. Good resistance to peeling.
15. Lack ductility.
16. Quite durable.
17. Easy maintenance of plastic surfaces (do not require any protective coat of paints).

## 13.5. SHORTCOMINGS OF PLASTICS

Following are the *shortcoming/limitations* of plastics:

1. Low heat resistance.
2. Not very hard.
3. Disintegrate gradually with the passage of time.
4. Exhibit high creep.
5. High coefficient of thermal expansion.

## 13.6. USES OF PLASTICS

The typical uses of plastics in *buildings* are:

1. Electrical conduits
2. Electrical insulators
3. Floor tiles
4. Jointless flooring
5. Decorative laminates and mouldings
6. Bath and sink units
7. Foams for thermal insulation
8. Overhead water tanks

9. Roof lights
10. Wall tiles
11. Corrugated and plain sheets
12. Cistern ball floats
13. Lighting fixtures
14. Paints and varnishes.
15. Water resistance adhesives etc.
16. Pipes to carry cold water etc.

## 13.7. TRADE NAMES AND TYPICAL APPLICATIONS OF SOME IMPORTANT PLASTICS

**A. Thermoplastics**

| S.No. | Material Type | Trade Name | Typical Applications |
|---|---|---|---|
| 1. | Acrylics | Lucite, Plexiglas | — Outdoor signs<br>— Lenses<br>— Transparent aircraft enclosures. |
| 2. | Nylons | Zytel, Plaskon | — Handles<br>— Bearings<br>— Covers<br>— Gears |
| 3. | Polyethylene | Alathon, Petrothene, Hi-fax | — Tumblers<br>— Ice trays<br>— Toys<br>— Flexible bottles |
| 4. | Polystrene | Styron, Lustrex, Rexolite | — Appliance housings<br>— Battery cases<br>— Wall tiles |
| 5. | Polyster (PET) | Mylar, Colanar, Dacron | — Automatic tyre cores<br>— Magnetic recording tapes<br>— Clothing |
| 6. | Vinyls | PVC, Tygon, Saran | — Phonograph records<br>— Floor coverings<br>— Garden house. |
| **B. Thermosetting** | | | |
| 7. | Phenolics | Bakelite, Resinox, Durez | — Auto distributors<br>— Telephones<br>— Motor housings |
| 8. | Silicones | DC resins | — Strips<br>— Laminates<br>— High temperature insulation |
| 9 | Epoxies | Araldite, Epon | — Adhesives<br>— Electrical mouldings<br>— Sinks |
| 10. | Polysters | Laminac, Selectron, paraplex | — Fans<br>— Chairs<br>— Helmets. |

## 13.8 COMMERCIAL FORMS OF PLASTICS

The important commercial forms of plastics used for civil engineering purposes are described below:

### 13.8.1. Plastic Sheets

**(*i*) PVC (Polyvinyl chloride) sheets:**

- PVC and its compounding ingredients are heated, masticated and then rolled into thin sheets between heated rollers. The sheets are then annealed and after cooling, they are rolled.
- PVC sheets (generally kept 1 mm for economic purposes) can resist the effects of grease, solvent etc.
- These should be laid only on dry surfaces, as they expand on being wetted.

**(*ii*) Fabric-backed sheets:**

- These are PVC sheets in which hession or felt is used for backing.
- These sheets are suitable for foot-traffic in dry conditions.

**(*iii*) Polythene sheets:**

- They are rough and water-resisting; available in wide rolls and in varying thicknesses.
- They are used for sheltering materials from rain on building site and to assist curing of concrete; also used as temporary covering for window and door openings etc.

**(*iv*) Corrugated plastic sheets :**

- They are light in weight and fire-resistant.
- Employed for roofing purposes in public buildings and factories.

### 13.8.2. Plastic Tiles

**(*i*) PVC tiles:**

- These tiles are used for flooring etc.

**(*ii*) Fabric backed PVC tiles:**

- These tiles are used for paving floors in dry conditions.

**(*iii*) Thermoplastic vinyl tiles:**

- These tiles are manufactured from a mixture of plasticised PVC, asbestos, limestone and pigments.
- They are suitable for heavy foot traffic in dry as well as in wet conditions.
- They can be used both on wooden and concrete sub-floors.

**(*iv*) Polystyene wall tiles:**

- They are suitable for lining of walls of high class buildings.

### 13.8.3. Plastic Laminates

These are formed by *impregnating sheets of fibrous materials such as paper, linen, canvas or silk with a synthetic resin and then compressing the sheets together with application of heat*. The synthetic resin may be phenolic resin, urea formaldehyde or a vinyl resin. The resin is usually dissolved in alcohol.

The material in roll form is immersed in the resin solution at atmospheric pressure and at room temperature and then run through a drier at 150°C. The rolls are next cut into sheets of given size, which are arranged into stacks. These stacks finally are compressed in a hydraulic press at about 170°C under pressure of 20 MN/m$^2$. The sheets are thus bonded to one another.

Some of the important *types of plastic laminates* are:

(*i*) Laminate bakelite
(*ii*) Melamine-surfaced phenolic paper laminates
(*iii*) Polyster paper laminate
(*iv*) Corrugated phenolic-resin bonded paper laminates.

**Properties:**

The laminated plastics have the following properties:

(*i*) Light and strong.
(*ii*) Machineable.
(*iii*) Resistant to wear, acids and alkalis.
(*iv*) Impervious to water and oil.
(*v*) Have a high dielectric constant.

**Uses:**

The laminated plastics are used for:

(*i*) Electric insulation.
(*ii*) Making silent gears.
(*iii*) Water lubricated bearings.
(*iv*) Pulley wheels.
(*v*) Pump parts.
(*vi*) Press tools.
(*vii*) Decorative purposes in wall panelling, translucent panelling table and counter parts.

**13.8.4. Plastic Panels**

- These are the sheets (3 m × 1.5 m × 25 mm) made from special craft paper impregnated with a phenolic resin.
- They are chiefly used for making light weight partitions.

**13.8.5. Plastic Pipes**

**(*i*) PVC pipes:**

- They have good ductility.
- They do not burst due to freezing of water.
- Used for distribution of water, mine drainage, gas distribution etc.
- Used at the places where the temperatures are moderate.

**(*ii*) Polythene plastic pipes:**

- They have inherent flexibility and excellent resistance to breakage from mechanical shock.
- Suitable for water supplies, and for use in chemical and food industries.

**13.9. MANUFACTURING PROCESSES**

**13.9.1. Casting**

- Moulds of different materials such as lead-antimony, sheet metal, clay or plaster of paris etc. are used for casting the plastic into the required shape without the application of any pressure.
- Both thermosetting and thermoplastics can be used in casting.
- Araldite resin and hardener liquids can be used for casting.

**13.9.2. Moulding**

Moulding is one of the most common methods used in the manufacture of plastic goods. Following are the different ways in which the method can be used:

**(*i*) Shaping:**

In this process of moulding, the hot plastic mass is spread over the shape and then allowed to cool and harden.

**(*ii*) Compression moulding:**

- The mould is filled with the required amount of moulding powder and then *heat and pressure are applied until cure is complete.*
- This process is employed mainly for *thermosetting resins.*
- The process can be speeded up by the use of *transfer moulds* in which the moulding powder is heated to the plastic stage and then forced into mould proper when the curing is completed.

This has the *advantage that only the exact amount of resin is fed to the mould and therefore no flash is formed*, thus eliminating the necessity of trimming the article. Also as the contents of the mould are in plastic condition when pressure is applied there is *no danger of uneven distribution of pressure and development of strains.*

**(*iii*) Injection moulding:**

- The pellets of thermoplastic are first compressed in the pressure chamber and then pushed into a heating chamber.
- The softened material in flowing state is then forced with pressure (about 960 N/mm$^2$) through a nozzle into a cold mould having cavity of the desired shape. Plastic articles of intricate shapes can be formed in these cavity moulds. The articles can be removed by opening the moulds (Fig. 13.2).

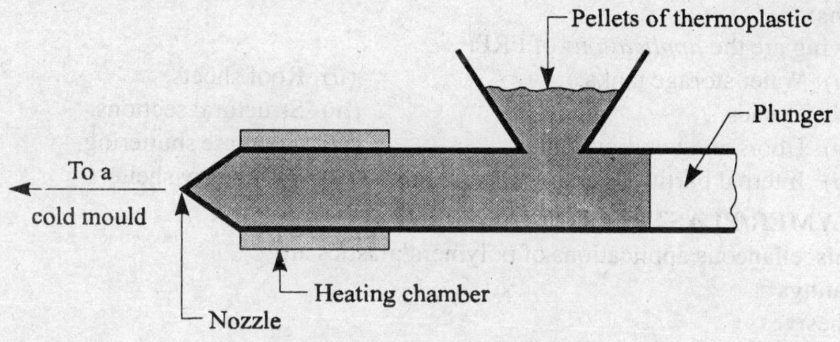


Fig. 13.2. Injection moulding.

**(*iv*) Extension moulding:**

The plastic mass is forced through a die, thus producing continuous lengths of tube or rod of the desired cross-section which can ultimately be cut upon machines as required.

**(*v*) Cold moulding:**

In this method, the articles are moulded in cold condition and are then baked in an oven or cured by steam.

### 13.9.3. Laminating

It is the process in which under the action of heat and pressure, layers of plastics or layers of other materials such as wood, fabric, metal etc. can be made into a composite, sand-witching a layer of plastics.

### 13.9.4. Blowing

- It is a process similar to glass blowing where various articles of plastics are made by blowing viscous mass of plastic material.
- It is used for fast and cheap production.
- This can be done manually or by automatic and semi-automatic machines. In automatic machines the temperature and blowing of plastics can be controlled for better quality products.

### 13.9.5. Calendering

It is the production of sheet of materials by rolling the plastics between multiple rollers.

### 13.9.6. Machining and Cementing

- Plastic sheets, rods and tubes can be machined with ordinary lathe machines to obtain the desired shapes and sizes. These can also be cemented or riveted with suitable glues and riveting materials. Thermoplastics can also be welded by using certain solvents.

## 13.10. FIBRE GLASS REINFORCED PLASTIC

- The fibre glass reinforced plastic (or FRP) is formed by using two materials in conjuction with each other to form a composite material of altogether different properties.

- In FRP, the glass *fibres provide stiffness and strength while resin provides a matrix to transfer load to the fibres.*

**Properties of FRP:**

Following are the properties which have made the FRP the most commercially successful composite material of construction:

(*i*) Aesthetic appeal
(*ii*) Dimensional stability.
(*iii*) Light weight.
(*iv*) Easy to repair.
(*v*) Durable.
(*vi*) Corrosion resistance.
(*vii*) Requires less energy for production.
(*viii*) Least maintenance required.
(*ix*) The tooling is inexpensive and fast.
(*x*) FRP products transmit a great deal of light.

**Applications:**

Following are the *applications* of FRP:

(*i*) Water storage tanks.
(*ii*) Roof sheets.
(*iii*) Domes.
(*iv*) Structural sections.
(*v*) Doors and window frames.
(*vi*) Concrete shuttering.
(*xii*) Internal partitions and wall panelling.
(*xii*) Temporary shelters

## 13.11. POLYMER/PLASTIC COATINGS

The miscellaneous applications of polymers/plastics are:

1. Coatings
2. Adhesives
3. Films
4. Foams

We shall discuss here only '*Coatings*'.

The surface of materials are applied coatings to serve one or more of the following *purposes*:

(*i*) To protect the object/item from corrosive or deteriorative reactions that the environment may produce.
(*ii*) To improve the appearance of the object/item.
(*iii*) To provide electrical insulation.

In coating materials, many of the ingredients are polymers and majority of which are of organic origin.

The organic coatings may be classified as:

(*i*) Paints
(*ii*) Varnishes
(*iii*) Enamels
(*iv*) Lacquers.

**Paints:**

Paint is a mechanical mixture of "*Vehicle*" and a "*pigment*". The vehicle binds together the pigment particles in the paint. It consists of *drying* oil and a *solvent thinner*. The "*pigments*" provide *both colour and binding power.*

The thin film of paint when applied over the surfaces gets dried up due to oxidation, polymerisation or evaporation of portions of its components and gives a film which has a considerable binding power.

**Varnishes:**

Varnishes are colloidal dispersions or solution of synthetic or natural resins in oil or in thinner or in oil and thinner both and *contain no pigments*. When varnish is applied on a material surface, the thinner evaporates and the oil resin mixture partially polymerises to form a clear dry film.

**Enamel:**

Enamel is a *mixture of both paint and a varnish*. The enamel dries at room temperature through oxidation and polymerisation or at elevated temperature where the film gets baked.

**Lacquer:**

Lacquer is a *finishing material* that dries quickly by evaporation of solvents and forms a film from its non-volatile constituents. It is solution or dispersion of cellulose nitrate or cellulose derivatives, resins and plasticizers in solvents and thinners.

The other organic coatings are:

(*i*) *Dispersion coatings* (finally divided insoluble resins dispersed in organic media);

(*ii*) *Emulsion coatings* (synthetic resins latex dispersed in water by means of dispersing agent); and

(*iii*) *Latex coatings.*

The organic protective coatings should have the following *characteristics*:

1. High hiding power.
2. Resist corrosion when applied on metals.
3. Film produced should be washable.
4. Should be able to resist the atmospheric conditions to which these are exposed.
5. Should be able to give the necessary gloss whenever required.

Organic coatings can be *easily applied* and are *flexible* and *cheap*. They can be applied by brush, spray or roller coaters. They can also be applied by hot spraying, by dipping articles in resin powders.

## 13.12. DEFORMATION OF PLASTICS

- Polymers undergo both elastic and plastic deformation just as metals do. As a class polymers have lower elastic moduli than metals. Also, as a class, plastics undergo permanent deformation more readily than the average metal.
- The deformation of plastics is affected by the *structure of the polymer*. Permanent deformation occurs as slip between adjacent molecules, because the *attractive forces are smaller between them within the molecules.*
- "Thermoplastic" resins (plastics) *deform easily under pressure and at high temperature, since the weak van der Waals' forces are easily overcome*. Consequently such a plastic when heated and injected into a mould under pressure takes on the shape of that mould easily. On cooling the plastic becomes rigid again. In "*thermosetting*" *resins (plastics) there is a continuous framework structure (developed by polymerisation) and slip cannot take place between the molecules.* The whole of the structure is one large molecule, since primary covalent bonds are present throughout the structure. These plastics do not increase in plasticity with increased temperature and take on a permanent set. They are widely used in processes requiring high temperatures. They are used for electrical outlets, telephone receivers and appliance handles whose processing makes use of the electrical or chemical insulating properties of the covalently bonded organic materials. Thermoplastics, on the other hand, have better formability and are economical because scrap resin can be recycled frequently.

## 13.13. TESTING OF PLASTICS

Plastics are tested for electrical and mechanical properties. The various characteristics measured are defined below:

**1. Insulation resistance after immersion in water :**

- A specimen immersed in distilled water for 24 hours is cleaned and dried.
- The specimen is then placed between pairs of electrodes and its resistance is measured at a potential difference of 500 V D.C., the temperature and humidity being under control.

**2. Dielectric strength :**

- This property measures the electrical breakdown through the material.
- Dielectric strength is measured as *voltage gradient as volts/mm to cause electrical breakdown.*

**3. Vieat softening point :**

It is the temperature at which a penetrator of circular section of 1 mm diameter will penetrate a specimen of thermoplastic to a depth of 1 mm under specified load and uniform rate of temperature rise.

**4. Melt flow index :**

- Plastic is extruded through a die by a loaded piston under controlled conditions of temperature. Number of grams extruded per 10 minutes is determined as melt flow index.

**5. Cross breaking strength :**

Refer Fig. 13.3.

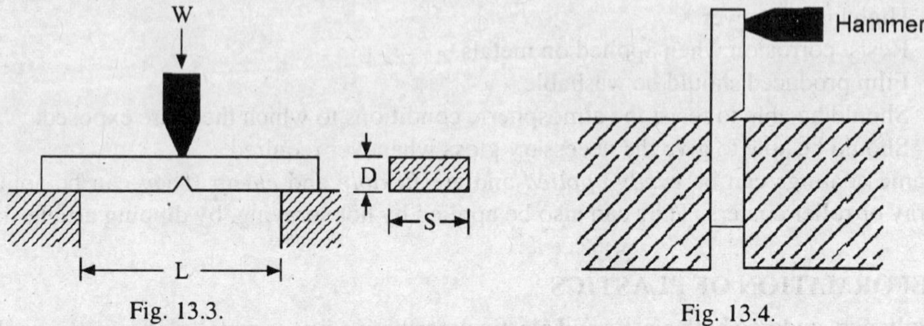


Fig. 13.3. Fig. 13.4.

- A plastic specimen of rectangular section ($B$ = width, $D$ = thickness) is supported like a simply supported beam over a span $L$ and a load ($W$) in the centre of the span applied to cause fracture.
- The cross breaking strength is measured as $\frac{1.5WL}{BD^2}$.

**6. Impact strength:**

Refer Fig. 13.4.

- Izod impact test in which a matched specimen of specified dimension $\left(\frac{1}{8}\text{" } thick \times \frac{1}{2}\text{" } wide \times 2\frac{1}{2}\text{" } long\right)$ held like a cantilever is struck by a swinging hammer at free end.
- The energy absorbed in fracture is the measure of impact strength.
- The impact strength is measured in kgf-cm or J/m. This figure is calculated on the basis of 1 m match length.

## 13.14. MACHINING OF PLASTICS

Two types of plastics are being used for fabrication. These are plane plastics in cast conditions or laminated plastics. Plastics present such problems as heating and edge fretting. During machining material is removed in fine particles and in some case gases may also evolve in chips and gases are harmful and hence must be removed immediately. Plastics may be subjected to almost all machining operations with the difference in cutting speed, feed, tool parameters etc. Brief suggestions are given below :

**1. Filing :** Both plastics can be filed and surface finishes of Rough (∇) and Fine (∇∇) types are obtainable.

- For rough filing use rasp or bastard file.
- For fine filing use smooth or dead smooth file.

**2. Sawing :**

Both plastics can be cut by a circular saw using cutting speed of 50 m/min and hand feed. Pitch of the saw should be close to 10 mm. However, the thickness of saw will vary with thickness of plastic to be cut.

- For thickness 0.5 to 4 mm use *3 mm thick saw.*
- For thickness of 5 to 8 mm use *4 mm thick saw.*
- For thickness 19 mm and above use *5 mm thick saw.*

**3. Drilling :**

Drilling a plastic may pose such problems as damaged hole periphery on surfaces where drill enters and exists and also stricking of drill in the hole.

By using sharp cutting edges and by carefully setting them on the surface before starting the operation, by putting a wooden support directly below the hole where drill will exist, and by frequently lifting the drill from hole the above problems are overcome. The drill may also be dipped in mineral oil for obtaining a smooth surface finish of hole.

- For plain plastic use : Point angle – 90° to 116°.
- However, if section is thin this angle may be 50°.
  : Flute angle - 60°
  : Cutting speed - 70 m/min
  : Feed - 0.2 to 0.3 mm/rev.
- For laminated plastics : Use cutting speed 60 to 90 m/min and feed 0.2 to 0.5 mm/rev.
- However, drilling may be done parallel or perpendicular to laminated plastics. If parallel use point angle of 70° and rake angle of 0-20°. If perpendicular use point angle 100-110°, rake angle 10 to 12°.

**4. Planing :** For both plain and laminated plastics use well sharpened and lapped carbide tipped tool at cutting speed of 10 to 20 m/min and feed of 0.2 to 0.8 mm/stroke.

**5. Turning :**

Carefully the tool must be set at the centre of workpiece, and the tool tip must be rounded to a radius of 1.5 to 2 mm.

- For rough surface (∇)

| | **Plain plastic** | **Laminated plastic** |
|---|---|---|
| Cutting speed | 80 to 200 m/min | 100 to 200 m/min |
| Feed | 0.3 to 0.5 mm/rev | 0.2 to 0.5 mm/rev |
| Depth of cut | 3 to 5 mm | 3 to 5 mm |

- For fine surface (∇∇)

| | | |
|---|---|---|
| Cutting speed | 200 to 300 m/min | 200 to 300 m/min |
| Feed | 0.1 to 0.3 mm/rev | 0.1 to 0.3 mm/rev |
| Depth of cut | 2 mm | 2 mm |

For both materials use clearance angle of 3° and rake angle of 15°. Chips should be removed by suction.

**6. Milling :**

Number of teeth on milling cutter must be $\frac{1}{2}$ to $\frac{2}{3}$ of the number of a corresponding steel job. Clearance angle of 20° and rake angle of 20 to 25° is recommended on carbide tipped tool. While feed for both types of plastics as 0.5 to 0.8 mm/rev is recommended the plain plastic should be cut with a cutting speed of 300 m/min and laminated plastic with a cutting speed of 120 to 250 m/min.

**7. Grinding :**

Grinding of plastics is possible with belt, sand paper and grinding wheels. For rough grinding grit 20 wheels and for finish grinding grit 60 silicon carbide wheels are used. For belt grinding $V = 6$ m/min, sand paper $V = 25$ m/min,

For surface grinding water may be used as coolant.

**8. Threading :**

Tapping, turning or milling can be used. In both turning and milling use HSS cutter with cutting speed 20 to 40 m/min.

Milling will result in better surface.

When using taps use wax and grease to lubricate with taps of wide flute.

Laminated plastics can be sheared upto thickness of 3 mm. However, if thickness greater than 3 mm is to be sheared the material is recommended to be warmed.

## 13.15. ADVANCED POLYMERIC MATERIALS

Some of the advanced polymeric materials are :

1. Ultrahigh molecular weight polyethylene (UHMWPE)
2. Liquid crystal polymers (LCPs).

**1. Ultrahigh molecular weight polyethylene :**

It is a linear polyethylene and its molecular weight is extremely high (nearly ten times greater than that of high density polyethylene).

**Properties :**

(*i*) It is electrically insulating and possesses excellent dielectric properties.
(*ii*) It possesses very good chemical resistance.
(*iii*) It offers outstanding resistance to wear and abrasion.
(*iv*) Its low-temperature properties are excellent.
(*v*) Its impact resistance is extremely high.
(*vi*) It provides a self lubricating and non-stick surface.
(*vii*) It offers a very low coefficient of friction.
(*viii*) It assimilates outstanding damping and energy absorption characteristics.
(*ix*) Due to its relatively low melting temperature, its mechanical properties diminish rapidly with increasing temperature.

**Uses :**

(*i*) Blood filters
(*ii*) Marking pen nibs
(*iii*) Pump impellers
(*iv*) Valve gaskets
(*v*) Bullet proof vests
(*vi*) Golf ball cores
(*vii*) Ice skating rink surfaces.

**2. Liquid crystal polymers :**

- These polymers belong to a group of chemically complex and structurally distinct materials and possess unique properties.
- They are composed of extended, rod shaped, and rigid molecules.
- In the melt; unlike other polymers (which are randomly oriented) LCP molecules become *alighed in highly ordered configurations.*
- Some of the liquid crystal polymers are rigid solids at room temperatures and exhibit the following properties/behaviour :

(*i*) High impact strengths, which are retained upon cooling to relatively low temperatures.
(*ii*) Excellent thermal stability.
(*iii*) Stiff and strong (tensile strength : 120 to 250 MN/m$^2$; tensile strength : 10 to 24 GN/m$^2$).

(*iv*) Chemically inert to wide variety of acids, solvents etc.
(*v*) Inherent flame resistance.
(*vi*) Combustion products are relatively nontoxic.

**Uses :**

1. These polymers are mainly used in liquid crystal displays (LCDs), on digital watches, laptop computers and other digital displays.
2. Interconnect devices, relays and capacitor housings etc.
3. Photocopiers and fibroptic components.

## 13.16. BEHAVIOUR OF POLYMERS

### 13.16.1. Thermal Behaviour

- The point of difference between thermoplastic (linear) and thermosetting (framework) polymers is that *thermoplastic polymers actually melt and lose all semblance of crystallinity at high temperatures*. Furthermore, thermosetting plastics, after completion of polymerisation, can eventually lose strength if they are exposed to high temperatures, due to degradation that takes place.
- The *major effect of increased temperature is an increased rate of chemical reactions*. The *temperature* at which polymers become *highly susceptible* to chemical reaction corresponds closely to the temperature at which *mechanical strength drops off.* This is *below the melting temperature in linear polymers*, and corresponds to the *start of degradation in framework polymers.*

### 13.16.2. Mechanical Behaviour

A polymer in molten state is amorphous and possesses random chain orientation. This structure of the molecules at higher temperatures, can be prevented at lower temperatures through a quench. Tension on such a mass produces most of the initial deformation and improves the alignment of the molecules. As a result, the stress-strain relationships are *not* like those of metals, because the *modulus of elasticity is increased when the stress is applied directly against the polymer chain after alignment has occurred.*

**Rate of deformation :**

- When the metals are exposed to stresses below their yield strength for extended periods of time, they may undergo creep. Atoms are moved locally at points of stress concentration. Whether only elastic deformation occurs, or whether both elastic and plastic deformation develop, depend on the factors namely *time and temperature.* More time or higher temperatures provide more opportunity for the atoms to establish new positions under applied stresses.
- *Polymers* are subject to the same, time-dependent phenomenon as metals are, and although molecular movements are more complicated than atom movements, simply because of size, the bonding forces are generally weaker so that high creep rates are developed.
- *Instantaneous stressing reveals only elastic responses, while prolonged exposure to even weak stresses allows plastic deformation to occur.*

**Stress relaxation:**

- Stresses are relaxed, with time, in those applications where they are initially developed from elastic elongation. The time required for adjustment of stresses is called the ***"relaxation time"***.
- As the relaxation of stresses is a continuous phenomenon, the *relaxation time* is defined mathematically as the *time it takes for the stress to be reduced by* $\frac{1}{e}$ *of its original value.*

**Anelasticity :**

- It has been observed that in most engineering materials a *time-dependent* elastic strain component exists (that is, elastic deformation will continue after the stress application and upon load release, some finite time is required for complete recovery). This *time-dependent* elastic behaviour is known as **anelasticity.**
- Anelasticity may arise because of the following :
  (*i*) Diffusion of thermal energy of interstitial atoms or substitutional solute atoms;
  (*ii*) Grain boundary effects, dislocations and thermal currents within the crystals.
  This effect is also a result of retarded elasticity and is called the **"elastic aftereffect".**
- The anelastic component for metals is normally small and often neglected. However, for some polymeric materials its magnitude is significant, in this case it is termed *viscoelastic behaviour.*

**Viscoelasticity :**

- *Viscoelasticity* is a third kind of deformation (other two being *elastic deformation and plastic deformation*), which is found in many polymers, such as plastics. Here, *after an applied stress that is within the material's elastic range is removed, the material will not completely recover to its original size.*
- An *amorphous polymer* may behave like a *glass* at low temperatures, a *rubbery solid* at intermediate temperatures, and a *viscous liquid* as the temperature if further raised. At *intermediate* temperatures, the condition is termed *viscoelasticity*.
- The *viscoelastic behaviour of polymeric materials is dependent on both time and temperature.* This behaviour can be measured and quantified by several experimental techniques. **Stress relaxation** measurements represent one possibility. A specimen, in these tests, is *initially strained rapidly in tension to a predetermined and relatively low strain level*. The *stress* necessary to maintain this strain *is measured as a function of time,* while *temperature* is held *constant. Stress is found to decrease with time due to molecular relaxation processes that take place within the polymer.*
  — **Relaxation modulus**, a *time-dependent elastic modulus for viscoelastic polymers*, is defined as :

$$E_r(t) = \frac{\sigma(t)}{\epsilon_0}$$

  where, $E_r(t)$ = Relaxation modulus,
  $\sigma(t)$ = Time-dependent stress, and
  $\epsilon_0$ = Strain level (maintained constant).

  — The *magnitude of the relaxation modulus is a function of temperature*; and to more fully characterise the viscoelastic behaviour of a polymer, isothermal stress relaxation measurements must be conducted over a range of temperatures.
- It has been observed that most of the polymers exhibit both an *elastic* as well as a *viscous* response to the external forces. Thus, the models for such viscoelastic behaviour comprise of viscous as well as elastic elements. The *spring illustrates purely elastic behaviour and the dashpot purely viscous behaviour.* If the spring is *Hookean* and the dashpot is *Newtonian*, the connections of these two elements in *series* and in *parallel* give two models namely **Maxwell** model and **Kelvin** or **Voigt** model respectively.
- Stress relaxation occurs with time if the stresses are developed by elastic deformation. The time required for the adjustment of stresses is called the ***relaxation time***. Relaxation time is taken as the *time at which the stress gets reduced to 1/e* (*i.e.*, *1/2.718*) *of its original value*. Under conditions of constant strain,

$$\frac{d\sigma}{dt} = -\frac{\sigma}{\lambda}$$

or

$$\sigma = \sigma_0 \, e^{-t/\lambda}$$

where $\sigma_0$ = Original stress,
$\sigma$ = Stress at time $t$, and
$\lambda$ = Relaxation time.

## 13.17. FIBRES

The fibre is a *filament or thread like piece of any material.* This term sometimes also refer to a raw material that can be drawn into threads.

Fibres may be of the following types:

1. Mineral fibres.
2. Animal fibres.
3. Vegetable fibres.

**1. Mineral fibres:**

*Mineral fibres* include *asbestos, glass fibre, slag wool and metal wool.* Asbestos due to its fibrous structure combined with inflammability is peculiarly suited for thermal and electrical insulation. Glass fibres and slag wool are employed for thermal insulation. The metal wool is used for fibres and cleaning applications.

**2. Animal fibres**

- *Animal fibres* are of two types: (*i*) Animal hair obtained from sheep, goats, pigs etc. (*ii*) Silk produced by the mulberry silkworm or from the larvae of other moths.
- The most important hair fibre is sheep wool. Its length varies from 2.5 to 20 cms. and its diameter from 0.0045 to 0.01 cm. Pig bristles are used in paint brushes. Wool waste for packing glands etc. is made from carpet yarn with fibres not less than 8 cm long.
- Silk from the mulberry silkworm possesses high strength while varieties obtained from other moths are of lower strength comparatively.

**3. Vegetable fibres**

*Vegetable fibres* consist mainly of *cellulose*. They may be seed hairs, such as cotton, or the inner bark of plants, such as flax, hemp, jute. The colours of cotton, flax and jute and hemp fibres are white, grey brown and brown respectively. The woody types are stiff and brittle whereas those with a high proportion of cellulose are flexible and elastic. *Cotton* is used for making batting, guncotton, cloth and *jute* is obtained from the stalks of the jute plant by *retting* (a process which involves steeping in water). The short fibres are used in paper making whereas longer ones are used in the manufacture of coarse woven fabrics such as burlap, carpet bindings, as a substitute for hemp in turn and small ropes, as a filler in cable, and as adulterant for other fibres. *Hemp* is the name given to the fibres obtained by retting the hemp plant. It is very strong and of great durability. Some grades of hemp are of dark brown colour and contain a mixture of cellulose and ligno cellulose. They are more hygroscopic than cotton, less affected by moisture and are disintegrated by bleaching. They are used mainly for turn and rope.

The following points in respect of **fibre polymers** are worth noting:

(*i*) The fibre polymers are capable of being drawn into long filaments having at least 100 : 1 length-to-diameter ratio. Most commercial fibre polymers are utilized in the *textile industry,* being woven or knit into cloth or fabric. The aramid fibres are employed in *composite materials.*

(*ii*) The molecular weight of fibre materials should be relatively high.

(*iii*) Fibre polymers must exhibit chemical stability to a rather extensive variety of environments, including acids, bases, bleaches, drycleaning solvents, and sunlight. In addition they must be relatively nonflammable and amenable to drying.

**Production/Manufacture of Synthetic Fibres:**

- There are, in general, following three methods of producing synthetic fibres:
  (*i*) Melt extrusion;
  (*ii*) Dry spinning;
  (*iii*) Wet spinning.

**(*i*) Melt extrusion :**

— In this method of producing synthetic fibres, a polymer is forced through a series of orifices, through a spinneret and solidified into thin filaments by stream of cool air.

— Polyamide (Nylon), Dacron and PVC are produced by melt extrusion.

**(*ii*) Dry spinning :**

— Dry spinning is accomplished by dissolving the polymer in a suitable solvent which is then extruded through a spinneret. As warm air is circulated in the system, the solvent evaporates and the filament solidifies.

— This method is used to produce the commercially known copolymers *Orlon* (cellulose acetate) and Acrilon (polyacrylonitrile).

**(*iii*) Wet spinning :**

— This method is used to produce *rayon fibres.*

— An alkaline viscous solution is forced through a spinneret into an acid bath where coagulation of fibres takes place resulting in solid filaments.

- After the production of fibres the next treatment is improving the chain molecules by *stretching the filament*. To increase crease resistance, water-proofing and fire proofing qualities, a number of chemical treatments are required. Like polymers metals and glass fibres are also produced and familiar example is of fibre glass which has many industrial applications (*e.g.*, safety helmets, racing cars, bullet proof bodies of automobiles and durable furniture items) because of its high strength.

## HIGHLIGHTS

1. The plastics are natural or synthetic resins, or their compounds, which can be moulded, extruded, cast or used as films or coatings.
2. Naturally occurring substances like coal, petroleum, cellulose, etc. form the raw materials for manufacturing plastics. An organic compound is used as a binder and is most important constituent. Binders may be resins or cellulose derivatives and are usually compounds of large molecular weight.
3. The moulding compounds used in the manufacture are: Binder, fillers, plasticizer, catalysts, pigment and lubricants.
4. Plastics can be classified basically into two groups: Thermoplastics and thermosetting plastics.
5. *Thermoplastics* become soft and plastic on heating without chemical changes taking place. These can thus be resoftened and remoulded without loss of any basic properties. *Thermosetting plastics* undergo chemical change in the presence of heat and catalyst and remain in hardened state even without cooling. They do not soften appreciably on heating.
6. Plastics have a relative high strength compared to their weight;
7. Commercial forms of plastics are:

   Plastic sheets, Plastic tiles, Plastic laminates, Plastic panels, Plastic pipes.
8. Manufacturing processes:

   Casting, Moulding, Laminating, Blowing, Calendering, Machining and cementing.
9. The fibre glass reinforced plastic (FRP) is formed by using two materials in conjunction with each other to form a composite material of altogether different properties.

## OBJECTIVE TYPE QUESTIONS

### A. Choose the Correct Answer :

1. PVC stands for
   (a) Polyvinyl chloride (b) Polythene vinyl chloride
   (c) Phosphorous-vanadium-carbide (d) Plastic very compact.
2. Fillers are added to moulding powders for plastics to
   (a) increase heat resistance (b) reduce cost
   (c) increase mouldability (d) reduce mould shrinkage
   (e) all of the above.
3. Bakelite is
   (a) highly inflammable (b) low resistance conductor
   (c) a semiconductor (d) incombustible.
4. Plasticizers are added to plastics to
   (a) crystallize into granules (b) impart colour
   (c) impart desirable properties of strength, toughness etc.
   (d) reduce melting temperature.
5. .................. is not a thermosetting plastic.
   (a) Cellulose (b) Phenol formaldehyde
   (c) Urea formaldehyde (d) Acrylic.
6. Plastic items can be manufactured by
   (a) impregnation (b) extrusion
   (c) casting (d) coating
   (e) any of the above.
7. Plastics can be broadly classified as
   (a) elastic and rigid (b) monomers and polymers
   (c) soft and hard (d) thermosetting and thermoplastic.

### ANSWERS

| | | | | |
|---|---|---|---|---|
| **1.** (a) | **2.** (e) | **3.** (b) | **4.** (c) | **5.** (d) |
| **6.** (e) | **7.** (d) | | | |

### B. Fill in the Blanks/Say 'Yes' or 'No'

1. A ............... can be broadly defined as any non-metallic material that can be moulded to desired shape.
2. ............... materials are those which soften on application of heat with or without pressure but they require cooling to set them to a shape.
3. ............... materials are those which require heat and pressure to mould them to a shape.
4. ............... have outstanding electric arc resistance.
5. ............... plastics are formed by impregnating sheet of fibrous materials such as paper, linen, canvas or silk with a synthetic resin and then compressing the sheets together with application of heat.
6. The trade name of Acrylics is Lucite.
7. Styron is the trade name of ...............
8. DC resins is the trade name of silicones.
9. The trade name of polysters is Laminac.
10. Paint is a mechanical mixture of vehicle and a ...............
11. The pigments provide both colour and binding power.
12. ............... is a mixture of both paint and a varnish.
13. The ............... is a filament or thread like piece of any material.
14. Vegetable fibres consists of mainly ...............
15. The most important hair fibres is sheep wool.
16. The plastics possess a limited number of colours.
17. The plastics are good conductors of electricity.
18. Thermoplastic materials soften on application of heat.
19. Thermosetting resins are stronger and harder than thermoplastic resins.
20. Tnermosetting resins can be repeatedly softened by heat and hardened by cooling.
21. Thermoplastic resins are usually supplied as granular materials.

22. Polythenes do not absorb moisture.
23. Polyvinyl chloride becomes soft beyond 80 C.
24. Silicones are used on high heat resistant insulations.
25. Polystrene has low resistivity.
26. Bitumen is the oldest known plastic.
27. Aminos are used for low voltage applications.
28. Melamines have excellent resistance to water.
29. Phenolics are made by reaction between phenol and formaldehyde.
30. Epoxy resins undergo very little shrinkage.
31. Laminated plastics cannot be machined.
32. Pig bristles are used in a paint brushes.
33. Fibre brands are made by mixing saw dust and fibres or tissues of wood with glue.
34. Hemp is the name given to the fibres obtained by retting the hemp plant.
35. One of the patented fibre brands is masonite whose thickness varies from 3 to 6 mm.

## ANSWERS

| | | | | |
|---|---|---|---|---|
| 1. plastic | 2. Thermoplastic | 3. Thermosetting | 4. Melamines | 5. Laminated |
| 6. Yes | 7. Polystrene | 8. Yes | 9. Yes | 10. pigment |
| 11. Yes | 12. Enamel | 13. Fibre | 14. cellulose | 15. Yes |
| 16. No | 17. No | 18. Yes | 19. Yes | 20. No |
| 21. Yes | 22. Yes | 23. Yes | 24. Yes | 25. No |
| 26. Yes | 27. No | 28. Yes | 29. Yes | 30. Yes |
| 31. No | 32. Yes | 33. Yes | 34. Yes | 35. Yes |

## THEORETICAL QUESTIONS

1. What are the special properties of plastics that make them useful engineering materials?
2. How do thermoplastics differ from thermosetting materials?
3. Explain briefly any three of the following plastics:
   Polyvinyl chloride, Silicones, Aminos, Phenolics, Expoxy resins.
4. What are laminated plastics? State their properties and uses.
5. What is a fibre?
6. Give the classification of fibres.
7. What are mineral fibres?
8. Write a short note on animal fibres.
9. Give a brief description of 'Vegetable fibres'.
10. What is a masonite?
11. How are fibre boards manufactured?
12. Name the different processes used in the manufacture of plastics.
13. Name the different methods of moulding plastics.
14. List the common properties of plastics which make them useful materials.
15. List the different uses of plastics in construction.
16. Mention uses of plastics.
17. Enumerate general properties of plastics.
18. What is polymerization?
19. Mention the shortcomings of plastics.
20. What are the properties of FRP?
21. Mentions the applications of FRP in building industry.
22. Write short note on "Deformation of plastics".
23. Explain briefly polymer/plastic coatings.
24. What are the factors which determine the mechanical behaviour of polymers?

# 14

# Glass

*14.1. Introduction. 14.2. Constituents of glass and their functions. 14.3. Properties of glass. 14.4. Glass furnaces – pot furnaces – tank furnaces – melting – fabrication of glass, 14.5. Classification of glass. 14.6. Uses of Glass. 14.2. The glass industry in India. Highlights – Objective Type Questions – Theoretical Questions.*

## 14.1 INTRODUCTION

- **Glass** is *any substance or mixture of substances that has solidified from the liquid state without crystallization*. Elements, compounds and mixture of wide varying composition can exist in the glass state, but the term "*glass*" as ordinarily used *refers to material which is made by the fusion of mixture of silica, basic oxides and a few other compounds that react either with silica or with the basic oxides.* No definite chemical compounds can be identified in glass. Many of its properties correspond to those of a supercooled liquid whose ingredients cannot be identified because they have not separated from the solution in crystalline form.
- *Glass* may also be defined as a *hard, brittle, transparent or transluscent material chiefly compound of silica, combined with varying proportions of oxides of sodium, potassium calcium, magnesia, iron and other minerals.*
- Glass is an amorphous substance having a homogeneous texure.

**Structure of glass:** The *glass is a random arrangement of molecules, the great majority of which are oxygen ions bounded together with the network forming ions of silicon, boron or phosphorous*. A glass made of silica alone has many desirable characteristics but unfortunately the high temperatures involved make it expensive, and difficult tó prepare. In order to reduce the temperature, required network-modifying ions are added. *Sodium, potassium* and *calcium* are the most common. The network-modifying ions increase the competition for the oxygen ions, thus loosening the Si-O bonds. Certain other ions may substitute for either the network-forming ions or network-modifying ions; *aluminium, zinc, beryllium, lead and iron* are a few of these intermediate irons.

## 14.2 CONSTITUENTS OF GLASS AND THEIR FUNCTIONS

The various *constituents of glass* and their *functions* are described below:

1. **Silica:**
   - It is the principal constituent of glass.
   - If silica alone is used in the manufacture of glass, it could be fused only at a very high temperature but it would give a good glass on cooling. However, it is imperative to add

some alkaline materials (sodium or potassium carbonate) and lime in suitable proportions to make the molten silica glass sufficiently viscous to make it amply workable and resistant against weathering agencies.

2. **Sodium or potassium carbonate:**
- It is an alkaline material and forms an essential component of glass.
- It is added in suitable proportion to *reduce the melting point* of silica and to *impart viscosity to the molten glass.*

3. **Lime:**
- It is added in the form of chalk.
- It is imparts durability to the glass.

In place of lime, sometimes, *lead oxide* is also added; it makes the glass *bright* and *shining*.

4. **Manganese dioxide:**
- It is added in suitable proportion to *correct the colour of glass due to the presence of iron in raw materials of glass.*
- It is also called 'Glass maker' soap.

5. **Cullet:**
- It is the old broken glass of the same type as that which is intended to be prepared.
- It is added in small quantity to provide body to the glass.

6. **Colouring substance :**

While manufacturing a coloured glass, a suitable colouring substance is added at fusion stage to provide the desired colour to the glass.

The various colouring substances for manufacturing glass of different colours are given below:

| | *Colour* | *Colouring substance* | | *Colour* | *Colouring substance* |
|---|---|---|---|---|---|
| 1. | *Black* | Cobalt, nickel and manganese oxide | 4. | *Violet* | Manganese dioxide |
| 2. | *Green* | Chromic oxide | 5. | *White* | Cryolite, tin oxide |
| 3. | *Red* | Cuprous oxide, selemium | 6. | *Yellow* | Cadmium sulphate. |

## 14.3 PROPERTIES OF GLASS

Following are the *properties* of glass:

1. No definite crystalline structure.
2. No sharp melting point.
3. Absorbs, refracts or transmits light.
4. Affected by alkalies.
5. An excellent electrical insulator at elevated temperatures.
6. Extremely brittle.
7. Available in beautiful colours.
8. Not affected by air or water.
9. Not easily attacked by ordinary chemical reagents.
10. Capable of being worked in several ways.
11. Can take up a high polish (and may be used as substitute for very costly gems).
12. Possible to weld pieces of glass by fusion.
13. As a result of advancement made in the science of glass production, it is possible to make glass lighter than cork or softer than cotton or stronger than steel.
14. Glass can be cleaned easily by any of the following methods:

(*i*) Applying methylated spirit. (*ii*) Rubbing finely powdered chalk.

(*iii*) Rubbing damp salt for cleaning paint spots.

(*iv*) Painting the glass panes with lime-wash and leaving it to dry and then washing with clean water.

**Commercial glass** must meet the following **requirements** :

(*i*) The material must melt at commercially obtainable temperature. Fused silica cools to a glass that is superior to ordinary glass, but the temperature required to melt it is so high that its production is expensive and its use restricted.

(*ii*) The molten mixture must remain in the amorphous or non-crystalline condition after cooling.

(*iii*) The fluidity of molten glass must persist to a sufficient extent to permit the formation of *desired shaped while the glass is cooling.*

(*iv*) The glass must be reasonably permanent in the use for which it is intended. Glass with a high proportion to sodium oxide is more readily attacked by water and acids than glass that contains less sodium oxide and more lime and magnesia. Glasses low in basic oxides are less readily attacked.

**Mechanical properties:**

Ranges of physical properties of glasses:

| | |
|---|---|
| Refractive index | 1.46 – 2.179 |
| Specific index | 2.125 – 8.120 |
| Compressive strength | 600 – 1200 $MN/m^2$ |
| Tensile strength | 27 – 700 $MN/m^2$ |
| Thermal conductivity | 0.0018 – 0.0028 cal/cm°C/sec. |
| Expansion co-efficient | $8 \times 10^{-7} - 140 \times 10^{-7}$ cm/cm/°C |
| Softening point | 500 – 1510°C |
| Annealing point | 350 – 890°C |
| Volume resistivity | $10^8 - 10^{18}$ $ohms/cm^3$ |
| Dielectric constant | 3.7 – 16.5 |

**Strength.** Glass is weak in tension and it always fails in tension no matter how stress is applied. Nevertheless, glass is four to six times as strong in compression as it is in tension and it should therefore be used under compressive loads. Strength depends upon the factors such as surface conditions, cross-section and amount of addition etc.

**Hardness and brittleness.** The hardness of glass cannot be measured by the Brinell or Rockwell machines because the test specimen fails when the surface is ruptured. Yet some glasses will scratch steel. The relative value of hardness can be evaluated by scratch or abrasion test. For *ordinary purposes the hardness of glass is a function of its tensile strength.*

The concept of brittleness has not been clearly defined. When glass is fixed it bends only a little, then shatters. There is no plastic deformation, and the curve of the *stress-strain diagram is straight until the test specimen breaks.*

**Thermal endurance.** The lower the co-efficient of expansion and the thinner the pieces the less likely it is to break under thermal shock. Glass does not break from thermal causes but from the tensile stresses set up by the temperature gradient.

**Electrical properties:**

**Conductivity.** Conductivity depends on composition, temperature and surface conditions. Sodium and potassium ions are largely responsible for imparting electrical conduction, but the function of the calcium ions or the mechanism of conduction in an alkali-free glass is not known. $Na_2O$ and $K_2O$ increase the conductivity. $Al_2O_3$ increases the conductivity slightly; ZnO, PbO, MgO, $Fe_2O$, BaO, $B_2O_2$ and CaO decrease the conductivity.

The presence of a film on the surface glass affects the surface resistivity to a marked degree. In high alkali glasses the surface conductivity may excced the volume conductivity. The presence of $CO_2$, $SO_2$, $H_2S$, solids and mineral soils also affects the surface conductivity.

In making conductivity tests on glass a precise value is seldom obtained because of the influence of the surrounding medium. Even oil immersion will not give the true value because it has been shown that such a test is a measure of the oil, which fails first, bringing about the failure of the glass. Care in eliminating the edge effects and other conditions show that glass has puncture values several times higher than that obtained in service. The practical limit must take the operating conditions into account.

In the usual method of testing glass for its electrical conductivity the glass specimen is placed between electrodes. Such a system consists of a capacitor with glass as the dielectric medium Upon charging such a capacitor, the initial charging current is high and then drops off. The charging current rapidly diminishes and straightens out, approaching a low but finite value. The discharge current behaves in like manner. If the charged capacitor is short circuited, the initial discharge current is high and drops off rapidly, approaching a low but finite value. The residual charge may be removed by short-circuiting the capacitor. It may be necessary to repeat this process several times. Because of this behaviour, the measurement of electrical resistance may be in error, and the time interval should be stated.

**Dielectric constant, power factor, dielectric loss.** The dielectrical constant K of glass varies from 3.7 to 16.5 and remains relatively constant over a wide frequency, the square of the voltage gradient, and to the product of dielectric constant and the power factor. For conditions of high frequency or high voltage or both, the value of K should be kept as low as possible. Both the dielectric constant and the power factor can be varied by changes in the glass composition by a ratio of more than 200 to 1. Two general glasses are available: those with high dielectric constant and low power factor, and those with low dielectric constant and low power factor.

**Dielectric strength**. Electrical breakdown in influenced by the characteristics of the dielectric, the thickness, the duration of the stress, temperature and edge effects, the surrounding medium and characteristics of the voltage, whether direct or alternating; thermal failure of dielectric is a result of internal heating, which causes it to become a conductor. Alternating currents cause failure at a lower voltage than direct current do.

## 14.4 GLASS FURNACES

For the manufacture of glass two types of *furnaces* are used:

1. Pot furnaces.
2. Tank furnaces.

### 14.4.1 Pot Furnaces

This type of furnace is used to *make small amounts of glass* such as optical glass that require careful control of ingredients and of melting and fining operations. The pots are fire clay crucibles ranging up to 0.9 metres in height and up to 1.5 metres in diameter. They are arranged on edges in a circular or oblong furnace where they are exposed to a steadily increasing temperature.

### 14.4.2 Tank Furnaces

In a *continuous tank* a constant level of molten glass is maintained by the addition of raw material or "batch" at the proper rates. The furnace is divided into **two compartments** : the large *melting compartment* and the smaller *working compartment* from which the molten glass is withdrawn. Molten glass flows from the melting to the working compartment through a small opening at the bottom of the dividing wall. The "*day tank*" is a smaller tank, designed to complete melting of a small charge in one day. It is particularly useful type of tank in cases where frequent changes in glass composition are required.

### 14.4.3 Melting

When the well-mixed raw materials are heated in the glass furnace, those substances with the lower melting points liquify and act as solvents for other substances, at the same time reacting with them and enabling them to react with each other. The system is very complicated one and the exact sequence of events cannot be determined. The temperature during melting is kept as high as practicable (1400 to 1500°C) in order to reduce the viscosity. Bubbles of gas (carbon dioxide, steam etc.) that are formed during the heating must escape, and their escape will be slower from a more viscous liquid. In order to remove the gases entirely, the glass must be held for sometime at the highest possible temperature. This is called **"plaining" or "fining"**. Additions are often made to the batch to lower its viscosity and to generate gas in large volumes at later stages in the melting process, thus sweeping out the small bubbles along with large ones. Salt cake is often used for this purpose. The under composed portion forms a liquid layer on the surface of the glass and reacts with the scum of siliceous material that tends to collect there. The working temperature *i.e.*, the temperature at which the glass has the proper viscosity for the shaping operations, is much lower than the highest temperature during melting, the actual working temperature being dependent on the kind of glass and the type of machine used. The range is roughly from 1000 to 1300°C or higher.

### 14.4.4 Fabrication of Glass

The various processes involved in the fabrication of glass are enumerated and described below:

| | |
|---|---|
| 1. Blowing | 2. Flat drawing |
| 3. Rolling | 4. Pressing into moulds |
| 5. Casting | 6. Spinning |

1. **Blowing:** In this process, a blow-pipe, 1.5 m to 2 m long and 12 mm diameter is used. One end of the blow-pipe is dipped in a molten mass of glass and a lump (of about 50 N weight) is taken out. The operator then blows vigorously from other end of the blow pipe (it can also be done with the help of a compressor); this blowing causes the molten mass to assume a cylindrical shape. It is then heated for ten seconds and is blown again. The blowing and heating are continued till the cylinder of required size is formed. It is then kept on an iron plate disconnected from the blow pipe and a cut is made on its top surface longitudinally by a diamond glass cutter so that by gravity the two ends of the cut cylinder will fall out, forming a rectangular thin sheet of glass.

2. **Flat drawing:** In this process, the molten glass in viscous form is drawn in the form of a plate by moving an iron bar side way through it. The plate is then passed over a large rotating roller which helps it in spreading out in a thin sheet.

3. **Rolling:** Rolling is carried out in **two** ways:

In one method the molten glass is poured on a flat iron casting table and it is then turned flat with the aid of a heavy iron roller.

In another method, the molten mass of glass is passed between heavy iron rollers and flat glass plate of unform thickness is obtained.

4. **Pressing into moulds:** In this process, the molten glass is pressed into moulds by machines. Such a glass is stronger and more durable than unpressed glass. This process is used for *hollow glass articles, ornamental articles* etc.

5. **Casting:** In this process molten glass is poured in moulds and then allowed to cool down slowly. This method is used to prepare large pieces of glass of simple design, mirrors, lenses etc.

6. **Spinning:** In this process, the molten glass is drawn into threads or is spun on a high speed wheel operated by power, to produce even finer than cotton threads.

The spun glass has *tensile strength equal to that of mild steel. It is very soft and flexible* and *does not fade, decay or shrink*. It is *not* attacked by acids, fire and vermins.

It is used for *producing insulation against heat, sound and electricity.*

## 14.5 CLASSIFICATION OF GLASS

As per composition and properties glass may be *classified* as:

1. Soda-lime or crown glass
2. Flint glass
3. Pyrex or heat-resistant glass.

**1. Soda-lime or crown glass :**

- It is the cheapest quality of glass.

- Its composition, like that of most glass, is not rigidly fixed, but can be varied both as to the amount of ingredients and chemical compounds used.

*Composition by weight:* Sand 75 parts; lime 12.5 parts; soda 12.5 parts; alumina 1 part; and waste glass 50 to 100 parts.

- It is easily fusible at comparatively low temperatures.
- It is available in clean and clear state.
- It is possible to blow or to melt articles made from this glass with the help of simple sources of heat.

**Uses:** Its principal uses are for *window glass*, *plateglass* and *container glass* (bottles, glass etc.)

**2. Flint glass :**

- It contains varying proportion of lead oxide to make it suitable for various purposes. Lead provides *brilliance* and *high polish* which makes the glass available for special purposes.

*Composition by weight :* Sand 100 parts; red lead 70 parts; potash 32 parts; waste glass 10 parts.

- It liquifies at a lower temperature than soda-lime glass and has better lustre.
- Owing to the ease with which lead compounds are reduced, the glass must be melted in an oxidizing atmosphere.

**Uses :** This is potash-lead glass used for better quality of *table-wares* and for *optical glass*. It is also used for *electric lamps, thermometers, electron tubes, laboratory apparatuses, containers for foods etc.*

**3. Pyrex or heat resistant glass :**

Both soda-lime and flint glasses are unable to withstand sudden temperature changes because of their large co-efficients of thermal expansion. The basic oxides that they contain make them susceptible to chemical attack by water and acids. Elimination of the basic oxides and inclusion of boron oxide produce a glass that is very resistant to thermal shock and to attack by water and acids. The temperature required to melt and fine such a glass is so high that it has to be heated in the electric arc. The familiar Pyrex glasses, which are used extensively for cooking utensils and laboratory wares, are **borosilicate glasses.**

*Composition by weight:* Silica 80 parts; boron oxide 14 parts, sodium oxide 4 parts; alumina 2 parts, with traces of potassium oxide, calcium oxide and magnesium oxide.

**Special Types of Glasses**

**1. Annealing glass:**

To prevent glass articles becoming too brittle and falling into pieces at the slightest shock, they are kept while still hot in an annealing furnace to cool very slowly. *The longer the annealing period, the better the quality of the glass.*

**2. Sheet glass :**

- It is roughly composed of 100 parts sand, 35 parts limestone or chalk, 40 part soda and 50 parts waste glass.
- It is made by blowing glass into hollow cylinder, splitting the cylinder and finally flattening it over a plane surface. It is manufactured in thickness varying from 1.5 to 5 mm and sizes upto 1.5 × 1 metre.

**Uses:** It is generally used for *doors and windows*.

**3. Plate glass:**

- Its composition is: White sand 100 parts; soda carbonate 33 parts; slaked lime 14 parts; manganese peroxide 0.15 parts and waste glass 100 parts.
- It is *made by pouring white hot glass over an iron table and rolling it to a uniform thickness under heavy roller.*

- It is thicker than sheet glass, and its thickness varies from 5 to 25 mm and sizes upto 275 cm × 90 cm.
- It is stronger and more transparent to sheet glass.
- In modern glass-fabrication, rolled plate glass is annealed because glass cooled normally is brittle.
- Plate glass includes *transparent, transluscent, opaque* and *structural glasses.*

**Uses** : (*i*) It is used for making looking-glasses, wind screens of motors, car skylights and glass houses.

(*ii*) It is also used for *sales counter* and *table tops* after being laminated with plywood or metal sheet.

**4. Fluted glass:**

When there are *corrugations* on one side of the plate glass then it is known as *fluted glass.* The other side is wavy but smooth. The light is admitted without the glare of the sun.

**Uses:** It is used in situations where it is desirable to *secure privacy without obstruction of light.* Horizontal ribs give more light in the middle and less at the sides, while upright ribs give more light at the sides and less in the middle. It is thus more *used for skylight roofs* and *for windows of railway stations and factories.*

**5. Ground glass:**

- It is made either by *grinding* (usually by sand blasting) one side, or by melting powdered glass upon it.
- This glass is mostly transluscent.
- This type of glass is also known as *frosted glass* or *obscured glass.*

**Uses:** This glass is used in situations where light is required without transparency. It is normally used for *window panes* and *bath room ventilators* etc.

**6. Wired glass:**

- It is glass with *wire netting or similar strengthening material embedded in it during manufacture.* This is why this glass is known as *reinforced glass.*
- It resists fire better than ordinary plate glass.
- In case the glass is fractured it does not fall into pieces.

**Uses:** It is used for *skylight* and *roofs*, also for *fire-resisting doors and windows.*

**7. Safety glass :**

- This type of glass is produced by sandwitching sheets of celluloid or other transparent plastic between two sheets of glass and sticking the whole combination together by means of colourless and heat resisting glue.
- Wired glass falls into the category of safety glass.

**8. Bullet-proof glass:**

- This glass is made of several layers of plate glass and alternate layers consist of vinyl-resin plastic. The outer layers of plate glass are made thinner than the inner layers. The special care is to be taken for heating and cooling of layers during manufacture.
- The thickness of this type of glass may vary from 15 mm to 75 mm or more.
- This glass will not allow bullet to fierce through it.

**Uses:** It is extensively used for *glazing bank teller, cages, cashier booths, jewellery stores, display cases* etc.

**9. Insulating glass:**

It is transparent glass unit composed of two or more plates, of glass separated by 6 to 13 mm of dehydrated captive air, hermetically sealed inside, is scientifically cleaned and dried.

Insulating glass provides a high resistance to heat-flow. *The sealed air makes the coefficient of heat transmission of the glass low and hence keeps the apartment cool in summer and warm in winter.*

**10. Foam glass :**

- It is prepared from powdered glass to which is added the desired quantity of carbon or any gas which makes the mass porous and light in weight.
- This glass is water-proof also.
- It can be easily cut and worked with common masonry tools.

**Uses:** It is used for *sound and heat insulation purposes. It is specially recommended for use in air-conditioning of buildings.*

**11. Glass blocks:**

- These consist of two halves so fused together as to *form a hollow inside.*
- They *provide insulation against heat, cold and noise* and *are easy to clean.*

**Uses:** They are widely used for constructing *wall partitions.*

**12. Soluble glass:**

- It is prepared by melting quartz sand, grinding and throughly mixing it with soda ash, sodium sulphate or potassium carbonate. The melting is carried out in glass tanks at a temperature between 1300°C to 1400°C and it takes about 7 to 10 hours. The resultant glass mass flows out from the furnace and it cools rapidly and breaks up into pieces, known as **silica lumps**.
- It is soluble in water, under normal conditions.
- The soluble glass in the form of silicate lumps is transported in containers and in the form of liquid, it is transported in barrels or glass bottles.

**Uses:** It is used for preparing *acid-resistant* cement.

**13. Ultra-violet ray glass:**

- It transmits ultra-voilet rays *effectively* even though it is not in the direction of the rays of sun.
- It is made from the raw mixture with minimum admixtures of iron, titanium and chrome.

**Uses:** It is used in the windows of schools, hospitals etc.

**14. Structural glass:**

- This type of glass is available in the form of *glass-crete square blocks, tiles* or *lenses* in thicknesses varying from 5 mm to 30 mm.
- These glass products are *hollow, light* and *transparent.*
- This type of glass can be *sawn, placed* and *drilled* like wood work, inspite of having general properties of glass.

**Uses:** Widely used for pavement lights, partitions, lantern lights; also used as roof covering material in industrial buildings, factories etc.

**Glass-Fibre or Glass-Wool**

- The usual *composition* of glass-fibres is that of a *soda-lime glass* but it may be varied for different purposes.
- The *glass-fibres are made by letting the molten glass drop through tiny orifices and blowing with air or steam to attenuate the fibres.*
- They have *very high tensile strengths,* upto about 2750 N/mm$^2$.
- Glass-fibre or glass-wool differs from mineral wool in that it is a glass made to a definite formulation with a uniformity not found in mineral wool.

## 14.6 USES OF GLASS

Besides other uses, some of the important uses of glass, based on the recent developments in the glass industry, are as follows:

1. The fibre glass reinforced with plastics can be used in the construction of furniture, cars, trucks, lampshades, bath room fittings etc.
2. Glass is used to form a rifle barrel which is lighter and stronger than conventional type.
3. Thousands of items in the body of a guided missile are made of glass.
4. Glass is used in the construction of noses of deep-diving vehicles.
5. Optical glass is finding wide application for the development and advancement of sciences of astronomy and bacteriology.
6. The glass linings are applied on equipments likely to be affected by the chemical corrosion such as valves, pipes, pumps etc.
7. Hollow glass blocks can be used for the construction of the walls and ceilings of the modern homes.
8. These days, it is possible to prepare the colour-changing glass; a window with such a glass will be transparent during the day and it will be a source of light at night.

## 14.7 THE GLASS INDUSTRY IN INDIA

Names/Addresses of some important glass manufacturing concerns in India are given below:

| S.No. | Name | Trade mark | Address |
|---|---|---|---|
| 1. | *Hindustan Pilkington Glass Works Ltd.* | "HPG" | Hindustan Building, 4, Chitranjan Avenue, CALCUTTA. |
| 2. | *Borosil Glass Works Ltd.* | "CORNING" Glass | 44, Khanna Construction House, Khan Abdul Gaffar Khan Road, Worli, MUMBAI. |
| 3. | *The Associated Glass Industries Ltd.* | "AGI" Make | Veradanagar, Kukatpally, HYDERABAD. |
| 4. | *Alembic Glass Industries (India) Ltd.* | | Alembic Road, BARODA-3 |
| 5. | *Vallabh Glass Works Ltd.* | | Vallabh Vidyanagar ANAND (Gujarat) |

## HIGHLIGHTS

1. *Glass* may be defined as a hard, brittle, transparent or transluscent material chiefly composed of silica, combined with varying proportion of oxides of sodium, potassium, calcium, magnesia, iron and other materials.
2. For the manufacture of glass two types of furnaces are used:
   (*i*) Pot furnaces (*ii*) Tank furnaces.
3. The various process involved in the fabrication of glass are:
   (*i*) Blowing; (*ii*) Flat drawing; (*iii*) Rolling
   (*iv*) Pressing into moulds (*v*) Casting; (*vi*) Spinning.
4. As per composition and properties the glass may be classified as:
   (*i*) Soda-lime or crown glass
   (*ii*) Flint glass
   (*iii*) Pyrex or heat resistance glass.
5. The *glas-fibres* are made by letting the molten glass drop through tiny orfices and blowing with air or stream to attenate the fibres.

## OBJECTIVE TYPE QUESTIONS

**A. Choose the Correct Answer:**

1. Bullet-proof glass
   (*a*) is always transluscent
   (*b*) is covered by steel sheet from inside
   (*c*) is made of thick glass sheets sandwiched by a layer of high tensile plastic
   (*d*) has iron wires embedded in its structure
   (*e*) none of the above.
2. Soluble glass is
   (*a*) concentrated brine (*b*) hydrofluoric acid
   (*c*) ethyl alcohol (*d*) acqueous sodium silicate.
3. Glass wool is used as
   (*a*) heat insulating material (*b*) sound absorbing medium
   (*c*) both (*a*) and (*b*) above (*d*) none of the above.
4. Silicate glass is least resistant to the action of
   (*a*) hydrochloric acid (*b*) sodium hydroxide
   (*c*) alcohols (*d*) hydrofluoric acid.
5. Which of the following glass resists heat passing through without affecting the normal passage of light?
   (*a*) Wired glass (*b*) Plate glass
   (*c*) Pyrex glass (*d*) Anti-actinic glass.
6. Common silicate glass is opaque to
   (*a*) infrared rays (*b*) ultra voilet rays
   (*c*) both (*a*) and (*b*) above (*d*) none of the above.
7. Which of the following glass is used in railway coaches?
   (*a*) Glass-crete (*b*) Wired glass
   (*c*) Anti-actinic glass (*d*) Pyrex glass.
8. Water glass is
   (*a*) water free from impurities (*b*) water formed by condensation on glass
   (*c*) solution of sodium silicate in water (*d*) glass made out of pure silica.
9. For manufacturing coloured glass, the colouring pigments are added in the glass at the stage of
   (*a*) cooling (*b*) fusion
   (*c*) melting (*d*) finishing.

### ANSWERS

| | | | | |
|---|---|---|---|---|
| **1.** (*c*) | **2.** (*d*) | **3.** (*c*) | **4.** (*d*) | **5.** (*d*) |
| **6.** (*c*) | **7.** (*c*) | **8.** (*c*) | **9.** (*b*). | |

**10. Fill in the Blanks/Say Yes or No:**

1. .............. is an amorphous substance having a homogeneous texture.
2. Glass does not have a sharp melting point.
3. Glass has a definite crystalline structure.
4. Glass is easily attacked by ordinary chemical reagents.
5. It is possible to weld pieces of glass by fusion.
6. It is possible to make glass lighter than cork.
7. Glass is not affected by alkalies.
8. The hardness of glass cannot be measured by the Brinell or Rockwell machines.
9. Glass does not break from thermal causes but from the tensile stresses set up by the temperature gradient.
10. Conductivity of glass depends on composition, temperature and surface conditions.
11. ................furnace is used to make small amounts of glass.

12. ................ furnace is particularly useful in cases where frequent changes in glass composition are required.
13. Soda-lime glass is the cheapest quality of glass.
14. The familar Pyrex glasses, which are used extensively for cooking utensils and laboratory wares, are ............... glasses.
15. Sheet glass is generally used for doors and windows.
16. When there are corrugations on one side of the glass then it is known as .............. glass.
17. ................ glass is also known as frosted glass or obscured glass.
18. Insulating glass does not provide resistance to heat flow.
19. ................ glass is prepared from powdered glass to which is added the desired quantity of carbon or any gas which makes the mass porous and light in weight.
20. Foam glass is water-proof.
21. Glass blocks are rarely used for constructing wall partitions.
22. Soluble glass is used for preparing acid-resistant cement.
23. ................ Structural glass can be sawn, planed and drilled like wood work.
24. Glass fibres have low tensile strength.
25. Borosil Glass works Ltd. is located in Calcutta.

## ANSWERS

| | | | | |
|---|---|---|---|---|
| **1.** Glass | **2.** Yes | **3.** No | **4.** No | **5.** Yes |
| **6.** Yes | **7.** No | **8.** Yes | **9.** Yes | **10.** Yes |
| **11.** Pot | **12.** Tank | **13.** Yes | **14.** borosilicate | **15.** Yes |
| **16.** fluted | **17.** Ground | **18.** No | **19.** Foam | **20.** Yes |
| **21.** No | **22.** Yes | **23.** Yes | **24.** No | **25.** No. |

## THEORETICAL QUESTIONS

1. What is 'glass'?
2. State the general properties of glass.
3. Describe the principal constituents of glass, giving the functions of each.
4. Give the classification of glass.
5. What are the requirements which a commercial glass must meet with.
6. Write a short note on the electrical properties of glass.
7. Explain briefly the various types of furnaces used in the manufacture of glass.
8. Describe the various processes of fabrication of glass.
9. Explain briefly the following:
   (*i*) Soda-lime glass; (*ii*) Flint glass; (*iii*) Heat-resistant glass.
10. Write short notes on the following:
    (*i*) Sheet glass; (*ii*) Wired glass; (*iii*) Structural glass.
11. How is soluble glass prepared?
12. What are glass blocks?
13. Give the important uses of glass.
14. What is difference between sheet glass and plate glass.
15. Why is lime added to the mixture during the manufacture of glass?

# 15

# Insulating Materials

*15.1 Introduction. 15.2 Classification of insulating materials. 15.3 Heat insulating materials – general aspects – requirements of heat insulating materials – classification and description of heat insulating materials. 15.4 Sound insulating materials – requirements of sound insulating materials – classification and description of sound insulating materials – Highlights – Objective Type Questions – Theoretical Questions.*

## 15.1 INTRODUCTION

*The materials which control transmission of heat and cold and offer resistance to reflection and transmission of sound and electricity are known* as **insulating materials.**

While designing and constructing public and residential buildings, much importance needs to given to heat, sound and electrical insulation. In outer to provide/promote comfortable living, safety, efficiency of work and auditory of building it is imperative to provide good thermal, accoustical and electrical conditions.

- **Heat/thermal insulation** is provided to control the transmission of heat or cold so as to provide a proper temperature inside the rooms of the building.
- **Sound insulation** is provided to control the disturbances due to noise and to remedy the acoustical (*acoustics* is a broad field which embraces music radio, sound reproduction and other fields; it includes those aspects of sound control associated primarily with man's normal physical environment) defects as to achieve the best sound effects.
- **Electrical insulation** is provided to separate the electrical conductors from other bodies to prevent escape of electricity.

## 15.2 CLASSIFICATION OF INSULATING MATERIALS

The insulating materials may be *classified* as follows:

1. Heat insulating materials (or heat or thermal insulators)
2. Sound insulating materials (or sound insulators)
3. Electrical insulating materials (or electrical insulators).

## 15.3 HEAT INSULATING MATERIAS

### 15.3.1 General Aspects

- The *function of a thermal or heat insulator is to resist the flow of heat through its body.*
- The heat insulating materials are required to grant *protection against heat and cold.* It may also be employed to prevent either the flow of heat from a heat furnace to the

surrounding atmosphere, or the ingress of heat from the environment to the plant operating at lower temperature.

- These materials are generally *porous* and their properties are governed not only by their porosity but also by nature of pores, their distribution, size and whether they are open or closed. *The materials with a great number of fine, closed and air-filled pores are the best heat insulating materials.*
- The bulk density of heat insulating materials is usually below 7000 N/m$^3$ and their coefficient of thermal conductivity does not exceed 0.75 kJ per in hr°C.
- The heat insulating materials should be protected against moisture (since the coefficient of thermal conductivity of water is about 25 times higher than that of air).

### 15.3.2 Requirements of a Heat Insulating Materials

The main requirements of a good heat insulating materials are:

1. Thermal stability
2. Chemical stability
3. Physical stability
4. Low thermal conductivity
5. Resistance to moisture
6. Low specific heat
7. Low specific gravity
8. Odourless
9. Resistance to vibration and shock
10. Non-inflammability
11. Porous and fibrous texture
12. Economical in its initial cost.

### 15.3.3 Classification and Description of Heat Insulating Materials

The heat insulating materials may be classified as follows:

A. *Organic heat insulators*

B. *Inorganic heat insulators*

**Organic insulators** are enumerated below:

1. Wool
2. Cattle hair
3. Eelgrass
4. Cotton wool
5. Corkboard
6. Silk
7. Wood pulp
8. Sugarcane fibre
9. Saw dust
10. Cardboard (corrugated)
11. Paper etc.

**Inorganic insulators** *include* :

1. Air (steel)
2. Slag wool
3. Mineral wool
4. Glass wool
5. Aluminium foil
6. Diatomaceous earth (powder)
7. Charcoal
8. Wood ashes
9. Gypsum (powder)
10. Slag
11. Asbestos etc.

Some important heat insulating materials are *described* below:

**1. Cork :**

- It is derived from the *bark of oak trees*. It is *ground, sized and baked in moulds*. When ground and baked, the natural resin in the cork binds the material into homogeneous mass which can be pressed into flexible sheets or boards etc. It is available in the form of *granulated cork, slab cork,* and *regranulated baked cork.*

- The structure of cork consists of an aggregation of minute air vessel, provided with thin, strong wall, so that if material is compressed it behaves more like a gas than an elastic solid; unlike the behaviour of spring, which exerts a pressure proportionate to the linear amount of compression. Cork, when compressed, exerts a pressure which increases in a more rapid manner and varies, approximately, *inversely as the volume.*

**Properties** : Following are the *properties* of cork:

1. Light in colour.
2. Porous in structure.
3. Specific gravity is about 0.24.
4. Not affected by moisture.
5. Thermal conductivity is low.
6. Can be easily compressed.
7. Resilient and reasonably elastic when dry.

**Uses:** Following are the *uses of cork*:

1. Cork sheets and boards are used for insulating walls and ceilings, both against heat and cold and also as a sound insulator.
2. Used as a non-conducting covering for pipes carrying steam or hot water.
3. Used as non-conducting material for scientific apparatuses.
4. Used in refrigeration and cold storage insulation.
5. Also used for bottle stoppers, vibration pads and floats for rafts and fish nets.

**2. Glass wool :**

- Glass wool is produced by *blowing high pressure jets of steam* or *air on molten streams of glass at a high temperature.* Molten glass is violently scattered in all directions, to give this product.
- Glass wool is a *form of fibrous glass* with short and fine fibres, scattered in various directions.
- It is available in the form of *loose fibres, mats, rigid quilts,* or *semi-rigid slabs or blocks* etc.

**Properties:**

1. Fibrous in structure.
2. Light in weight.
3. Has good tensile and dielectric strength.
4. Low thermal conductivity.
5. Quite durable.
6. Acts as an excellent insulating material because of the presence of large pockets of air in it.
7. Not affected by low temperatures and has been used successfully at temperatures as low as – 212°C.

Glass fibres have the following *characteristics* :

- Do not catch fire.
- Not easily affected by heat.
- Not spoiled by insects and moisture.

**Uses:** 1. Mostly used for insulation of *pipes, bends, valves* etc.

2. Used for panel insulation for all types of industrial equipment.
3. Can be used for thermal and sound insulation of aircrafts.

4. *Glass wool blocks* can be used in the *construction of partition walls for thermal insulation purposes.*
5. Used in boilers, ovens, cylinder or pipe insulation.

**3. Rock wool:**

- It is produced from flint rock containing some calcareous matter. In the absence of such a natural rock, flint and lime are mixed in the requisite proportions and melted in a furnace at temperature of about 1700°C. This molten material is then formed into small globules by means of a steam jet. These globules are then drawn into very fine fibres by hurling them in a large container. These fibres of wool are then formed into boards or blankets (to be used as insulators). It can also be pressed, rolled and secured between fabric of wire-netting of brass or copper.
- It is available in the following forms:

*Loose fibres, matresses, mats, boards or felts, rigid or semi-rigid slabs, quilts.*

**Properties:**

1. Soft and flexible
2. Resilient and woody consistency
3. Heat and sound proof (due to the presence of milions of minute dead air cells)
4. Specific gravity is about 0.48.

**ses :** (*i*) Employed for heat and sound insulation pruposes.

(*ii*) Also used as an electric insulator.

**4. ›ag wool:**

- It is an aggregate of fine filaments of slag produced by blowing air through a stream of blast furnace slag.
- It is available in the form of loose fibres.

**Uses:** It is used for *heat insulation in high temperature furnaces.*

**5. Asbestos:**

- Asbestos is a mineral fibre composed of hydrous silicate of magnesia with a small amount of iron oxide and alumina.
- Asbestos sheets or boards consist of natural asbestos fibres mixed with a binding agent (usually cement) and then rolled in the form of sheets or boards. There are available in the market under the trade name 'Salamander'.

**Properties:**

1. White, grey or brown in colour.
2. Flexible and can resist high temperature.
3. Fire-proof.
4. Unaffected by acids and fumes.
5. Resistant to corrosion and vermin effect.
6. Excellently resists heat and electricity.

**Uses:** Employed for heat and sound insulation of buildings. Also used for insulation of furnaces.

**6. Thermocol:**

- Thermocole is one of the trade names of *Polystrene*. This product was developed (in U.S.A) during second world war. It was made by *direct extrusion of the foam form raw materials.*

**Properties:**

(*i*) It has a very attractive, natural, snow white colour.

(*ii*) Very light in weight (density : 150 to 300 N/m$^3$). The foam is very light because it contains over 98% (by volume) air, trapped in 3 to 6 millions closed cells per litre.

(*iii*) Compressive strength = 0.07 to 0.1 MN/$m^2$; cross breaking strength = 0.14 to 0.18 MN/$m^2$.

(*iv*) Very low value of thermal conductivity.

(*v*) Highly resistant to moisture.

(*vi*) Odourless, chemically stable and resistant to fungus attack.

(*vii*) Fully resistant to water, salt, soaps, bleaching agents and HCl (35%), $HNO_3$ (upto 50%), $H_2SO_4$ (upto 95%), caustic soda, caustic potash, strong ammonia, alcohols and silicon oil.

(*viii*) Not resistant to organic solvents like benzene, paint thinners and saturated aliphatic hydrocarbons like petroleum and gasoline.

(*ix*) Very good shock-protecting properties.

(*x*) Ability of being moulded into well-fitting, contoured cases.

**Uses.** 1. Thermocole (with operational range of – 200°C to 80°C) is an excellent material for cold insulation in refrigerators, cold storages, air conditioning, chilled pipe lines and chemical processes.

2. It is used for industrial insulation and insulation for buildings, against extremes of climate.
3. In the form of specially made flexible sheets, thermocole can be used on intermediate concrete floors in multistorey buildings, to *reduce impact sound transmission.*
4. It is used for *packing* electronic goods like transistor, radios, tape recorders and calculating machines, clocks, medicine bottles, cameras etc.
5. It is also used for air-dropped packaging, decorative and gifts packaging and edge protecting packaging.

**Reflecting paper:**

- Reflecting paper (also known as *building paper*) is a strong tough paper which is lined with aluminium or copper foil on the exposed side, which reflects back heat waves coming from a source and this keeps the walls and the enclosed rooms cool.
- Sometimes, reflecting coatings of *varnishes, paraffins, gums or synthetic resins* are applied to various grades of paper of fibrous materials.

**Properties:**

1. Strong and tough in nature.
2. Heat-resistant.
3. Possesses adequate dielectric strength.

**Uses:** Used for *heat insulation purposes.*

**8. Gypsum:**

- It is a hydrated sulphate of calcium ($CaSO_4.2H_2O$) occuring in monoclinic crystals.
- It seldom occurs in nature in pure state; contains impurities such as alumina, calcium carbonate, magnesium carbonate and silica upto 6 percent.
- When it is burnt in kilns, 'Plaster of Paris' is obtained.
- After mixing with asphalt and casting into slabs, it is burnt in a kiln to form very strong sheets which possess very good insulating properties.

**Properties:**

1. Crystalline and fibrous in structure.
2. Controls the setting time of cement.
3. Gypsum boards are good insulators of heat.

**Uses:** Employed for heat insulation purposes. Ceiling panels made of gypsum are used for *suspended ceilings.*

**9. Aluminium foils :**

- These are very thin foils or sheets of aluminium and are also known as 'Alfoils'.
- These are available in the form of paper-backed foils, separated layers of foils and some rigid materials faced with foils.

**Properties:**

1. Light in weight.
2. Low thermal conductivity.
3. Possess smooth and shining surface.
4. Low emissivity (which decreases the radiation losses).
5. Resistant to ordinary atmospheric gases.

**Uses :** Used as *heat insulator in refrigerators.*

**10. Expanded blast furnace slag:**

- It is obtained during the production of iron steel. It is collected in the form of liquid slag which collects on top of the molten iron.
- It is highly resistant to corrosion and attack by insects and microscopic organisms.
- It is also a good fire-resistant. But it has high water absorption properties and can be used only in situation where there is no risk of moisture penetration.

**Uses:** It *can be used in the roofs and in floors above the level of damp proof course.*

**11. Light weight concrete:**

- Light weight concrete, also known as *cellular concrete,* consists of *crushed slag or grits and cement mixed with aluminium powder.* Bubbles of hydrogen gas are then liberated through the concrete mix which form cells and render the plaster porous or form a foam like cellular concrete, known as *light weight concrete.*
- It is made in the form of cellular concrete.

**Uses:** Employed for lining walls and roofs for *heat and sound insulation of buildings.*

**12. Vermiculite:**

- It is the geological name given to a group of hydrated, laminar minerals which are aluminium-iron-magnesium silicates and resemble mica in appearance.
- It possesses excellent fire resisting properties.
- Vermiculite concrete can be made by mixing vermiculite, portland cement and concrete. By varying the density, different strength and thermal properties can be obtained.

**Uses:** 1. *Low density vermiculite* is used for insulating roofs and wall cavities as a loose filler material.

2. *Vermiculite* can be used for protection of steel girders and stanchions. It can be bonded with bitumen and used as a composite thermal insulation and water-proofing material.
3. *Vermiculite concrete* can be used for *in situ* roofing and also for making blocks, tiles and slabs.

**13. Coconut fibres:**

The fibres obtained from the outer layers of coconut are hard and elastic. The felted fibres are sandwiched between paper and covered on both sides with a layer of bitumen. Such a material in completely *water repellent:*

**Use:** (*i*) Employed for underfloor insulation and floating floors.

(*ii*) Also used for internal insulation of walls.

**14. Cellulose:**

- It is made by converting paper waste or other wood to fibre form by addition of some chemicals like borax, boric acid, aluminium sulphate, etc.

Mostly cellulose insulating material is obtained by shredding and pulverizing waste paper and mixing it with dry chemicals.

**Uses:** It is used as *loose fill* in the insulation of ceilings and walls of residential and commercial buildings, both for new construction and retrofit purposes.

Table 15.1 shows the density and thermal conductivity of some of the common building and insulating materials. It has been observed, in general, that materials of low density provide better thermal insulation than those having higher density.

**Table 15.1. Density and Thermal Conductivity of Some of the Common Building and Insulating Materials**

| S.No. | Material | Density (kN/m$^3$) | Thermal conductivity (kJ/mh°C) | S.No. | Material | Density kN/m$^3$ | Thermal conductivity (kJ/mh°C) |
|---|---|---|---|---|---|---|---|
| 1. | *Glass fibre* | 0.48 | 0.121 | 11. | *Sand-lime bricks* | 18.4 | 3.89 |
| 2. | *Hair felt* | 0.80 | 0.142 | 12. | *Sandstone* | 20 | 4.69 |
| 3. | *Glass cellular* | 1.68 | 0.255 | 13. | *Limestone* | 21.8 | 5.53 |
| 4. | *Saw dust* | 1.92 | 0.214 | 14. | *Cement concrete* | 22.4 to 22.8 | 5.19 |
| 5. | *Compressed straw slab* | 3.68 | 0.31 | 15. | *Asphalt* | 22.4 | 4.40 |
| 6. | *Fibre board* | 2.4 to 4 | 0.19 to 0.23 | 16. | *Glass* | 25.1 | 3.79 |
| 7. | *Wood-wool slab* | 4 | 0.297 | 17. | *Granite* | 26.4 | 10.55 |
| 8. | *Asbestos cement sheet* | 15.2 | 1.05 | 18. | *Plastering* | 12.8 to 16 | 2.09 |
| 9. | *Artificial stone* | 17.6 | 4.77 | 19. | *Terrazo* | 24.3 | 5.77 |
| 10. | *Ordinary bricks* | 17.6 | 2.93 to 6.03 | 20. | *Timber* | 4.8 to 27 | 0.519 |

## 15.4 SOUND INSULATING MATERIALS

- *The materials which offer resistance to reflection and transmission of sound are called* **sound insulating materials.**
- The sound absorbent materials can be incorporated in building structures either in compressed state or in suspended state or in free state.

### 15.4.1 Requirements of Sound Insulating Materials

A good sound insulating material should fulfil the following *requirements* :

1. It should be able to absorb noise to the desired extent.
2. It should be such that it can be cleaned, washed or painted easily to maintain a clean appearance.
3. It should be resistant to attacks of vermin, insects, termite and dry rot.
4. It should be fire-resistant.
5. It should be able to withstand weathering effects.
6. It should be lighter so that it can be handled and fixed easily.
7. It should be economical in its initial cost.

8. In its finished form, it should offer pleasant appearance.
9. It should be economical in its initial cost.

### 15.4.2 Classification and Description of Sound Insulating Materials

The sound insulating/absorbent materials may be classified as follows:

**1. Soft-materials:**

These materials *have sufficient porosity and are good sound absorbers.*

*Examples : Asbestos, rockwool, glass, silk, Hanfelt.*

**2. Semi-hard materials:**

They are *stiff enough to stand rough handling* and *can also serve as building panels.*

*Examples: Mineral wool boards, cane fibres.*

**3. Hard materials:**

These are hard materials which have been made porous during manufacture. They also serve as protective surfaces.

*Examples: Porous tiles of masonry.*

Some of the *common types of sound absorbent materials* are *described* below:

**1. Acoustical plaster:**

- It is made from the mineral called *perlite.*
- It is also called *fibrous plaster.*
- It possesses an absorbent coefficient of 0.30 at 500 cycles per second.
- It is available in different colours and is usually applied in two coats to a finished thickness of about 12 mm. The rough finish of such a surface has slightly better sound absorption effect as compared to a smooth finish.
- The acoustic plaster boards are also available. They can be fixed on the wall and 'then' coefficient of absorption varies from 0.15 to 0.30.

**2. Acoustic pulp:**

- It is mainly composed of asbestos and cellulose fibres mixed with certain binders and preserving chemicals. When water is added it becomes plastic and can be applied to wall and ceiling surfaces to a thickness of upto 20 mm.
- It is applied in layers of 6 mm thickness.
- It is easily shaped and finished (being plastic).

**3. Unifil accoustical plaster:**

- It is manufactured from vermiculite, other constituents being gypsum and lime or portland cement. Water is added to make it plastic for application.
- It is an inert, featherweight and a granular substance.

**Uses:** It is adopted to energy type of architectural treatment and is used mainly for *interior finishes.*

**4. Acoustical boards or tiles:**

- They are usually made of either compressed cane or wood fibre or mineral wool.
- They are pre-finished at the factory and can be painted or coloured to give desirable decorative appearance and light reflection characteristics.
- They have uniform physical and sound absorption characteristics.
- They can be made in different sizes and shapes to suit the requirements.
- The accoustical tiles are, however, relatively costly than other absorbent materials.
- Pre-fabricated boards may be of unperforated or perforated type.

*Acoustic tiles are most suitable for rooms in which small area is available for acoustical treatment.*

Some of the common types of boards and tiles made of material other than mineral fibre are:

(*i*) Corkoustic
(*ii*) Celotex building board
(*iii*) Acousti-lux
(*iv*) Heraklith
(*v*) Econacoustic tiles
(*vi*) Unifil acoustical tiles
(*vii*) Acousti-celotex
(*vii*) Muffletone tiles.

**5. Glass fibres:**

- Glass fibres of 0.015 mm diameter are used for the purpose of sound insulation.
- When faced on one or both sides with a non-metallic flexible material such as paper, muslin, or glass cloth it is known as *fibre glass blanket*. Such *blankets* are extensively used in broadcasting, television and film studios (available from 12 mm to 100 mm in thickness)
- Fibre glass boards and tiles can be manufactured into a lightly decorative material having uniform texture and excellent light properties.

**6. Quilts and mats:**

- Quilts and mats are *prepared from glass wool or mineral wool and are fixed in the form of acoustic blankets.*
- Their absorption coefficients depend on the density, thickness, perforation, nature of backing, frequency of sound and mode of fixing.

**Note:** For electrical insulating materials please refer Art. 20.4.

## HIGHLIGHTS

1. In order to create a comfortable and healthy environment, it is required to insulate the building to reduce excessively low or high temperatures and to minimize outside noise disturbance.
2. A good *heat insulating material* should have a low thermal conductivity value, be a poor absorber of moisture, possess adequate strength and easy to apply and maintain.
3. Thermal insulation can be achieved by forming air spaces between structural components. Coating the surfaces with reflective material helps reduce transfer of heat.
4. The materials commonly employed for purpose of thermal insulation are:
   Aerated concrete, blast furnace slag, gypsum, asbestos, vermiculite, cork, rock wool, cellulose, fibre glass, cellular plastic. The material is applied in the form of sheets, board and blankets.
5. *Sound insulation* is carried out to minimize the outdoor/indoor noise. A good sound insulating material should be able to absorb the noise to the desired level, should be resistant to moisture, vermin and insects; should be incombustible, lighter in weight, easy to maintain and also offer pleasant appearance.
6. The materials commonly employed for sound insulation are:
   Asbestos, cellulose fibre, cork, wood, gypsum, fibre, fibre glass, perlite, vermiculite etc.

## OBJECTIVE TYPE QUESTIONS

**Fill in the Blanks/Say 'Yes' or 'No':**

1. The heat insulation materials are required to give protection against .............. and ..............
2. The heat insulating materials should be protected against moisture.
3. .............. is derived from the bark of oak tree.
4. Thermal conductivity of cork is high.
5. Cork is not affected by moisture.
6. Glass fibres catch fire easily.
7. .............. wool is a form of fibrous glass with short and fine fibres, scattered in various directions.
8. .............. wool is prepared from flint rock containing some calcarious matter.

9. The specific gravity of rock wool is about 0.48.
10. .............. wool is an aggregate of fine filaments of slag produced by blowing air through a stream of blast furnace slag.
11. .............. is a mineral fibre composed of hydrous silicate of magnesia with a small amount of iron oxide and alumina.
12. Asbestos is resistant to corrosion and vermin effects.
13. Thermocole is one of the trade names of 'Polystrene'.
14. Thermocole has very poor shock-protecting properties.
15. .............. is a hydrated sulphate of calcium occuring in a monoclinic crystals.
16. Light weight concrete is also known as .............. concrete.
17. Vermiculite possesses excellent fire resisting properties.
18. Acoustical plaster is made from the mineral called ..............
19. Acoustic pulp is applied in layers of 6 mm thickness.
20. Accoustic tiles are most suitable for rooms in which small area is available for accoustical treatment.

## ANSWERS

| | | | | |
|---|---|---|---|---|
| 1. heat, cold | 2. Yes | 3. cork | 4. No | 5. Yes |
| 6. No | 7. Glass | 8. Rock | 9. Yes | 10. Slag |
| 11. Asbestos | 12. Yes | 13. Yes | 14. No | 15. Gypsum |
| 16. cellular | 17. Yes | 18. pertile | 19. Yes | 20. Yes |

## THEORETICAL QUESTIONS

1. How are heat insulating materials classified?
2. What are the requirements of a good heat insualting material?
3. List the different types of thermal insulating materials.
4. Write short notes on the following:
   Rockwool, Gypsum, Thermocole, Light weight concrete, Vermiculite.
5. Give the classification of sound insulating materials.
6. State the requirements of a good sound insulating material.
7. Write short notes on the following :
   Acoustical plaster, Accoustical tiles, Quilts and mats.

# 16

# Fly-ash, Gypsum and Gypsum Plaster

*16.1. Fly-ash – general aspects – applications of fly-ash. 16.2. Gypsum. 16.3. Gypsum plaster. Highlights – Objective Type Questions – Theoretical Questions.*

## 16.1 FLY-ASH

### 16.1.1 General Aspects

**Composition:**

- The **fly-ash** *is the residue from the combustion of pulverized coal collected by the mechanical or electrostatic separators from the flue gases of thermal power plants.* The fly-ash obtained from electrostatic precipitators is finer than portland cement, whereas that obtained from cyclone separators is comparatively coarser and may contain large amounts of unburnt fuel.
- It consists mainly of spherical glassy particles ranging from 1 to 150 μm in diameter, of which the bulk passes through a 45-μm sieve.
- The fly-ash, like portland cement, contains oxide of calcium, aluminium and silicon, but the amount of calcium oxide is considerably less.

The **constituents** of fly-ash are: *Silicon dioxide* ($SiO_2$) = 30 to 60%;
*Aluminium oxide* ($Al_2O_3$) = 15 to 30%; *Unburnt fuel* (carbon) = up to 30%;
*Calcium oxide* (CaO) = 1 to 7%; *Magnesium oxide* (MgO) = small amounts;
*Sulphur trioxide* ($SO_3$) = small amounts.

In fly-ash, the carbon content *should be as small as possible*, whereas *the silicon content should be as high as possible.*

The composition of fly-ash varies with the *type of fuel burnt, load on boiler* and *type of separator* etc.

**Disposal of fly-ash:**

Presently, the fly-ash is disposed off in the following two ways:

1. **Dry system:** Here the fly-ash is transferred into an overhead silo or a bunker at the plant *penumatically.* The use of mechanical means (*e.g.* screw conveyors etc.) for the removal of fly-ash is restored to only if the quantity of flash to be handled is small.
2. **Wet system:** In this system of disposal of fly-ash, the *fly-ash is mixed with water and sluiced to the settling ponds or dumping areas near the plant.* The system, however, will work satisfactorily if the following *conditions* are satisfied: (*i*) The water supply is available continuously; (*ii*) Large areas of wasteland for ponding are available; and (*iii*) The filled-

up ponds are emptied regularly.

The present system of fly-ash disposal causes the following *problems:*

1. The region's ecology is disturbed.
2. Wastage of costly land (due to dumping of fly-ash) in the vicinity of thermal power stations.
3. The fly-ash causes health hazard to the people living near the thermal power stations.
4. The transport of fly-ash to the selected ash ponds entails heavy expenditure.
5. Soil, vegetation, under ground-water resources etc. get polluted.

### 16.1.2 Applications of Fly-ash

The fly-ash may be used in the following ways:

1. The fly-ash may be used in **concrete** as an *admixture* or *in-part replacement of cement.* The fly-ash is generally used in the following three ways:

- As a *part replacement of cement.*
- As a *part replacement of fine aggregate.*
- As a *simultaneous replacement of cement and fine aggregate.*

2. **Cellular concrete blocks:**

These blocks are light in weight and are produced by *autoclaving a set mix of a fine silicous material such as fly-ash and binder in the form of lime.*

***Advantages:***

(*i*) Low thermal conductivity.

(*ii*) Stable against temperature and humidity variations.

(*iii*) Better sound insulation.

(*iv*) Better strength to weight ratio.

(*v*) Since the cellular concrete blocks are machine finished and uniform in size, therefore, less quantity of cement mortar is required. Further as the blocks are smooth and uniformly coloured, plaster can be avoided completely.

3. **Fly-ash building bricks:**

- To manufacture fly-ash building bricks, the *fly-ash*, *sand* and *lime* are mixed approximately in the ratio of 80 : 13 : 7. The bricks are made by using *hydraulic press and the semi-dried bricks are cured in a steam chamber at an appropriate temperature and pressure.*
- As compared to conventional bricks, these bricks are:

— *superior in shape, technical specifications, impermeability* and *compressive* strength;

— *light in weight* (20 percent); and

— *cheaper* (about 10 to 15 percent).

## 16.2 GYPSUM

- The chemical formula for gypsum is $CaSO_4.2H_2O$ (hydrated sulphate of calcium).
- It very seldom occurs in nature in pure state; it contains impurities such as alumina, calcium carbonate, magnesium carbonate and silica. The main gypsum producing regions in India are Bikaner, Jodhpur, Jabalpur and Barmer districts of Rajasthan, Trichurapalli in Tamil Nadu, Nellore in Andhra Pradesh, Hillar, Bhavnagar and Kutch in Gujarat, Dehra Dun and Garhwal in Uttar Pradesh and Uri in Jammu and Kashmir.
- It is a white crystalline substance sparingly soluble in water.
- It is soluble in hydrochloric acid but insoluble in sulphuric acid.
- As a binding material, it sets and hardens quickly.
- The physical form of gypsum may be crystals as selenite, fibrous as stain spar and massive as alabaster.

**Properties of gypsum products/items:**

Following are the *properties of gypsum products/items*:

1. Good sound absorbers.
2. Possess small bulk density.
3. They are incombustible.
4. They have poor strength in wet state.
5. They develop high creep under load, especially in moist surroundings.

In order to inprove the moisture resistance and water-resisting properties, sometimes, the water-proof paints or pastes are applied on the surface of gypsum.

**Uses:** Following are the *uses of gypsum*:

1. It is used as a filler in paint, paper and rubber industries.
2. It is used in the *manufacturing of cement* to increase its setting time.
3. It is also used to prepare *"Plaster of Paris" and gypsum boards*. These boards are popularly known as the *gypboards* and are formed by mixing gypsum with asphalt.

The *gypboard* claims the following *advantages* over other conventional materials:

(*i*) Easy to nail, screw and cut with ordinary carpenter's tools.
(*ii*) Provides a very good final finish.
(*iii*) It is cheaper than plywood partitioning systems.
(*iv*) Very light in weight.
(*v*) It is fire-resistant.
(*vi*) It possesses good sound insulating properties.
(*vii*) Its surface is smooth and hence finishing processes such as painting, veneering, wall papering etc. can be easily carried out.

## 16.3 GYPSUM PLASTER

- On heating gypsum between 120-180°C, it loses about 14.7% of the water of crystallisation in the form of steam and forms *calcium sulphate hemihydrate*, known as **plaster of Paris**.
- On heating further, the remaining water is also expelled and *calcium sulphate anyhydrite* is formed, which is known as second settle plaster. This plaster is mixed with a small amount of accelerator (*i.e.* alum potassium sulphate, raw gypsum) to make **anhydrous** or **Keene's gypsum plaster.**

**Manufacture:**

The manufacture of gypsum plaster involves the following steps:

— *Finishing of raw gypsum*;
— *Calcination;* and
— *Addition of suitable retarders.*

Finishing process consists of crushing and grinding of gypsum to 80% passing 150 micron IS sieve. Among the various processes employed for calcination of gypsum, the **open pan process** is foremost, and 80% gypsum plaster is manufactured by this method.

**Properties:**

1. It is practically unaffected by bacteria.
2. It is light in weight (thus ideally suited for use in multi-storeyed buildings).
3. The plaster shows good adhesion to fibrous materials.
4. It is slightly soluble in water (hence it cannot be adopted for damp conditions and external work).
5. It sets by natural process of crystallization, hence it can be applied with ease and without wastage.

6. It sets with little change in volume and with negligible shrinkage on drying.
7. Since it dries up quickly, the finishing coat can be applied soon after the undercoat has set, usually the same day.
8. A *porous gypsum is an important sound absorbing material while dense and hard plasters are efficient reflectors of sound upto 90% of sound energy incident at their surfaces.*

**Uses:** 1. It is used as a heat insulating material to protect wood or metal columns and beams from high temperature.
2. It is used for ornamental plaster work.
3. The *gypsum plaster boards* which are cheap, light in weight, fire-proof and easy to work are used for *ceiling,* for *internal lining of walls and for partition walls.*
4. **Gypsum blocks** are very useful in the construction of non-load bearing partitions in dry conditions because of good thermal insulation and small dead load and therefore facilitate speedy construction. These *blocks are also used for the protection of columns, beams and elevated shafts against fire.*
5. Slotted gypsum plaster tiles 13 mm thick measuring 300 mm × 300 mm size can be prepared using calcined gypsum and saw dust mix in proportion of 65 : 35 parts, by weight. These **acoustic tiles** are used in the *large offices and anditoria to reduce the echo.* The effectiveness of tiles depends on their surface porosity and extent to which pores are interconnected.
6. The plaster of Paris is used in pottery, artwork, dentistry etc.

## HIGHLIGHTS

1 The *fly-ash* is the residue from the combustion of pulverized coal collected by the mechanical or electrostatic separators from the flue gases of thermal power plants.

In fly-ash, the carbon content should be as small as possible, whereas the silicon content should be as high as possible.

The fly-ash may be used in concrete as admixture or in part replacement of cement. It may also be used in the form of cellular concrete blocks and fly-ash building bricks.

**2.** Gypsum ($CaSO_4 . 2H_2O$) is a white crystalline substance sparingly soluble in water.

Gypsum items/products are good sound abserbers, incombustible, possess small bulk density and have poor strength in wet state.

**3.** Manufacture of gypsum plaster involves finishing of raw gypsum, calcination, finally regrinding and mixing it with suitable retarders. Gypsum plaster is practically unaffected by bacteria, shows good adhesion to fibrous materials, light in weight and slightly soluble in water.

Gypsum plaster is used to protect wood or metal columns and beams from high temperature; used in pottery, artwork, dentistry; used for internal lining of walls and for partition walls etc.

## OBJECTIVE TYPE QUESTIONS

**Fill in the Blanks/say 'Yes' or 'No' :**

**1.** The fly-ash resembles a pozzolana.
**2.** The fly-ash obtained from electrostatic precipitators is finer than portland cement.
**3.** In fly-ash the carbon content should be as high as possible.
**4.** In fly-ash the silicon content should be as small as possible.
**5.** The composition of fly-ash varies with the type of fuel burnt, load on boiler and type of separator etc.
**6.** In fly-ash the constituent silicon dioxide varies from - to - percent.
**7.** .........................concrete blocks are produced by autoclaving a set mix of a fine siliceous material such as fly-ash and builder in the form of lime.
**8.** The chemical formula of gypsum is ....................
**9.** Gypsum is a white crystalline substance sparingly soluble in water.
**10.** Gypsum products possess large bulk density.

11. Gypsum items have poor strength in wet state.
12. Gypsum products are good sound absorbers.
13. Gypsum boards are popularly known as ...........
14. Gypsum plaster is practically unaffected by bacteria.
15. Gypsum plaster adheres poorly to fibrous materials.

## ANSWERS

| | | | | |
|---|---|---|---|---|
| 1. Yes | 2. Yes | 3. No | 4. No | 5. Yes |
| 6. 30, 60 | 7. Cellular | 8. $CaSO_4.2H_2O$ | 9. Yes | 10. No |
| 11. Yes | 12. Yes | 13. gypboards | 14. Yes | 15. No. |

## THEORETICAL QUESTIONS

1. What is fly-ash?
2. What are the constituents of fly-ash?
3. How is fly-ash disposed off?
4. Describe briefly dry and wet systems of fly-ash disposal.
5. List the problem caused by the present system of fly-ash disposal.
6. Explain briefly the applications of fly-ash.
7. Write short notes on "Cellular concrete blocks" and "Fly-ash building bricks".
8. What is Gypsum?
9. What are the properties of gypsum products/items?
10. Enumerate the uses of gypsum.
11. How is gypsum plaster manufactured?
12. What is plaster of Paris?
13. Give the properties of gypsum plaster.
14. Describe the uses of gypsum plaster.
15. Write short notes on "Gypsum blocks" and Acoustic tiles (gypsum plaster).

**17**

# Elastomers and Composite Materials

*17.1.* ***Elastomers.*** *Introduction – Properties of rubber – Uses of rubber – Types of rubber – Forms of rubber – Plasticization – Compounding – Calendering – Vulcanization – Extruding – Moulding – Reclaimed rubber – Properties of elastomers. 17.2.* ***Composite Materials.*** *General aspects – Classification – Particle reinforced composites – Fibre-reinforced composites – Structural composites – Wood and timber. Highlights – Objective Types Questions – Theoretical Questions.*

## 17.1. ELASTOMERS

### 17.1.1. Introduction

**Elastomer** commonly known as *rubber* is a *linear polymer which possesses elasticity and good resistance to corrosive fluids.*

The properties of rubber depend upon the type of *fillers and adhesives used.*

### 17.1.2. Properties/Characteristics of Rubber

Rubber finds a wide field of application because of the following *properties*:

1. It is elastic; because of this property a rubber band can be stretched to 9 or 10 times its original length and when the load is removed it regains its original length.
2. It is *strong* and *tough*. Because of this property it can be put to use even under abnormal conditions.
3. It is *highly impermeable to both water and air* and therefore, it can be used to retain water as in rubber bottles, hoses etc.
4. It exhibits a great resistance to abrasion, tearing and cutting over a wide range of temperature – 7 to 115°C.
5. It is a bad conductor of heat.
6. It can contain liquids and gases.
7. The synthetic rubber offers great resistance to acids, petroleum products etc.
8. Its properties such as *hardness, strength, abrasion, resistance* etc. can be modified to the desired extent by compounding techniques.
9. When two fresh surfaces of milled rubber are pressed together, they coalesce to form a single piece. This property (known as *tackiness*) of rubber makes the manufacturing of composite articles such as a tyre, from a separate piece very simple.
10. The *plasticity of rubber makes it pliable and amenable to all manufacturing processes.*

11. *By vulcanizing the rubber, its mechanical properties can be considerably improved.*
12. Rubber insulation and other rubber products require shaping prior to vulcanisation because the vulcanised rubber cannot be shaped by mechanical pressing once it has acquired its characteristic plasticity. By vulcanizing rubber in moulds, products of rather intricate shape can be obtained.
13. The electrical insulating properties of pure rubber are inferior to those of vulcanised rubbers.
14. Ordinary electrical insulating rubbers, have the following electrical characteristics under normal conditions:
    Volume resistivity ............ $1 \times 10^{14}$ to $1 \times 10^{15}$ ohm cm.
    Loss tangent...................... 0.01 to 0.03 at 50 Hz
    Dielectric constant................... 2.5 to 5.
15. The electric strength of organic rubbers strongly depends on the kind of current involved, the degree of stretch, and the time during which the voltage remains applied. When left unstretched and subjected to a short-time 50 Hz test voltage, rubber will have an electric strength within the following limits, depending on the pure-rubber constant.
    For a 20 to 25% rubber content .................. 20 to 30 kV/mm
    For a 30 to 35% rubber content .................. 30 to 45 kV/mm
    The D.C. electric strength of rubber is 2 to 2.5 times the electric strength at 50 Hz.
16. Although rubber is practically water and gas tight its electrical characteristics are affected by moisture, especially for rubbers compounded with considerable quantities of the substances which increase the sensitivity to moisture. Only specially compounded rubbers can maintain their electrical characteristics nearly unchanged when kept continuously in contact with moisture.
17. For *normal rubbers the maximum operating temperature is usually considered to be 55°C; for rubbers of great heat resistance it is equal to 65°C.* For butyl rubbers the working temperature can be as high as 90°C.

**17.1.3. Uses of Rubber**

The following are some of the *uses of rubbers*:

1. Thermal insulation
2. Lining for reservoir
3. Rain water and flexible tubing
4. Belting of all types
5. Tyres and tubes
6. Adhesive
7. Hose pipes, printing rollers etc.
8. Mounting material
9. As a buffer.
10. Gasketing material.

**17.1.4. Types of Rubber**

There are generally *two* types of rubber :

1. Natural rubber
2. Synthetic rubber.

**1. Natural rubber :**

- *Natural rubber* is an elastic material present in the *latex* of certain plants. More than 95 per cent of this rubber is obtained from the latex of rubber trees. Hevea prasiliensis, grows on plantations, mainly in Sri Lanka and Malaya Peninsula. Small quantities of rubber are produced in Brazil from uncultivated trees and from the *guagle* shrub in Mexico and south western United States.

- *Latex* is a milky colloidal fluid containing from 30 to 40 percent of the rubber, the remaining being mainly water and a small amount of protein and resinous material. It oozes from superficial cuts in rubber trees and is collected in containers. A plantation grown tree continues to yield for about 40 years and give enough latex to make from 1.5 kg to 3 kg of rubber each year. Latex is treated in two ways to obtain rubber goods. The crude rubber is either co-agulated from it by acids or heat and then processed; or the latex itself is mixed with appropriate compounding materials and then precipitated directly with appropriate compounding materials and then precipitated directly from solution in the shape to be used, such as rubber glove
- The recovery of rubber from *guayule* is carried out in different manner. The entire plant, which may contain as much as 20 per cent by weight of rubber, is harvested after 4 years growth, ground up and soaked in water. The latex comes at the surface and is skinned off; the woody material becomes water logged and sink to the bottom. The resin content of *guayule rubber* is 18 to 20 percent compared to about 4 percent for hevea rubber. This resin may be extracted by means of solvents and can be recovered and utilized.
- Natural rubber is a polymerised form of isoprene (2-methyl-1, 3-butadiene). The polyhydro-carbon chain consists of 2000-3000 monomer links. The polymerisation occurs by a biochemical reaction in which a particular type of enzyme acts as a catalytic agent.

$$\underset{\text{Isoprene}}{(-CH_2-\underset{\underset{CH_3}{|}}{C}-CH=CH_2-)_n} \qquad \underset{\text{Butadiene}}{CH_2=CH-CH=CH_2}$$

**Crude rubber:**

- It is usually obtained from latex by coagulation with organic acids, washing and coagulum with water as it passes between rolls, and finally drying the washed sheet as it comes from the rolls.
  - — If the *rough rolls* rotating at *high speeds* are used and the rubber is then *hung in the air to dry*, the product is known as pale **"crepe rubber".**
  - — If *smooth rolls* rotating at the *same speed* are used and the rubber is then *dried in a smoke house*, the product is known as **smoke sheets.** It is tougher than pale crepe.
- Crude rubber is tough, strong, elastic substance made up of 92% or more of a hydro-carbon chain polymer $(C_5H_8)$.
- Crude rubber does not possess the properties that cause the familiar manufactured articles to be considered so valuable. At the temperature of very hot summer weather, pure rubber becomes soft and sticky, during very cold weather it becomes hard and brittle. In either of these conditions, it is useless for the purposes to which it is commonly put. However, its properties are greatly improved by addition of other materials followed by suitable heat or other treatment.

***Physical properties of crude natural rubber :***

The *physical properties of crude natural rubber* are indicated below :

1. At low temperature rubber becomes stiff and when it froze, it attains fibrous structure.
2. Raw rubber when heated to 130°C becomes soft and plastic. The plasticity can be varied within certain ranges by chemicals.
3. Co-efficient of cubical expansion is $670 \times 10^{-8}$. When rubber is extended heat is produced. This is called **Joule's effect.** When rubber is stretched to 82% it generates 680 calories/gm of heat.
4. Heat of combustion of raw rubber is 10547 calories/gm.
5. Specific gravity of raw rubber at O°C is 0.95 and is 0.934 at 20°C.
6. Specific heat of raw rubber at room temperature is 0.502.

**2. Synthetic rubber:**

- Rubber *produced by artificial chemical process* is known as **"synthetic rubber".**
- Synthetic rubbers, or "*elastomers*" are derived from such raw materials as coke, limestone, petroleum, natural gas, salt, alcohol, sulphur, ammonia and coal tar.
- Elastomers are not strictly speaking synthetic rubber, for rubber has never been synthesized. They are rubber like materials which have many of the characteristic properties of rubber and some of which resemble rubber in their chemical nature.
- The processing of synthetic rubbers involves approximately the same steps and equipments as that of crude rubber. Their properties while similar to those of rubber, are capable of wider variation. Some elastomers are more resistant than rubber to sunlight, others have greater solvent resistance; and some have greater elasticity.

Details for some of the commonly used synthetic rubbers are as under :

***(i)* Neoprene (Polychloroprene) rubber:**

- It is one of the most successful synthetic rubber, and it is in some ways superior to natural rubber. Its basic unit is a chlorinated butadiene whose formula is :

$$CH_2 = \overset{\overset{Cl}{|}}{C} - CH = CH_2$$

- The raw material for this product is acetylene, which makes the product more expensive than some other synthetic rubbers.
- It has *good mechanical properties, considerable resistance to oils,* and *excellent resistance to oxidation in the air.*
- It can be vulcanised by heat alone but generally zinc oxide is incorporated to control and facilitate vulcanisation.

**Uses :**

(*i*) Shoe heels

(*ii*) Protective clothing

(*iii*) Gaskets

(*iv*) Cable insulation

(*v*) Solid tyres

(*vi*) Oil seals

(*vii*) LPG, oils, petrol hoses.

***(ii)* Butyl rubber:**

- It is made from isobutylene and isoprene by addition polymerisation as shown below :

$$\underset{\text{Isobutylene}}{CH_2 = \underset{\underset{CH_3}{|}}{\overset{\overset{CH_3}{|}}{C}}} + \underset{\text{Isoprene}}{CH_2 = C - C + CH_2} \xrightarrow[\text{Polymerisation}]{\text{Addition}} \underset{\text{Butyl rubber}}{\underset{\underset{CH_3}{|}}{\overset{\overset{CH_3}{|}}{CH_2}}} - CH_2 - C = \overset{\overset{CH_3}{|}}{C} - CH_2 -$$

- Its chief properties are *extremely low impermeability to gases and high resistance to acids, alkalies and ozone.*
- It can be used for service temperature from –50°C to 100°C.

**Uses:**

(*i*) Automobile tubes
(*ii*) Tubeless tyres
(*iii*) Life jackets and life rafts
(*iv*) Cable insulation
(*v*) Lining of steam hoses
(*vi*) Lining of diaphragms etc.

**(*iii*) Silicone rubber:**

- The silicone rubbers are based on the chains of silicone atoms rather than carbon atoms. They are therefore, expensive than carbon based rubbers. By hydrolysis and condensation of dimethyldichlorosilane; we get silicone oil with $n$ = upto 2000 approx. By further increasing the value of $n$ between 5000 to 10000 a gummy material is obtained. By oxidation, cross-linking takes place to produce silicone rubber.

$$n\,Cl-\overset{CH_3}{\underset{CH_3}{Si}}-Cl \xrightarrow[+\text{ Condensation}]{\text{Hydrolysis}} O(-\overset{CH_3}{\underset{CH_3}{Si}}-O)_n$$

Dimethyldichlorosilane — Silicone oil

**Properties :**

(*i*) These rubbers can withstand wide range of service temperature between –100°C to 300°C.

(*ii*) It possesses excellent electrical properties and unique resistance to sunlight, water and other gases.

(*iii*) It is colourless, tasteless and non-toxic.

(*iv*) It does not stain, corrode or deteriorate materials with which it comes in contact.

(*v*) It performs well for sealing purpose between –70°C to 250°C.

**Uses :**

(*i*) Rubber tubing for cables
(*ii*) Gas masks
(*iii*) T.V. corona shield
(*iv*) Medical tubing
(*v*) Sealing gas rings in jet engines
(*vi*) Rubber rolls
(*vii*) Sealant for expansion joint
(*viii*) Aircraft and missile wire insulation
(*ix*) Gaskets for oven and refrigeration doors etc.
(*x*) Hot air duct in aircraft
(*xi*) Nuclear power cable.

**(*iv*) Styrene butadiene rubber (SBR or Buna-S) :**

- It is also known as *bunas rubber*.
- This type of rubber is *produced by addition polymerisation* of butadiene and styrene.
- Chemically it is represented as under :

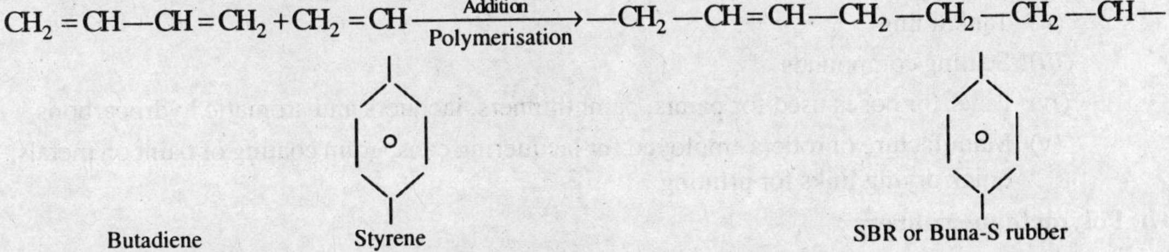


- Most of its properties are similar to natural rubber.
- It is cheaper and more resistant to weathering.

**Uses :**

(*i*) Conveyor belts
(*ii*) Car tyres
(*iii*) Cable insulation
(*iv*) Hose pipes
(*v*) Moulded articles.

**(*v*) Nitrile rubber (Butadiene acrylonitrile NBR, Buna—N Rubber) :**

- Nitrile rubber (closely related to natural rubber) is produced by the *addition polymerisation of butadiene and acrylonitrile (15 to 40 percent).*

$$\underset{\text{Butadiene}}{CH_2 = CH - CH = CH_2} + \underset{\text{Acrylonitrile}}{\underset{\displaystyle CN}{CH_2 = \overset{}{\underset{|}{CH}}}} \xrightarrow[\text{Polymerisation}]{\text{Addition}} \underset{\text{NBR-Buna N rubber}}{—CH_2 — CH = CH_2(—CH_2 — \underset{\underset{\displaystyle CN}{|}}{CH})_n}$$

- It has excellent resistance to oils and greases.
- It is extremely resistant to organic liquids.
- It can be used for temperatures from –50°C to +100°C.
- *Nitrile rubbers* are more expensive than SBR, butyl and natural rubbers. They have *good resistance to oils and organic solvents, and possess better weathering resistance.*

**Uses:**

(*i*) Tank lining
(*ii*) Ingredient for contact adhesives
(*iii*) Hoses, gaskets and seals
(*iv*) Filler for phenol resins.

**(*vi*) Polysulphides (Thiokol)**

- Thiokol rubber is made from *ethylene dichloride and sodium polysulphide by condensation polymerisation.*

$$\underset{\text{Ethylene dichloride}}{Cl — CH_2 — CH_2 — Cl} + \underset{\text{Sodium polysulphide}}{Na_2S_4} \xrightarrow[\text{Polymerisation}]{\text{Condensation}} \underset{\text{Thiokol}}{(—CH_2 — CH_2 — \underset{\underset{\displaystyle S}{\|}}{\overset{\overset{\displaystyle S}{\|}}{S}} — S—)_n} + 2NaCl$$

- Thiokols are extremely resistant to atmospheric weathering, oils, greases.
- They have obnoxious smell which is typical of sulphides.

**Uses:**

(*i*) Moulded rubber articles
(*ii*) Tank lining
(*iii*) Sealing compounds
(*iv*) Liner for hoses used for paints, paint thinners, lacquers and aromatic hydrocarbons
(*v*) Manufacture of rollers employed for lacquering cans, grain coating of paint on metals, quick drying links for printing.

**(*vii*) Polymethane rubbers :**

- When an isocyanate reacts with an alcohol, a methane is produced.
- Polymethane (depending upon the nature of starting material and fillers) can be produced in the following various forms :

— *Flexible foams* : mattresses, acoustic thermal insulation, packaging for delicate equipments;
— *Rigid foams* : thermal insulation;
— *Elastomers* : sealants, gaskets;
— *Liquids* : for paints, adhesives.

- These rubbers have *higher tensile strengths, tear and abrasion resistance than other rubbers.*
- These are *relatively hard and offer good resistance to oxygen and ozone.*

**Uses:**

(*i*) Oil seals, diaphragms
(*ii*) Industrial chute lining
(*iii*) Tyres of fork lift trucks
(*iv*) Heels and soles of shoes.

**(*viii*) Fluorocarbon rubbers :**

- Rubber based upon *fluorine containing monomers are known as* ***fluorocarbon rubbers.***
- Most compounds can be moulded either by compression, transfer or injection as with other rubbers. Curing temperature is between 150 to 170°C and cure time is 5 to 30 minutes.

**Uses :**

(*i*) Gas masks
(*ii*) Nuclear power cables
(*iii*) Sealant for expansion joints
(*iv*) T.V. corona shield
(*v*) Insulation in frying pan set
(*vi*) Gaskets for refrigeration doors.

The important *characteristics and typical applications for five commercial elastomers* are given in the Table 17.1.

**Table 17.1. Important characteristics and typical applications of commercial elastomers**

| *S. No.* | *Chemical type* | *Trade (Common) name* | *Elongation (%)* | *Useful temperature range (°C)* | *Major application characteristics* | *Typical applications* |
|---|---|---|---|---|---|---|
| 1. | **Natural polyisoprene** | *Natural Rubber (NR)* | *500-700* | *–60 to 120* | Excellent physical properties; good resistance to cutting, grouping, *and abrasion; low heat, ozone, and oil resistance; good electrical properties.* | Pneumatic tyres and tubes; heels and soles; gaskets |
| 2. | **Styrene butadiene copolymer** | *GRS Buna S (SBR)* | *450-500* | *–60 to 120* | *Good physical properties; excellent abrasion resistance; not oil, ozone, or weather resistant; electrical properties good, but not outstanding.* | *Same as natural rubber* |

| | | | | | | |
|---|---|---|---|---|---|---|
| 3. | **Acrylonitrile butadiene copolymer** | *Buna A Nitrile (NBR)* | *400-600* | *–50 to 150* | *Excellent resistance to vegetable, animal and petroleum oils; poor low-temperature properties; electrical properties not outstanding.* | *Gasoline, chemical and oil hose; seals and O-rings, heels and soles* |
| 4. | **Chloroprene** | *Neoprene (CR)* | *100-180* | *–50 to 150* | *Excellent ozone, heat and weathering resistance; good oil resistance, excellent flame resistance; not as good in electrical applications as natural rubber* | *Wire and cable; tank linings; belt; hoses, seals, and gaskets* |
| 5. | **Polysilo-xane** | *Silicone (VMQ)* | *100-180* | *–115 to 315* | *Excellent resistance to high and low temperatures; low strength; excellent electrical properties.* | *High and low-tempe-rature insulation; seals, diaphragms; tubing for food and medical uses.* |

### 17.1.5. Forms of Rubber

The important forms of rubber are described below:

1. **Gautta percha rubber:**
   - It is a variety of natural rubber, prepared from the leaves of tress known as the dichopsis gutta and palaginum gutta (mostly grown in Malaya peninsula).
   - It becomes soft and sticky at a temperature of about 100°C.
   - As compared to other varieties of rubber, it absorbs less water.

**Uses:** It is extensively used for preparing *ropes of submarine* and as an insulating material in electrical works.

2. **Foam rubber:**

It is prepared by adding the chemically producing gases in the liquid latex and stirring the mixture till foam is formed. It is then converted into solid form and is given the desired shape.

**Uses:** It is widely used for packing pads, pillows etc.

3. **Sponge rubber:**

It is prepared by adding sodium bicarbonate during the process of vulcanization. The evaporation of moisture leaves pores which result in sponge rubber.

**Uses:** It is used as a *heat and sound insulating material.*

4. **Guayle rubber:**
   - It is a variety of natural rubber available in North America.
   - It is prepared from the branches of guayule.
   - It contains 70 percent of hydrocarbon, 20 percent of resin, 10 percent of insoluble materials, cellulose, liquir etc.

5. **Smoked rubber:**
   - It is a variety of crude rubber.
   - The rubber pieces after coagulation are dried in room filled with smoke at a temperature of about 40°C to 50°C.
   - It is so named as drying is carried out in a smoke-room.

6. **Polybutadiene rubber:**

It is a variety of synthetic rubber (produced by The Indian Petrochemicals Corporation Ltd. near Baroda); commercially known as *Cisrub*.

- It has high abrasion resistance and strength.

**Uses:** It is nicely used in producing beltings, floor tiles, automotive moulded goods, hoses, tyres, seals, gaskets etc.

### 17.1.6. Plasticization

In order to make crude rubber workable, it is kneaded to a plastic mass by passing it through heated rollers which travel towards each other at different speeds. By this process the rubber is subjected to both compression and shearing stresses which cause the rubber to twist, crack or tear and become plastic. Other ingredients can also be added into the rubber mass during the process.

### 17.1.7. Compounding

The rubber available in plastic state (as mentioned in Art 17.1.6) is *compounded* with other ingredients to *modify its properties*. A number of different substances which can be mixed with rubber to give it definite properties, are classified as follows:

(i) *Vulcanates:*

- These are the substances (*e.g.* sulphur etc.) which are added to crude rubber to reduce its plasticity while maintaining in elasticity.
- These are essentially required for carrying out the process of vulcanization.

(ii) *Plasticizers:*

These are the materials (e.g. vegetable oils, waxes, stearic acid, etc.) which are used to *soften the crude rubber*.

(iii) *Accelerators:*

- These are the substances (*e.g.* lime, magnesia, litharge, white lead etc.) which are used to *reduce the time required for the completion of the process of vulcanization.*
- With their use, the quantity of sulphur required for vulcanization is also reduced.

(iv) *Antioxidants:*

These are the materials (*e.g.* phenols, amines, waxes etc.) which when added in small quantities to rubber decrease the rate of its deterioration by agencies like light, heat and by the presence of copper or manganese, salts or soaps etc.

(v) *Fillers, reinforcing agents and colouring agents:*

- The addition of **fillers** is made *to modify the properties of rubber and to lower the cost of rubber product.*
- A **reinforcing agent** is one which when added to rubber increases its tensile strength. Materials like zinc oxide and carbon black serve both as fillers and reinforcing agents. Some other reinforcing agents are magnesium carbonate, calcium carbonate and barium sulphate.
- **Colouring agents** are added to obtain the rubber in desired shade.

**Example:** *Ferric oxide* — red colour; *Lithophone* — white colour; *Lead chromate* — yellow colour etc.

### 17.1.8. Calendering

A calendering machine is used to obtain rubber in the *form of sheets of definite thickness*. The rubber mass is made to pass between the large steel rollers of the machine, at carefully controlled temperature and pressure. The rollers press the rubber mass into sheets of uniform thickness and desired finish.

### 17.1.9. Vulcanization

- *Vulcanization* is the process in which *plasticity of the rubber compound is reduced without affecting its elasticity.*

- In this process 1 to 3% of sulphur is added to the rubber compound which is then heated in closed moulds which are heated by steam under a pressure of about 3 bar. Consequently, the sulphur gets chemically combined with rubber and results in remarkable changes in the properties of the rubber compound.
- Vulcanization normally results in *increased elasticity and strength, increased durability and increased resistance to adverse effect of weather and chemical agents.*
- Vulcanized rubber has *low electrical and thermal conductivity and is resistant to abrasion and action of chemical reagents and weather.*
- The process of vulcanization, in the manufacture of rubber goods, is carried out after calendering and extrusion. When the rubber goods are to be manufactured by the process of moulding, vulcanization is performed simultaneously with moulding.
- Fabricating operations and lathe cutting are performed after the vulcanization process.

**17.1.10. Extruding**

- By this process rubber goods of *constant cross-section* like tubes, insulation on wire etc. can be manufactured.
- In this process, this rubber compound is fed into the extrusion die by means of a screw fed and the extruded products are either coiled or laid in lengths in large flat trays and vulcanized in open steam heaters.

**17.1.11. Moulding**

- In this process the *rubber is forced into the mould* (of heavy metallic construction) under *high pressure* (rubber is not melted and poured like metal).
- For getting non-porous uniform product, the process of moulding is performed in heated moulds in which the rubber compound is foiced under high pressure.

**17.1.12. Reclaimed Rubber**

- Reclaimed rubber is the rubber *recovered from worn-out rubber articles and rubber wastes from factories and then treated suitably for re-use.*
- For reclaiming rubber, the metal and fabric contents of the worn-out rubber are first removed and then it is *heated in a closed iron vessel which contains an alkali solution.* This treatment makes the rubber *free from the remaining fabric contents and free sulphur*. The treated rubber is washed and dried.

**Advantages:**

1. It is durable.
2. It is cheap.
3. Can be prepared speedily.
4. Has a uniform composition.

**Disadvantages:**

1. Elasticity is low.
2. Low resistance to friction.
3. Low tensile strength.

**Uses:** It may be used for making the articles like shoe soles and heals, water-proof foot-wear, electrical goods etc.

**17.1.13. Properties of Elastomers**

The *properties of elastomers* are described below:

**1. Tensile strength, hardness and compression set:**

- In natural and synthetic rubber, the uncoiling of molecular chains around myriads of bond centres allows extensibilities of as much as 800%. Above this point the ordinary laws of mechanics of rigid bodies apply.

- The stress-strain curve becomes steep and the rupture point rapidly reaches. Typical stress-strain curves for elastomers are shown in Fig. 17.1.

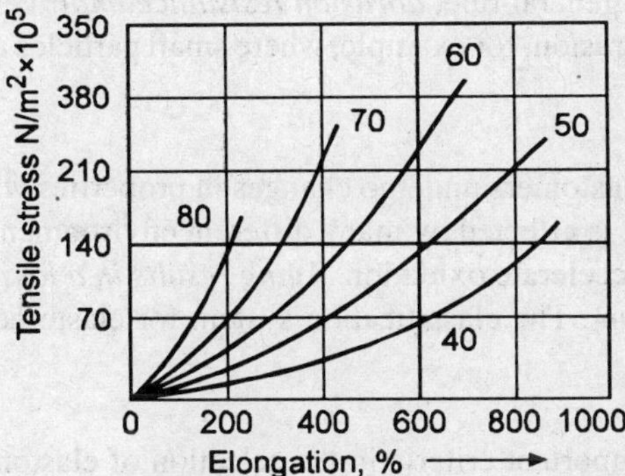


Fig. 17.1. Typical stress strain curves for elastomers of various hardness.

The *modulus increases with hardness.* There is, however, a wide range of modulus at a given hardness level which is achieved by the use of different fillers and extenders.

— The number 40 compound has a low modulus. The modulus for elastomers is based on original cross-sectional area and the load required to stretch a test piece to a given elongation. Thus, the number 60-durometer compound has a higher tensile strength and lower elongation than 40-durometer compound.

— Compounds in the 50 to 70 durometer range have the best *balance of properties, good impact, abrasion, tear and cut resistance.* Applications of elastomers in this category are: *Belt covers, automobile tyre treads, hose covers, roll covers, automotive bumper, blocks and many general moulded items.* All general purpose rubbers such as styrene-butadiene, natural, neoprene and acrylonitrile-butadiene are included in this category.

— The 80-hard elastomer has a much higher modulus, lower tensile strength and considerably lower elongation. Applications for rubbers in this range include *flat gas belts, spacers, stiff tubing, solid industrial tyres, press pads, soling and floor tile.*

- The properties of rubber compounds are evaluated by using both the *tension set and compression set.*

— **Tension set** refers to the *amount of deformation not recovered after a tensile load.* The length increase is expressed as a percent of the original length. Values depend on the formulation as well as on the amount and length of time of deformation.

— **Compression set** is the *amount (percent) that a standard test piece of rubber fails to return to its original thickness after being subjected to a constant compressive load or a constant deflection for a fixed period of time and temperature.* The hardness of the rubber is used as a guide in determining the extent of the deflection. Elastomers that give satisfactory compression set characteristics are : (*i*) Natural rubber (NR), (*ii*) Styrene butadiene (SBR) and (*iii*) Acrylonitrile-butadiene (NBR).

*Compression set characteristics* are especially important in applications such as seals and gaskets which must maintain tight joints. Moulded gaskets for these applications are designed with a double row of ribs on both sides.

**2. Tear resistance :**

- Elastomers as a class have *low tear strength.*
- If tear strength is extremely low, small nicks or cuts can cause catastrophic failure and under flexing, failure will quickly occur. Also, elastomers with poor tear strength generally have low abrasion resistance.

**3. Abrasion resistance :**

Since several elastomer applications involve friction and wear, abrasion resistance is usually an important consideration. As a general rule, *abrasion resistance improves with increase in hardness.* However, for some kinds of abrasion, for example, where small particles are involved, relatively soft-elastomers perform better.

**4. Deterioration resistance :**

Owing to oxidation all elastomers undergo changes in properties with time. This deterioration, which is referred to as *ageing*, is affected by many different environmental factors such as sunlight, heat and ozone, all of which accelerate oxidation. *Aging results in a loss of resilience, a decrease of hardness and evèntual cracking*. The classification system for elastomers is based in part on their dry-heat resistance (ageing).

**5. Oil resistance :**

- Oil resistance is an important criteria in the selection of elastomers because many of their applications involve an environment of hydrocarbon fluids such as oil and gasoline.
- Resistance to oils is based on how much the elastomer *swells* in the presence of oil.

## 17.2. COMPOSITE MATERIALS/COMPOSITES

### 17.2.1. General Aspects

- A **composite material** is *a combination of two or more materials having compositional variations and depicting properties distinctively different from those of the individual materials of the composite.* The composite material is generally better than any of the individual components as regards their strength, heat resistance or stiffness.
- Composites include the following :
  (*i*) Multiphase metal alloys;
  (*ii*) Ceramics;
  (*iii*) Polymers.

**Examples** of composites are :

(*i*) **Pearlitic steels.** The *pearlitic steels* have a microstructure consisting of alternating layers of $\alpha$ ferrite and cementite; the *ferrite phase is soft and ductile,* whereas *cementite is hard and very brittle.* The combined mechanical characteristics of the pearlite, reasonably high ductility and strength, are superior to those of either of the constituent phases.

(*ii*) **Wood.** It is one of the examples of the composites that occur in nature. It consists of strong and flexible cellulose fibres surrounded and held together by a stiffer material called *lignin.*

*Plywood* is the composite of thin sheets of wood with grains of alternate sheets perpendicular to each other and bonded together by a polymer in between them.

(*iii*) **R.C.C.** It has steel rods embedded in the concrete mix, which, itself, is a composite of cement, sand aggregate and water. The resulting R.C.C. structure can take loads which, otherwise, cannot be carried by the concrete alone. Steel rods can be of different shape, size and provided in various directions.

(*iv*) **Vehicle tyres.** The vehicle tyres are rubber reinforced with woven cords.

- Each of the materials in composites serves one or more specific functions. The properties of composites are affected by the following :
  (*i*) The size and distribution of the constituents in relation to each other;
  (*ii*) The bond strength between them;
  (*iii*) The shape, size, amount and properties of each material.

- The base material surrounding other materials is normally present in higher percentage and is called **matrix.** Other materials which reinforce the properties of base material are called **reinforcements.** Cohesion between the matrix and reinforcement is essential and may take place in any or combination of the following ways :
  (*i*) Chemical reaction at the interfaces of the constituents.
  (*ii*) Mechanical keying between the matrix and the reinforcement.
  (*iii*) Mechanical bonding between the matrix and reinforcement by vander Waals' forces acting between the surface molecules of the various constituents.
- The selection of matrix and reinforcement is made in such a way that their *mechanical properties complement each other,* while *deficiencies are neutralised.* In particular cases there may be more than one type of reinforcement present at the same time. The *effect of reinforcement is to increase both the tensile strength and tensile modulus of the composites.*

### 17.2.2. Classification

The composite materials may be classified as follows : Refer Fig. 17.2.

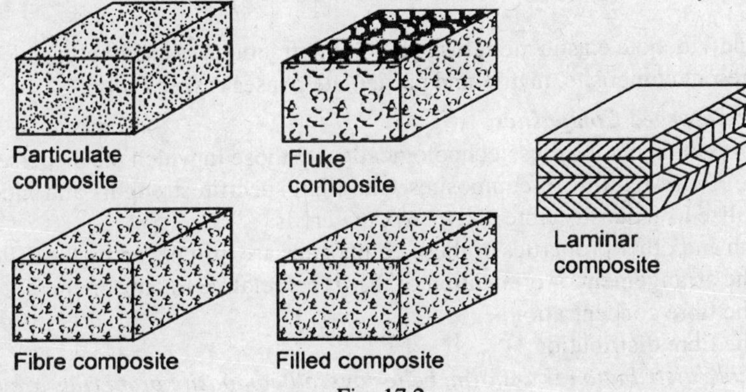


Fig. 17.2.

**1. Particle reinforced :**

(*a*) Large particle
(*b*) Dispersion strengthened.

**2. Fibre-reinforced :**

(*a*) Continuous (aligned)
(*b*) Discontinuous (short)
  (*i*) Aligned
  (*ii*) Randomly oriented.

**3. Structural:**

(*a*) Laminates
(*b*) Sandwich panels.

#### 17.2.2.1. *Particle reinforced Composites*

**(*a*) Large-particle composites :**

- These composites are utilized with all three types of materials, *viz., metals, polymers and ceramics.*

**Examples :**

(*i*) *Automobile tyres*—containing 15 to 30% volume of spherical particles of carbon with 20-50 μm diameter in vulcanized rubber which improves the tensile strength, toughness and resistance against corrosion.

(*ii*) *Concrete*—consisting of cement (the matrix), and sand and gravel (the particulates).

(*iii*) **Cements,** *e.g., cemented carbide*—consisting of extremely hard particles of a refractory carbide ceramic (such as tungsten carbide or titanium carbide) embedded in a matrix of a metal such as cobalt or nickel (load is shared between ceramics and metals). These composites are widely used for *cutting tools* for hardened steels, *drills, electrical contacts, magnets* and *rocket nozzles* etc. Other applications of cemented carbides are *burner nozzles, gauges and plugs used for inspection of materials, grinding balls and liners in grinding mills.* The *cermets are prepared by powder metallurgy techniques.*

- In the case of *two-phase composites*, the following two mathematical expressions have been formulated for the dependence of the elastic modulus on the volume fraction of the constituent phases:

$$E_c \text{ (upper bound)} = E_m V_m + E_p V_p \quad \text{...}(i)$$

$$E_c \text{ (lower bound)} = \frac{E_m E_p}{V_m E_p + V_p E_m} \quad \text{...}(ii)$$

where, $E$ and $V$ denote elastic modulus and volume fractions respectively; and $c$, $m$, $p$ represent composite, matrix and particulate phases respectively.

**17.2.2.2. *Fibre-reinforced Composites***

The most important composites, technologically, are those in which the disperse phase is in the form of the fibre. Fibre-reinforced composites with high specific strengths and moduli have been produced that utilize low-density fibre and matrix materials.

The strength and other properties of these composites are influenced by the following :

(*i*) The arrangement or orientation of the fibres relative to one another;

(*ii*) The fibre concentration;

(*iii*) The fibre distribution.

When the *fibre distribution is uniform, better overall composite properties are realized.*

- **The fibre phase.** The fibres, on the basis of diameter and character, are grouped into three different classifications : *Whiskers, fibres and wires.*

**Whiskers :**

— These are very thin crystals having extremely large length-to- diameter ratios (crystal size ranges from 0.5 to 2.0 microns in diameter to 20 mm length). As a consequence of their small size, they have a high degree of crystalline perfection and are virtually free from flaws, which accounts for their exceptionally high strengths (tensile strength of whisker is approx. 21 GN/m$^2$ compared to 3 GN/m$^2$ for carbon fibre); they are strongest known materials.

— They are *difficult and costly to manufacture* (as such whiskers are not employed extensively as a reinforcement medium).

— Whisker materials include graphite, silicon carbide, silicon nitride and aluminium oxide.

— Boron and carbon whiskers are used in polymeric matrix. Alumina whiskers are used to reinforce the metal nickel.

**Fibres :**

— Fibres are either *polycrystalline or amorphous and have small diameters.*

— The fibrous materials are generally either polymers or ceramics.

**Wires :**

— Wires are used as a radial steel reinforcement in automobile tyres, in filament-wound rocket castings, and in wire-wound high- pressure hoses.

— Fine wires have relatively large diameter; typical materials include steel, tungsten and molybdenum.

- **The matrix phase :** In fibrous composites, the matrix phase may be a metal, polymer, or ceramic. The metals and polymers, in general, are employed as matrix materials because some *ductility is desirable.* In case of ceramic-matrix composites, the reinforcing component is added to *improve fracture toughness.*

*The matrix phase, in case of fibre-reinforced composites, serves the following purposes/functions:*

(*i*) *It binds the fibres together* and acts as a medium by which an externally applied stress is transmitted and distributed to the fibres; only a very small proportion of an applied load is sustained by the matrix phase.

(*ii*) It protects the individual fibres from surface damage as a result of mechanical abrasion or chemical reactions with the environment.

(*iii*) The matrix separates the fibres and, by virtue of its relative softness and plasticity *prevents the propagation of brittle cracks from fibre to fibre*, which could result in catastrophe failure.

In order to minimize fibre pull-out it is essential that *adhesive forces between fibre and matrix be high.*

Some ***important fibre-reinforced composites are :***

- *Polymer-matrix composites*
  — Glass fibre-reinforced polymer (GFRP) composites
  — Carbon fibre-reinforced polymer (CFRP) composites
  — Aramid fibre-reinforced polymer composites.
- *Metal-matrix composites*
- *Ceramic-matrix composites*
- *Carbon-carbon composites*
- *Hybrid composites.*

### 17.2.2.3. Structural Composites

**(*a*) *Laminar composites :***

- These composites are made from the *two dimensional* sheets or panels that have a *preferred high strength direction* (such as found in wood and continuous and aligned fibre reinforced plastics).
- The layers are sacked and subsequently cemented together such that the orientation of the high strength direction varies with each successive layer (*e.g.*, in *plywood*, adjacent wood sheets are aligned with the *grain direction at right angles to each other*).

**(*b*) *Sandwich panels:***

- These composites are composed of *two strong outer sheets or faces* (typical face materials include aluminium alloys, fibre-reinforced plastics, titanium steel and plywood), separated by a *layer of less-dense material* or "*core*" (typical core materials include; foamed polymers, synthetic rubbers, inorganic cements and wood) which has lower stiffness and lower strength.
  The core, structurally performs the following *two functions :*
  (*i*) It separates the face and resists deformation perpendicular to the face plane.
  (*ii*) It provides a certain degree of shear rigidity along planes which are perpendicular to the faces.
- Sandwich panels find wide *applications* in the following :
  (*i*) Roofs, floors and walls of building;
  (*ii*) In aircraft for wings, fuselage and tailplane skins.

**Tensile Modulus of Composite :**

Consider a composite cylinder made up of continuous fibres, all parallel to the axis of the cylinder in a matrix. Each component in the composite shares the applied force. Thus,

Total force = force on matrix + force on fibre

= stress on matrix × area of matrix + stress on fibre × area of fibre

[∵ Force = stress × area]

$$\therefore \frac{\text{Total force}}{\text{Total area}} = \left[\text{stress on matrix} \times \frac{\text{area of matrix}}{\text{total area}}\right] + \left[\text{stress on fibre} \times \frac{\text{area of fibre}}{\text{total area}}\right]$$

∴ Stress on composite = (stress on matrix × % area of matrix) + (stress on fibre × % area of fibre) ...(1)

Strain on composite = strain on matrix = strain on fibre ...(2)

From eqns.(1) and (2), we get

Tensile modulus of composite = modulus of matrix × % area of matrix + modulus of fibre × % area of fibre ...(3)

**Example 17.1.** *Determine the Young's modulus of a composite containing 65 vol % of glass fibre ($E_f$ = 70 GN/m²) in a matrix of epoxy resin ($E_m$=3 GN/m²) under isostress condition.*

**Solution :** Volume fraction of glass, $V_f$ = 65% or 0.65

∴ Volume fraction of matrix, $V_m$=1 – 0.65 = 0.35

We know that Young's modulus of composite is given by

$$E_c = E_f V_f + E_m V_m$$

$$E_c = 70 \times 0.65 + 3 \times 0.35 = \mathbf{46.55\ GN/m^2}\ \textbf{(Ans.)}$$

**Example 17.2.** *Calculate the volume ratio of aluminium and boron in Al-Boron composite which can have the Young's modulus equal to that of iron. The Young's modulus of Al, iron and boron are 71, 210, 440 GN/m² respectively.*

**Solution :** *Given :* $E_{al} = 71\ \text{GN/m}^2$; $E_i = 210\ \text{GN/m}^2$; $E_b = 440\ \text{GN/m}^2$.

**Volume ratios :** $V_{al}$; $V_b$:

We know, $E_i = E_{al}V_{al} + E_b V_b$

$$210 = 71 \times V_{al} + 440\ V_b \quad \ldots(i)$$

Also $V_{al} + V_b = 1$

∴ $V_{al} = 1 - V_b$

Substituting the values of $V_{al}$ in eqn. (*i*), we get

$$210 = 71(1 - V_b) + 440\ V_b = 71 - 71\ V_b + 440\ V_b$$

∴ $V_b = \mathbf{0.377}$ and $V_{al} = 1 - 0.377 = \mathbf{0.623}$ **(Ans.)**

**Example 17.3.** *A tensile load of 120 N is applied to an aluminium-boron composite of 1 mm² cross-sectional area. The volume of the parallel fibres is 30 percent. What is the stress in the fibres, when the load axis is (i) parallel to the fibres, and (ii) perpendicular to the fibres? The Young's modulus of aluminium and boron are 71 GN/m² and 440 GN/m² respectively.*

**Solution :** Given : Load, $P = 120\text{N}$; Area, $A = 1\text{mm}^2$; $V_f = 30\%$; $E_{al} = E_m = 71\ \text{GN/m}^2$; $E_b = E_f = 440\ \text{GN/m}^2$.

**Stress in the fibres, $\sigma_f$ :**

(*i*) *When the load axis is parallel to the fibres :*

When the load is applied parallel to the fibres, the elastic strain ∈ is the same in both the fibres and the matrix. If $\sigma_f$ and $\sigma_m$ are the stresses in the fibres and the matrix respectively; then,

$$E_f = \frac{\sigma_f}{\epsilon_f}$$

and $$E_m = \frac{\sigma_m}{\epsilon_m}$$

Since $$\epsilon_f = \epsilon_m$$

$\therefore$ $$\frac{E_f}{E_m} = \frac{\sigma_f}{\sigma_m} = \frac{440}{71} = 6.2 \qquad \left(\text{or } \sigma_m = \frac{\sigma_f}{6.2}\right)$$

Also $$0.3\,\sigma_f + 0.7\sigma_m = 120$$

$$0.3\,\sigma_f + 0.7 \times \frac{\sigma_f}{6.2} = 120$$

or $$0.3\sigma_f + 0.113\,\sigma_f = 120$$

$\therefore$ $$\sigma_f = 290.55 \text{ N/mm}^2 = \mathbf{290.55\ MN/m^2\ (Ans.)}$$

*(ii) When the load axis is perpendicular to the fibres :*

When the load is applied perpendicular to the fibres, the stress in the fibre and the matrix is the same, **120 MN/m²** $\left(\because \sigma = \frac{P}{A} = \frac{120}{1} = 120\text{N / mm}^2 \text{ or } 120 \text{ MN / m}^2\right)$ **(Ans.)**

**Example 17.4.** *A continuous and aligned glass reinforced composite consists of 40 vol % of glass fibres having modulus of elasticity of 69 GN/m² and 60 vol % of a polyester resin having modulus of elasticity of 3.4 GN/m². if the area of the composite is 210 mm² and a stress of 50 MN/m² is applied in the longitudinal direction, then compute the magnitude of the load carried by each of the fibre and the matrix phases.*

**Solution :** *Given* : $V_f = 40\% = 0.4;\ V_m = 60\% = 0.6;\ E_f = 69 \text{ GN / m}^2;\ E_m = 3.4 \text{ GN / m}^2;$

$$\text{A}_\text{c} = 210 \text{ mm}^2;\ \sigma_\text{c} = 50 \text{ MN / m}^2$$

$P_f;\ P_m$:

Let us first find the ratio of fibre load to matrix load using the following eqn.,

$$\frac{P_f}{P_m} = \frac{E_f V_f}{E_m V_m} = \frac{69 \times 0.4}{3.4 \times 0.6} = 13.5 \qquad (\text{or } P_f = 13.5\ P_m)$$

Also, load sustained by the composite, $P_c = \sigma_c \times A_c$

$$= 50 \times (210 \times 10^{-6}) \text{ MN} = 10500 \text{ N}$$

Again, $$P_f + P_m = P_c$$

$$13.5\ P_m + P_m = 10500$$

$\therefore$ $$P_m = \mathbf{724.14\ N\ (Ans.)}$$

and $P_f = 13.5 \times 724.14 =$ **9775.86 N (Ans.)**

**Example 17.5.** *Calculate the tensile modulus of a composite consisting of 60% of volume of glass fibre having tensile modulus of 84 kN/mm² in a polyester matrix of tensile modulus 5.25 kN/mm².*

**Solution :** *Given* : $V_f = 60\% = 0.6; E_f = 84 \text{ kN/mm}^2; E_m = 5.25 \text{ kN/mm}^2$

**Tensile modulus of composite, $E_c$ :**

Tensile modulus of composite = (modulus of fibre × % area of fibre) + (modulus of matrix × % area of matrix)

*i.e.,* $$E_c = 84 \times 0.6 + 5.25 \times (1 - 0.6) = \mathbf{52.5 \text{ kN/mm}^2} \textbf{ (Ans.)}$$

### 17.2.3. Wood and Timber

- *Wood is a natural polymer composite.*
- The principle polymeric molecules are those of *cellulose* [Fig. 17.3.(*a*)] in which the molecular range upto 2,000,000 g/mole (more than 12,000 mers). It is composed of cells resembling long tubes with tapered ends [Fig. 17.3(*b*)]. Its properties are mainly dominated by the life cycle and growth conditions of the trees as well as its molecular chains which are both strongly oriented and crysalline.

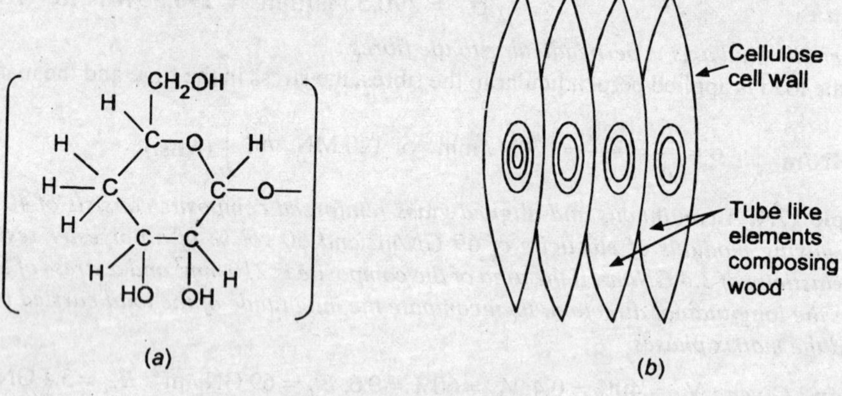


Fig. 17.3. Cellulose structure.

- During the growth process passage ways for flowing of sap and moisture are provided by the cell cavities. *Wood* contains about *50 to 60 percent cellulose* ($C_6H_{12}O_5$) a natural product having three to four thousands $C_6H_{12}O_5$ units in the chain and are *highly crystalline due to the polar nature of the hydroxyl groups on the molecules*. It also contains 20 to 35 percent *lignin*—a carbohydrate compound and smaller amount of other carbohydrates like pentosanes, resins, gums and traces of mineral matter. At elevated temperatures, the assembly of molecules decomposes before softening. In order to reduce the molecular weight as well as neutralise the polar influence of the hydroxyl groups cellulose can be treated chemically for acquiring the mechanical processing.

## HIGHLIGHTS

1. *Elastomer* commonly known as *rubber* is a *linear polymer* which possesses elasticity and good resistance to corrosive fluids.
2. There are generally two types of rubber : Natural rubber and synthetic rubber.
3. The crosslinking process in elastomers is called **vulcanization,** which is achieved by a non-reversible chemical reaction, ordinarily carried out at elevated temperature.
4. A **composite material** is a combination of two or more materials having compositional variations and depicting properties distinctly different from those of the individual materials of the composite.

5. Composites include the following :
   (*i*) Multiphase metal alloys (*ii*) Ceramics
   (*iii*) Polymers.
6. The composite materials may be *classified* as under :
   (*i*) Particle-reinforced (*ii*) Fibre- reinforced
   (*iii*) Structural.

## OBJECTIVE TYPE QUESTIONS

**Fill in the Blanks/Say Yes or No.**

1. By ................ the rubber, the mechanical properties can be considerably improved.
2. Rubber is weak and brittle.
3. The plasticity of rubber makes it pliable and amenable to all manufacturing processes.
4. The electrical insulating properties of pure rubber are ................ to those of vulcanised rubbers.
5. At low temperature rubber becomes stiff and when it is frozen, it attains fibrous structure.
6. Butyl rubber is copolymer of isobutylene and ................
7. A ................ material is a combination of two or more materials having compositional variation and depicting properties distinctively different from those of the individual materials of the composite.
8. ................ are composed of thin layers of the materials in which the successive layers have different orientations of the grain or fibre.

## ANSWERS

**1.** Vulcanizing. **2.** No. **3.** Yes. **4.** inferior. **5.** Yes.
**6.** isoprene. **7.** composite. **8.** Laminates.

## THEORETICAL QUESTIONS

1. Enumerate important properties of rubber.
2. What is natural rubber?
3. What are the characteristics of natural rubber?
4. What is synthetic rubber?
5. Enumerate various types of synthetic rubbers.
6. Write short notes on the following synthetic rubbers:
   Neoprene, Butyl rubber and Silicon rubber.
7. Write short notes on:
   (*i*) Plasticization (*ii*) Compounding
   (*iii*) Calendering (*iv*) Vulcanization
   (*v*) Reclaimed rubber.
8. List the uses of rubber.
9. Enumerate the properties and uses of rubber.
10. Explain briefly any two of the following synthetic rubbers:
    Neoprene rubber, Butyle rubber, Silicone rubber, Nitrile rubber, Polysulphides.
11. Enumerate important forms of rubber.
12. Explain briefly the following, as applied to rubber:
    (*i*) Plasticization (*ii*) Compounding
    (*iii*) Calendering (*iv*) Vulcanization.
13. What is a composite material?
14. How are composite materials classified?
15. State the advantages and disadvantages of reclaimed rubber.
16. What is composite material? Give three examples of composites.
17. What is the distinction between matrix and dispersed phase in a composite material?

**18**

# Lubricating, Belting and Packing Materials

*18.1. Lubricating materials — Introduction – types of lubricants. 18.2. Belting materials. 18.3. Packing materials. Objective Type Questions – Theoretical Questions.*

## 18.1 LUBRICATING MATERIALS

### 18.1.1 Introduction

When one solid surface is moved relative to another solid surface with which it is in contact, there is a frictional resistance to motion. If the surfaces are perfectly smooth, the frictional resistance arises primarily from the operation of molecular forces of cohesion. With practical surfaces which are not perfectly smooth, the interference of minute projections on the surface contributes to the resistance. This frictional resistance causes surface wear and dissipation of energy as heat. To hold energy dissipation and surface wear to a minimum in all sorts of machines, lubricant is used to keep the moving surfaces apart.

In lubrication, a *lubricant performs the following functions*:

1. It reduces the co-efficient of friction between two rubbing surfaces of machinery as a result of which the loss of energy due to friction is reduced.

2. It does not allow the direct contact between the rubbing surfaces by keeping its own film in between these rubbing surfaces. In this way, injuries to surfaces due to seizure and wear are reduced.

3. It acts as a coolant or heat transfer medium by removing the frictional heat generated due to rubbing of surfaces.

4. It acts as a seal in internal combustion engines by sealing the piston and cylinder wall at the compression rings so that there is no leakage of gases at high pressure in combustion chamber and in this way power loss is averted by lubrication in internal combustion engines.

### 18.1.2 Types of Lubricants

— Most lubricants are "*oils*" or "*greases*".

— However, in special circumstances other fluids like *water, air* etc. and solids such as *graphite* may perform the function of lubricants.

— *Synthetic lubricants* are also used occasionally.

**1. Mineral lubricating oils:**

The mineral lubricating oils are obtained from the residual mass left during crude petroleum distillation. In this treatment, gasoline and gas oil fractions are recovered from distillate. Paraffinic crudes generally give a higher yield of lubricating oils than napthenic crudes. The residue is subjected to vacuum distillation. In several cases steam is also introduced into the system. Bubble towers are used for fractionation of oil into two or three fractions of different viscosities.

The mineral oils are generally associated with many impurities such as dissolved waxes, asphaltic substances, and resinous materials. Different methods are used to remove these undersirable components. The two most important methods are: (*i*) *Dewaxing*, (*ii*) *Solvent refining*.

**Fatty oils:**

Fatty oils from animal and vegetable origin are sometimes used alone, but are frequently mixed with mineral oils. The fatty oils exhibit poor keeping quality and undergo decomposition. Fatty oils however, exhibit more oiliness than mineral oils of the same viscosity. Non-drying oils such as olive oil, caster oil, rape seed oil, lard, and fish oil are mainly used for lubricating purposes. They are generally compounded with mineral oil to the extent of about 2 to 10 per cent in order to increase the oiliness of mineral oils. *Olive oil is employed for lubricating textile machinery*, as it can be easily washed out from the cloth without leaving any stain. *Rape seed oil is used for making lubricants for railway engines*, while *lard is used for producing lubricants for internal combustion engines. Caster oil is usually compounded with heavy mineral oils for making extreme-pressure lubricants.*

**Synthetic lubricating oils:**

The need by both industry and military for lubricants that will perform satisfactorily at conditions which are too severe for petroleum lubricants has led to the development of various synthetic lubricants. These lubricants are termed synthetic, because they are not obtained directly from petroleum. Various applications require oils which will have, for example, a lower point, or better viscosity-temperature characteristics, or a higher degree of resistance to oxidation or thermal decomposition than petroleum derived lubricants do.

The synthetic lubricants presently in use are classified as: (*i*) Dibasic acid esters; (*ii*) Organo-phosphate esters; (*iii*) Silicate esters; (*iv*) Silicon polymers; (*v*) Polyglycol ethers and related compounds; and (*vi*) Fluorinated and chlorinated hydrocarbon compounds.

The *applications* of synthetic lubricants are given below:

— **Dibasic acid esters.** Instrument oils, low volatility grease bases, special hydraulic fluids, jet engine lubricants.

— **Phosphate esters.** Fire resistance hydraulic fluids, low-volatility high-lubricity grease bases, special low-temperature lubricants.

— **Silicones**. Condensation pump lubricants, low-volume grease bases for highly loaded bearings, damping fluids, devices requiring minimum viscosity changes with temperature.

— **Silicate esters.** Heat transfer fluids, high temperature hydraulic fluids, low volatility, low viscosity grease bases. Components for low viscosity hydraulic fluids.

— **Polyglycol ether compounds.** Synthetic engine oil, special hydraulic fluids, foaming and drawing lubricants, low temperature grease bases, vacuum pump lubricants, components of other synthetic lubricant formulations.

— **Trifluoro vinyl chloride polymers.** Non-flammable, extreme oxidation resistant lubricant for plant processes or devices handling reactive materials. Heat transfer fluid.

**2. Greases:**

*Lubricating grease is a solid to a semi-solid dispersion of a thickening agent in liquid lubricant. Other ingredients imparting special properties may be included.*

Grease are normally used under conditions of lubrication for which oil is not as suitable or convenient. *Greases perform better than oils under conditions requiring:*

— High bearing loads and shock loads.

— Slow journal speed.

— Temperature extremes.

— Cleanliness or avoidance of splash or drip.
— Minimum attention.
— A seal against external contaminants.
— Large bearing clearances.

**Calcium soap greases.** The water proof nature of calcium soap grease is its primary advantage. Because most of these greases are stabilized with water, they cannot be used for any length of time at 70°C without separation of soap and oil. The calcium soap fibres are very short and give the grease a buttery texture. Calcium greases are widely used as cup and pressure-gun greases for lubrication of plain bearing which operate at normal temperatures and average loads, but they are not suitable for use at very high pressures.

**Sodium soap greases.** Sodium greases are fibrous, and they are useful for *higher temperature lubrication* than most greases. Their chief disadvantage is that they are water soluble and therefore they should not be used where moisture is present. The length of fibre is varied by the manufacturing process. The long-fibre soda greases are used for *lubrication of plain bearings* and other *sliding surfaces* that operate at temperatures in excess of 65°C. They are not suitable for *high-speed antifriction* bearings because the fibres tend to wrap around cages. For *anti-friction bearings, short fibre greases are used.*

**Aluminium soap greases.** These are extremely short fibred and either smooth or stringy in texture. They are not suitable for use much above 70°C because the grease will deteriorate structurally if heated near the melting range and cooled. Aluminium greases are highly adhesive or tacky and find application as lubricants for cams, chains, and oscillating surfaces to which other gases will not cling.

**Mixed soap greases.** Soaps have been mixed to produce greases having general purpose characteristics. The most common combination is calcium sodium. Resultant properties vary widely depending upon the proportions of calcium and sodium used.

**Barium soap greases.** Barium soap greases approach the characteristics of a multi-purpose grease. They are resistant to water, they melt at fairly high temperatures, and they can be used upto almost 140°C. They are smooth or fibrous, depending upon manufacturing procedure, and are used as chasis and antifriction bearing lubricants.

**Lithium soap greases.** These are presently considered to be superior to other soap-base greases as multi-purpose lubricants. They melt at about 175°C and maintain a desirable structure through a wide range of temperatures. They are suitable for lubrication at 140°C and as low as –38°C. Lithium soap greases are buttery to stringy in nature. *They are wear resistant, and some of them have especially good resistance to breaking down upon continuous subjection to shear stress.*

**Pure petroleum greases.** A considerable amount of refinery residuum of asphaltic material is sold as grease, even though it is not a grease by definition. Residuum grease is extremely adhesive and is furnished in a range of viscosities or consistencies as a lubricant for heavy open gears and wire rope. Additives such as graphite, mica or fats, are sometimes included in the grease to enhance properties for particular applications.

Petroleum find limited application as lubricants because of low melting points, these being in range of 45 to 70°C. They are produced by fractional distillation of still residues from the steam distillation of paraffin-base petroleum or by mixing heavy petroleum lubrication oils with low melting point paraffin wax.

**3. Solid lubricants:**

*Solid lubricants are solid materials having strong adherence to metals, very low shear strength, and good elastic properties.* These materials *should be stable at high temperatures, chemically inert, and have a high conductivity.* Some solid lubricants are useful in bearings subjected to temperatures above those at which oil or grease may be used. They are used in *applications requiring lubricants that are chemically inert and also on bearings adjacent to parts that must not be contaminated with foreign materials.*

*A disadvantage of solid lubricants is that they do not flow to cover ruptures in the lubricant film.* The most common solid lubricants are *laminar solids and solid organic compounds.*

(*a*) **Laminar solid**. These materials have lattice structures consisting of layers. Atoms within a layers are strongly bonded, and between layers weakly bonded. Each of slipping between lattice layers accounts for low shear strength. *Graphite, molybdenum disulphide, tungsten disulphide, talc, mica and silver sulphate and borax are examples of laminar solids* which are used as solid lubricants. Some materials perform much more satisfactory if they are bonded into a film on the surface to be lubricated.

- **Graphite** is the most commonly used solid lubricant. Investigators state that water or certain gases or vapours must be present along with graphite to obtain low wear. It is considered that the foreign substances act as separating molecules. Graphite is satisfactory for use as a lubricant to nearly 370°C; above this temperature graphite will begin to oxidize appreciably in air.
- **Molybdenum disulphide** has found considerable use as a solid lubricant for last several years. It is thermally stable and retains lubricity upto 400°C in air. Molybdenum disulphide adheres well to metal surfaces, and is used as an extreme pressure agent in grease or liquid lubricant. Structurally, the molybdenum atoms are oriented in a plane between two planes of sulphur atoms. The sulphur-to-sulphur bond of adjacent molecules is considered to be very weak.

(*b*) **Organic compounds.** These consist of dry soaps, waxes and fats or fatty acids. Soaps are in general use as lubricants in wire drawing. High melting point waxes are not extremely used, probably because of poor adherence of metals. An example of fatty acid useful as solid lubricant in powdered form is stearic acid.

## 18.2 BELTING MATERIALS

The belts are usually made from the following materials:

(*i*) Leather
(*ii*) Rubber
(*iii*) Cotton
(*iv*) Balata
(*v*) Canvas
(*vi*) Steel.

**Leather belts.** These have the *best pulling capacity*. They are used very rarely and at some especially important cases because of high cost of leather. The high speed belts are made by cementing together leather strips. The leather belts are oak-tanned and chrome-tanned. The chrome-tanned belts cannot be easily affected by oils, acids, fumes and moist. The various thicknesses of the belt are obtained by combining a number of hides of different thicknesses. Thus we get single, double and triple belts depending upon the number of hides joined. Weight of leather for belting is 980 N/m, ultimate strength for leather belting is 20 MN/m$^2$. The maximum velocity recommended for leather belt is limited to 40 m/s.

Leather belt should not be allowed to get dry and brittle. Whenever found drying, it is best treated with a dressing of a mixture of neat's foot oil and castor oil. Specially water-proofed belts can be employed in moist places.

**Rubber belts.** Rubber belts consist of layers of fabric impregnated with rubber comoposition. To achieve better flexibility rubber linings are placed between layers of fabric. Usually from three to ten plies in thickness of belts are available. The commercial rubber belting have width from 20 mm to 500 mm and thickness from 2.5 mm to 13.5 mm. The ultimate tensile strength of rubber belting is 37 to 44 MN/m$^3$. Modulus of elasticity is 80 to 120 MN/m$^2$. These belts stand up better in atmospheric conditions, therefore they are preferred for outdoor service. If this belt is kept in contact with oil or gease it is destroyed very soon. As compared with leather belts for the same power transmission, rubber belts are cheaper but have shorter life. If the belt fork is available, then it can be made to operature under varying loads. Maximum velocity recommended is 20-30 m/s.

**Cotton belts**. These belts may be either woven in a loom at one operation or may be made by doubling and sewing together several thicknesses on one piece of woven canvas. The strength and thickness of the belting is measured from the number of plies used. The thickness of the belt is 4.5 to 8.5 mm and width 30 to 250 mm. The ultimate tensile strength of the woven cotton belts is 35 to 45 $MN/m^2$ and modulus of elasticity 30 to 60 $MN/m^2$. Such belts are used for rough and short service. To prevent the edges being damaged by shifting forks or guides the edges of the belts are woven in the leather.

**Balata belts**. Balata is a gum just like rubber. It is used in the manufacture of fabric belts. Balata belts are acid proof and water proof. They cannot stand temperatures higher than about 40°C at which it starts softening. The thickness, width, ultimate strength and modulus of elasticity of balata belts are same as those of rubber belts.

**Canvas belts.** These are made 4, 6, 8 or 10 ply and of width ranging from 0.04 to 1.6 m. The strength of the canvas belts is of the same tune as that of leather belts.

**Steel belts.** These belts are prepared from specially prepared homogeneous steel in various thicknesses upto about 1 mm and are run on pulleys with camber which are covered with thin layer of cork. They have the *advantages* that there is no permanent stretch and centrifugal tension is much lower for a given speed than in case of a leather belt, so that higher belt speeds are used. Their *disadvantages* are that shafts must be perfectly aligned and pulley runs perfectly true.

## 18.3 PACKING MATERIALS

These materials are used for packing into joints or between surfaces of machine parts in contact. These are:

(*i*) Gasket
(*ii*) Lead wool
(*iii*) Glass wool
(*iv*) Hessian
(*v*) Burlap
(*vi*) Cork
(*vii*) Rubber
(*viii*) Rubberized fabric
(*ix*) Asbestos.

**Gasket:**

There are three main types of gaskets used in automobiles and an aircraft engines, namely:

1. For *high temperature applications such as cylinder-head joints and for inlet and exhaust manifold and pipe joints.*
2. For *medium temperature joints such as valve and rocker arm, cover joints, water-pump turning gear and oil-pump joints etc.*
3. For *low temperature joints, such as fuel pump, gear box and back axle cover ones.*

— For *high temperature joints "compressed asbestos"*, with or without wire gauge reinforcement, is generally employed, as for inlet manifolds and exhaust joints. For cylinder-head joints the most widely used form of gasket is the "*copper-asbestos*" one having two thin soft copper layers with the asbestos between these gaskets are made to exact shape of the cylinder-head joint, the holes for the studs, cylinder bores and water connections being punched out, and all edges are generally sealed with the thin copper sheet, so that no asbestos is exposed.

— For *exhaust manifold joints compressed asbestos with a suitable high temperature resistant binder is usually* employed, although copper-asbestos gaskets are also used for this purpose. When asbestos gaskets are employed they are stamped or cut to shape from sheet material coated with graphite upon both faces.

— For *medium temperature* purposes compressed asbestos and rubber coated asbestos are favoured. *Synthetic rubbers* are also in use for such applications. These *gaskets are suitable for use on valve cover joints, rocker arm covers, but not cylinder-head joints.* They *will withstand the effects of oil, steam, water, glycerine, ethylene glycol, etc., at temperatures upto 200°C.*

— For joints that are not exposed to temperatures above about 100°C "*compressed cork sheet*" and "*rubberized fabric*" are frequently employed. The covers of automobile head lamps are usually provided with compressed cork ring joints backed with fabric for strength purpose; this is necessary precaution since the cork possesses very little tensile strength.

— Corrugated cupro-nickel steels are now widely used for *internal combustion engines* and *steam joints subjected to high pressures and temperatures. This type of gasket is highly resistant to corrosion and corrugations seal surface irregularities.*

**Glass wool:**

Glass wool or glass silk or glass fibre is made as very fine fibres of diameter about 0.0025 cm by allowing molten glass to drop through very fine orifices and blowing with either air or stream. It is used as *packing material for heat, sound and electrical insulation and for making fire proof and acid-resistant fibres.*

**Lead wool:**

Lead wool is a lead in finely divided fibrous form, used for cold packing into water-pipe joints in most situations, where molten lead cannot be used.

**Hesian** is jute fabric.

**Burlap** is course jute, hemp or flax textile.

*Other materials have already been discussed in the book.*

## OBJECTIVE TYPE QUESTIONS

**Fill in the Blanks/say Yes or No:**

1. A ........................ reduces the coefficient of friction between two rubbing surfaces of the machinery.
2. A lubricant acts as a coolant or heat transfer medium by removing the frictional heat generated due to rubbing of surfaces.
3. Most lubricants are oils or greases.
4. Synthetic lubricants are used on a large scale.
5. The mineral lubricating oils are obtained from the residual mass left during crude petroleum distillation.
6. Fatty oils exhibit less oiliness than mineral oils of the same viscosity.
7. Caster oil is usually componded with heavy mineral oils for making extreme-pressure lubricants.
8. ........................ is a solid to semi-solid dispersion of a thickening agent in liquid lubricant.
9. Calcium greases are most suitable for use at very high pressures.
10. Sodium soap greases are not suitable for high-speed antifriction bearings.
11. Aluminium soap greases are very suitable for use above 70°C.
12. Solid lubricants are solid materials having weak adherence to metals, very high shear strength, and poor elastic properties.
13. Solid lubricants do not flow to cover ruptures in the lubricating film.
14. ................ is the most commonly used solid lubricant.
15. Graphite is satisfactory for use as a lubricant to nearly 370°C.
16. ................ belts have best pulling capacity.
17. The maximum velocity recommended for leather is limited to 80 m/s.
18. Leather belt should not be allowed to get dry and brittle.
19. The ultimate tensile strength of rubber belting is 37 to 44 MN/$m^2$.
20. Maximum velocity recommended for rubber belting is 40-60 m/s.
21. Cotton belts are used for rough and short service.
22. Balata belts cannot stand temperatures higher than about 40°C.
23. The strength of the canvas belt, is much more than that of leather belt.
24. Glass wool is used as a packing material for heat, sound and electrical insulation and for making fire proof and acid-resistant fibres.
25. Lead wool is a lead in finely divided fibrous form, used for cold packing into water-pipe joints in most-situations, where molten lead cannot be used.

## ANSWERS

| | | | | |
|---|---|---|---|---|
| **1.** lubricant. | **2.** Yes. | **3.** Yes | **4.** No. | **5.** Yes |
| **6.** No. | **7.** Yes. | **8.** Grease. | **9.** No. | **10.** Yes. |
| **11.** No. | **12.** No. | **13.** Yes. | **14.** Graphite. | **15.** Yes |
| **16.** Leather. | **17.** No. | **18.** Yes. | **19.** Yes. | **20.** No. |
| **21.** Yes. | **22.** Yes. | **23.** No. | **24.** Yes. | **25.** Yes. |

## THEORETICAL QUESTIONS

**1.** What are the functions performed by a lubricant, in lubrication?

**2.** Enumerate various types of lubricants.

**3.** Explain briefly the following lubricants:
(*i*) Mineral lubricating oils (*ii*) Greases
(*iii*) Solid lubricants.

**4.** Name the materials of which the belts are usually made.

**5.** Explain briefly any three of the following belting materials:
Leather, Cotton, Balata, Canvas and Steel.

**6.** Enumerate the commonly used packing materials.

**7.** Explain briefly the following packing materials:
Gasket, Lead wool, Rubber and Asbestos.

# 19

# Cutting Tool Materials

*19.1. Characteristics of an ideal cutting-tool material. 19.2. Types of tool materials – carbon steels – medium alloy steels – high speed steels – stellites – cemented carbides – ceramics – diamonds – abrasives – Objective Type Questions – Theoretical Questions.*

## 19.1 CHARACTERISTICS OF AN IDEAL CUTTING TOOL MATERIAL

An ideal cutting-tool must possess the following *characteristics:*

1. The material must *remain harder than work material at elevated temperature.* (*Hot hardness*)
2. The material must *withstand excessive wear* even though the relative hardness of the tool-work materials changes. (*Wear resistance*)
3. The material must have sufficient *strength* and *ductility* to withstand shocks and vibrations and to prevent breakage. (*Toughness*)
4. The *coefficient of friction* at the chip tool interface must remain low for minimum wear and reasonable surface finish.
5. The cost and *easeness of fabrication* should be within reasonable limits.

### 19.2 Types of Tool Materials

- While selecting proper tool material the type of service to which the tool will be subjected should be given primary consideration. No one material is superior in all respects, but rather each has certain characteristics which limits its field of application.
- The *principal cutting tool materials* are:

1. Carbon steels.
2. Medium alloy steels.
3. High speed steels.
4. Stellites.
5. Cemented carbides.
6. Ceramics.
7. Diamonds.
8. Abrasives

1. **Carbon steels:**

- *Carbon tool steels are characterised by the low stability of the supercooled austenite.* Therefore, they have a high critical cooling rate and a low hardenability. Through hardening can be achieved only in parts upto 12 or 15 mm in thickness or diameter. Consequently steel may be recommended for *small-sized tools,* which are quenched in oil or molten salts, and for comparatively large tools (15 to 30 in mm diameter) in which the cutting section is only the surface layer (files, core drills, short reamers etc.)
- When large tools are hardened (of diameter over 30 mm) the layer with a high hardness is so thin, even upon quenching in water, that the tools are not fit for cutting purposes.

**Advantages:**

(*i*) *Cheapness.*

(*ii*) *Low hardness* (BHN=170 to 180).

(*iii*) Good machinability.

(*iv*) Formability in the annealed state.

(*v*) Retain a tough unhardened core due to low hardenability.

This last factor *improves their resistance to breakage under vibration and impacts.*

**Disadvantages:**

(*i*) *Narrow range of hardening temperature.*

(*ii*) *Necessity for rapid quenching in water or aqueous alkali solutions (salt).*

This leads to *considerable deformation* and *warping* and *even to the formation of cracks.*

Therefore, tools of complex form with sharp changes in section and with a large length-to-diameter ratio should not be made of carbon steels. *Warping and crack forming may be reduced somewhat by quenching in water only to 200 – 250°C with subsequent retarded cooling in oil.* Stepped quenching (martempering) is advisable for small-size tools. Good results may be obtained in applying induction hardening to certain types of tools.

**Uses:** Carbon steels are only applicable for tools operating at low cutting speeds (about 12 m/min) since their hardness is substantially reduced at temperatures above 190-200°C.

2. **Medium alloy steels:**

- The high carbon medium alloy steels have a carbon content akin to plain carbon steels, but in addition there is, say, upto *5 percent alloy content consisting of tungsten, molybdenum, chromium and vanadium.* Small additions of one or more of these elements *improve* the performance of the carbon steels in respect of *hot hardness, wear resistance, shock and impact resistance and resistance to distortion during heat treatment.*
- The alloy carbon steels broadly occupy a midway *performance position* between *plain carbon* and *high speed steels.*
- They lose their required hardness at temperature from 250 to 350°C.

3. **High speed steels:**

- *High speed steels are distinguished for their high red-hardness, their* capability to retain their structure (martensite), hardness, and wear resistance at the high temperature generated on the cutting edges when machining at high cutting speeds. High-speed steels are designed for the manufacture of high production tools *with high wear resistance which must retain their cutting properties at temperatures upto 600-620°C.*
- High speed steels are obtained by alloying tungsten, chromium, vanadium, cobalt and molybdenum with steel. This alloying produces metals which remain hard at temperatures at which normal steel becomes quite soft.

(*a*) **18-4-1 High speed steel.** A common analysis is: *18 percent tungsten, 4 per cent chromium and 1 percent vandium,* with a carbon content of 0.6 - 0.7 percent. This alloy is termed 18-4 -1, while an increase of vanadium to 2 per cent provides 18 - 4 - 2 steel.

(*b*) **Cobalt high speed steel.** This is sometimes called *super high-speed steel.* Cobalt is added from 2 to 15 percent to increase hot hardness and wear resistance. One analysis of this steel contains *20 percent tungsten, 4 percent chromium, 2 percent vanadium, and 12 percent cobalt.*

(*c*) **Molybdenum high speed steel.** This class of high speed steels contains a *lower percentage of tungsten,* this being compensated by the addition of molybdenum.

This steel containing *6 percent molybdenum, 6 percent tungsten, 4 per cent chromium and 2 percent vanadium have excellent toughness and cutting ability.*

**4. Stellites:**

- Stellite is a *non ferrous alloy* with range of elements: *cobalt = 40 to 48 percent; chromium = 30 to 35 percent and tungsten = 12 to 19 percent.* In addition to one or more carbide foming elements, carbon is added in amounts of 1.8 to 2.5 percent.
- They *cannot be forged to shape,* but may be deposited directly on the tool shank in an oxy-acetylene flame, alternately, small tips of cast stellites can be brazed into place.
- Stellites *preserve hardness upto* 1000°C and can be operated on steel *at cutting speeds 2 times higher than for high speed steel.*
- These materials are not widely used for metal cutting, since they are very brittle, however, they are *used extensively in some non-metal-cutting application,* such as in rubbers, plastics where loads are gradually applied and the support is firm and where wear and abrasion are problems.

**5. Cemented carbides:**

- These are *so named because they are composed principally of carbon mixed with other elements.*
- The basic ingredient of most cemented carbide is *tungsten carbide* which is extremely hard. Pure tungsten powder is mixed under high heat, at about *1500°C,* with *pure carbon* (lamp black) in the ratic of 94 percent and 6 percent by weight. The new compound, tungsten carbide, is then mixed with cobalt until the mass is entirely homogeneous. This homogeneous mass is *pressed, at pressures from 100 to 420 MN/m²,* into suitable blocks and then heated in hydrogen. *The blocks are cut and ground into specified shapes and then sintered at high temperature heating in the presence of hydrogen.* Boron, titanium and tantalum are also used to form carbides.
- The amount of cobalt used regulates the toughness of the tool. A typical analysis of a carbide suitable for *steel machining is*:

| | | |
|---|---|---|
| Tungsten carbide | = | 82 percent |
| Titanium carbide | = | 10 percent |
| Cobalt | = | 8 percent |

- The *carbide tools* are made by *brazing or silve-soldering* the formed inserts on the ends of the commercial steel holders.
- The most important properties of cemented carbides *are very high heat and wear resistance.* Cemented carbides tipped tools *can machine metals even when their cutting elements are heated to a temperature of 1000°C. They can withstand cutting speed 6 or more than 6 times higher than tools of high speed steel.*
- Cemented carbide is the hardest manufactured material and has extremely high compressive strength. However, it is very brittle, has low resistance to shock, and must be very rigidly supported to prevent cracking.

**Types of cemented carbides.** The two types are:

1. Tungsten type cemented carbides
2. Titanium-tungsten type cemented carbides.

*"Tungsten type cemented carbides"* are *less brittle* than tilanium-tungsten type; they contain

92 to 98 percent tungsten carbide and 2 to 8 percent cobalt. These cemented carbides are *designed chiefly for machining brittle metals such as cost iron, bronze*, but they may *also be used for non-ferrous metals and alloys, steel etc.*

"*Titanium-tungsten type cemented carbides*" are *more wear resistance*. They contain 66 to 85 per cent tungsten carbide, 5 to 30 percent titanium carbide and 4 to 10 percent cobalt. These cemented carbides are designed for *machining tougher materials chiefly for various steels.*

**6. Ceramics:**

- Ceramic tools are made by *compacting aluminium oxide powder in a mould at about 28 MN/m$^2$ or more.* The part is then sintered at 2200°C. This method is known as *cold pressing. Hot pressed ceramics are more expensive due to higher mould costs.*
- Ceamic tool materials are made in the form of *tips* that are to be clamped on metal shanks.
- Ceramic tools have *very low heat conductivity and extremly high compressive strength, but they are quite brittle and have a low bending strength.* For this reason, these materials *cannot be used for tools operating in interrupted cuts, with vibrations, as well as for removing a heavy chip. But they can withstand temperatures upto 1200°C* and can be used at cutting speeds 4 times that of cemented carbides.
- Ceramic tools are *chiefly used for single-point tools* in semi-finish turning of cast iron, plastics, and other work, but only when they are *not subject to impact loads.*
- To give increased strength to ceramic tools often ceramic with a metal bond, known as "*cermets*" is used.
- Because of the high compressive strength and brittleness the tips are given a 5 to 8 degrees *negative rake for carbon steel and zero rake for cast iron and non-metallic materials to strengthen* their cutting edge and are *well supported by the tool holder.*
- Ceramic tools are generally used *without a coolant since they have very low heat conductivity.*

**7. Diamonds:**

- The diamond (hardest-known material) can be run at *cutting speeds about 50 times greater than that for high speed steel tools, and at a temperature upto 1650°C.*
- In addition to its hardness the diamond is *incompressible, is of a large grain structure, readily conducts heat, and has a low co-efficient of friction.*
- Diamonds are suitable for *cutting very hard materials* such as glass, plastics, ceramics and other abrasive materials and for *producing fine finishes.*
- The maximum depth of cut recommended is 0.125 mm with feeds of, say, 0.05 mm.

**8. Abrasives:**

*Abrasive is a class of mineral used to sharpen the edges of cutting tools, and to reduce or polish metallic or other sufaces.*

Abrasive particles held together by a bonding material comprise the cutting edges in grinding wheels known as *abrasive wheels.*

There is fairly small group of generally used abrasives, some *natural*, some *artificial*. Of the *natural abrasives* perhaps the best known are *emergy, corundum and diamond dust.*

'*Emery' is tough* and *durable*, containing about 70 per cent aluminim oxide, a valuable abrasive which does the actual cutting and is known as crystalline fused alumina. While aluminium oxide occurs in nature, it is also produced synthetically by fusing the mineral bauxite in the electric arc furnace. *Corundum contains about 90 percent aluminium oxide.*

*Diamond dust* is slowly forging ahead as abrasive. When diamond dust is used in a "loose" condition it is mixed with oil or grease into a paste-like consistency.

'*Carborundum*" is a trade name for *silicon carbide*, one of the *most important modern artificial abrasives.*

## OBJECTIVE TYPE QUESTIONS

**Fill in the blanks:**

1. Carbon tool steels are characterised by the low stability of the supercooled .......................
2. Carbon steels have narrow range of ....................... temperature.
3. Carbon steels are only applicable for tools operating at ....................... cutting speeds.
4. High speed steels retain their cutting properties at temperatures upto .......................
5. The analysis of 18-4-1 high speed steel is: 18 percent ....................... 4 percent ....................... and 1 percent .......................
6. Cobalt high speed steel is sometimes called ....................... high speed steel.
7. Stellite is a ....................... alloy.
8. Stellites preserve hardness upto .......................
9. The basic ingredient of most cemented carbides is ....................... which is extremely hard.
10. The carbide tools are made by ....................... or silver .......................
11. Ceramic tools are made by compacting ....................... powder.
12. Ceramic tools have very ....................... heat conductivity.
13. ....................... tools are generally used without coolant.
14. ....................... is the hardest known material.
15. Carborundum is trade name for .......................

## ANSWERS

**1.** Austenite. **2.** Hardening. **3.** Low. **4.** 600 – 620°C.
**5.** Tungsten, chromium, vanadium. **6.** Super. **7.** Non-ferrous. **8.** 1000°C.
**9.** Tungsten carbide. **10.** Brazing, soldering.
**11.** Aluminium oxide. **12.** Low. **13.** Ceramic. **14.** Diamond.
**15.** Silicon carbide.

## THEORETICAL QUESTIONS

1. What are the characteristics which an ideal cutting-tool must possess?
2. Enumerate the principal cutting tool materials.
3. Explain briefly the following cutting tool materials:
   High speed steels, Stellites, Cemented carbides, Diamond.

# 20

# Electrical Engineering Materials

*20.1. Classification of electrical engineering materials. 20.2. Conductors – resistance – specific resistance – superconductivity – classification of electrical conductors. 20.3. Semiconductors – characteristics – atomic structure – intrnisic semiconductor extrinsic semiconductor – applications of semiconductors – 20.4. Insulators – characteristics of a good electrical insulating material – classification of electrical insulating materials. 20.5. Magnetic materials – introduction – terms connected with magnetic materials – classification of magnetic materials – magnetic domains – paramagnetism – ferromagnetism – diamagnetism – principal ferromagnetic elements – magnetically soft materials – magnetically hard materials. Objective Type Questions – Theoretical Questions.*

## 20.1 CLASSIFICATION OF ELECTRICAL ENGINEERING MATERIALS

The electrical engineering materials are classified as follows:

1. Conductors
2. Semiconductors
3. Insulators (or Dielectrics)
4. Magnetic materials.

## 20.2 CONDUCTORS

**Conductors** *may be defined as the materials which have free valence electrons in plenty for electric conduction.* **Examples:** *Silver, copper, aluminium etc.*

### 20.2.1 Resistance

*It is defined as the property of a substance due to which it opposes the flow of electricity through it.* The practical unit of resistance is ohm and a conductor is said to have a resistance of one ohm if it allows one ampere current to flow through it when one volt is applied across its terminals.

### 20.2.2 Specific Resistance

Specific resistance or resistivity of the material may be defined as "*the resistance between the opposite faces of a cm. cube of that material*". It is usually represented by $\rho$.

### 20.2.3 Superconductivity

At very low temperature some metals acquire *zero electrical resistance and zero magnetic induction*; the property known as *superconductivity*. Some of the important superconducting elements are:

(*i*) Aluminium,

(*ii*) Zinc,

(*iii*) Cadmium,

(*iv*) Mercury, and

(*v*) Lead.

The characteristic temperature at which a metal becomes superconducting depends on (*i*) the strength of magnetic field, and (*ii*) upon whether the field is applied externally or is the result of current used to measure the resistance.

An interesting feature to note is that metals such as copper, silver, and gold which are very good conductors do not show superconducting properties whilst metals and compounds which are superconducting are rather bad conductors at room temperature.

The *superconductivity can be destroyed by the application of a strong magnetic field.* When the magnetic field exceeds a certain critical value the superconducting state disappears, the magnetic field penetrates the material and electrical resistance is restored. *The disappearance of superconductivity by means of a strong magnetic field is the principle on which switching elements like the "Cryotrons" operate.*

Superconductivity is known to be sensitive to structure but its mechanism is still under reasearch.

**Applications of superconductors:**

Some of the *important applications* of superconductors are as follows:

1. The *Cryotron.*

2. *Electrical machines:* It is possible to manufacture electrical generators and transformers in exceptionally small sizes having efficiency of 99.99%.

3. *Electromagnet:* It is possible to produce superconducting solenoids which do not produce heat during operation. By the use of superconductivity, it is possible to design electromagnets for use in laboratories for lower temperature devices (like *maser*).

4. *Power cables.*

### 20.2.4 Classification of Electrical Conductors

Materials serving as electrical conductors can be mainly divided into two groups:

1. Materials of *low resistivity* (and of high conductivity)
2. Materials of *high resistivity.*

The *former are employed for making conductors for all kinds of windings needed in electrical machines, apparatus and devices as well as for transmission and distribution of electric energy.* Important *examples* being *copper, aluminium and steel.*

The latter find their use in the making of *thermocouples, resistance heating devices etc.* Materials (alloys) like manganin, constantan, nichrome, carbon etc. belong to this category.

Low and high resistivity materials are discussed in detail below:

#### 20.2.4.1 Low resistivity materials

1. **Copper:**

**Properties:**

(*i*) Temperature co-efficient ($\alpha$) = 0.0038; resistivity ($\rho$) = 1.79 $\mu$ ohm cm.

(*ii*) Melting point = 1084°C; boiling point = 2595°C.

(*iii*) Density = 8.94 gm/c.c (20°C).

(*iv*) Possesses *high resistance to corrosion.*

(*v*) Conductivity decreases when hard drawn into wires; annealing is therefore necessary before the material can be used in electrical equipments.

(*vi*) *Mechanical properties* are improved by rolling or drawing. This property makes the copper suitable for low voltage overhead distribution lines.

(*vii*) By addition of one percent cadmium the tensile strength of copper is increased by 50 percent but conductivity however decreases by 15 percent. Cadmium addition makes copper suitable for the use in long span lines of thin cross-section which require conductors of high mechanical strength.

(*viii*) High purity copper is obtained by electroytic refining. Traces (0.1%) of iron, silicon or phosphorus seriously reduce the conductivity of copper.

(*ix*) It can be soldered easily thus simplifying the joining operations.

(*x*) It has *lowest resistance per unit volume.*

(*xi*) It has *low contact resistance.*

**Uses:**

(*i*) *Soft copper is used to make an extensive variety of winding wires and conductors for producing cables.*

(*ii*) Hard copper is used to make commutator segments and contact wire.

(*iii*) Copper conductors having a steel core are also employed for long span transmission lines, where a combination of high conductivity, small sag and minimum cross-section is desired. In such conductors, an insulating tape over the wire has to be provided in order to prevent the corrosive action of steel on copper.

2. **Aluminium:**

**Properties:**

(*i*) Temperature co-efficient ($\alpha$) = 0.004; Resistivity ($\rho$) = 2.69 $\mu$ ohm cm.

(*ii*) Melting point = 657.6°C; boiling point = 2057°C

(*iii*) Density = 2.703 gm/cm$^3$ (at 20°C).

(*iv*) It is malleable and ductile and can be drawn into wire 0.1 mm diameter.

(*v*) *Offers high resistance to corrosion.*

(*vi*) It is much lighter than copper for the same mass.

(*vii*) It poses no problems in cable manufacture and installation.

(*viii*) Pure aluminium is softer than copper and can be rolled into thin sheets down to 5 to 7 microns in thickness (so called *foil*).

(*ix*) Aluminium oxidizes more quickly in normal atmospheric conditions and acquires a thin film of oxide, $Al_2O_3$ which effectively protects it from further oxidation. In sea coast regions, aluminium is more heavily oxidized than copper.

(*x*) Aluminium cannot be soldered easily but fluxes have been devised to make soldering easy. Machanical clamping and screwing methods have also been developed.

(*xi*) It has lowest resistance per unit weight.

(*xii*) It has high contact resistance.

**Uses:**

(*i*) The steel reinforced aluminium conductors (A.C.S.R.) finds extensive use in long span transmission lines.

(*ii*) Aluminium conductors are particularly suitable for operation in very high ambient temperature.

3. **Iron and steel:**

(*i*) Steel is used as conductor rail in traction on account of its cheapness and rigidity.

(*ii*) Steel possesses *high resistance per unit weight and volume.*

(*iii*) Galvanised steel and iron wires which are usually used for earth conductors in low voltage distribution systems may also be used for phase conductors in rural areas where cheapness is the main consideration. Such lines generally have large voltage drops due to high resistance and inductance.

(*iv*) Manganese addition has a hardening effect on steel and manganese steel (about 1.3% manganese) has the further property of being practically non-magnetic.

(*v*) Steel alloyed with chromium and aluminium is employed for starting rheostats where lightness combined with robustness, compactness and good heat dissipation are important considerations, as in some traction vehicles.

(*vi*) Cast iron is employed in the manufacture of resistance grids to be used in the starting of large d.c. motors.

**Low resistivity copper alloys:**

Copper base alloys containing tin, cadmium, beryllium and certain other metals generally called '*bronze*' are also used as high conductivity materials. *All bronzes possess high mechanical strength as compared to copper, but have higher resistivity.* The most widely used as conductors are cadmium copper and beryllium bronzes. *Cadmium copper is employed for making contact wires, commutator segments and beryllium bornze is used for making current carrying springs, brush holders, sliding contacts, knife switch blades etc.*

*Another type of copper-base alloy is 'brass' an alloy of copper with zinc which has greater mechanical strength and wear resistance than copper*, but considerably lower conductivity (higher resistivity). Brass is fairly resistant to corrosion and finds its wide use in the *manufacture of electrical apparatus and instruments as a current carrying and structural material.*

### 20.2.4.2 High resistivity materials

**1. Nichrome or Brightray B:**

*Composition:* Ni = 60%; Cr = 15%; Fe = 25%

*Properties:* (*i*) $\rho$ = 1.10 $\mu$-ohm cm, $\alpha$ = 0.0002

(*ii*) Melting point is 1350°C.

(*iii*) Specific gravity is 8.24.

**Uses:** Employed in making of tabular heaters and electric irons.

**2. Nichrome V or Brightray C:**

*Composition:* Ni = 80%; Cr = 20%

*Properties:* (*i*) $\rho$ = 40 $\mu$-ohm cm, $\alpha$ = 0.0001

(*ii*) Specific gravity is 8.4

(*iii*) Melting point is 1400°C.

(*iv*) Possesses high resistance to oxidation and temperatures effects.

**Uses:** In the making of heating elements and furnaces.

**3. Constantan or Eureka:**

*Composition:* Ni = 40%; Cu = 60%

*Properties:* (*i*) $\rho$ = 49 $\mu$-ohm cm, $\alpha$ = 0.00002.

(*ii*) Melting point is 1300°C.

(*iii*) Specific gravity is 8.9.

(*iv*) Heat resisting properties poor.

**Uses:** It is used for making *starter and field regulator resistances; wires for resistance boxes and thermocouples.*

**4. Manganin:**

*Composition:* Cu = 84%; Mn = 12%; Ni = 4%

*Properties:* (*i*) $\rho$ = 44 $\mu$-ohm cm, $\alpha$ = + 0.00015.

(*ii*) Melting point is 1020°C.

(*iii*) Specific gravity is 8.19.

**Uses:** It is used in *instruments, shunts and standard resistance coils.*

5. **German silver or Nickle silver or electrum:**

*Composition:* Cu = 50%; Zn = 20%; Ni = 30%

*Properties:* (*i*) $\rho$ = 30 $\mu$-ohm cm, $\alpha$ = 0.0004

**Uses:** It is used in electrical measuring instruments.

6. **Nirosta:**

*Composition:* Cr = 20%; Ni = 10%; Si = 0.05%; C = 0.15%; Mn = 0.25%

*Properties:* (*i*) $\rho$ = 73.00 $\mu$-ohm cm.

(*ii*) Melting point is 1400°C

(*iii*) Density is 7.86.

**Uses:** It is particularly suitable where exceptional corrosion resistance and heat resistance are of significant importance.

7. **Carbon:**

(*i*) $\rho$ = 1000 to 7000 $\mu$-ohm cm, $\alpha$ = – 0.0002 to – 0.0008.

(*ii*) Melting point is 3500°C.

(*iii*) Density varies from 1.7 to 3.5.

(*iv*) Carbon is used in *automatic voltage regulators* for making the *pressure sensitive pile resistors*.

## 20.3 SEMICONDUCTORS

**Semiconductors** *are solid materials, either non-metallic elements or compounds, which allow electrons to pass through them so that they conduct electricity in much the same way as metal.*

### 20.3.1 Characteristics

Semiconductors possess the following *characteristics:*

(*i*) The resistivity is usually high.

(*ii*) The temperature co-efficient of resistance is always *negative.*

(*iii*) The contact between semiconductor and a metal forms a layer which has a *higher resistance in one direction than the other*.

(*iv*) When some suitable *metallic impurity* (*e.g.* Arsenic, Gallium etc.) is added to a semiconductor, its *conducting properties change appreciably*.

(*v*) They exhibit a rise in conductivity in the increasing temperature, with the decreasing temperatures their conductivity falls off, and at low temperatures semiconductors become dielectrics.

(*vi*) They are usually metallic in appearance but (unlike metals) are *generally hard and brittle*.

Both the resistivity and the contact effect are as a rule very sensitive to small changes in physical conditions, and the great importance of semiconductors for a wide range of uses apart from rectification depends on the *sensitiveness*.

Of all the elements in the periodic table, *eleven* are semiconductors which are listed below:

| *S.No.* | *Element* | *Symbol* | *Group in the periodic table* | *Atomic no.* |
|---|---|---|---|---|
| 1. | Boron | B | III | 15 |
| 2. | Carbon | C | IV | 6 |
| 3. | Silicon | Si | IV | 14 |
| 4. | Germanium | Ge | IV | 32 |
| 5. | Phosphorous | P | V | 15 |
| 6. | Arsenic | As | V | 33 |
| 7. | Antimony | Sb | V | 51 |
| 8. | Sulphur | S | VI | |

9. Selenium Se VI
10. Tellurium Te VI
11. Iodine L VIII

Examples of semi-conducting compounds are given below:

(*i*) Alloys: $Mg_3 Sb_2$, ZnSb, $Mg_2Sn$, CdSb, AISb, InSb, GeSb

(*ii*) Oxide: ZnO, $Fe_3O_4$, $Fe_2O_3$, $Cu_2O$ CuO, BaO, CoO, NiO, $Al_2O_3$, $TiO_2$, $UO_2$, $Cr_2O_3$, $WO_2$, $MoO_3$.

(*iii*) Sulphides: $Cu_2S$, $Ag_2S$, PbS, ZnS, CdS, HgS, $MoS_2$.

(*iv*) Halides: Agl, Cul.

(*v*) Selenides and Tellurides.

*PbS is used in photo conductive devices, BaO in oxide coated cathodes, caesium antimonide in photomultipliers etc.*

### 20.3.2 Atomic Structure

To understand how semiconductors work it is necessary to study briefly the structure of matter. All atoms are made of electrons, protons and neutrons. Most solid materials are classed, from the stand point of electrical conductivity, as conductors, semiconductors or insulators. *To be a conductor, the substance must contain some mobile electrons-one that can move freely between the atoms.* These free electrons come only from the valence (outer) orbit of the atom. Physical force associated with the valence electrons bind adjacent atoms together. The inner electrons below the valence level, do not normally enter into the conduction process.

Conductivity depends on the number of electrons in the valence orbit. Electron diagrams for three typical elements, aluminum, phosphorus and germanium are shown in Figures. 20.1, 20.2, 20.3.

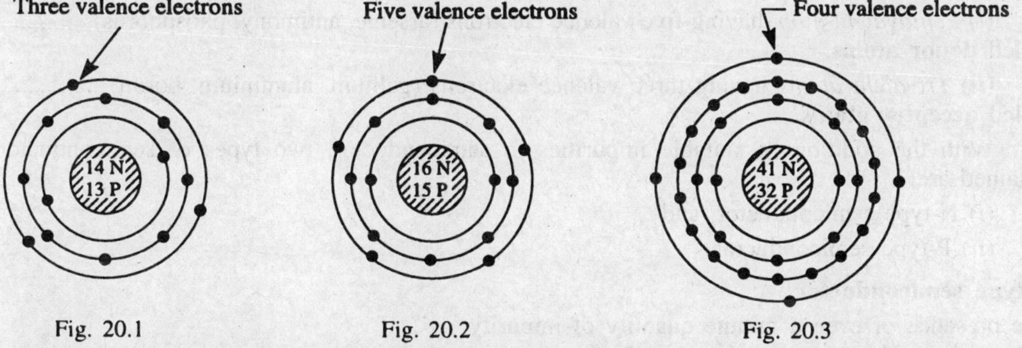


Fig. 20.1 Fig. 20.2 Fig. 20.3

These elements can also be used in semiconductor manufacture.

The *degree of conductivity* is determined as follows:

1. Atoms with *fewer than four valence* electrons are *good conductors*.
2. Atoms with *more than four valence* electrons are *poor conductors.*
3. Atoms with *four valence electrons* are *semiconductors.*

Fig. 20.1 shows *aluminium* which has *three* valence electrons. When there are less than four valence electrons they are loosely held so that at least one electron per atom is normally free; hence aluminium is a *good conductor*. This ready availability of free electrons is also true of copper and most other metals.

Fig. 20.2 shows *phosphorus* with *five* valence electrons. When there are more than four valence electrons, they are tightly held in orbit so that normally *none* are *free*. Hence phosphorus and similar elements are *poor conductors* (insulators).

Germanium (Fig. 20.3.) has *four* valence electrons. This makes it neither a good conductor nor a good insulator, hence its name "*semiconductor*". Silicon also has four valence electrons and is a semiconductor.

**Note:** *The energy level of an electron increases as its distance from the nucleus increases. Thus an electron in the second orbit possesses more energy than electron in the first orbit; electron in the third orbit have higher energy than in the second orbit and so on. It follows, therefore, that electrons in the last orbit will possess very high energy. These high energy electrons are less bound to the nucleus and hence they are more mobile. It is the mobility of last orbit electrons that they acquire the property of combining with other atoms. Further it due to this combining power of last orbit electrons of an atom that they are called* **valence electrons**.

### 20.3.3 Intrinsic Semiconductor

*A pure semiconductor is called* **intrinsic semiconductor**. Here no free electrons are available since all the covalent bonds are complete. *A pure semiconductor, therefore behaves as an insulator.* It exhibits a peculiar behaviour even at room temperature or with rise in temperature. *The resistance of a semiconductor decreases with increase in temperature.*

When an electric field is applied to an intrinsic semiconductor at a temperature greater than 0° K, conduction electrons move to the anode and, the holes (when an electron is liberated into the conduction band, a positively charged hole is created in valence band) move to cathode. Hence semiconductor current consists of movement of electrons in opposite direction.

### 20.3.4 Extrinsic Semiconductor

In a pure semiconductor, which behaves like an insulator under ordinary conditions, if small amount of certain *metallic impurity is added it attains current conducting properties*. The impure semiconductor is then called *impurity semiconductor* or *extrinsic semiconductor*. The process of adding *impurity (extremely in small amounts, about 1 part in $10^8$) to a semiconductor to make it extrinsic (impurity) semiconductor is called* **Doping**.

Generally following *doping agents* are used:

(*i*) *Pentavalent atom* having five valence electrons (arsenic, antimony, phosphorus) .............. called **donor atoms.**

(*ii*) *Trivalent atoms* having three valence electrons (gallium, aluminium, boron) ................ called **acceptor atoms.**

With the addition of suitable impurities to semiconductor, two types of semiconductors obtained are:

(*i*) N-type semiconductor, and

(*ii*) P-type semiconductor.

**N-type semiconductor:**

The presence of even a minute quantity of impurity, can produce N-type semiconductor. If the impurity *atom has one valence electron* **more** than the semiconductor atom which it has substituted, this *extra electron* will be loosely bound to the atom. For example, an atom of germanium possesses four valence electrons; when it is replaced in the crystal lattice of the substance by an impurity atom of antimony (Sb) which has five valence electrons, the fifth valence electron (free electron) produces extrinsic N-type conductivity *even at room temperature*. Such an impurity induced into a semiconductor is called *donor impurity* (or donor). The conducting properties of germanium will depend upon the amount of antimony (*i.e.* impurity) *added.* This means that controlled conductivity can be obtained by the proper addition of impurity. Fig. 20.5. shows

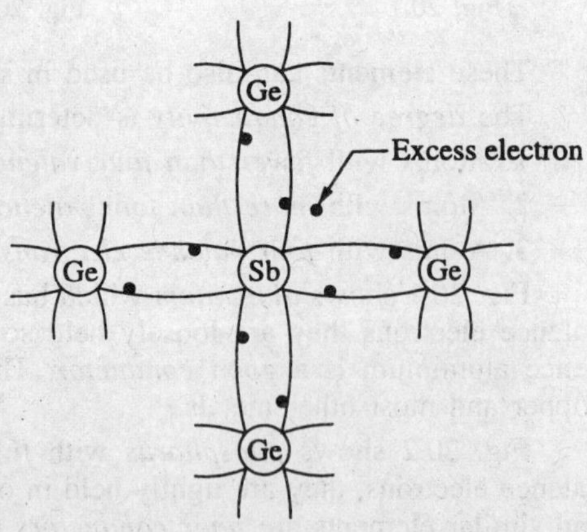

Fig. 20.4. N-type semiconductor.

extrinsic semiconductor (N-type) at T = 0°K showing the loosely bound excess electron contributed by the donor atom.

**P-type semiconductor:**

P-type extrinsic semiconductor can be produced if the impurity atom has *one valence electron* **less** than the semiconductor atom that it has replaced in the crystal lattice. This impurity atom cannot fill all the interatomic bonds, and the free bond can accept an electron from the neighbouring bond; leaving behind a vacancy or hole. Such an impurity is called an *acceptor impurity* (or accepter). Fig. 20.6 shows structure of *P*-type semiconductor (Germanium and Boron) at 0°K.

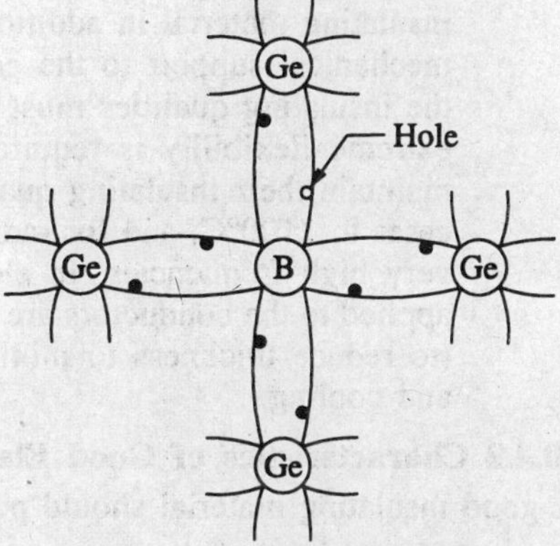

Fig. 20.5. P-type semiconductor.

### 20.3.5 Applications of Semiconductors

1. Copper oxide and selenuim, were the first materials to be used to serve as rectifiers.

2. Germanium and silicon came into commercial use somewhat later after copper-oxide and selenium. Germanium rectifiers found application earlier than silicon ones. One of the reasons for this is that it is easier and simpler to produce germanium monocrystals, although the process involves considerable technological difficulties. The Germanium and silicon semiconductors find wide use in both *high frequency and commercial-frequency circuits* particularly as non-controlled rectifiers (diodes) and controlled rectifiers (for example, silicon controlled rectifiers).

3. *Non-linear resistors*. These are also called *varistors*. These are the semiconductors whose resistance is marked by dependence on the applied voltage, due to which the current rises non-linearty with rise in voltage.

These are made mainly from *silicon carbide* obtained by electrically heating a mixture of quartz sand with carbon to temperature of about 2000°C. This is commonly known as *synthetic (electrical) carborundum.*

4. *Temperature-sensitive resistors*. These are also called *Thermistors*. They possess a *negative temperature resistivity of high absolute value.*

They are made from oxides of certain metals such as copper, manganese, cobalt, iron and zinc. Thermistors are produced in the form of discs, short rods, beads etc. by ceramic techniques.

5. *Photo-conductive and photo-voltaic cells*. These are prepared from the materials which possess high sensitivity to light. The materials are sulphides, selenides and tellurides. Sometimes germanium and silicon are also used for the purpose.

These days considerable use is made of silicon photo-voltaic cells, more commonly called *solar cells* which serve to convert solar radiant energy into electric energy for spacecraft power supply, etc.

Photo-voltaic cells find wide *applications* in the following:

(*i*) Automatic control systems.

(*ii*) Television circuits.

(*iii*) Sound motion picture recording and reproducing equipment.

## 20.4 INSULATORS (OR DIELECTRICS)

### 20.4.1 Introduction

- **'Electrical insulating materials'** are defined as *materials which offer a very large resistance to flow of current, and for that reason they are used to keep the current in its proper path along the conductor.*
- A large number of substances and materials may be classified as insulators, many of which have to be employed in practice, as no single substance or material can satisfy all

the requirements involved in the numerous and varied applications of insulators in electrical engineering. Such requirements involve consideration of physical properties, reliability, cost, availability, adaptability to machining operations etc. Thus in applications, the insulating material in addition to its function as an insulator may have to act as a rigid mechanical support to the conductor and may be installed out of doors, in which case the insulating qualities must be retained under all atmospheric conditions. In other cases extreme flexibility is required. Again, in electric heaters the insulating materials must maintain their insulating qualities over a wide range of temperatures extending in some cases to 1100°C, and for radio purposes the insulating qualities must be maintained upto very high frequencies. In electrical machines and transformers the insulating materials applied to the conductors are required to be flexible, to have high specific electric strength (to reduce thickness to minimum) and ability to withstand unlimited cycles of heating and cooling.

### 20.4.2 Characteristics of Good Electrical Insulating Material

A good insulating material should possess the following *characteristics:*

1. Large insulation resistance.
2. High dielectric strength.
3. Uniform viscosity—it gives uniform electrical and thermal properties.
4. Should be uniform throughout—it keeps the electric losses as low as possible and electric stress uniform under high voltage differences.
5. Least thermal expansion.
6. When exposed to arcing should be non-ignitable.
7. Should be resistant to oils or liquids, gas fumes, acids and alkalies.
8. Should have no deteriorating effect on the material, in contact with it.
9. *Low dissipation factor (loss tangent)*
10. High mechanical strength.
11. High thermal conductivity.
12. Low permittivity.
13. High thermal strength.
14. Free from gaseous insulation to avoid discharges (for solids and gases)
15. Should be homogencous to avoid local stress concentration.
16. Should be resistant to thermal and chemical deterioration.

### 20.4.3 Classification of Electrical Insulating Materials

The insulating materials can be *classified* in the following *two ways:*

1. Classification according to substances and materials.
2. Classification according to temperature.

**1. Classification According to Substances and Materials:**

(*i*) **Solids** (Inorganic and organic):

Mica, wood, slate, glass, porcelain, rubber, cotton, silk, rayon, terelyne, paper and cellulose materials.

(*ii*) **Liquids** (Oils and Varnishes):

Linseed oil, refined hydrocarbon mineral oils, spirit, synthetic varnishes etc.

(*iii*) **Gases:**

Dry air, carbon dioxide, argon, nitrogen etc.

**2. Classification According to Temperature:**

| Class | Insulating materials included | Assigned limiting insulating temperature |
|---|---|---|
| Y (Formely O) | Cotton, silk, paper, cellulose, wood, etc. neither impregnated nor immersed in oil. Materials of Y class are *unsuitable for electrical machines and apparatus as they deteriorate rapidly and are extremely hygroscopic.* | 90°C |
| A | Materials of class Y impregenated with natural resins, cellulose esters, insulating oils etc. Also included in this list are laminated wool, varnished paper. | 105°C |
| E | Synthetic resin enamels, cotton and paper laminates with formaldehyde bonding etc. | 120°C |
| B | Mica, glass fibres, asbestos with suitable bonding substances, built up mica, glass fibre, and asbestos laminates. | 130°C |
| F | Materials of class B with bonding materials of higher thermal stability. | 155°C |
| H | Glass fibre and asbestos materials, and built up mica, with silicon resins. | 180°C |
| C | Mica, ceramics, glass, quartz without binders or with silicon resins of higher thermal stability. | above 180°C |

## 20.5 MAGNETIC MATERIALS

### 20.5.1 Introduction

**Magnetic materials** *are those materials in which a state of magnetisation can be induced.* Such materials when magnetised create a magnetic field in the surrounding space.

- In a material all the molecules contain electron which orbit around the nucleus; these orbits are therefore equivalent to circulating currents and so develop an m.m.f. Each m.m.f. due to an individual orbit, in most molecules, is neutralised by an opposite one. But in magnetic materials such as iron and steel *there are a number of unneutralised orbits*, such that a resultant axis of m.m.f. exists which produces a magnetic dipole.
- In *unmagnetised specimens,* because of mutual attraction and repulsion among the dipoles, the molecular *m.m.f. axes live long continuous closed paths, and no external magnetic effect can be detected.*
- In *magnestised specimens* the dipole line up parallel with the exciting m.m.f. and when the exciting m.m.f. is removed, a *number of dipoles may remain aligned in the direction of the external field and thus exhibit permanent magnetism.*
- The readiness of a material to accept magnetism is expressed by its permeability for all materials except a few magnetic ones, the permeability is that of free space and is constant. It is denoted by $\mu_0$ ($= 4\pi \times 10^{-7}$ H/m)

  For magnetic materials the permeability ($\mu$) equals $\mu_0$ times the relative permeability $\mu_r$ *i.e.* $\mu = \mu_0\mu_r$.

  *The relative permeability varies with ttle degree of magnetisation of the material and may have a value* as high as 2500.

### 20.5.2 Terms Connected with Magnetic Materials

1. *Magnetic force.* It is the force exerted by one magnet on another either to attract it or repel it.

2. *Unit pole strength.* It is defined as the strength of that pole which when placed in vacuum

at a distance of one metre from a similar and equal pole, repels it with a force of one newton.

3. *Magnetic flux density.* It is usually represented by letter *B* and is defined as the flux or lines of force passing per unit area through any substance through a plane at right angles to the direction of magnetic flux.

It is measured in Wb/$m^2$ (or *T i.e.* Tesla).

4. *Magnetic field strength.* It may be defined in the following two ways:

(*i*) Field strength at any point within a magnetic field is the number of lines of force passing through a unit area round the point considered and held perpendicular to the lines.

(*ii*) Field strength at any point within a magnetic field is the force exerted by a unit North pole at that point.

5. *Relative permeability.* It is the ratio of the flux density (*B*) produced in that material to the flux density produced in vacuum by the same magnetising force (*H*). It is denoted by $\mu_r$.

6. *Absolute permeability:* Absolute permeability of the material is the ratio of flux density in that material to the magnetising force producing that flux density and is denoted by $\mu$; $\mu = \mu_0\mu_r$ where $\mu_0$ is the permeability of free space having a value of $4 \pi \times 10^{-7}$ H/m.

7. *Magnetic potential.* The magnetic potential at any point within a magnetic field is measured by the work done in carrying a unit North pole from infinity to that point against the force of magnetic field.

8. *Intensity of magnetism.* It is defined as the pole strength per unit area of the bar or magnetic moment per unit volume of the bar. It is denoted by latter *I*.

9. *Susceptibility.* It is defined as the ratio of intensity of magnetisation (*I*) to magnetising force (*H*).

10. *Magnetomotive force.* It is that force which drives or tends to drive the flux through a magnetic circuit. In short it is written as m.m.f. It is the product of number of conductors (*N*) and current (*I*) passing through them *i.e.* m.m.f. = *NI*.

11. *Magnetic reluctance.* It is that property of the material which opposes the production of magnetic flux in it.

12. *Coercive force.* It may be defined as the demagnetising force which is necessary to neutralise completely the magnetism in an electromagnet after the value of magnetising force becomes zero.

13. *Remanance.* It is that property of magnetic flux density which still persists in magnetic material even when the magnetising force is completely removed. It is expressed in Wb/$m^2$.

14. *Retentivity.* It is that property of magnetic material which is measured by its maximum value of the residual induction.

15. *Core loss, incremental.* The core loss in a magnetic material when subjected simultaneously to a D.C. biasing magnetising force and an alternating magnetising force.

16. *Core loss, specific.* The power expended per unit mass of magnetic material in which there is symmetrical, sinuosidally varying induction of a specified maximum value, B, at a specified frequency, *f*.

17. *Core loss, total.* The power expended in a magnetic specimen in which there is cyclically *alternating induction.*

**Note:** Measurements of core loss are normally made with sinuosidally alternating induction, or the results are corrected for deviations from the sinuosidal condition.

18. *Curie temperature.* The temperature above which a ferromagnetic material becomes paramagnetic.

19. *Cyclically magnetised condition.* A magnetic material is in a cyclically magnetised condition when, after having been subjected to a sufficient number of identical cycles of demagnetising force, it follows identical hysteresis loops on successive cycles which are not necessarily symmetrical with respect to the origin of the axes.

20. *Hysteresis loop, normal.* A closed curved obtained with a ferromagnetic material by plotting (usually to rectangular co-ordinates) corresponding D.C. value of magnetic induction $B$ for ordinates and magnetising force H for abscissa when the material is passing through a complete cycle between equal definite limits of either magnetising force $\pm H_m$ or magnetic induction $\pm B_m$. In general, the normal hysteresis loop has inversion symmetry with respect to the origin of the $B$ and $H$ axes, but this may not be true for special materials.

21. *Hysterisis loss, incremental.* The hysteresis loss occurring in a ferromagnetic material when subjected simultaneously to a constant biasing magnetic force and cyclic excitation.

22. *Hysteresis loss, normal.* (*i*) The power expended in a ferromagnetic material as a result of hysteresis, when the material is subjected to an SCM (symmetrically cyclically magnetised) excitation. (*ii*) The energy loss per cycle in magnetic materials as a result of magnetic hysteresis when the induction is cyclic (but not necessarily periodic).

23. *Magnetic hysteresis.* The property of a ferromagnetic magnetic material revealed by the lack of correspondence between the changes in induction resulting from increasing magnetising force and from decreasing magnetising force.

24. *Induction, intrinsic.* The vector difference between the magnetic induction in a magnetic material and the magnetic induction that would exist in a vaccum under the influence of the same magnetising force.

25. *Induction normal.* The maximum induction, in a magnetic material, that is in symmetrically cyclically magnetised condition.

**Note.** Normal induction is a magnetostatic parameter usually measured by ballistic methods.

26. *Induction curve, normal.* A curve of a previously demagnetised specimen depicting the relation between normal induction and corresponding ascending values of magnetising force. This curve starts at the origin of $B$ and $H$ axes.

27. *Magnetic field of induction.* A state of a region such that a conductor carrying a current in the region would be subjected to mechanical force and an e.m.f. would be inducted in an elementary loop rotated with respect to the field in such a manner as to change the flux linkage.

28. *Aging co-efficient.* The percentage change in specific magnetic property resulting from a specified aging treatment.

29. *Magnetic aging.* The change in magnetic properties of a material resulting from metallurgical change. This term applies whether the change results from a continued normal or specified accelerated aging condition.

### 20.5.3 Classification of Magnetic Materials

All materials possess magnetic properties to greater or lesser degree and these are determined by the facts that (*i*) a magnetic field exert forces and torques on the bodies, (*ii*) a body placed in a

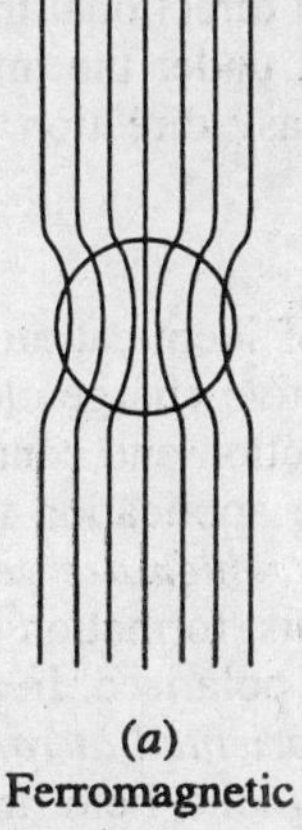

(*a*)
Ferromagnetic

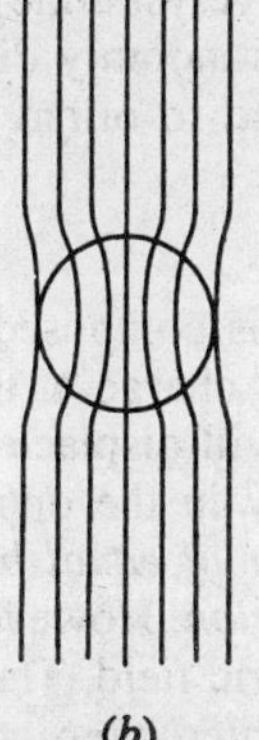

(*b*)
Paramagnetic

(*c*)
Diamagnetic

Fig. 20.6

magnetic field distorts the field. The magnetic properties of the materials are characterised by their relative permeabilities. In accordance with the value of relative permeability the materials may be classified in the following three ways:

1. **Ferromagnetic materials.** The relative permeabilities of these materials are *much greater than unity* and are dependent on the field strengths. They attract the lines of force strongly (Fig. 20.6 *a*). The principal ferromagnetic elements are *iron, cobalt, nickle*. Gadolinium however also comes under this classification. These have high susceptibility.

2. **Paramagnetic materials.** These have relative permeabilities *slightly greater than unity* and are magnetised slightly. They attract the lines of, forces weakly (Fig. 20.6 *b*). Aluminium, platinum and oxygen belong to this category.

3. **Diamagnetic materials.** The relative permeabilities of these materials are *slightly less than unity*. They repel the lines of force slightly (Fig. 20.6 *c*). The example are bismuth silver, copper and hydrogen.

Fig. 20.7 shows the arrangement of atomic magnetic moment.

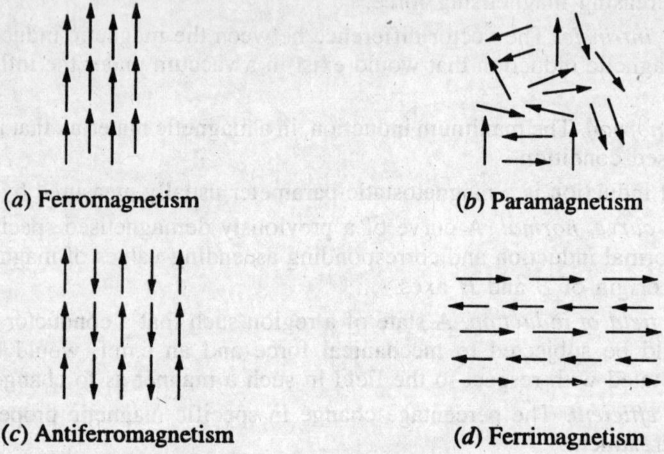


Fig. 20.7. Arrangement of atomic magnetic moment.

### 20.5.4 Magnetic Domains

In a ferromagnetic material the *unpaired electrons* bring about *alignment in neighbouring atoms* and this causes certain microscopic regions called *magnetic domains* to develop saturated magnetic moment. The magnetic domains have actually been observed. The boundaries between such domains are regions at which the direction of spin of electrons changes. A single crystal of metal can show magnetic directionality but a polycrystalline mass cannot exhibit directional magnetism, since the domains in such a material are randomly distributed. However, under the influence of a strong magnetic field these domains tend to orient themselves along easy direction nearest to the field and the mass becomes magnetic.

### 20.5.5 Paramagnetism

Let us take the case of a material which is composed of large number of identical atoms. Each atom can be visualised as a positive point charge at the centre of a negative charge cloud. Now if an electric field is applied the nucleus will displace slightly in one direction and centres of the negative electron cloud will move slightly in the opposite direction. The application of electric field produces a dipole (A *dipole is a pair of equal but opposite charges which are very close.*) This process is known as *induced polarisation*. However in certain materials, formation of dipoles takes place even in the absence of an electric field. These are permanently polarised. In a material which is permanently magnetised, the *dipoles interact weakly*. They are *oriented at random and have a low net magnetisation.* When these materials are placed in a magnetic field, the dipoles orient themselves up in the direction of magnetic field and the substance shows magnetisation

This phenomena is known as *paramagnetism and the materials exhibiting this property are known as* **paramagnetic substances.**

### 20.5.6 Ferromagnetism

The outstanding characteristic of a ferromagnetic material is that it is very strongly attracted by the magnet. Importance of iron as a magnetic material has led to the name ferromagnetism. Magnetic properties even exist in the absence of applied magnetic field.

Ferromagnetism arises out of the electronic structure within the atoms. According to *Pauli exclusion principle, not more than two electrons can occupy each energy level of an isolated atom* and the same holds true for atoms in a crystal structure. The two electrons having the same energy levels have *"spins" in opposite direction*, and since each spinning electron can be considered equivalent to a moving charge, each electron acts as an extremely small magnet having north and south poles. A material having even number of electrons in general have as many electrons spinning in one direction as in the other and the sum total effect is magnetically neutral in structure. In case of a material with unfilled subvalence shell, *more electrons spin in one direction than in other resulting in a net magnetic moment* (Fig. 20.8). In materials like α-iron, cobalt and nickle, these moments are quite strong and atoms are sufficiently close together so that there is *spontaneous magnetic alignment of adjacent atoms.*

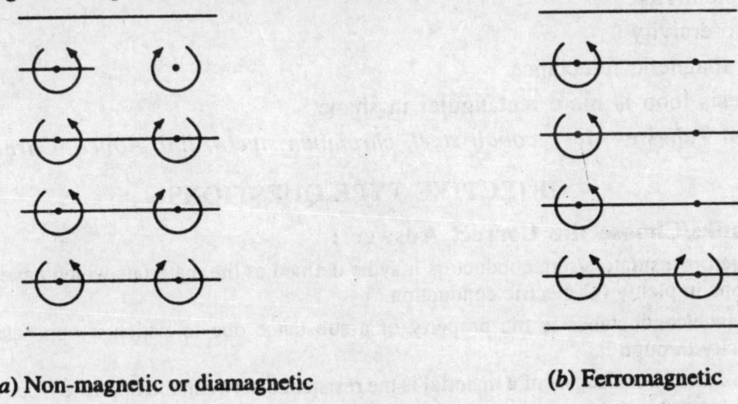

(*a*) Non-magnetic or diamagnetic (*b*) Ferromagnetic

Fig. 20.8

### 20.5.7 Diamagnetism

*Diamagnetism is the property of material due to which it, when placed in a magnetic field, becomes weakly magnetised in a direction opposite to the magnetisation of the external field.* Consequently purely diamagnetic substances are repelled by a magnetic field. Diamagnetism refers to the magnetic moment induced in the individual atoms and molecules by an applied field, the process being similar to that by which currents are induced in a conducting loop of wire when the flux linking the loop is changed. *The induced moment is directly opposite to the applied field.*

### 20.5.8 Principal Ferromagnetic Elements

Here follows the discussion of *ferromagnetic elements*:

1. *Iron.* It has the highest susceptibility comparatively. This property added with low cost makes iron most suitable for commercial purposes. Its magnetic properties are considerably effected by the presence of traces of carbon, oxygen and nitrogen; therefore, these undesirable elements should be reduced to the possible attainable limit. High purity means high permeability and reduced hysteresis loss. Its permeability is to the tune of 2,000.

2. *Nickel.* It is also one of the important ferromagnetic elements. When it is heated, it remains ferromagnetic upto 395°C and beyond that it gets converted into paramagnetic. Its properties as ferromagnetic material are considerably improved when alloyed with iron and cobalt. Its permeability is 300.

3. *Cobalt.* Its permeability is 250 and increases with the temperature upto 300°C and loses

its magnetic properties at about 1130°C. When in pure state it does not prove to be an important ferromagnetic material. Its cost is adequately high.

### 20.5.9 Magnetically Soft Materials

The magnetically soft materials are characterised as follows:

1. They have high permeability.
2. The magnetic energy stored is not high.
3. They have negligible co-ercive force (due to which these are not suitable for making permanent magnets).
4. They have low remanance.

**Examples:** *Pure or ingot iron, Cast iron, Carbon steels, Silicon steel, Manganese and nickel steel, Permalloy, Mumetal, Perminvar.*

### 20.5.10 Magnetically Hard Materials

Magnetically hard materials are used for *making permanent magnets* and have the following *characteristics:*

1. They possess high vlaue of *BH* product.
2. High retentivity.
3. High co-ercivity.
4. Strong magnetic reluctance.
5. Hysteresis loop is more rectangular in shape.

**Examples:** *Tungsten steel, cobalt steel, chromium steel, Alini, Alnico, Curnife, Hypernic.*

## OBJECTIVE TYPE QUESTIONS

**Fill in the Blanks/Choose the Correct Answer :**

**1.** Conductors/insulators/semiconductors may be defined as the materials which have free valence electrons in plenty for electric conduction.

**2.** Resistance/conductance is the property of a substance due to which it opposes the flow of electricity through it.

**3.** ........................ resistance of a material is the resistance between the opposite faces of a cm cube of that material.

**4.** Several metals possess nil resistivity/conductivity at absolute zero of temperature, the phenomenon known as superconductivity.

**5.** Mechanical stressing of crystal structure decreases/increases conductivity of metal.

**6.** At very low temperature some metals acquire zero electrical resistance and zero magnetic induction, the property known as ........................

**7.** ........................ can be destroyed by the application of a strong magnetic field.

**8.** ........................ force is the force exerted by one magnet on another either to attract it or repel it.

**9.** ........................ permeability is denoted by $\mu_r$.

**10.** ........................ is defined as the ratio of intensity of magnetisation to the magnetisation force.

**11.** ........................ force is that force which drives or tends to drive the flux through a magnetic circuit.

**12.** ........................ is that peoperty of the material which opposes the production of magnetic flux in it.

**13.** ........................ force is defined as the demagnetising force which is necessary to neutralise completely the magnetism in an electromagnet after the value of magnetising force becomes zero.

**14.** ........................ is the magnetic flux which still presists in magnetic material even when the magnetising force is completely removed.

**15.** Match the following:

1. Ferromagnetic materials (*a*) $\mu_r < 1$

2. Paramagnetic materials (*b*) $\mu_r >> 1$
3. Diamagnetic materials (*c*) $\mu_r > 1$

**16.** A ....................... is a pair of equal but opposite electric charges which are very close.

**17.** The outstanding characteristic of ....................... material is that it is very strongly attracted by the magnet.

**18.** ....................... arises out of the electronic structure within the atoms.

**19.** ....................... is the property of material due to which it, when placed in a magnetic field, becomes weakly magnetised in a direction opposite to the magnetisation of the external field.

**20.** Nickel has permeability to the tune of
(*a*) 200 (*b*) 300 (c) 400 (*d*) 500.

**21.** Cobalt loses its magnetic properties at about
(*a*) 830°C (*b*) 930°C (*c*) 1030°C (*d*) 1130°C.

**22.** Nikel when heated remains ferromagnetic upto
(*a*) 195°C (*b*) 930°C (*c*) 395°C (*d*) 495°C.

**23.** Magnetically soft material have ............ remanance.

**24.** Magnetically hard materials have ............ retentivity.

**25.** .................... steel is used in making permanent magnets for dynamos, motors etc.

**26.** The saturation flux density of cobalt steel is
(*a*) 1.4 $Wb/m^2$ (*b*) 2.4 $Wb/m^2$ (*c*) 3.4 $Wb/m^2$ (*d*) 4.4 $Wb/m^2$.

**27.** Coercivity of chromium steel is
(*a*) 5500 AT/m (*b*) 6500 AT/m (*c*) 7500 AT/m (*d*) 8500 AT/m.

**28.** Alnico is more expensive/cheaper than Alni.

**29.** Saturation flux density of Alnico is
(*a*) 1.2 $Wb/m^2$ (*b*) 1.8 $Wb/m^2$ (*c*) 2.2 $Wb/m^2$ (*d*) 2.8 $Wb/m^2$.

**30.** Hypernic contains nickel and iron in the ratio of
(*a*) 40 : 60 (*b*) 50 : 50 (*c*) 60 : 40 (*d*) 70 : 30.

**31.** Cunife is suitable for small-size/large-size magnets.

**32.** When a ferromagnetic material is magnetised small changes in dimensions occur, the effect being known as ......................

**33.** ....................... temperature is the rising temperature at which the given material ceases to be ferromagnetic.

**34.** The dielectric strength of a material is usually increased/decreased by moisture contaminations, elevated temperatures, aging and mechanical stresses.

**35.** The combined effect of the volume and surface resistance represents ...................... resistance of the material.

**36.** Conductors/semiconductors/insulators are solid materials, either non-metallic element or compounds, which allow electrons to pass through them so that they conduct electricity in much the same way as a metal.

**37.** Resistivity of semiconductors is usually low/high/nil.

**38.** Semiconductors show a rise in conductivity in the decreasing/increasing temperature.

**39.** The temperature co-efficient of resistance of a semiconductor is always positive/negative.

**40.** Semiconductors are usually .................... in appearance.

**41.** Atoms with fewer than four valence electrons are
(*a*) Poor conductors (*b*) Good conductors (*c*) Semiconductors (*d*) Insulators.

**42.** Atoms with four valence electrons are conductors/semiconductors.

**43.** The energy level of an electron increases/decreases at its distance from the nucleus increases.

**44.** It is the mobility of first orbit/last orbit electrons that they acquire the property of combining with other atoms.

**45.** A pure semiconductor is called intrinsic/extrinsic semiconductor.

**46.** A pure semiconductor behaves as an insulator/conductor.

**47.** The resistance of a semiconductor decreases/increases with increase in temperature.

**48.** Semiconductor current consists of movement of electrons in the same/opposite direction.

49. The impure semiconductor is called extrinsic/intrinsic semiconductor.
50. The process of adding impurity to a semiconductor is called ......................
51. Pentavalent atoms having five valence electrons are called ...................... atoms.
52. Trivalent atoms having three valence electrons are called .................. atoms.

## ANSWERS

| | | | | | |
|---|---|---|---|---|---|
| 1. | Conductor | 2. | Resistance | 3. | Specific |
| 4. | Resistivity | 5. | Decreases | 6. | Superconductivity |
| 7. | Superconductivity | 8. | Magnetic | 9. | Relative |
| 10. | Susceptibility | 11. | Magnetomotive | 12. | Magnetic reluctance |
| 13. | Coercive, | 14. | Remanance, | 15. | 1 (*b*), 2 (*c*), 3 (*a*) |
| 16. | Dipole | 17. | Ferromagnetic | 18. | Ferromagnetism |
| 19. | Diamagnetism | 20. | (*b*) | 21. | (*d*) |
| 22. | (*c*) | 23. | Low | 24. | High |
| 25. | Tungsten | 26. | (*b*) | 27. | (*c*) |
| 28. | Expensive | 29. | (*a*) | 30. | Small size |
| 31. | (*b*) | 32. | Magnetostriction | 33. | Curie |
| 34. | Decreased | 35. | Insulation | 36. | Semiconductors |
| 37. | High | 38. | Increasing | 39. | Negative |
| 40. | Metallic, | 41. | (*b*) | 42. | Semiconductors |
| 43. | Increases | 44. | Last orbit | 45. | Intrinsic |
| 46. | Insulator | 47. | Decreases | 48. | Opposite |
| 49. | Extrinsic | 50. | Doping | 51. | Donor |
| 52. | Acceptor. | | | | |

## THEORETICAL QUESTIONS

1. What are the general properties of conductors ?
2. Give the properties and uses of the following conducting materials:
   Copper, Aluminium, Nickel and Tin.
3. Give the desirable properties of high resistivity materials.
4. Give the properties and uses of the following high resistivity materials:
   Tungsten, Carbon, Nichrome, Constantan and Chromel
5. Explain the term 'superconductivity'.
6. What is a semiconductor ?
7. What are the characteristics of semiconductors?
8. List the applications of semiconductors.
9. Differentiate between semiconductors and insulators.
10. How are insulating materials classified ?
11. Give the description of the following solid insulating materials:
    Mica, Rubber, Asbestos, Ceramic and Slate.
12. Define the following terms as applied to magnetic materials:
    Magnetic flux, Relative permeability, Magnetic potential and Coercive force.
13. How are magnetic materials classified?
14. Describe the following principal ferromagnetic elements:
    Iron, Nickel and Cobalt.
15. What are magnetic domains?
16. State the characteristics of magnetically soft materials.
17. State the characteristics of magnetically hard materials.

21

# Material Science of Metals

*21.1. Meaning of material science. 21.2. Structure of atoms and molecules – concept of atom– crystal structure – crystal structure for metallic elements – bonds in solids. 21.3. Deformation of metals – deformation – slip – twinning imperfections (defects) in crystals – cold working – hot working. 21.4. Heat treatment – definition – objects – constitutents of iron and steel – iron-carbon equilibrium diagram – theory of heat treatment of steel – heat treatment processes. 21.5. Testing of materials – introduction – classification of tests – non-destructive tests – destructive tests. 21.6. Corrosion – factors influencing corrosion – general types of corrosion – various types of corrosion– control and prevention of corrosion. 21.7. Manufacturing processes. Objective Type Questions – Theoretical Questions.*

## 21.1. MEANING OF MATERIAL SCIENCE

The term *'Material science'* consists of two words : *Materials and science. Materials* means engineering materials and those are limited to only solid materials. In general the word *'Science'* defines the knowledge arranged under general truth and principles and it naturally covers today a wide range of subjects but in material *science,* 'science' refers to the physical sciences relating to physics and chemistry. In material science since we confine out attention to solid materials only so the subject is related to solid-state physics and solid-state chemistry. In general the material "science" refers to that branch of applied science which is concerned with *investigating the relationship existing between the structures of materials and their properties* and their inter-disciplinary study of materials for practical purposes.

## 21.2 STRUCTURE OF ATOMS OF MOLECULES

### 21.2.1 Concept Of Atom

All substances are made up of atoms. Each atom consists of the following :

1. Nucleus. 2. Electrons

Fig. 21.1 shows the structure of an atom. Central nucleus is surrounded by orbital electrons which move in concentric spherical shells.

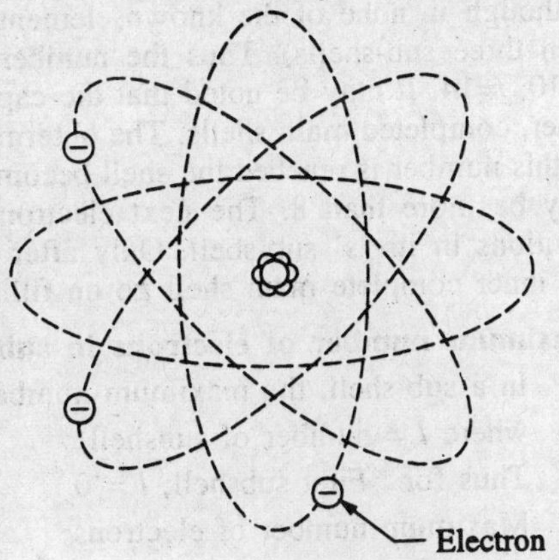

Fig. 21.1 : Concept of atom.

*Nucleus* : The nucleus is at the centre of the atom

and comprises of protons and neutrons. Its diameter is $\frac{1}{10,000}$th of the atom as a whole. Almost the entire mass of a given atom is concentrated in its nucleus. Due to the presence of protons nucleus is positively charged. The number of protons in the nucleus is equal (numerically) to the atomic number of the element. A neutron is an uncharged particle and has same mass as the hydrogen nucleus.

*The weight of the atom and its radioactive properties are associated with the nucleus. The chemical properties and spectrum on the other hand depend on the planetary electrons.*

*Electrons* : An electron is a negatively charged particle present in an atom. The number of electrons which surround the nucleus of a neutral atom is equal to the number of protons within the nucleus, *i.e.*, the atomic number. The electrons move about the nucleus. The orbit or shell nearest to the nucleus known as the K-orbit, contains a maximum of two, the next (L-orbit) eight and next (M-orbit) eighteen electrons and so on. In fact, the number of electrons in any orbit is equal to $2n^2$ where $n$ is the serial number of the orbit taking first orbit nearest to the nucleus, with the exception that the *outermost orbit cannot have more than eight.* In a given atom all orbits may not be complete.

Fig. 21.2 (*a, b, c*) illustrates atomic structures of some elements *viz.* lithium, sodium and argon

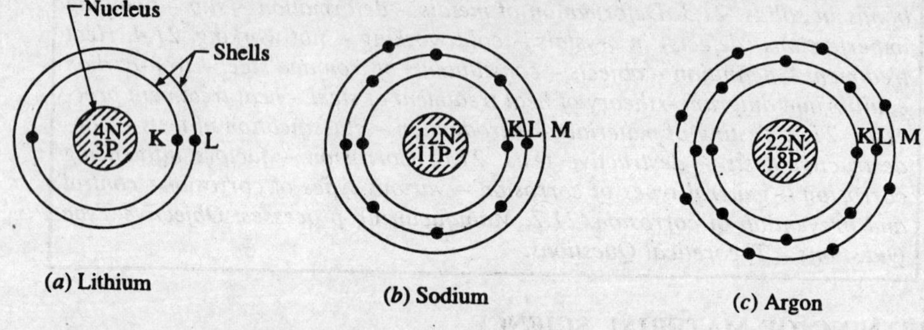

(*a*) Lithium (*b*) Sodium (*c*) Argon

Fig. 21.2

The main electrons shells (as stated above) are designated by the letters K, L, M, N, O, P and Q and by the corresponding principal quantum numbers, $n$ = 1, 2, 3, 4, 5, 6 and 7. The K shell, for example, has the principal quantum number $n$ = 1. The *sub-shells in each main shell are designated by small letters, s, p, d and f with quantum numbers* $l$ = 0, 1, 2, and 3 respectively. The number of sub-shells in a given main shell is the same as its principal quantum number, $n$ (although in none of the known elements do any of the last three shells O, P, Q, contain more than three sub-shells). Thus the number of electrons that can occupy the subshells, $s$=2; $p$=6; $d$=10; $f$=14. It may be noted that the capacities thus obtained, as described above, apply only to inner, completed main shells. The outermost main shell never has more than 8 electrons. As soon as this number is reached the shell becomes stable even though the sum of its sub-shell capacities may be more than 8. The next electrons to be added start to fill the next main shell, taking positions in its 's' sub-shell. Only after both spaces in this '*s*' sub-shell have been filled does the inner complete main shell go on filling up.

**Maximum number of electrons in sub-shells :**

In a sub-shell, the maximum number, of electrons is = 2 (2$l$ + 1)

where $l$ = number of subshell

Thus for : First subshell; $l$ = 0

Maximum number of electrons $= 2\,(2l + 1)$

$= 2\,(2 \times 0 + 1) = 2$

The subshell is known as *s*-sub-shell. The symbols 2*s*, 3*s*, 4*s* .... denote that *s*-sub-shell of the first, second, third and fourth energy levels respectively.

Second sub-shell $l = 1$

Maximum number of electrons = 2 (2 × 1 + 1) = 6

The sub-shell is known as *p*-sub-shell. The symbols 2*p*, 3*p*, 4*p* ... denote the *p*-sub-shell of the second, third and fourth energy levels respectively.

Third sub-shell $l = 2$

Maximum number of electrons = 2 (2 × 2 + 1) = 10

This sub-shell is known as *d*-sub-shell. The symbols 2*d*, 3*d*, 4*d* ........ denote the *d*-sub-shell of the third, fourth and fifth energy levels respectively.

Fourth sub-shell: $l = 3$

Maximum number of electrons = 2 (2 × 3 + 1) = 14

The sub-shell is known as *f*-sub-shell. The symbols 4*f*, 5*f* ... denote the *f*-sub-shell of the fourth and fifth energy levels respectively.

The following points are worth noting :

(*i*) The maximum number of electrons in the first principal energy level is 2. Therefore it will have only one sub-shell (1*s*) containing two electrons.

The first energy level *cannot have any other sub-shell.*

(*ii*) The maximum number of electrons in second principal energy level are 8. Therefore it will have two sub-shells namely 2*s* (containing two electrons), and 2*p* (containing 6 electrons).

(*iii*) The maximum number of electrons in the third principal energy level are 18. Therefore, it will have three sub-shells namely 3*s* (containing two electrons), 3*p* (containing 6 electrons) and 4*d* (containing 10 electrons).

(*iv*) The maximum number of electrons in the fourth principal energy level are 32. Therefore it will have four sub-shells namely 4*s* (containing two electrons) 4*p* (containing 6 electrons), 4*d* (containing 10 electrons) and 4*f* (containing 14 electrons).

### 21.2.2. Crystal Structure

Materials can broadly classified as crystalline and non-crystalline solids. In a crystal, the arrangement of atoms is in a *periodically repeating pattern* whereas no such regularity of arrangement is found in a non-crystalline material. A crystalline solid can either be a single crystal, where the entire solid consists of only one crystal, or an aggregate of many crystal separated by *well defined boundaries*, in the latter form, the solid is said to be *poly-crystalline.*

#### 21.2.2.1 Unit cell and space lattice

The metallic crystals can be considered as consisting of tiny blocks which are repreated in three dimensional pattern. *The tiny block formed by the arrangement of a small group of atoms is called the* **unit cell.** It is that *volume of solid* from which the entire crystal can be constructed by translational repetition in three dimensions. (Fig. 21.3).

If each atom in a lattice is replaced by a point; then each point is called a **lattice point** and the arrangement of the points is referred to as the (three dimensional) **lattice array.** Thus a **space lattice** *is, defined as an array of points in three dimensions in which every point has surroundings identical to that every other point in the array* (Fig. 21.4). The distance between the atoms-points is called **inter-atomic** or **lattice spacing.** A space lattice is a conventional geometrical basis by which crystal structures can be described. Every *point of a space lattice has identical surroundings.*

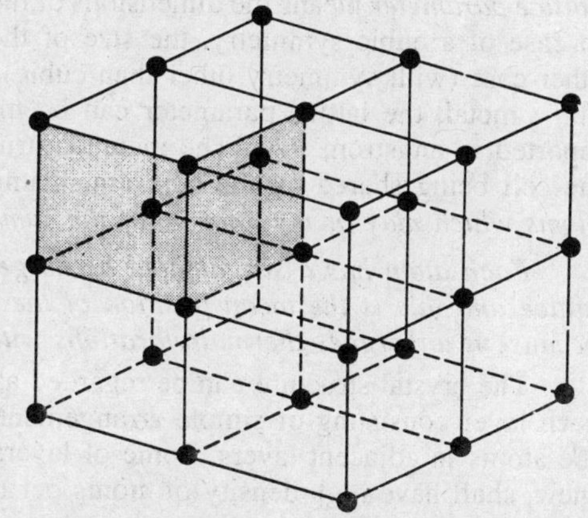

Fig. 21.3 : Unit cell.

The *length of the side of a unit cell is the*

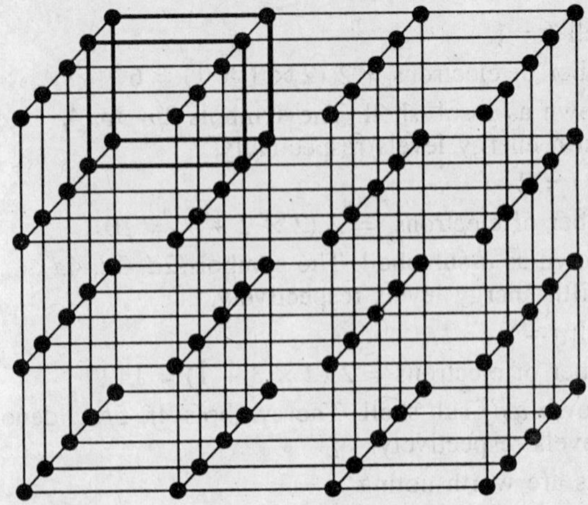

Fig. 21.4 : Space lattice.

*distance between the atoms of the same kind*. In the case of pure metals whose crystals have simple cubic structure [Fig. 21.5 (*a*)], it is equal to the basic distance '*a*' only. However in case of crystals of chemical compound like the cubic crystals of sodium chloride (NaCl), it is twice the distance '*d*' between two adjacent atoms as shown in [Fig. 21.5 (*b*)].

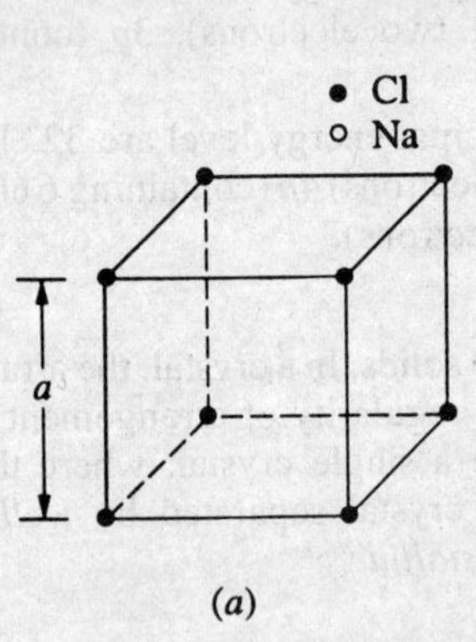


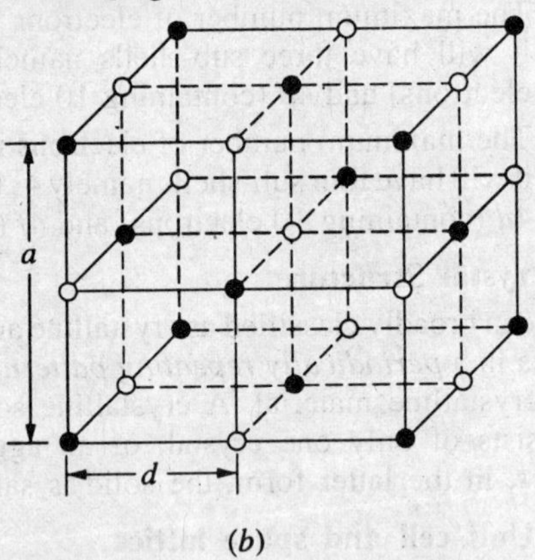


Fig. 21.5

### 21.2.2.2 Lattice parameter and crystallographic planes

*Lattice parameter* means the dimensions of the unit cell in any of the crystallographic arrangements. In case of a cubic symmetry, the size of the lattice is fixed by the length of the cubic unit. In other case (with symmetry other than cubic) more than one lattice parameter has to be specified. For a metal, the lattice parameter can be measured by X-ray diffraction studies and results are reported in angstrom' (Å). The metallic structure is built up of series of unit cells, each face of the cell being shared by the adjacement units. *A crystal thus consists of a regular assembly of atoms which may or may not be of the same species.*

*Each atom has a site defined by its geometrical centre of the relationship with the crystal lattice and this is the mean position of the atom. The position of the atom is less well defined, because it undergoes thermal vibrations within the space available to it between its neighbours.*

The crystal structure can be regarded as an array of atoms built of layers one over the other, each layer consisting of simple arrangement of uniform rows of atoms, which mate or key with the atoms in adjacent layers. Some of layers, depending on the lattice arrangement and angle of view, shall have high-density of atoms per unit area and other may not be so dense. *Such layers of atoms are called crystallographic planes* and the planes and the *interatomic forces in a solid vary greately with the geometry of these planes.*

### 21.2.3 Crystal Stuructures for Metallic Elements

The crystal of most metals havė highly symmetrical structure with closed packed atoms. The most common types of space lattice (or unit cells) with which metallic elements crystallise are given below.

1. Body centred cubic structurę (BCC)
2. Face centred cubic structure (FCC)
3. Hexagona close-packed structure (HCP).

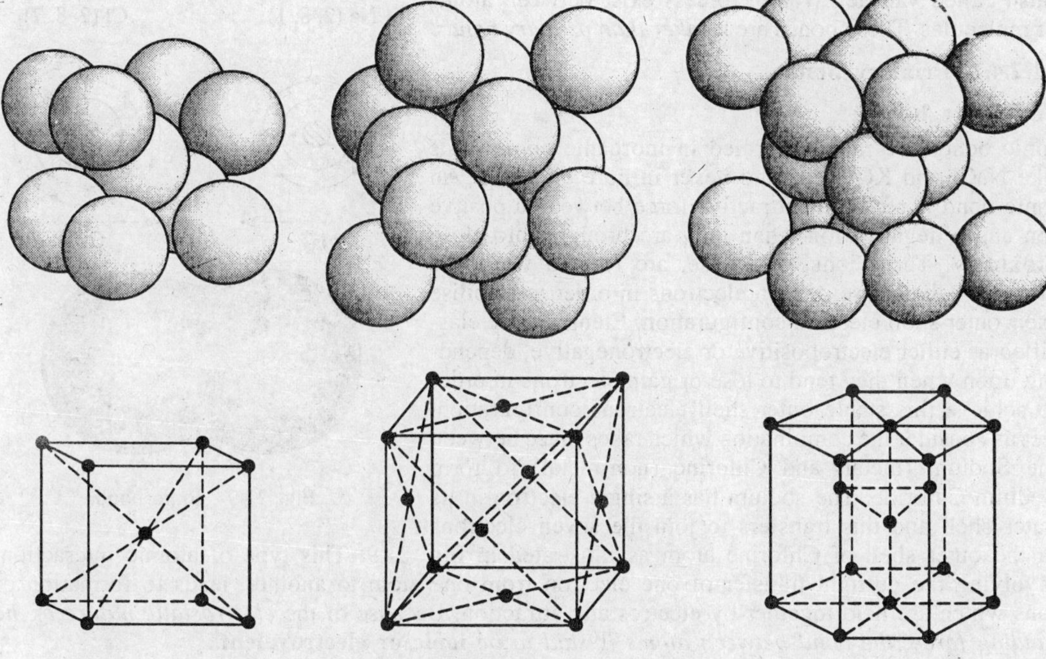

Fig. 21.6 : B.C.C. Fig, 21.7 : F.C.C. Fig. 21.8 : H.C.P.

**1. Body centred cubic structure (BCC).** Refer Fig. 21.6. Here the atoms are located at the corners of the cube and one atom at its centre. This type of unit cell is found in metals like *lithium, sodium, potassium, barium, vanadium, molybdenum* etc.

**2. Face centred cubic structure (FCC).** Refer Fig. 21.7. Here the atoms are located at the corners of the cube and one atom at the centre of each face. This type of unit cell is found in metals like *copper, silver, gold, calcium, aluminium, lead* etc.

**3. Hexagon close packed strucure (HCP).** Refer Fig. 21.8. Here the unit cell has an atom at each of the twelve corners of the hexagonal prism, one atom at the centre of each of the two hexagonal faces and three atoms in the body of the cell. This type of unit cell is found in metals like *zinc, lithium, magnesium beryllium* etc.

### 21.2.4. Bonds in Solids

The atoms and *molecules in a solid state* are more closely packed as compared to gaseous and liquid states and are held together by *strong mutual forces of attraction.* These interatomic forces are electrostatic in nature and depend upon the electronic structure of matter. When atoms come closer and finally unite to form molecules their electrons rearrange themselves in such a way as to achieve a *stable configuration*. This arrangement of electrons gives rise to different types of bonds due to which atoms are held together.

Bonds can be broadly classified as :

1. Primary bonds

2. Secondary bonds.

**Primary bonds (or chemical bonds):** These are the *strongest bonds* between atoms which can be further divided as follows :

(*i*) Ionic (or electrostatic) bond.

(*ii*) Covalent (or atomic or homopolar) bond.

(*iii*) Metallic bond.

**Secondary (or molecular) bonds :** Attraction forces (also called van der Waals' forces) exist between atoms or molecules. These bonds are *weaker than primary bonds.*

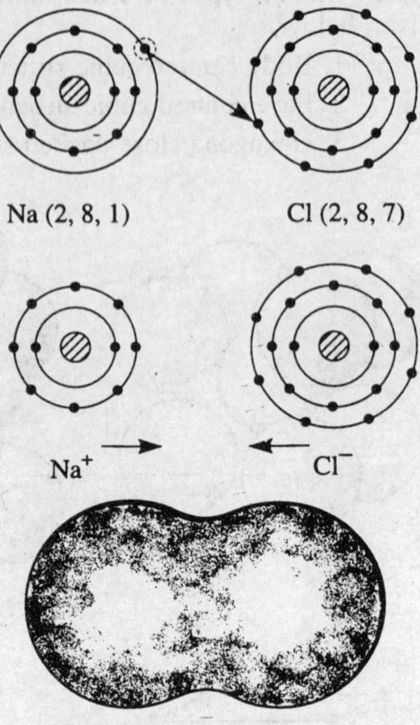

Fig. 21.9 : Ionic bond.

### 21.2.4.1. Primary bonds

#### 1. Ionic bond :

Ionic bonds are mainly formed in inorganic compounds like NaCl and KOH etc. and never in pure elements. An ionic bond is really the attractive force between a positve ion and a negative ion when they are brought into close proximity. These ions, of course, are formed when the atoms involved lose or gain electrons in order to stabilise their outer shell electron configuration. Elements are classified as either electropositive or electronegative, depending upon when they tend to lose or gain electrons in order to achieve this stable outer shell electrons configuration. Let us consider the combination which takes place between the Sodium (metal) and Chlorine (non-metal) to form Sodium Chloride. The sodium has a single electron in its outer shell and this transfers to join the seven electrons in the outer shell of Chlorine atom as illustrated in Fig. 21.9. This type of atomic interaction, involving the outright transfer of one electron from one atom to another, leads to formation of ions which are held together by electrostatic attraction. Because of the *electrostatic nature of the binding force, the bond between atoms is said to be* **ionic or electrovalent.**

*Characteristics of Ionic Compounds:*

(*i*) Ionic bonds are undirectional.

(*ii*) They are generally crystalline in nature.

(*iii*) They are rigid.

(*iv*) They have high melting and also high boiling point because strong electrostatic forces bind them.

(*v*) They are generally non-conductors of electicity but their melts and solutions conduct electricity.

(*vi*) They are usually highly soluble in water and other polar solvents but insoluble in organic solvents.

#### 2. Covalent bond:

*The covalent bond is formed by sharing of electrons between atoms rather than by transfer of electrons.* Only a few solids are held together by covalent bonds. Covalent bonding alone is not sufficient to build three dimensional solids. The majority of solids incorporating covalent bonds are bound also either by ionic or van der Waals' bonds.

An excellent example of covalent bonding is found in the Chlorine molecule (Fig. 21.10 *a*). Here the outer shell of each atom possesses seven electrons. Each chlorine atom would like to gain an electron and thus form a stable octet. This can be done by sharing of two electrons between pairs of chlorine atoms thereby producing stable diatomic molecules. The nature of covalent bonds in the molecules of hydrogen ($H_2$) and hydrochloric acid (HCl) are illustrated in Fig. 21.10 (*b*).

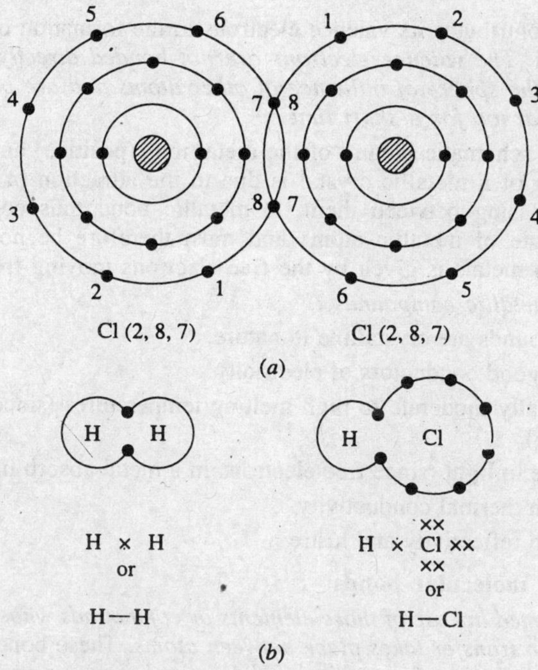


Fig. 21.10 : Covalent bond.

*Characteristics of covalent compounds:*

(*i*) Covalent compounds are directional in nature.

(*ii*) They can be solids, liquids or gases.

(*iii*) They are soft and have low melting and boiling points.

(*iv*) They are generally insulators electrically.

(*v*) They are soluble in non-polar solvents such as benzene and carbon tetrachloride and insoluble in water.

**3. Metallic bond:**

*A metallic bond results from the sharing of variable number of electrons by a variable number of atoms.* The atoms in metal and alloys are held together by such bonds. This type of bond is characteristic of the elements having small number of valence electrons, which are loosely held, so that they can be easily released to the comon pool. The bonding takes place when each of

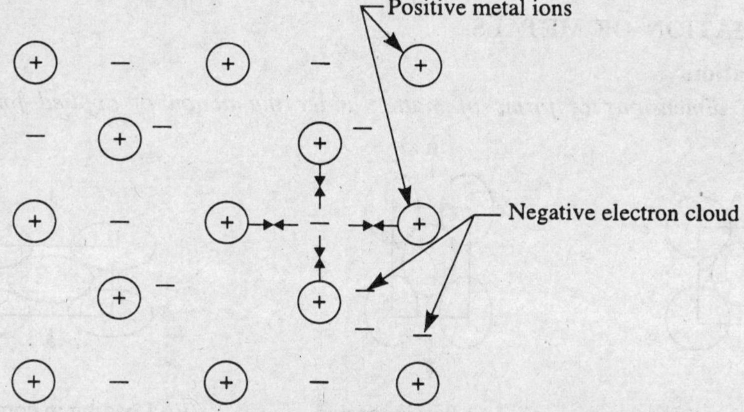


Fig. 21.11 : Metallic bond.

the atoms of the metal contributes its valence electrons to the formation of an electron cloud that pervades the solid metal. *The valence electrons are not bonded directly to an individual atom but they move freely in the sphere of influence of other atoms and are bound to different atoms at different times and that too for a short time.*

Fig. 21.11 shows a schematic picture of the metal ions (positive) and that of electron cloud (negative). The cohesion of a metallic crystal is due to the attraction of the positive nuclei and the valence electrons passing between them. A metallic bond thus conceived can exist only between a large aggregate of metallic atoms and must therefore be non-directional. The high electrical conductivity to metals is given by the free electrons moving freely in an electric field.

*Characteristic of metallic compounds :*

(*i*) Metallic compounds are crystalline in nature.

(*ii*) They are very good conductors of electricity.

(*iii*) They have usually moderate to high melting temperatures (since the metallic bonds are not very strong).

(*iv*) The are opaque to light (since free electrons in a metal absorb light energy).

(*v*) They have high thermal conductivity.

(*vi*) They have high reflectivity and lustre.

#### 21.2.4.2 Secondary or molecular bonds

*Molecular bonds are formed in case of those elements or compounds whose electron configuration is such that little electron transfer takes place between atoms.* These bonds are formed as a result of week Van der Waals' forces of attraction which exist between various atoms. These forces are *due to the electrostatic attraction between the nucleus of one atom and the electrons of the other.* This is largely but not completely neutralised by the electrostatic repulsion of the nucleus of one atom by the nucleus of the other. The *resultant weak attraction between the two atoms is called van der Waals' force.*

Three types of intermolecular bonds known are :

1. Dispersion bond
2. Dipole bond
3. Hydrogen bond.

*Characteristics of moleculer bonds:*

1. Moleculer bonds can be both crystalline and non-crystalline.
2. They have low melting point (because of weak molecular bond)
3. They are usually transparent to light.
4. They are good insulators (since they do not have valence electrons).

## 21.3 DEFORMATION OF METALS

### 21.2.1 Deformation

The *change in dimensions of forms of matter under the action of applied forces is called*

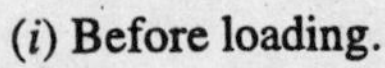

(*i*) Before loading.

(*ii*) Loading in tension.

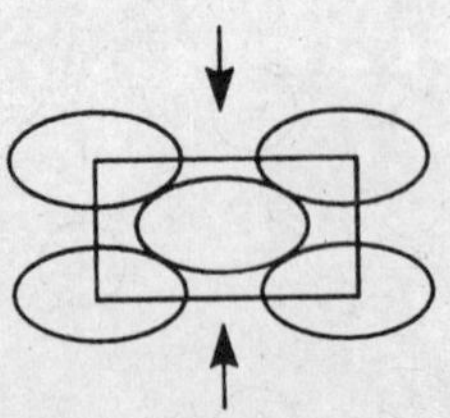

(*iii*) Loading in compression.

Fig. 21.12 : Elastic deformation.

*'deformation'.* It is caused either by the mechanical action of external forces or by various physical and physio-chemical processes. To form various metallic shapes the deformation of metals is necessary. The deformed or mechanically worked metals are much superior to cast metals from which they are produced.

**Types of metal deformation :**

The metal deformations are of the following *two types :*

1. Elastic deformation, and
2. Plastic deformation.

**1. Elastic deformation:**

Elastic deformation *is the deformation which disappears when the load is removed.* It precedes (take place before) plastic deformation. This deformation occurs when a stress is applied on a metal piece.

Fig. 21.12 (*i*) (*ii*) (*iii*) shows the form of the atoms before loading, after loading in tension and compression reapectively. When a tensile load is applied, the piece becomes slightly longer whereas a compressive load shortens the piece. On removal of load (tensile or compressive) the metal piece or specimen regains its original shape (dimensions). Within the elastic range, the strain is result of a slight elongation of the unit cell in the direction of the tensile load, or a slight contraction in the direction of compressive load.

For elastic deformation, the strain is nearly proportional to stress. The *ratio between stress and strain under elastic deformation is known as modulus of ealsticity or Young's modulus* (*E*). It is a characteristic of the type of the metal and its magnitude depends upon the force of attraction between atoms of a metal.

*Shear stresses* are also set up in crystal structures in engineering applications. These produce displacement of one plane of atoms relative to the adjacent plane of atoms. The elastic strain is defined as the *tangent of the shear angle.* The ratio between shear stress and shear strain is known as *shear modulus of rigidity (G).*

**2. Plastic deformation:**

It is the *deformation which persists even after the load is removed.* Plastic deformation is observed at stresses exceeding the elastic limit. In contrast to elastic deformation, which depends primarily on stress in the simplest cases, plastic deformation is typically a *function of stress, temperature,* and the *rate of straining.* Plastic deformation is associated with the displacement of the atoms within the grains and causes ultimately *permanent changes in shape of the material.* Tensile, compressive and torsional stresses can cause plastic deformation.

Plastic deformation process is generally applied in important metallurgical operations of shaping. The operations inlcude rolling of boiler plates, drawing of wire, extrusion of telephone cables, stamping of automobile parts etc.

In metals the plastic deformation generally takes place by the process of **"slipping"**.

**Comparison Between Elastic and Plastic Deformation**

| S. No. | Elastic deformation | Plastic deformation |
|---|---|---|
| 1. | It appears and disappears with the application and removal of *stress-temporary deformation.* | It exists even after the removal of *stress-permanent deformation.* |
| 2. | It is the begining of the progress of deformation. | It takes place after the elastic deformation has stopped. |
| 3. | In elastic deformation, the stain reaches its maximum value after the stress has reached its maximum value. | In plastic deformation, the strain occurs simultaneously with the application of stress. |
| 4. | It takes place over a short range of stress-strain curve. | It takes place over a wide range of stress-strain curve. |

### 21.3.2 Slip

Slip is defined as a *shear deformation that moves atoms by many inter-atomic distances in one crystal plane over the atoms of another crystal plane.* Fig. 21.13 shows the slip process under a shear load (in a single crystal). Because of plastic deformation, the crystal is divided into layers or slip blocks which are displaced in reference to each other and are separated by thin layers in which a considerable displacement of atoms has taken place. These *intermediate layers with strongly distorted lattices are called* **"slip planes".** Slip occurs more readily along certain crystal directions and planes than along other. This is shown in Fig. 21.14 where a single crystal HCP metal was deformed plastically.

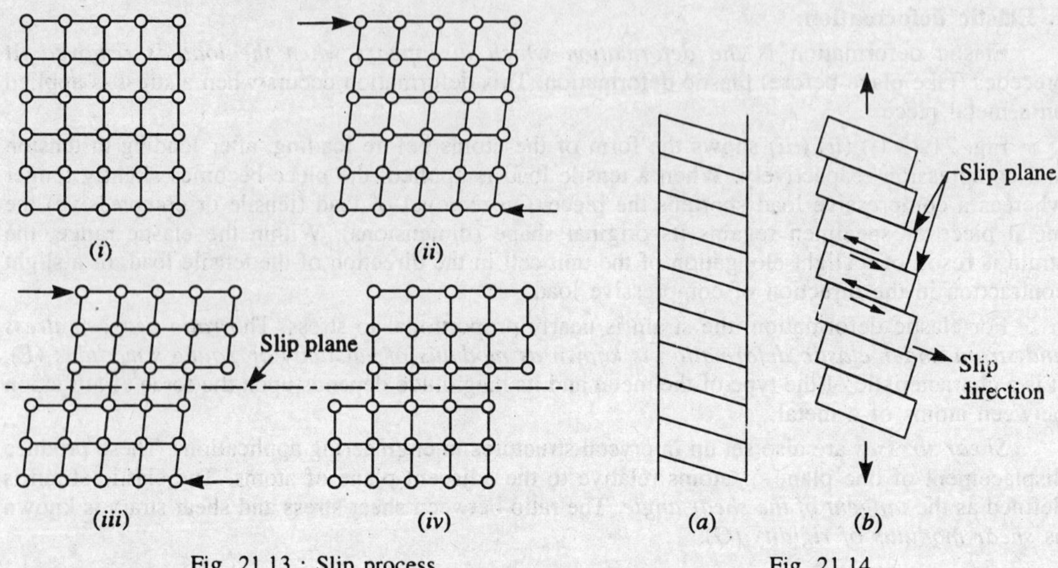


Fig. 21.13 : Slip process. Fig. 21.14

### 21.3.3. Twinning

**Twinning** *is that process by which a portion of the crystal takes up an orientation which makes that portion a mirror image of the parent crystal.*

Deformation by twinning is most common in HCP metals and its effects on others is to move parts of the grains such that they acquire favourable orientation for the slip to occur. *Twinning is produced suddenly and is accompanied with sound.* The well known cry of tin is the result of twinning.

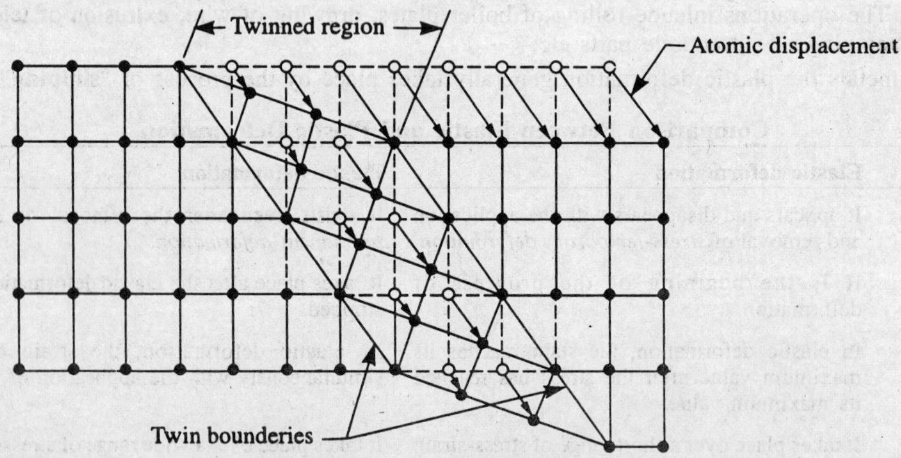


Fig. 21.15 : Twinning.

Fig. 21.15 shows the classical atomic picture of twinning. The twinning plane is at righı angle to the paper. If a shear stress is applied, the crystal will turn about the twinning plane. The region to the left of the twinning plane is undeformed. To the right of this plane, the planes of atoms have sheared in such a way to make the lattice a mirror image across the twin plane. The *mechanisms of twinning* may thus be described as the simple sliding of one plane of atoms over the next, the extent of movement of each plane being proportional to its distance from the twinning plane.

In case of slip, the offset is a multiple of interatomic spacing, but in *twinning* the *offset* produced by sliding of one plane against its neighbour is a *fraction* of unit slip. This causes a difference in the orientation between the twinned and untwinned regions in the crystal and on polishing and etching these can be easily differentiated under the microscope.

Twins may be of the following two types :

1. **Mechanical twins :** These are produced by mechanical deformation. These are produced in BCC or HCP metals under conditions of rapid rate of loading (shock loading) and decreased temeprature.
2. **Annealing twins :** These are produced as a result of annealing following plastic deformation.

### 21.3.4. Imperfections (Defects) in Crystals

Imperfections are found in all cyrstals unless special means are used to reduce them to a low level.

These may be classified as follows :

(a) Point defects,

(b) Line defects, and

(c) Other defects.

**(*a*) Point defects :** Point defects are imperfect point like regions in a crystal. The typical size of a point defect is one or two atomic diameters.

The various point defects are discussed below :

1. **Vacancy :** Refer Fig. 21.16. A vacancy is the simplest point defect and involves a *missing atom within a metal.* These defects may come up as a result of *imperfect packing during the original crystallisation.* They may also arise from thermal vibrations of the atoms at high temperatures.

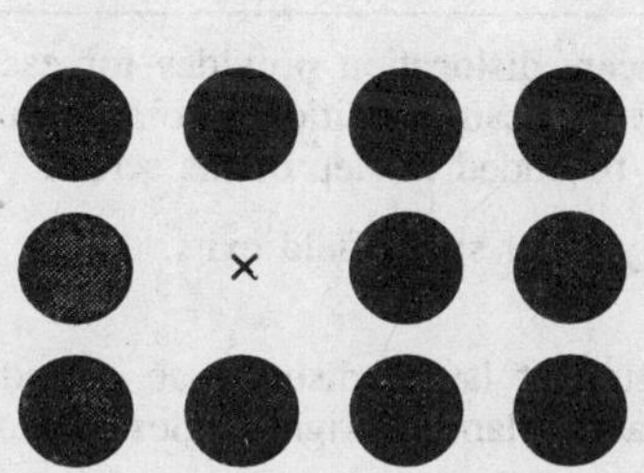

Fig. 21.16 : Vacancy.

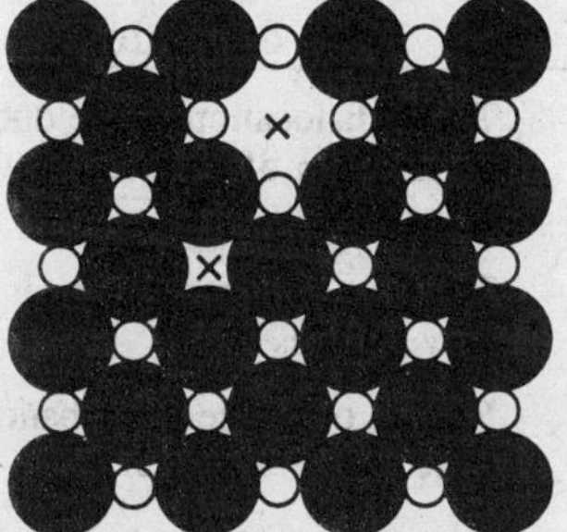

Fig. 21.17 : Schottky imperfections.

2. **Schottky imperfections :** Refer Fig. 21.17. These are closely related to vacancies but are found in compounds which must maintain a charge balance. They involve vacancies of pair of ions of opposite charges.

   Both *vacancies and schottky defects facilitate atomic diffusion.*
3. **Interstitialcy :** Refer Fig. 21.18. It is the addition of an extra atom within a crystal structure particularly if the atomic packing factor is low. This results in atomic distortion. The foreign atom may form added alloying agent or simply an impurity. The vacancy and interstitialcy are therefore inverse phenomena.

4. **Frenkel defect :** Refer Fig. 21.19. An ion dislodged from the lattice into an interstitial site is called *Frenkel defect*. The interstitialicies and Frenkel defects are less in number than vacancies and Schottky defects, because additional energy is required to force the atom into the new position.

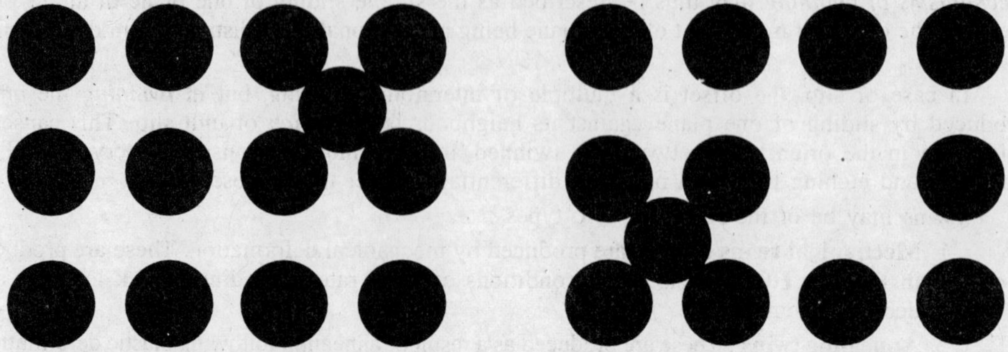

Fig. 21.18 : Interstitialcy defect. Fig. 21.19 : Frenkel defect.

**(b) Line defects (Dislocations)** : *A linear disturbance of the atomic arrangement, which can move very easily on the slip plane through the crystal* is known as **dislocation**. The dislocation may be caused *during growth of crystals* from the melt or from a vapour or they may occur during a slip.

These are basically of the following two types :

1. Edge dislocation, and
2. Screw dislocation.

Several actual dislocations are the combinations of the above two types of dislocations and are termed as *complete dislocations* or *mixed dislocations.*

**Comparison between an edge dislocation and a screw dislocation:**

The comparison between an edge dislocation and a scew dislocation is spelled out in table given below :

| S. No. | Edge disloation | Screw dislocation |
|---|---|---|
| 1. | These dislocations arise due to introduction or elimination of an extra row of atoms. | A scew dislocation provides for easy crystal growth because additional atoms and units cells can be added to step of the screw. |
| 2. | Tensile, compressive and shear stress fields may be present. | Only shear stress field exits. |
| 3. | Region of lattice disturbance extends along an edge inside a crystal. | Region of lattice disturbance extends in two separate planes at right angles to each other. |
| 4. | Burger's vector is always perpendicular to the dislocation line. | Burger's vector is parallel to the dislocation line. |
| 5. | These dislocations are formed during deformation and during crystallisation. | These dislocations are also formed during deformation and during crystallisation. |
| 6. | An edge dislocation (pure) can glide and climb. | A screw dislocation can glide only. |

**(c) Other defects**

**Surface defects :** *Surface defects are the two dimensional regions in a crystal.* They arise from a change in the stacking of atomic planes on or across a aboundary.

They are of the following two types :

1. External defects 2. Internal defects

**1. External defects :** The external type is just what its name implies, the defects or imperfections represented by a boundary. The external surface of the material is an imperfection itself, because the atomic bonds do not extend beyond it. The surface atoms have neighbours on one side only, while atoms inside the crystal have neighbours on either side of them. Since these surface atoms are not entirely surrounded by others, they possess higher energy than those of internal atoms.

**2. Internal defects :** These defects are discussed below :

(*i*) **Grain boundaries :** *Grain boundaries are those imperfections which separate crystals or grains of different orientation in polycrystalline aggregation during nucleation or crystallisation.*

In materials like copper there may be crystals of various orientations. These individual crystals are called *grains.* In one grain atoms are arranged with one orientation and one pattern. However there is a transition zone between two adjacent grains which are not aligned with either grain.

(*ii*) **Tilt boundary :** Tilt boundary in reality is a series of aligned dislocations which tend to anchor dislocations movements normally contributing to plastic deformation. Little energy is associated with this type of boundary.

(*iii*) **Twin boundaries :** These are the imperfections which separate two orientation that are mirror image of one another. The volume of the material which has an orientation that is mirror image of the matrix orientation is called a *twin.*

(*iv*) **Stacking fault :** It is a surface imperfection that arises from the stacking of one atomic plane out of sequence on another while the lattice on *either* side of the fault is perfect.

**Impurities :** These are the foreign atoms occupying position between the atoms of an ideal crystal. Foreign atoms may also occupy lattice sites from which regular atoms are missing.

**Exciton and Photons :** These are energy quanta associated with excitation and vibration of crystals.

### 21.3.5 Cold Working

*A metal is said to be cold worked, if it is mechanically processed below the recrystallisation temperature of the metal.* Cold working produces an improved surface finish and closer dimensional tolerance and because of this characteristic cold working processes are generally used in making end-use products. Since recrystallisation does not take place in cold working, the grains are permanently distorted. *Residual stresses are set up during cold working. As their presence is undesirable a suitable heat treatment is generally necessary to neutralise these stresses and restore the metal to its original structure.*

The various cold working operations are :

1. Drawing 2. Squeezing. 3. Bending. 4. Shearing. 5. Hobbing. 6. Shot peening. 7. Cold extruding.

When there is excessive cold work the metal may fracture before reaching the desired shapes and size; and in order to avoid it cold working operations are carried out in *several steps.*

### 21.3.6. Hot working

*When plastic deformation of metal is carried out at a temperature above the recrystallisation temperature the processes performed on metals are termed as* **hot working.** Hot working process can be considered as simultaneous combination of cold working and annealing. Any work hardening effect caused by plastic deformation is neutralised immediately by the effect of high temperature. Hot working process facilitates metal shaping with low power requirements though it is expensive to handle hot materials. There is also loss of metal by scaling and fine dimensionsal tolerance cannot be achieved in this process. Hot working increases the density since any pores or cavity in the cast metal disappear during hot working. Grain structure becomes more refined.

The various hot working operations are :

1. Rolling 2. Forging. 3. Pipe welding 4. Hot piercing 5. Hot drawing 6. Hot spinning 7. Hot extruding.

**Comparison of Hot Working and Cold Working Processes**

| S.No. | Cold working | Hot working |
|---|---|---|
| 1.. | Cold working is done at a temperature below the value required for recrystallisation, so no appreciable recovery takes place during deformation. | Hot working is done at a temperature above recrystallisation temperature, so it can be regarded as a simultaneous occurence of deformation and recovery process. |
| 2. | Hardening is not eliminated as working is done at a temperature below recrystallisation, so this is always accompanied by strain hardening. | Hardening due to plastic deformation is completely eliminated by recovery and recrystallisation. This is true, however, only if the rate of crystallisation is higher than rate of deformation. |
| 3. | Cold working decrases the value of elongation, reduction of area and impact values. | Mechanical properties like elongation, reduction of area and impact value are improved. |
| 4. | Crystallisation does not take place, so refinement of crystals is out of question. Grains are elongated. | Refinement of crystals occurs. |
| 5. | Uniformity of material is lost and properties are affected a lot. | Promotes uniformity of material by facilitating diffusion of alloys, constituents and breaks brittle film of hard constituents or impurity *e.g.,* cementite in steel. |
| 6. | Chances of crack propagation is more. | Cracks and unoxidised blow holes are sometimes welded up; alternatively, serious cracks or faults are usually shown up at an early stage. |
| 7. | Cold working increases ultimate temsile strength, yield point, hardness and fatigue strength while resistance to corrosion is decreased. If severely worked, yield point may coincide with ultimate strength value. | Ultimately tessile strength, yield point, fatigue strength, hardness and resistance to corrosion etc. are *not affected* if hot working is done properly. |
| 8. | Internal and residual stresses are produced. | Internal and residual stresses are not produced. |

## 21.4. HEAT TREATMENT

### 21.4.1 Definition

"**Heat treatment**" *is defined as an operation or combination of operations, involving heating and cooling of a metal or alloy in its solid state with the object of changing the characteristics of the material.*

### 21.4.2 Objects

Heat treatment is generally employed for following purposes :

1. To improve machinability.
2. To change or refine grain size.
3. To relieve the stresses of the metal induced during cold or hot working.
4. To improve mechanical properties *e.g.,* tensile strength, hardness, ductility, shock resistance to corrosion etc.
5. To improve mechanical and electrical properties.
6. To increase resistance to wear, heat and corrosion.
7. To produce a hard surface on a ductile interior.

### 21.4.3. Costituents of Iron and Steel

The different microscopic constituents of iron and steels which commonly occur are as follows:

1. Ferrite
2. Cementite
3. Pearlite
4. Martensite
5. Austenite
6. Troostite
7. Sorbite.

The other constituents comprise the three allotropic form of nearly pure iron, graphite and slag.

**1. Ferrite:** Iron which contains little or no carbon is called *ferrite*. It is very soft and ductile and is known as *'aplha'* iron by the metallurgists. Ferrite is present to some extent in a great range of steels, particularly those low in carbon content, and it is also present in soft cast iron. *Ferrite does not harden when cooled rapidly.* It forms smaller crystals when cooled from a bright red heat at a rapid rate.

**2. Cementite :** This is a definite carbide of iron ($Fe_3C$) which is *extremely hard,* being harder than ordinary hardened steel or glass. Cementite increases generally with the proportion of carbon present, and the hardness and also the brittleness of cast iron is believed to be due to this substance.

It contians 6.6. percent carbon and occurs either in the form of a network or in globular or massive form, depends on the analysis of the steel and the heat treatment to which it is subjected. It is *magnetic below 250°C. Its presence in iron or steel decreases the tensile strength but increases the hardness and cutting qualities.*

**3. Pearlite :** Pearlite is the name given to *a mixture of about 87.5 percent ferrite and 12.5 percent cementite.* It comprises of alternate layers of ferrite and cementite in steel. Under high magnification the ferrite and cementite can be seen to be arranged in alternate laminations or plates. When seen in the microscope the surface has appearance like mother of pearl, hence the name pearlite. The thickness of alternate plates and the distance between them is governed by the rate of cooling, slow cooling produces a coarser structure than rapid cooling. *Pearlite is a eutectoid of steel.*

It has been found that the proportion of pearlite increases from nothing in the case of pure carbonless iron upto 100%, or saturation, for steel containing 0.90% of carbon, thus a 0.3 percent carbon steel will consists of about 33 percent pearlite and rest ferrite. It is the characteristic of *soft steels that they contain ferrite and pearlite, and the hardness increases with the proportion of pearlite. Hard steels are mixtures of pearlite and cementite.*

**4. Martensite :** It is a *hard brittle mass of fibrous or needle like structure and is the chief constituent of hardened steel.* The vickers pyramid numeral is anything upto 900 for an original carbon content of 0.9 percent. It has been found that martensite is produced by the rapid quenching of high carbon steel from a slightly higher temperature than the maximum temperature of critical interval. It *is not as tough as austenite. It differs from austenite in being magnetic.*

**5. Austenite :** It is a solid solution of iron-carbon which is stable only within a particular range of composition and temperature, and is *non-magnetic.* On cooling below 700°C it is completely transformed into ferrite which is magnetic and cementite to form the eutectoid pearlite, together with free ferrite or free cementite, depending on whether the carbon content is less or greater than 0.87 percent respectively.

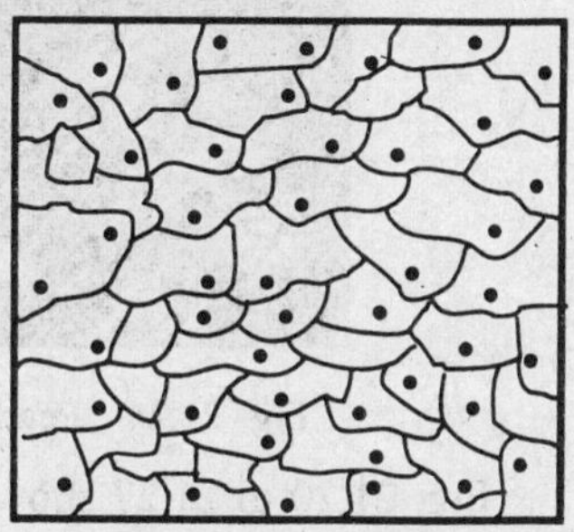

It is formed when carbon steel with more than 1.1. percent carbon is quenched rapidly from about 1000°C. The amount of

Fig. 21.20 : Pure iron.

austenite increases with the proportion of carbon 0 at 1.1. percent carbon upto 70 percent for 1.6 and 1.8 percent carbon. *Austenitic steels cannot be hardened by usual heat treatment methods and are non-magnetic.*

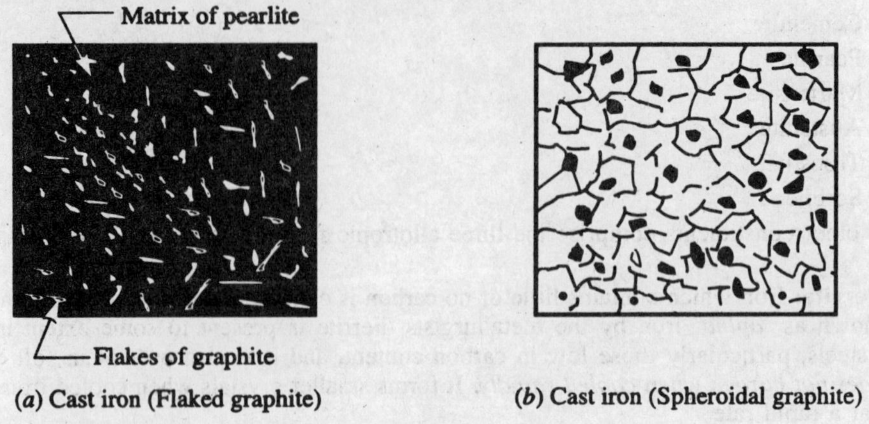


(*a*) Cast iron (Flaked graphite) (*b*) Cast iron (Spheroidal graphite)

Fig. 21.21

**6. Troostite :** It is a structure in steel (consisting of very finely divided iron carbide in what is known as "*alpha-iron*") produced either by tempering a martensitic steel at between 250 and 450°C, or by quenching steel at a speed insufficient to suppress the thermal change point fully. The structure produced by the latter method should be more accurately termed very fine pearlite.

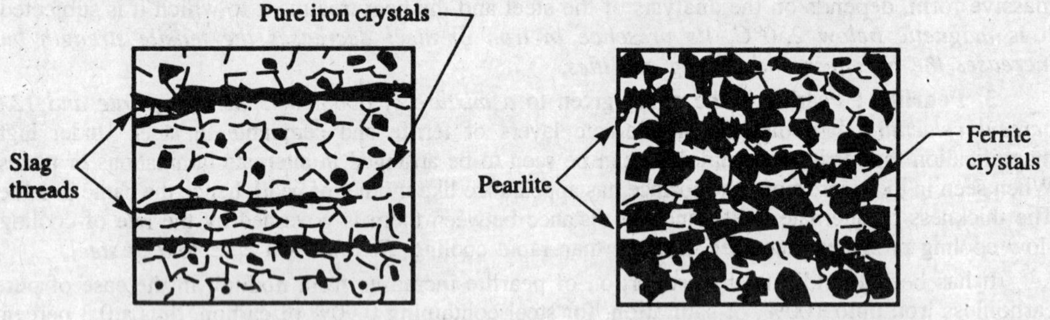


Fig. 21.22 : Wrought iron. 21.23 : Mild steel (0.25 – 0.3% carbon).

**7. Sorbite :** It is a strucutre which consists of evenly distributed carbide or iron particles in a mass of ferrite, formed when a fully hardened steel is tempered between 550 and 650°C. *A sorbite structure is characterised by strength and a high degree of toughness.*

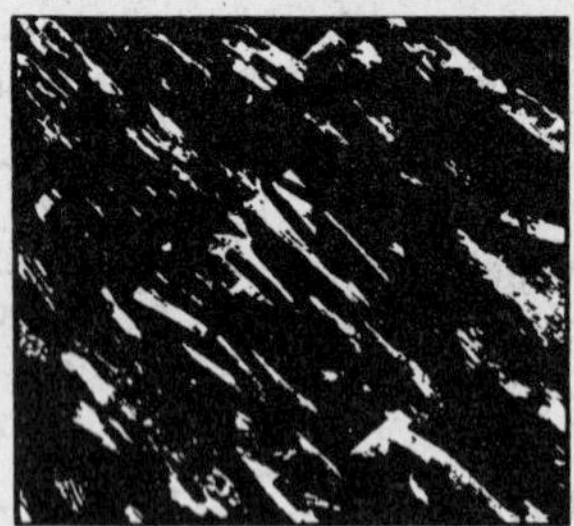

Fig. 21.24 : Pearlite.

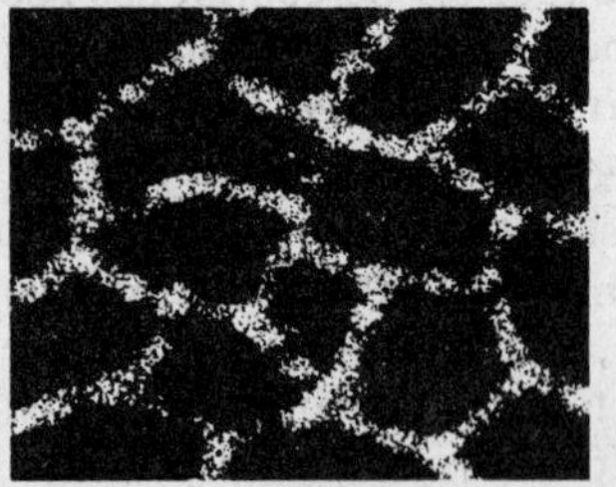

21.25 : High carbon steel.

Fig. 21.20 to 21.27 shows microstructures of constituents of iron and steel and some of the ferrous metals.

Fig. 21.26 : Ferrite and pearlite.

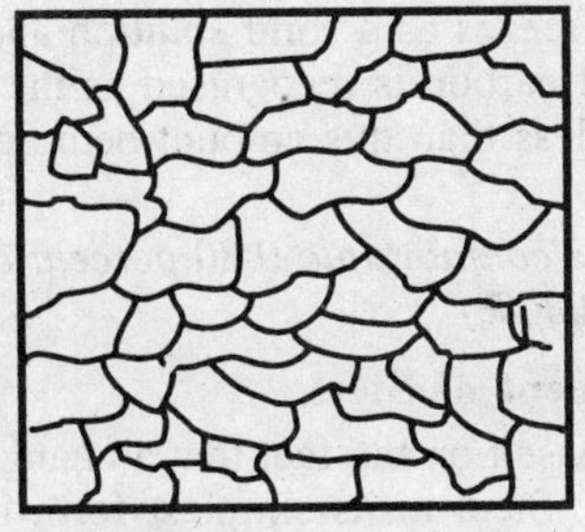

Fig. 21.27 : Austenite.

Fig. 21.28 : Martensite.

### 21.4.4. Iron-Carbon Equilibrium Diagram

Fig. 21.29 shows the *iron carbon equilibrium diagram.* Any iron-carbon alloy in the molten state may be considered to be solution of $Fe_3C$ in iron. When this solution cools and solidifies, it depends upon the amount of carbon present whether the alloy will solidify as a solid solution or form a *eutectic*. (In this system the two constituent metals are soluble in all proportions in liquid state but completely insoluble in the solid state). If the carbon does not exceed 2.0% the alloy will solidfy as a solid solution. A solid solution in γ-iron is known as *austenite,* and this name applies to all such solutions regardless of the amount of carbon, which may range from almost nil upto the saturation point 2.0%. The saturated solution is sometime known as 2.0% austenite. If the alloy contains more than 2.0% of carbon it solidifies with the formation of eutectic, which freezes at 1130°C and contains 4.3% of carbon. In other words everything to the left of the line

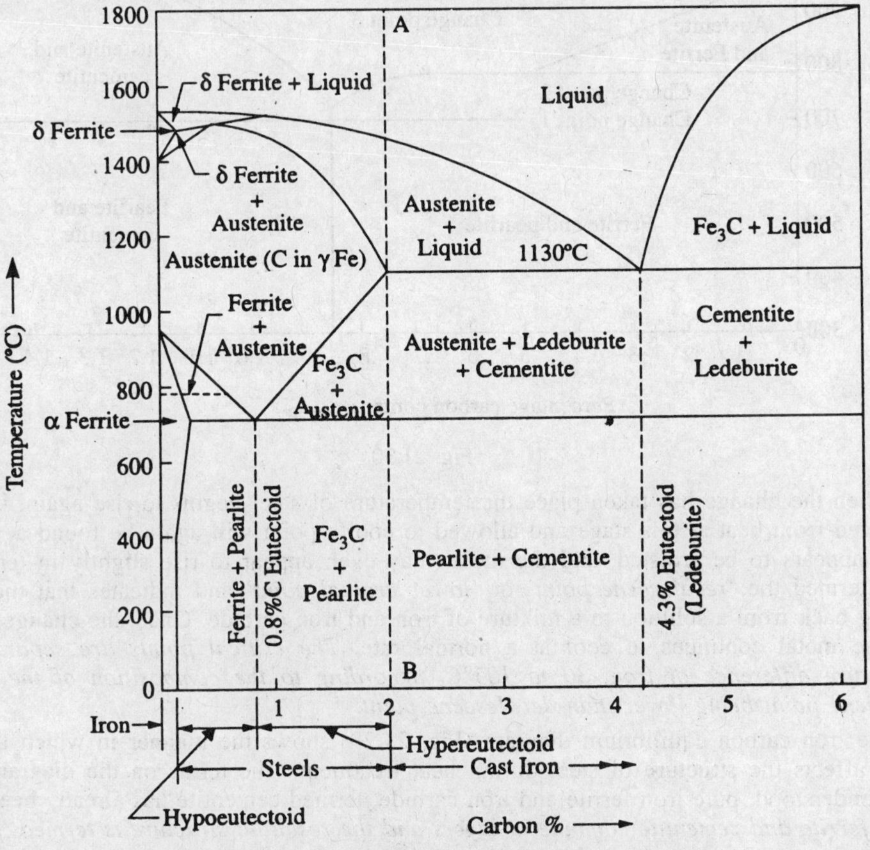

Fig. 21.29 : Iron-carbon equilibrium diagram.

*AB* in the diagram (Fig. 21.29) freezes as a solid solution and everything to the right of it freezes selectively. The limit of 2.0% of carbon is recognised as the division between steel and cast iron, an iron-carbon alloy containing less than this amount being defined as steel and any that contains more being defined as cast iron.

**Note :** *Hypoeutectoid steels contain upto 0.80 percent carbon whereas hypereutectoid steels higher percentage (more than 0.80%).*

### 21.4.5 Theory of Heat Treatment of Steel

The theory of heat treatment is based on the fact that a change takes place in the internal structure of steel at specific temperatures. Steel in its simplest form is composed of pure iron to which a small percentage of carbon has been added. At normal temperatures the steel consists of the pure iron, known as '*ferrite*' combined with iron carbide, which is called '*cementite*'.

The hard brittle iron carbide does not become fully combined with the iron however, until a temperature of approximately 800°C is reached. If the steel is heated progressively it will be found that at this point the temperature of metal does not increase, although heat is for the time being, used up in bringing about a structural and chemical change in the metal. The carbon is in fact, going into what is termed as '*solid solution*' with the iron, although the steel is well below its melting point. This is termed the '*decalescene point*' or '*upper critical point*' by metallurgist.

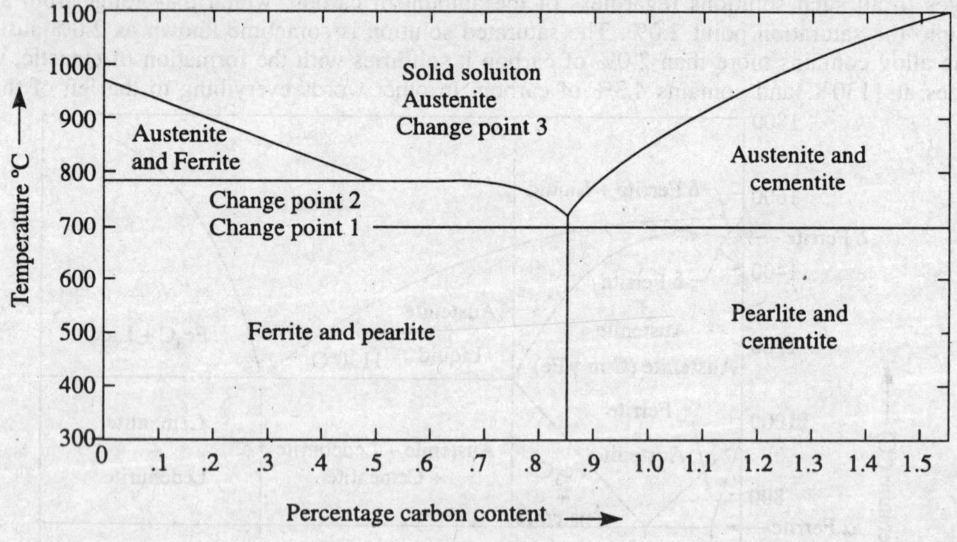


Fig. 21.30

When the change has taken place the temperature of steel begins to rise again, if the steel is removed from heat at this stage and allowed to cool a point will again be found at which the cooling appears to be arrested, and the metal may even appear to rise slightly in temperature. This is termed the '*recalescene point*' or '*lower critical point*' and indicates that the metal is changing back from a solution to a mixture of iron and iron carbide. Once, the change has taken place the metal continues to cool at a normal rate. *The critical points are separated by a temperature difference of from 30 to 105°C, according to the composition of the steel, the recalescene point being lower than decalescene point.*

The iron carbon equilibrium diagram (Fig. 21.29) shows the manner in which the carbon content affects the structure of steel during heat treatment. The terms on the diagram may be simply understood, pure iron ferrite and iron carbide, termed cementite has already been referred to. The *ferrite and cementite combine in layers and the resultant structure is termed 'perarlite'*. If the heating process is continued beyond the critical temperature, the pearlite dissolves to form 'austenite' when the steel contains 0.89% of carbon, a true pearlite structure is obtained, with less carbon some free iron remains uncombined, whereas as excess of carbon over 0.89% results in uncombined cementite.

If a piece of steel is heated above its upper critical point until it becomes cherry red *i.e.*, Austenitic condition, which is equivalent to about 900°C, and is then plunged into cold water to cool it, the sudden cooling traps the carbon in its solid solution state and an extremely hard, needle-shaped structure known as '*martensite*' is formed.

### 21.4.6. Heat Treatment Processes

Refer Fig. 21.31.

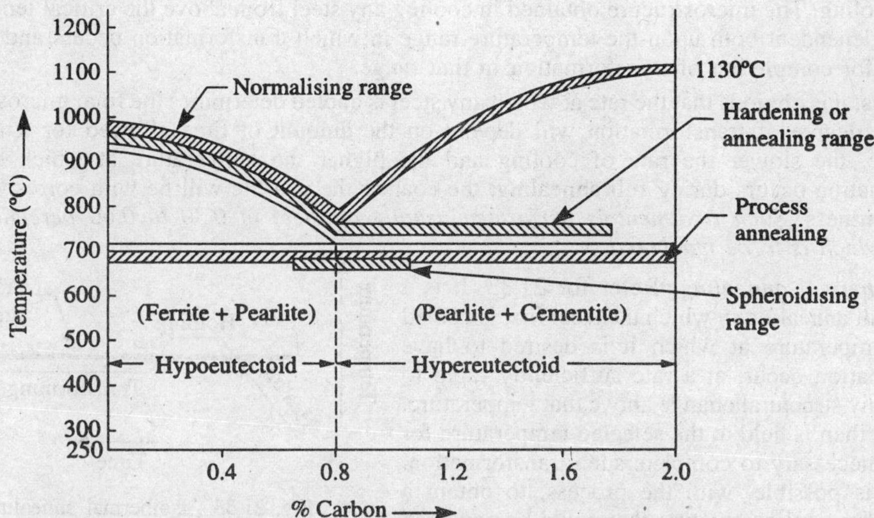


Fig. 21.31 : Heat treatment processes.

The various heat treatment processes are enumerated and discussed below :

1. Annealing
2. Normalising
3. Quenching/hardening
4. Tempering
5. Surface hardening
   (*i*) Case hardening (by carburising).
   (*ii*) Nitriding.
   (*iii*) Cyaniding.
   (*iv*) Flame hardening

### 21.4.2.1 Annealing

The objects of annealing are as follows :

(*i*) To soften the metals.

(*ii*) To improve machinability.

(*iii*) To refine grain size due to phase recrystallisation.

(*iv*) To increase ductility of metal.

(*v*) To prepare steel for subsequent treatment.

(*vi*) To modify electrical and magnetic properties.

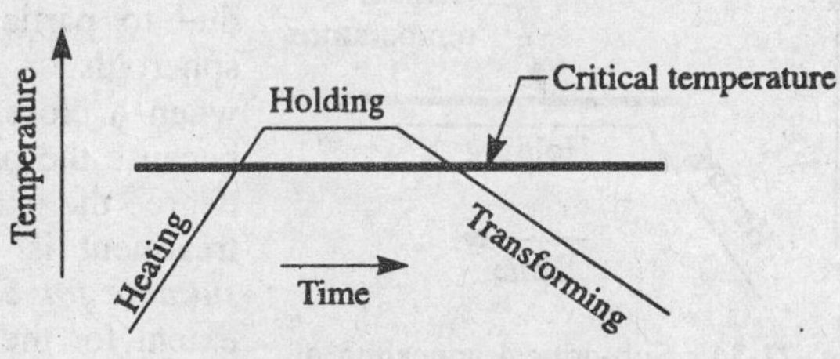


Fig. 21.32 : Full annealing.

(*vii*) To relieve internal stresses.

(*viii*) To remove gases.

(*ix*) To produce a definite micro structure.

*Full annealing :* Refer Fig. 21.32. If it is desired to refine the grain structure and produce a lamellar pearlite, a full annealing cycle should be used. This consists of heating the steel to a temperature above the transformation range, holding for one to two hours, and then cooling at a predetermined rate to obtain the desired microstructure. Grain refinement is accomplished in this instance by the recrystallisation of the steel in passing through the critial range both in heating and in cooling. The microstrucure obtained in cooling any steel from above the critical temperature range is dependent both upon the temperature range in which transformation occurs and the time required for completion of transformation in that range.

Thus, it is obvious that the rate at which any steel is cooled determines the final microstructure, since the degree of transformation will depend on the amount of time allowed for it to occur. Therefore, the slower the rate of cooling and the higher the temperature at which complete transformation occurs during full annealing, the coarser the pearlite will be with correspondingly lower hardness. *Such treatment is performed usually on steel of 0.30 to 0.60 percent carbon content which is to be machined.*

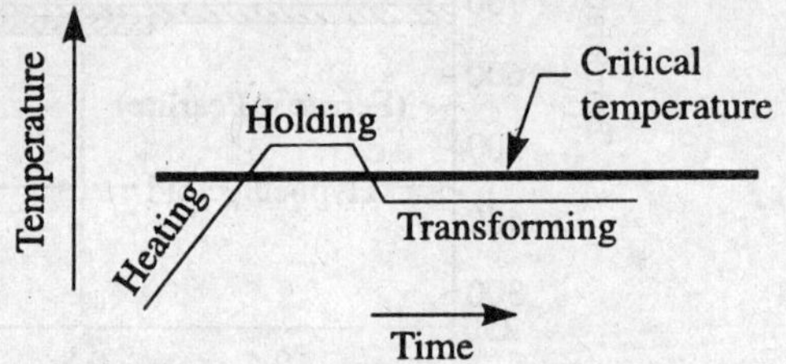


Fig. 21.33 : Isothermal annealing.

*Isothermal annealing:* Refer for 21.33. It is a type of full annealing in which the steel first is cooled to the temperature at which it is desired to have transformation occur, at a rate sufficiently rapid to prevent any structural change above that temperature. The steel then is held at the selected temperature for the time necessary to complete such transformation. Thus, it is possible, with the process, to obtain a more uniform microstructure that could be expected by continuous cooling. However, since it is necessary to drop the temperature rapidly to prevent any transformation above the desired temperature, there are definite limitations as to the mass that can be so treated. It is applicable, therefore only to *small sections* and would be suitable for large bars or large load in batch type furnances since it would be impossible to cool them at a rate sufficiently rapid to prevent some transformation.

A modified application of isothermal annealing is possible however, in which the charge is heated in one furnance and transferred to another, which has been set at a temperature somewhat lower than the desired temperature of transformation in order that the temperature of the charge will drop rapidly to that required. The selection of the temperature of the second furnace will be governed by the temperature to which the charge first is heated, the mass of the charge and the desired transformation temperature. Suitable handling equipment must be available to transfer the entire charge rapidly, since any undue delay might result in portions of the charge being cooled to low a temperature. Continuous furnaces also are applicable to this type of cycle.

*Process of sub-critical annealing:* Refer Fig. 21.34. Another type of annealing called *process or sub-critical annealing* consists of heating the steel to a temperature first under lower critical point and holding at this temperature for the proper time (usually 2 to 4 hours) followed by air cooling. This type of annealing results in softening the steel due to particle coagulation of the carbide to form the spheroids or small globules of carbide. It is not suitable when a close control of hardness or structure is desired, because the prior strucure of steel determines to a marked degree the extent of spheroidisation which will occur. The treatment is quite satisfactory for *rendering bars more suitable for cold sawing or shearing* and is used to great extent for these purposes. Since the temperature to which the bars are heated is somewhat lower than in full annealing there is less scaling and warping can be controlled.

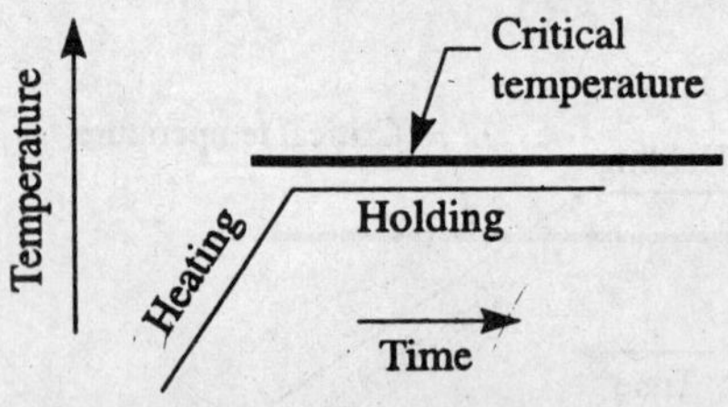


Fig. 21.34 : Sub-critical annealing or process annealing.

*Spheroidization:* Refer Fig. 21.35. It is a type of annealing which causes practically all carbides in the steel to agglomerate in the form of small globules or spheroids. There may be wide range of hardness with such a structure for any grade of steel since the size of the gloubles has a direct relation to hardness. Spherodizing may be accomplished by heating to a *temperature just below the lower critical* and *holding for sufficient period of time*. A more desirable and commonly used method for spheroidizing is to heat to temperature just *above the critical* and *cool very slowly* (about 6°C per hour) through the critical range or to heat to a temperature within the critical range but not above the upper critical and cool slowly.

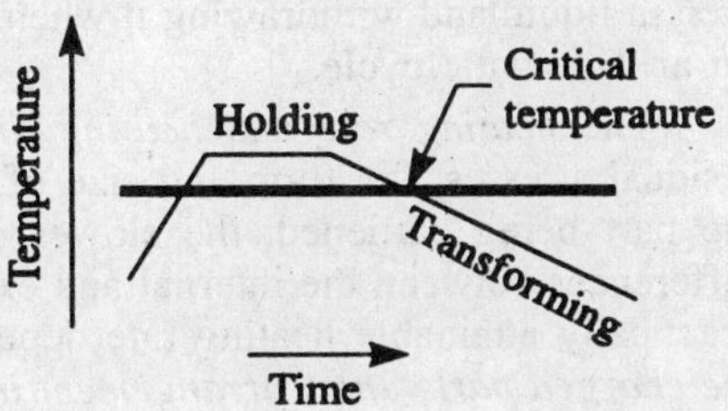


Fig. 21.35 : Spheroidize annealing.

This treatment is used *for practically all steels containing over 0.6 percent carbon that are to be machined or cold formed.*

**21.4.6.2 Normalising**

The *objects* of *normalising* are :

(*i*) To eliminate coarse grain structure obtained during forging, rolling and stamping.

(*ii*) To increase the strength of medium carbon steel.

(*iii*) To improve machinability of low carbon steel.

(*iv*) To improve the structure of welds.

(*v*) To reduce internal stresses.

(*vi*) To achieve desired results in mechanical and electrical properties.

Refer Fig. 21.31. Normalising is the term applied to the process of heating the steel approximately 4°C above the critical temperature followed by cooling below this range in *still air. This is one of the simplest treatments. The steel produced by this treatment is harder and stronger but less ductile than annealed steel having the same composition.*

This treatment is frequently applied to castings, forgings etc. to refine grain structure and to relieve stresses set up in previous operations. It is commonly applied after cold working, overheating or any other operation resulting in non-uniform heating or cooling. It may be used to efface the effects of previous heat treatments.

**21.4.6.3 Quenching**

Refer Fig. 21.36 *Quenching* is the method of *rapid cooling of a metal in a bath of liquid during heat treatment*. As earlier mentioned that if a piece of steel is heated above its upper critical temperature and plunged into water to cool it an *extremely hard, needle-shaped structure* known as *"martensite"* is formed. In other words, sudden quenching of steel greatly increases its hardness.

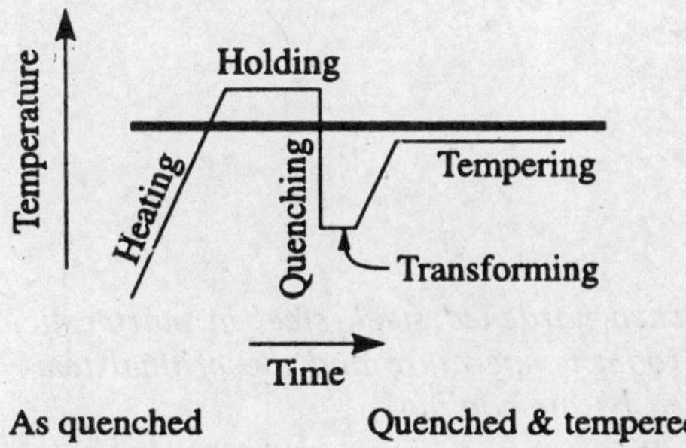


Fig. 21.36 : Thermal cycles (*a*) quenching (*b*) quenching and tempering.

The rapidity with which the heat is absorbed by the quenching bath has a considerable effect on the hardness of the metal. Clear, cold water is very often used, while the addition of salt still increases the degree of hardness. Oil however, gives the best balance between hardness, toughness and distoriton for standard steels. Special soluble oils are used in many plants instead of ordinary fish oil, linseed oil or cotton seed oil.

In order to increase the cooling rate, the parts may be moved around the quenching both, either by hand, or by passing them through the tank in a basket attached to mechanical conveyor. Large parts may be lowered into the tank by a crane and kept moving while cooling. It is often cheaper and more efficient, however to circulate the cooling liquid around the hot part. Modern quenching equipment is often highly mechanised, a rapid conveyor taking the load from the furnace to the quenching tank at 75 metres/min., lowering the charge into the tank, moving it then in liquid and withdrawing it when cool, the whole operation being controlled by push-buttons on an automatic cycle.

The *heating rate* and *heating time* depend on the composition of the steel, its structure, residual stresses, the form and size of the part to be hardened. The more the intricate and large the part being hardened, the slower it should be heated to avoid stresses due to temperature differences between the internal and external layers of the metal, warping and even cracking. The practically attainable heating rate depends upon the *thermal capacity of the furnace, the bulk of the charged parts, their arrangement in the furnace,* and other factors. The heating rate is usually reduced, not by reducing the furnace temperature but by preheating the articles.

The heating time for carbon tool steels and medium-alloy structural steels should be from 25 to 30% more than for carbon structural steels. The heating time for high-alloy structural and tool steels should be from 50 to 100% higher.

**Quenching media :** The quenching media in general use are : Water, brine, air, oils, molten salt. '*Water*' is probably the most widely used as it is simple and effective, it cools at the rate of 982°C per second. It tends however to form bubbles on the surface of the metal being quenched and causes soft spots, so a brine solution is often used to prevent this trouble. "*Brine*" is a very rapid cooling agent, and may tend to cause distortion of the parts, as will water. "*Oil*" is therefore used when there is any risk of distortion, although it is more suitable for alloy steels than for plain carbon steels. Where the risk of distortion is great, quenching must be carried out in air blast. Since the rate of cooling is then lower, more hardening elements must be added to the steel, forming an *air blast.* Since the rate of cooling is then lower, more hardening elements must be added to the steel, forming an air hardening alloy. The air blast must be dry, since any moisture in the air will crack the steel. High speed steels are often quenched in molten salt to harden them.

**Note :** Hypo-eutectoid steel containing very little carbon, say less than 0.25%, cannot be easily hardened by sudden quenching because of large amount of soft ferrite which it contains and all of which cannot be retained in solution even on very quick cooling. The hardening capacity of steel increases with carbon content.

#### 21.4.6.4. Tempering

Martensitic structures formed by direct quenching of high-carbon steel are hard and strong, but unfortunately are also brittle. They cannot be plastically deformed and have very little toughness, and although strong then are unable to resist impact loads and are extremely sensitive to stress concentration. Some of the hardness and strength must be sacrificed to obtain suitable ductility and toughness. This is done by *tempering* the martensitic steel. Thus tempering process is carried out to :

(*i*) Increase toughness

(*ii*) Decrease hardness,

(*iii*) Stabilise structure

(*iv*) Relieve stresses, and

(*v*) Change volume.

The *process of tempering consists of heating quenched hardened steel, steel in martenstic condition, to some predetermined temperature between room temperature and the critical temperature of the steel for a certain length of time, followed by air cooling.*

**1. Method of tempering :** Tempering of steel may be carried out in liquid baths such as oil, salt or lead, the bath being heated to the correct temperatures and steel immersed in bath for the determined length of time after which it (steel) is removed and allowed to cool to room temperature. Air tempering furnaces are fully automatic and the hot air (heated by gas or electric means) is circulated around the parts to be tempered.

**2. Theory of tempering :** The steel which has been hardened by quenching is considered to be in an unstable condition. And perhaps in most quenching operations some austenite is retained at room temperature. The martensitic structure of hardened steel is much different from the structure of normal pearlite formed in steels that have been slow cooled from the austenitic state. The martensite which is fresh born has a tetragonal atomic arrangement and is known as *alpha martensite.* In this condition, martensite is eager to change to a more stable structure (more nearly pearlitic) and undergoes this change when offered an opportunity (*such as when temperature is raised during the tempering operation*). When alpha martensite is heated to app. 90°C, beta martensite (with body centered cubic lattice) is formed. If beta martensite is further heated it precipitates carbon in the form of cementite which has been held in supersaturated solution in the martensite. The size of the precipitated cementite particles is very small and their size is not revealed by microscopic examination when the tempering is low and around 200°C; but if the temepring temperatures is raised beyond 200°C the size of minute cementite particles continue to grow in size and finally becomes microscopic.

Upon reheating to a temperature of 260°C any retained austenite found in hardened carbon or low alloy steels may be transferred to martensite or a tempered form of martensite. When austenite changes to martensite upon tempering it is accompanied by an expansion (which may be very marked) and such a change induces internal stresses and may be removed as partially by increasing the temperature. Changing the retained austenite to martensite by reheating to 260°C will effect a change in original martensite which results in a loss of maximum hardness.

Any retained austenite at room temperature may be transformed to martensite by cold treating (*i.e.*, cooling to subnormal temperatures 21°C to 33°C). Cold treating does not cause any loss of hardness of the original martensite as that which occurs during tempering and yields maximum hardness. The cold treating of hardened steel may be carried out after tempering treatment.

The results from tempering depend on the *time* of treatment. The longer the time of treatment (at a given tempering temperature) better are the results. It is recommended that for getting satisfactory results at least one hour be allowed at any tempering temperature. Some tempering operations consume several hours.

**Temper colours :** Tempering can be judged by the temper colours which appear on the bright red surface, and experienced eyes are generally guided by those colours while heating steel materials for, tempering. The following are the colours formed on steel in the process of tempering:

| Colour | Tempering temperature |
|---|---|
| Very pale yellow | About 220°C |
| Straw yellow | About 238°C |
| Brown yellow | About 260°C |
| Bright purple | About 277°C |
| Dark purple | About 288°C |
| Pale yellow | About 321°C |
| Blue tinged with green | About 322°C |

**Effects of tempering temperature on the machanical properties of steel :** Tempering process decomposes the martensite into a ferrite-cementite mixture and thus the properties of steel are strongly affected.

(*i*) At low tempering temperature (upto 200° and 250°C),

(*a*) bending and true tensile strength are increased, and

(*b*) hardness changes to a small extent.

(*ii*) If the temperature is further increased the following results :

(*a*) Hardness, true tensile strength, proportional limit and yield point are reduced

(b) Reduction of area and relative elongation are increased.

(iii) Tempering at 250° to 400°C reduces the impact strength of steel. Therefore the temperature range 250° to 400°C should be avoided in assigning tempering temperatures.

(iv) The properties after structural improvement *i.e.*, hardening followed by high tempering are always higher than those of annealed steel. This is due to the difference in structure of the ferrite-cementite mixture.

### 21.4.6.5 Surface Hardening

The various methods of surface hardening are discussed below :

**Case hardening :** The objects of case hardening are :

1. To obtain a hard and wear resistance surface in machine parts with enrichment of the surface layer with carbon to concentration of 0.75 to 1.2%.
2. To obtain a tough core.
3. To obtain close tolerances in machining parts.
4. To obtain a higher fatigue limit and high mechanical properties in the core.

*"Case hardening" consists in heating a steel in the presence of a solid, liquid or gas, rich in carbon in order to enable the surface to be hardened, while retaining a tough ductile core.*

There are three methods of adding carbon to the surface of the metal:

1. Pack hardening
2. Liquid carburizing.
3. Gas carburizing.

**1. Pack hardening:** Refer Fig. 21.37. This method is the oldest. The articles to be carburised are packed in metal boxes or pots surrounded by a suitable compound which is rich in carbon.

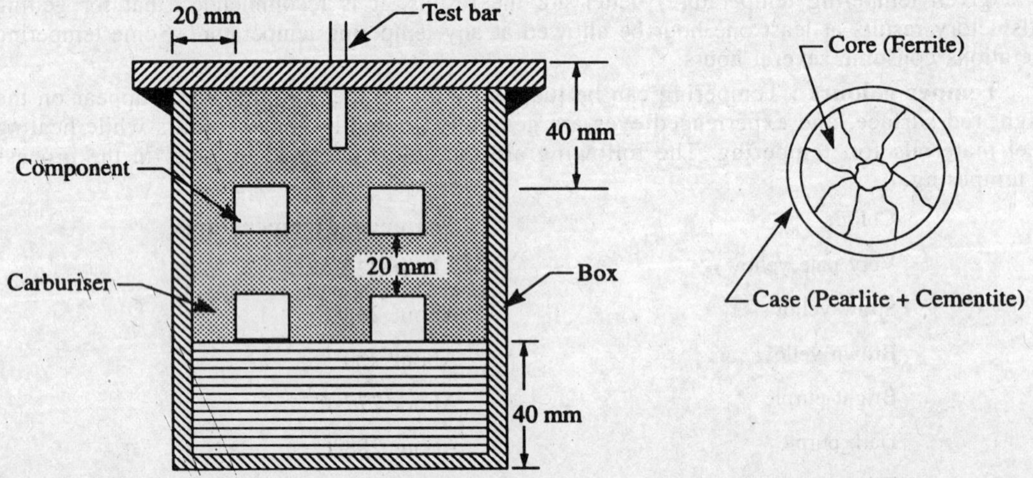


Fig. 21.37 : Pack hardening.

The boxes are sealed with clay to exlcude air, and are placed in an oven or furnace, where they are heated to a temperature between 900 to 920°C, depending on the composition of steel. The carbon from the carburizing compound soaks into the surface of the hot steel to depth which depends on the time that the box is left in the furnace, so that the low-carbon steel is converted into high-carbon steel in the form of thin case. The internal section of the steel, and any parts which have been protected by tinning, however, remain unaffected, the result being a piece of steel with a dual-structure. The steel is allowed to cool slowly in the box.

The steel is then removed from the box and reheated to a temperature just above its critical point, or approximately 915-925°C for fine grain steel, followed by quenching in water, brine or oil. This hardens the skin and at the same time refines the core. Smaller articles and thin sections are heated to a lower temperature in order to avoid distortion. The steel is usually given a second heat treatment at about 760-780°C, in order to improve the ductility and impact resistance of the core and case.

Small parts and single jobs are often carburized by heating them in forge, and covering them with a carburizing power when the metal has reached a bright red heat. The carburizing compound melts and flows over the surface of the metal, which is then returned to the forge and maintained at a bright red heat for sufficient time to allow the carbon to penetrate the surface, quenching then follows as usual.

Many commercial 'carburizing' compounds are available in suitably mixed form. Among the ingredients, combined in different percentage are *powdered charred leather, wood charcoal and horn.* Wood charcoal is very largely used, although its value varies with the type of wood. *Hickory* gives the best results, and a normal rate of penetration gradually decreases and ceases after eight hours. Wood charcoal gives the slowest rate of penetration of any of the carburizing materials.

**2. Liquid carburizing :** Where a fairly thin case is required a more economical process is to carburize the parts in a liquid both. This consists of a container filled with a molten salt, such as sodium cyanide, which is heated by electrical immersion elements or by a gas burner. Salt-bath carburizing reduces distortion of the parts to the minimum, while equal heating is assured. The parts leave the bath with a clean, bright finish, the scaling experienced during pack hardening being avoided.

Heat treatment following liquid bath carburizing is much the same as that used after pack hardening, although for cheaper classes of work the parts may be quenched immediately after removal from the salt-bath. It is more usual to reheat the parts to about 760°C and quench them again. Even in this case a considerable saving in time results from the fact that quenching can follow carburizing, instead of waiting for a red-hot box to cool before the parts can be removed for heat treatment.

**3. Gas carburizing :** It is another method of introducing extra carbon into the suface of the steel, in this case by heating the metal in a furnace into which a gas which is rich in carbon, such as methane, propane, butane, is introduced. It is necessary to maintain a continuous flow of carburizing gas into the furnace, and to extract the spent gas. The first cost of gas carburizing eqipment is high, and the process is economical only for large outputs. The horizontal rotary type of gas carburizing furnace has a retort of muffle which revolves slowly, so that the parts are rotated in the stream of gas; this is suitable for smaller parts such as ball and roller bearings, chain links, pins, axles and so on. Larger parts are usually carburized in a vertical rotary furnace, in which gas is given a swirling rotary motion so that it circulates around the parts.

**Nitriding:** The nature of the nitriding process used to obtain a case-hardened product is very different from that of the carburizing process. *Nitrogen, instead of carbon, is added to the surface of the steel. Carbon does not play any part in the nitriding operation but influences the machinability of steel. The temperatures used in nitriding are much lower than those used in carburizing and below the critical temperature of the steel.*

*Simple carbon steels, which are often used for carburizing are not used for nitriding. Steels used in the process are special alloy steels.* With the nitriding developing rather thin cases, a high core hardness is required to withstand any high crushing loads. High tempering temperatures call for a steel with a higher carbon content in order to develop this increase in core hardness. In addition to higher carbon content, various alloying elements are called for in the steel to bring about an increase in the formation of these nitrides. Aluminium seems to display the strongest tendency in the formation of these nitrides. Chromium, molybdenum, vanadium and tungsten, all being nitrides formers, are also used in nitriding steels. Nickel in nitriding steel hardens and strengthens the core and toughens the case but with slight loss in its hardness.

*Nitriding Operation* : In nitriding process, *nitrogen is introduced to the steel by passing ammonia gas through a muffle furnace containing the steel to be nitrided.* The ammonia is purchased in tanks as a liquid and introduced into the furnace as a gas at slightly greater than atmospheric pressure. With the nitriding furnace operating at a temperature of 480°C to 540°C, the ammonia gas partially dissociates into nitrogen and hydrogen gas mixture. The dissociation of ammonia is shown by the following equation.

$$2NH_3 = 2N + 3H_2$$

The operation of the nitriding cycle is usually controlled so that the dissociation of the ammonia gas is held to approximately 30% but may be varied from 15% to 95%, depending upon operating conditions. The gas mixture leaving the furnace consists of hydrogen, nitrogen, and undissociated ammonia. The undissociated ammonia, which is soluble in water, is usually discharged into water and disposed of in this manner.

The free nitrogen formed by this dissociation is very active, uniting with the iron and other elements in the steel to form nitrides. These nitrides are more or less soluble in the iron and form a solid solution or, more likely, are in a fine state of dispersion, imparting hardness to the surface of the steel. From the surface the nitrides diffuse slowly, and the hardness decreases inwardly until the unaffected core is reached. The depth of penetration depends largely upon the length of time spent at the nitriding temperature. Diffusion of these nitrides is much slower than diffusion of carbon in the carburizing operations so a much longer time is required to develop similar penetration.

In general, the surface hardness is much greater after nitriding than it is after carburizing and hardening. The maximum hardness obtained from a carburized and hardened case runs around 67 Rockwell C: whereas, it is possible to obtain surface hardness value in excess of 74 Rockwell C by nitriding. The surface hardness of the nitrided cases cannot be measured by the case; it is usually measured by the Rockwell superficial scale such as 15N and 30N scale. The hardness value on the Rockwell C scale may be estimated from a hardness conversion table.

*In order to obtain localized nitriding of parts or to stop nitriding on some surface areas, the best method is to use tin as a protecting agent against the nitriding action.* Tin, in the form of a paste or paint made from tin powder, or tin oxide mixed with glycerine or shellac, may be applied to the areas to be protected against nitriding. A thin tin electroplate may be applied as a satisfactory stop off method. Although tin melts at a lower temperature than used in nitriding, sample protection is provided by thin layer of tin that is held to the surface tension. Care should be exercised to apply the tin paint or electroplate to a clean surface and to avoid a thick layer of tin which may run or dip into surface where protection is not desired.

*All heat treatments, such as the quench hardening of steel in the nitriding process, are carried out before the nitriding operation.* After rough machining operations, the steel is heated to about 950°C, held for the necessary length of time at this temperature in order to have the alloying elements go into solution in the austenite, thereby imparting core strength and toughness after quenching and finish machining to produce a sorbitic structure which has a rough case and eliminates any brittleness resulting from any free ferrite. This hardening, thus reduces distortion during nitriding. After tempering all oxide film and the traces of decarburization are removed. Any decarburization left on the surface of the steel to be nitrided will usually result in failure of the nitrided surfaces by peeling or spalling off. The steel is then nitrided and allowed to cool slowly to room temperature in the nitriding box or chamber. No quenching is required; the steel develops its maximum hardness without necessitating a further quenching operation.

**Merits and demerits of nitriding :**

**Merits :** Nitriding (a hard surfacing operation) is associated with the following *merits.*

1. Greater resistance to wear and corrosion.
2. Less warping or distortion of parts treated.
3. Greater surface hardness.
4. Greater fatigue strength under corrosive conditions.
5. Higher endurance limit under bending stresses.
6. Better retention of hardness at elevated temperatures.

**Demerits :** The *demerits* of 'Nitriding' are so follows :

1. Medium used is expensive.
2. High furnace costs due to the long time of treatment.
3. Necessity of using special alloy steels.
4. Necessity of using high alloy containers to resist the nitriding.

**Cyaniding :** Cyaniding is a process of superficial case-hardening in which the steel is heated in a molten cyanide salt at about 850°C followed by quenching. Both carbon and nitrogen are absorbed in this process.

In carbon nitriding steel is heated in a gaseous mixture of ammonia and hydrocarbons whereby both carbon and nitrogen are absorbed.

**Flame hardening :** It is process of surface hardening by which steel or cast iron is raised to high temperature by a flame and then almost immediately quenched. The process of hardening is used for local hardening of such components as *wheel teeth.*

Fig. 21.38 shows a flame hardening of gear teeth. A flame from an oxy-acetylene or similar burner is played on to the teeth so as to raise the temperature rapidly above the hardening temperature. Hardening results when the austenised surface is quenched by spray (usually) that follows the flame.

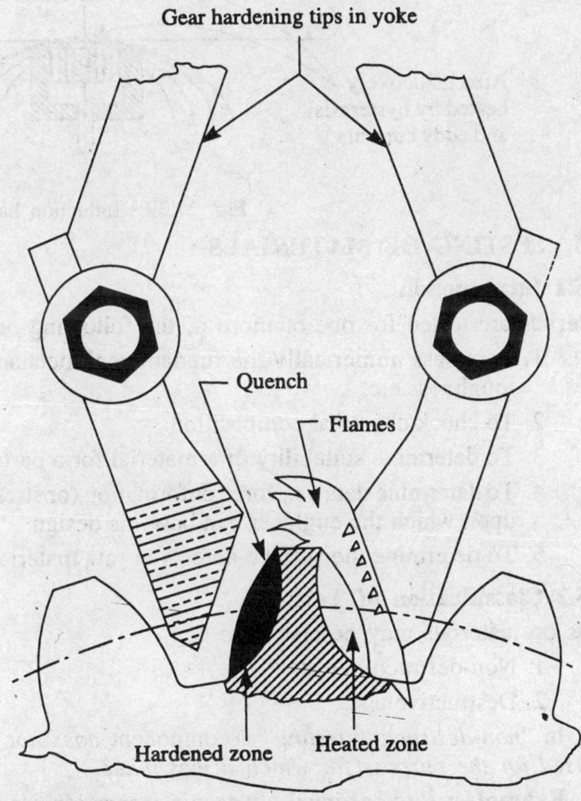

Fig. 21.38 : Flame hardening.

The *advantage* of this process is that there is *much lesser distortion than in ordinary methods. It is quick,* and *the hardening is restricted to parts which are affected by wear.*

**Induction hardening :** This process of surface heating is based upon inductive heating in which a high-frequency current is first transformed from high to low voltage and the heavy low voltage, current is passed through the inductor blocks which surround the bearing journal to be hardened without actually touching it. The inductor block current induces current in the surface of the metal the block surrounds and it is this induced current which heats the surface to be hardened. When the area in question has been thus subjected to an accurately controlled high-frequency current for proper length of time, the electrical circuit is opened and simultaneously the heated surface is quenched by a spray from a water jacket built into the inductor block. Fig. 21.39 shows the scheme of induction hardening.

The induction hardening is at present *extensively used for producing hard surfaces on crankshafts, camshafts, axels and gears. The principal* **advantages** *are listed below :*

1. The time required for this heat-treatment operation is less thereby increasing the labour productivity.
2. Deformation due to heat-treatment is considerably reduced.

3. The articles which are induction heated have no scale effect.
4. The hardening of the surface can be easily controlled by controlling the current.
5. The depth of hardness can be easily controlled by varying frequency of supply voltage.

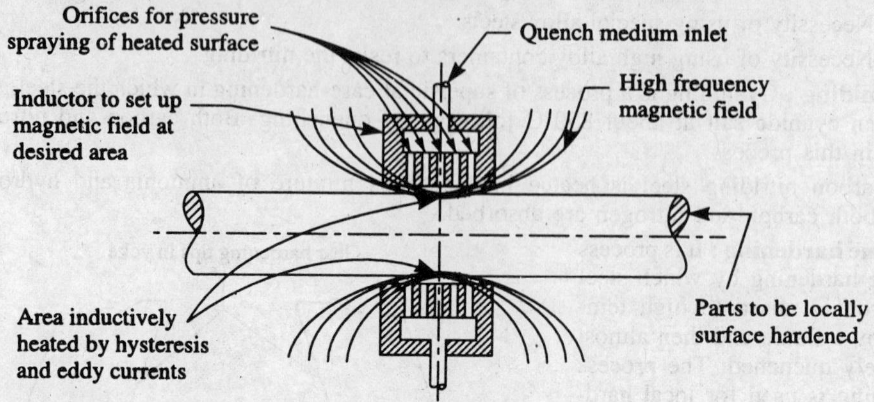


Fig. 21.39 : Induction hardeing.

## 21.5. TESTING OF MATERIALS

### 21.5.1 Introduction

Materials are tested for one or more of the following *purposes* :

1. To assess numerically the fundamental mechanical properties of ductility, malleability, toughness etc.
2. To check chemical composition.
3. To determine suitability of a material for a particular application.
4. To determine data *i.e.,* force deformation (or stress) values to draw up sets of specifications upon which the engineer can base his design.
5. To determine the surface defects in raw materials or processed parts.

### 21.5.2. Classification of Tests

Tests on materials may be classified as :

1. Non-destructive test.
2. Destructive test.

In "*non-destructive testing*" a component *does not break even after being tested, so it can be used for the purpose for which it was made.*

**Examples:** *Radiography, ultrasonic inspection etc.*

In "*destructive testing*" the component or *specimen either breaks or remain no longer useful for further use.*

**Examples:** Tensile test, impact test, torsion test etc.

### 21.5.3. Non-Destructive Tests

**Non-destructive** *tests may be defined as those which in a specific context would not damage the material being examined to an extent such that it is rendered useless for future for which it was originally meant.*

**Importance:** Although non-destructive tests do not provide direct measurement of mechanical properties, yet they are extremely useful in *revealing defects in components* that could impair their performance when put in service. These tests make components *more reliable, safe and economical.*

The various methods used for non-destructive testings are as follows :

1. X-ray radiography

2. Gama radiography.
3. Magnetic particle inspection.
4. Ultrasonic testing.
5. Electrical method.
6. Damping tests.

### 21.5.4. Destructive tests (Mechanical tests)

The component or specimen, after being destructively tested, either breaks or remains no longer useful for further use. Examples of destructive or mechanical tests are : *tensile test, impact test, torsion test, fatigue test etc.*

### 21.5.4.1. Importance of mechanical tests

Structures, machines and products of various kinds are usually subjected to load and deformation. Therefore, the properties of materials under the action of load and deformation so produced under various environments become an important engineering consideration. The microscopic properties of materials under applied forces or loads are broadly classed as "*mechanial properties*". They *are a measure of the strength and lasting characteristics of a material in service and are of great importance particular to the design engineer. Unfortunately these properties cannot be desired from the structural or bonding considerations alone since most of them are structure-sensitive, are much more affected by crystal imperfections and other factors* such as composition, grain size, heat treatment etc. Therefore, mechanical properties do not depend on them in all situations. A great number of mechanical properties are, therefore, best evaluated by mechanical testing of the materials like metals and alloys.

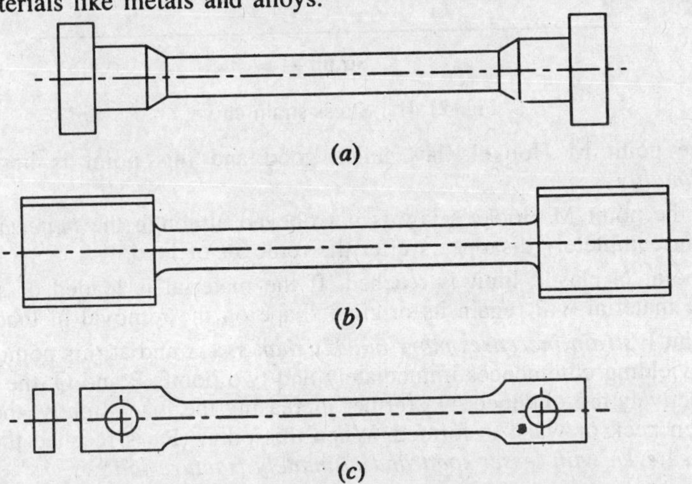


Fig. 21.40 : Specimens for the tensile test.

The following important mechanical tests give valuable information about metals and alloys as given below :

| S. No. | Name of the Test | Information Supplied About |
|---|---|---|
| 1. | **Tensile test** | Tensile strength, yield point, elastic limit, Young's modolus, ductility, toughness etc. |
| 2. | **Impact test** | Toughness of a material under shock loading conditions. |
| 3. | **Hardness test** | Wear resistance, indentation resistance, scratch resistance or cutting ability of a material. |
| 4. | **Fatigue test** | Behaviour of a material under repeatedly applied stress and its endurance limit. |
| 5. | **Creep test** | Behaviour of a material under a steady load over a long period of time and creep limit of a material. |

### 21.5.4.2. Tensile test

The tensile test is one of the most widely used of the mechanical tests. There are may variations of this test to accommodate the widely differing character of materials such as metals, elastomers, plastics and glass. The *tensile test on a mild steel test piece* is described below :

The tensile test is carried out on a bar of uniform cross-section throughout the gauge length. The specimen is mounted in the jaw of a testing machine with which a gradually increased loading can be applied. The extension or elongation of the gauge length is recorded continuously and finally a graph is drawn between the load and extension or between the stress and strain which is of the type shown in Fig. 21.41.

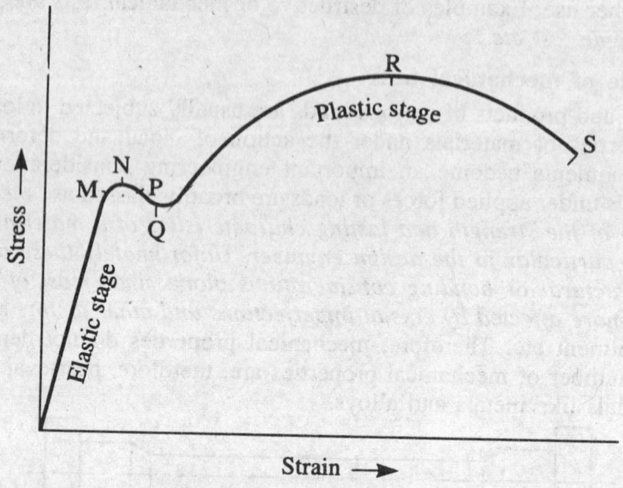


Fig. 21.41 : Stress-strain curve.

— Upto the point M Hooke's law holds good and this point is known as *"limit of proportionality"*.

— Beyond the point M Hooke's law is not obeyed although the *material remains elastic i.e.,* strain completely disappears after the removal of load.

— At the point N elastic limit is reached. If the material is loaded or stressed upto this point the material will regain its original shape on the removal of load.

Upto the point P *strain increases more quickly than stress* and at this point the *metal yields.* In the mild steel yielding commences immediately and two points P and Q, the *upper and lower yield points* respectively are obtained. On further increasing the load slightly, the strain increases rapidly till R when neck or waist is formed. When this point (R) is reached the deformation or *extension continues even wtih lesser load and ultimately fracture follows.*

### 21.5.4.3. Impact test

**Significance of impact test:**

— *An impact test signifies toughness of material that is ability of material to absorb energy during deformation. Static tension tests of unnotched specimens do not always reveal the susceptibility of metal to brittle fracture.* This important factor is determined by impact test. *Toughness takes into account both the strength and ductility of the material.*

— Several engineering materials have to withstand impact or suddenly applied load while in service. *Impact strengths are generally lower as compared to strengths achieved under slowly applied loads* of all types of impact tests, the notched bar tests are most extensively used. Therefore, the impact test measures the *energy necessary to fracture standard notch bar by applying an impulse load. The test measures the notch toughness of material under shock loading.*

— *Values obtained from these tests are not of much utility to design problems directly and are highly arbitrary. Still it is important to note that it provides a good way of comparing toughness of various materials or toughness of the same material under different conditions.* This test can also be used to *assess the ductile briltle transition temperature of the material occuring due to lowering of temperature.*

**Impact tests :** A pendulum type impact testing machine is generally used for conducting notched bar impact tests. The following types of impact tests are performed on this machine.

1. Izod test. 2. Charpy test.

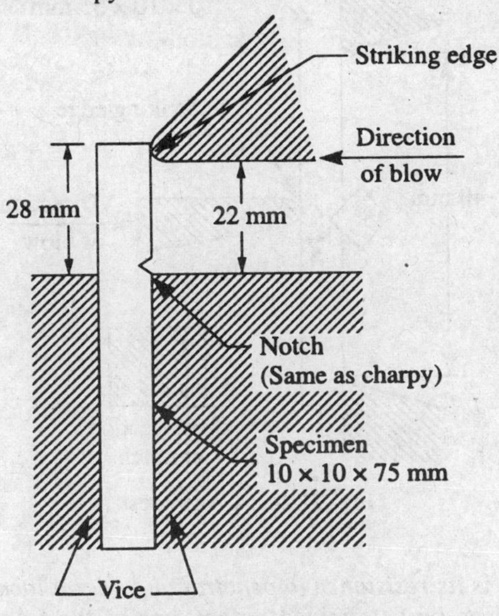


Fig. 21.42 : Izod test.

**Izod test :** Refer Fig. 21.42. This test uses a cantilever test piece of 10 mm × 10 mm section specimen having standard 45° notch 2 mm deep. This is broken by means of a swinging pendulum which is allowed to fall from a certain height to cause an impact load on the specimen. The angle rise of the pendulum after rupture of the specimen or energy to rupture the specimen is

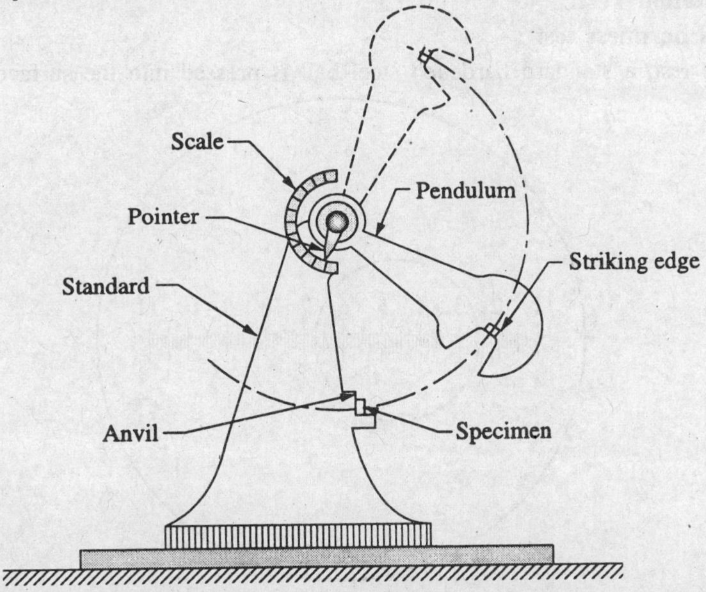


Fig. 21.43 : Pendulum type impact testing machine.

indicated on the graduated scale by a pointer. The energy required to rupture the specimen is the function of the angle of rise.

**Charpy test :** This test is more common than Izod test and it uses simply supported test piece [Fig. 21.44] of 10 mm × 10 mm section. The specimen is placed on supports or anvil so that the blow of striker is opposite to the notch.

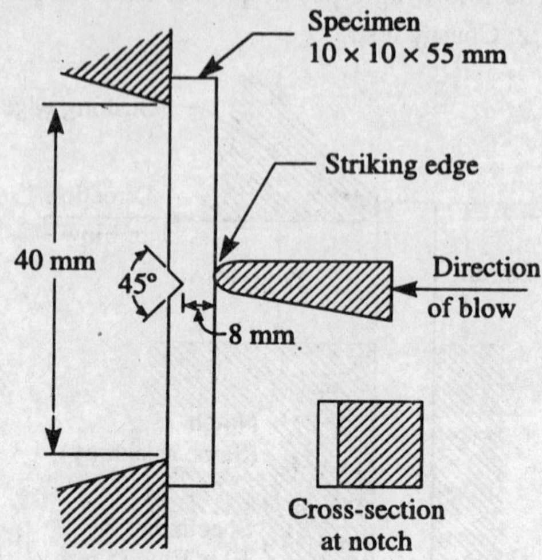


Fig. 21.44 : Charpy test.

#### 21.5.4.4 Hardness tests

The **hardness** of a material is its *resistance to penetration under a localised pressure or resistance to abrasion.* The hardness can be determined by any one of the following tests :

(*a*) Indentation or penetration test.

(*i*) Brinell (*ii*) Vicker's (*iii*) Rockwell

(*b*) Rebound test

(*c*) Scratch test.

(*a*) **Indentation Test:**

(*i*) **Brinell hardness test :**

— In this test, a standard hardened steel ball is pressed into the surface of the specimen

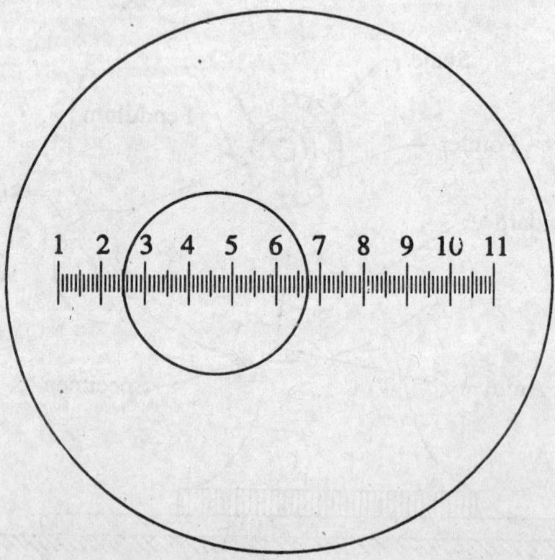

by a gradually applied load which is maintained on the specimen for a definite time. The impression (Fig. 21.45) so obtained is measured by a microscope and the brinell hardness number (B.H.N) is found out by the following equation :

$$\text{B.H.N.} = \frac{\text{load}}{\text{area of impression / indentation}}$$

$$= \frac{2P}{\pi D (D - \sqrt{D^2 - d^2})}$$

where $P$ = load (kgf)

$D$ = diameter of ball (mm)

$d$ = diameter of indentation (mm).

— The test is carried with a *hardened steel or carbide ball of 10 mm diameter. A 3000 kgf load is used for testing ferrous alloys and alloys of similar hardness. When brass soft alloys are tested, a 500 kgf load is used.* The time of loading is specified between 10 and 30 seconds, depending upon the alloy being examined.

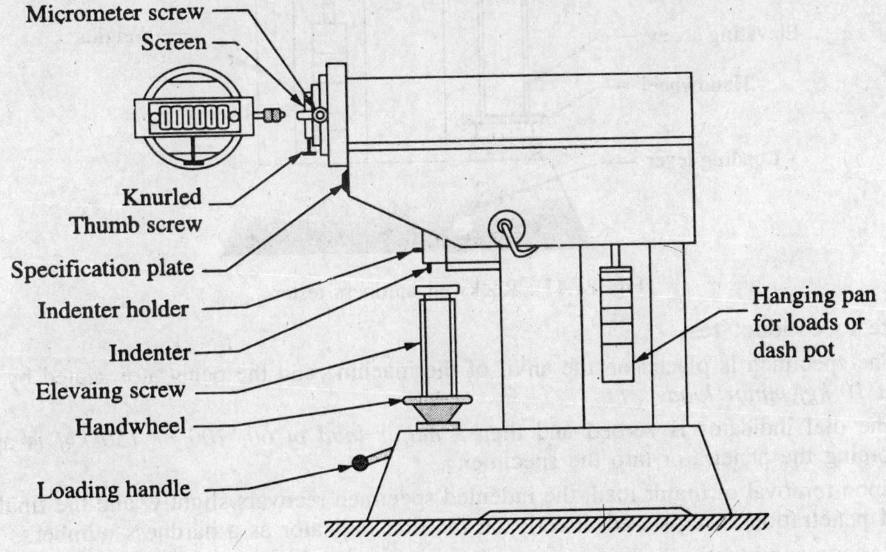


Fig. 21.46 : Brinell hardness tester.

(*ii*) **Vicker's hardness test :** Here a polished square based pyramid diamond tool with an angle of 136° between the faces is used. This tool under gradual load makes an impression on the specimen. The load when divided by the area of indentation in mm$^2$ gives what is known as pyramid hardness number (D.P.N.).

This is given by the expression,

$$\text{D.P.N.} = \frac{2P \sin \frac{\theta}{2}}{d^2},$$

where P = load in kgf.

$\theta$ = angle between the opposite faces.

$d$ = mean length of two diagonals in mm.

The loading on the plunger carrying the diamond is applied by a cam, which is rotated by a drum on which a wire carrying a weight is wound. The load and time during which it is applied are thus mechanically controlled.

(*iii*) **Rockwell hardness test:**

— The Rockwell hardness test is *probably the most widely used method* of hardness testing.

— Rockwell testers use much smaller penetrators and loads than does the Brinell tester. Four sizes of hard balls from 1/16″ to 1/2″ in diameter are available as well as a cone shaped diamond. For testing metallic materials the 1/16″ ball and the diamond penetrator are most commonly used. The penetrator chuck is mechanically connected to a dial indicator which responds to vertical motion of the penetrator (Fig. 21.47). Since the penetrators are small the specimen should be ground smooth and clean.

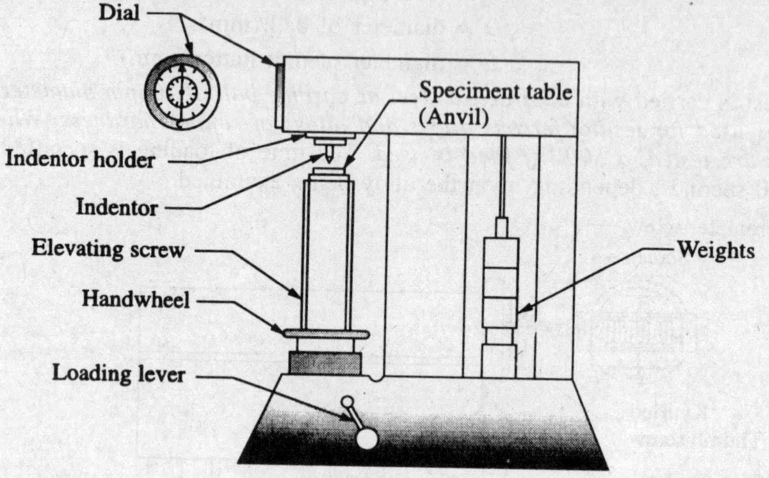


Fig. 21.47 : Rockwell hardness tester.

**Procedure to conduct test**

- The specimen is placed on the anvil of the machine and the penetrator seated by means of *10 kgf minor load.*
- The dial indicator is zeroed and then a *major load of 60, 100 or 150 kgf* is applied, forcing the penetrator into the specimen.
- Upon removal of major load, the indented specimen recovers slightly, and the final depth of penetration is registered directly on the dial indicator as a hardness number.
- Various combinations of penetrator and major load are based and designed by a series of letters. The two commonest scales are the HRB scales and the HRC scales, respectively standing for the 1/16″ ball with 100 kgf load, and the diamond penetrator with the 150 kgf major load. In general very hard materials are tested with the diamond penetrator.

Mild steel might have HRB reading 90; hardened alloy steel might have an HRC of 55. These are stated as HRB 90 and HRC 55.

(*b*) **Rebound Test (Shore scleroscope):** This test was carried out by A.F. Shore in 1910. In this test a steel cylinder hammer is dropped from a height of 25 cm through a glass tube on the surface to be tested. The height of the rebound is used as a measure of hardness of surface. The surface to be tested should be smooth, free from oil and tube should be truly vertical. Shore's hardness number for steel is 1/6 to 1/7 of Brinell hardness number (B.H.N).

(*c*) **Scratch Test:** A Martens in 1820 defined scratch hardness as the load in grammes under which a 90° diamond will produce a scratch 0.01 mm wide. According to Moh's test, iron and steel range in hardness from 4 to 9 depending upon the heat treatment and the alloy content.

**21.5.4.5 Fatigue test**

*"Fatigue failure" occurs as a result of repeated application of small loads which are individually incapable of producing detactable plastic deformation.* Eventually these repeated loads cause a macrocrack to open and spread across the piece. Stress intensification occurs and ultimately a

sudden, brittle fracture results. Ferrous metals and alloys have a limiting value of repeated stress which can be applied and reversed for an indefinitely large number of cycles without causing

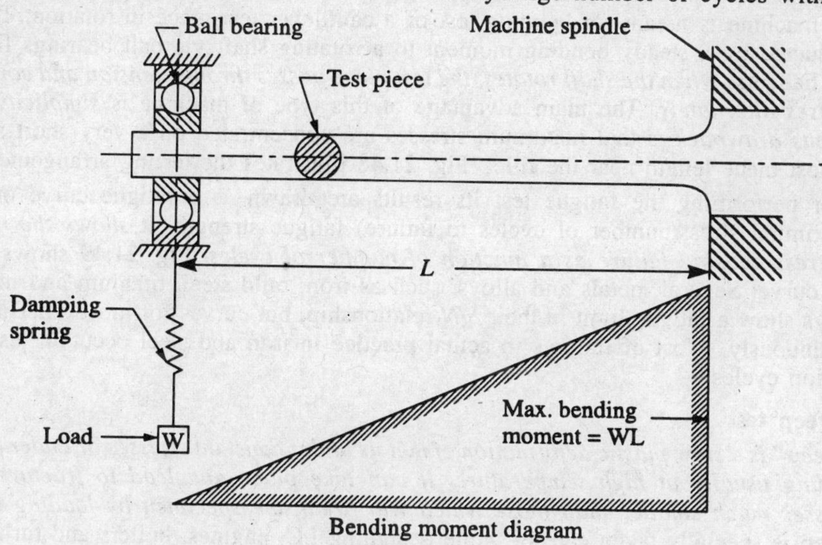


Fig. 21.48

failure. This stress is known as *fatigue limit.* Non-ferrous metals and alloys do not have limiting stress values below which failure will not occur if the cycle is repeated often enough. For these materials, fatigue strengths are usually given as stresses for which failure can be expected to occur at $10^7$, $10^8$ or some other specified number of cycles of loading. The results of fatigue tests are usually plotted on semilog plots.

— Fatigue failure starts at the point of highest stress. This point may be determined by the shape of a part; for instance, by stress concentrations in a groove. It *can also be caused by surface finish, such as tool marks or scratches, and by internal voids such as shrinkage cracks and cooling cracks in castings and weldments and defects introduced during mechanical working and by defects, stresses introduced by electroplating.*

It must be remembered that surface and internal defects are stress raisers, and the point of highest actual stress may occur at these rather than at minimum cross-section of highest nominal stress. Thus, processing methods are extremely important as they affect fatigue behaviour.

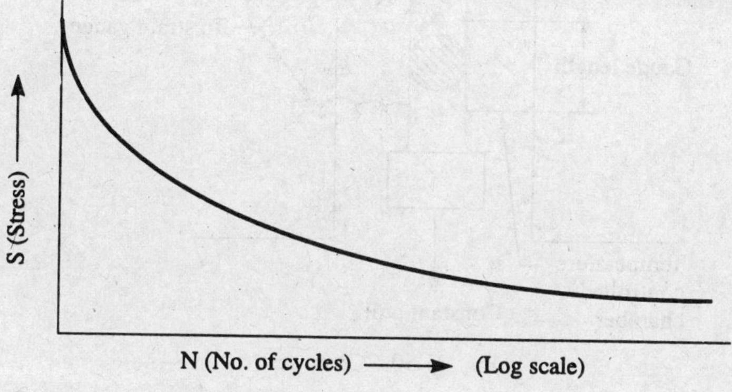


Fig. 21.49 : S/N curve.

**Procedure to conduct test:**

- To carry out fatigue test various machines have been developed; the tests are conducted for long periods under varying loads at high frequencies.

- For laboratory investigations a rotary bending test based '*wholer system*' is often used. The machine is meant for bending test of a cantilever test piece in rotation. The test is conducted by a steady bending moment to a rotating shaft via ball bearings fitted with load hangers. *When the shaft rotates, the test piece passes through tension and compression stresses alternately.* The main advantage of this type of machine is *simplicity* whereas *serious drawback* is that maximum stresses are concentrated on a very short section of the test piece length near the fillet. Fig. 21.48 illustrates the testing arrangement.
- After performing the fatigue test its results are drawn as a fatigue curve or the *S/N* (maximum stress/number of cycles to failure) fatigue strength. It *shows the magnitude of stress causing failure as a function of number of cycles.* Fig. 21.49 shows a typical *S/N* curve. Several metals and alloys, such as iron, mild steel, titanium and magnesium alloys show a fatigue limit in their *S/N* relationship, but curves for most other metals fall continuously. Most of failures in actual practice in iron and steel occur in less than 10 million cycles.

### 21.5.5.6 Creep test

- *"Creep" is a slow plastic deformation of metals under constant stresses or under prolonged loading usually at high temperature. It can take place and lead to fracture at static stresses much smaller than those which will break the specimen by loading it quickly.* Creep is specially taken care of while designing I.C. engines, boilers and turbines.
- The creep at room temperature is known as *low temperature creep* and can occur in load pipes, roofings, glass as well as in white metal bearings. The creep at high temperature is known as *high temperature creep*. It mainly depends upon the *metal, service temperature to be encountered and stress involved.* For studying its effects, the specimens are put under a constant load; the creep is measured during various time intervals and results then plotted to get a creep curve.

**Procedure to conduct test:**

Refer Fig. 21.50. The installation for creep test is shown in the figure.

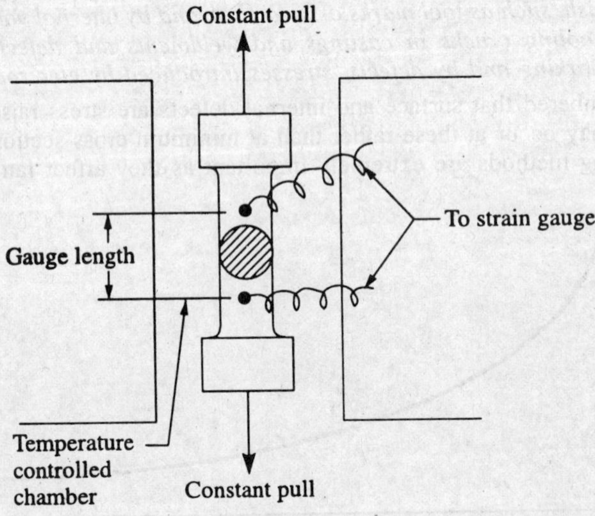


Fig. 21.50 : Creep test.

— The specimen to be tested is placed in the electric furnace where is heated to a given temperature and is constantly subjected to a load.

— The strain variations are measured with strain gauges.

— The plot with time is shown in Fig. 21.51. The various stages are explained below:

**Primary stage**. Primary creep represents a transient stage in which the resistance of the metal increases due to its deformation. It forms part of the total extension reached in a given time affecting clearance. This stage is of great interest to the designer.

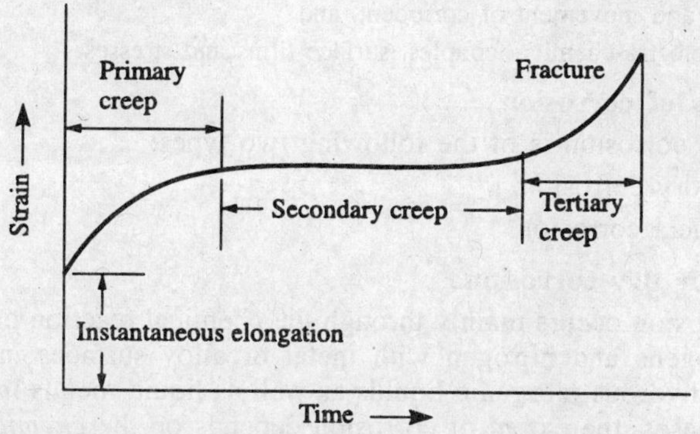


Fig. 21.51

**Secondary stage.** This stage of curve indicates the period of extension during which the creep occurs at a more or less constant rate having its minimum creep rate. Secondary creep is the result of balance between process of strain hardening and recovery. *It is the important part the curve and covers the whole of estimated service life of the alloy.*

**Tertiary stage**. In this stage, the rate of extension or strain rate accelerates rapidly leading to rupture finally. In this stage, the use of alloy should be avoided. *Tertiary creep is considered to be due to necking.*

## 21.6 CORROSION

*Corrosion is a gradual chemical or electro-chemical attack on a metal by its surroundings such that the metal is converted into an oxide, salt or some other compound.* It may be brought about by an almost unlimited number of corrosive media such as air, *industrial atmospheres, soil, acids, bases and salt solutions.* It may also occur at elevated temperatures in media which are inert when near or below room temperatures.

The metals may be corroded as a result of electro-chemical or chemical reactions between a metal surface and the environment. Many corrosion phenomena are essentially electro-chemical in nature and involve the presence of an electrolyte in contact with metal. The electrolyte is usually an aqueous solution of salt, acid, or alkali and consequently this type of corrosion can be called *wet corrosion.* All other corrosion phenomena involving chemical reaction between metal and gas or a liquid which is not an electrolyte can be classified by *dry corrosion.* There is however, no sharp dividing line between these two general types of attack on metal, since metal gas reactions often affect the course and extent of the electrochemical process occurring at liquid metal interface.

### 21.6.1 Factors Influencing Corrosion

The corrosion of a metal is dependent on:

(*i*) Chemical nature of metal,
(*ii*) Environment
(*iii*) Internal structure, and
(*iv*) Nature of engineering applications for which it is used.

**Metallurgical factors:**

(*i*) Composition,
(*ii*) Distribution of secondary phases,
(*iii*) Voids,
(*iv*) Inclusions, and

(*v*) Dissolved gases.

**Environmental factors:**

(*i*) Concentration

(*ii*) Temperature and movement of corrodent, and

(*iii*) Presence of electro-chemical couples, surface films and stresses.

### 21.6.2 General types of corrosion

As a chemical action corrosion is of the following two types:

1. Chemical or dry corrosion.
2. Electro-chemical corrosion.

#### 21.6.2.1 Chemical or dry corrosion

Chemical or dry corrosion occurs mainly through the chemical reaction of gases such as oxygen, sulphur vapour, halogens and nitrogen with metal or alloy surfaces in immediate proximity. Some organic and anhydrous inorganic liquids as well as liquid metals may attack directly solid metals. In all these cases, the extent of corrosion depends on the *chemical affinity between the corrosive environment (either gas or liquid) and solid metals and also upon the ability of metal to form a protective film.*

**1. Oxidation corrosion:**

This type of corrosion involves the reaction of metals with oxygen at high temperatures, usually in the absence of the moisture. The nature of the oxides that is formed plays an important role in the oxidation process.

(*i*) *The oxide may be unstable (as in case of silver, gold oxide) and oxidation does not occur.*

(*ii*) *The oxide may be volatile (as in the case of molybdenum oxide) and is removed as a vapour in the gas stream. In such a case oxidation occurs at a constant but relatively high rate.*

(*iii*) *The oxide formed may form a layer or layers at the metal surface.*

**2. Corrosion by other gases:**

The corrosive effect of other dry gases like $CO_2$, $SO_2$, $Cl_2$, $F_2$ on a metal depends mainly on their chemical affinity for the metal involved. And just as in the case of oxidation by air, the degree of the attack in such cases also depends on metal surface. As soon as protective or non-porous films are formed the intensity of attack decreases. When, however, no protective film is formed on the metal surface the whole metal is gradually destroyed.

**3. Liquid metal corrosion**

Liquid metal corrosion is caused due to chemical attack of a flowing liquid metal at high temperatures on a solid metal or alloy. This type of corrosion has been found to occur in devices used for nuclear power. The corrosion reaction essentially *involves dissolution of a solid metal by a liquid metal or intergranular penetration of the liquid metal into the solid metal. The usual cause of these phenomena is the existence of a temperature gradient within the solid-liquid system.*

#### 21.6.6.2 Electrode potential

When a piece of metal is partly dipped in a salt solution a potential difference is spontaneously set up between the metal surface and solution. This is called *electrode potential* and its magnitude depends upon the *nature of the metal, the concentration solution and the temperature.* If the electrolyte in the solution is other than the salt of electrode metal, then also there exists an electrode potential. If two such electrodes are electrically connected to complete the circuit it forms a *galvanic cell* sending out an electric current. There is no method of measuring the absolute value of an electrode potential. Hence all electrode potentials are determined under a standard condition with reference to a standard hydrogen electrode whose electrode potential is assumed to be zero. The electrochemical series (Table 21.1) is obtained by arranging metals in the order of their standard electrode potential. The sign of an electrode potential varies according to convention.

**Table 21.1. Electro-chemical series**

| Metal | Iron<br>Noble | Electrode potential, volts<br>Cathodic (protected) end |
|---|---|---|
| Gold | $Au^{+++}$ | + 1.50 |
| Platinum | $Pt^{++++}$ | + 0.86 |
| Silver | $Ag^{+}$ | + 0.80 |
| Mercury | $Hg^{++}$ | + 0.70 |
| Copper | $Cu^{++}$ | + 0.34 |
| Hydrogen | $H^{+}$ | 0.00 (Reference) |
| Lead | $Pb^{++}$ | – 0.13 |
| Tin | $Sn^{++}$ | – 0.14 |
| Nickel | $Ni^{++}$ | – 0.25 |
| Cobalt | $Co^{++}$ | – 0.28 |
| Cadmium | $Cd^{++}$ | – 0.40 |
| Iron | $Fe^{++}$ | – 0.44 |
| Chromium | $Cr^{++}$ | – 0.44 |
| Zinc | $Zn^{++}$ | – 0.76 |
| Aluminium | $Al^{++}$ | – 1.66 |
| Beryllium | $Be^{++}$ | – 1.85 |
| Magnesium | $Mg^{++}$ | – 2.37 |
| Sodium | $Na^{+}$ | – 2.71 |
| Potassium | $K^{+}$ | – 2.92 |
| Lithium | $Li^{+}$ | – 2.96 |

**Base anodic (corroded) end**

When current is drawn from such a cell, as the Daniel cell, certain irreversible effects occur adjacent to the electrodes that tend to oppose the direction of current flow. This is known as the *polarisation* of electrodes; it causes the solution potential of the anode to become more cathodic and that of the cathode to become more anodic. As a result the potential difference between electrodes steadily decreases, whereas the amount of current in the closed circuit diminishes. *Polarisation at the electrodes is due to concentration polarisation, overvoltage, or presence of surface film on the electrodes.*

### 21.6.2.3 Wet or electro-chemical corrosion

This type of corrosion occurs:

(*i*) *Where the liquid with metal is conducting.*

(*ii*) When there exists a difference of potentials either between two dissimilar metals or between different areas on the surface of the metal or alloy.

It progresses due to the existence of *anodic* and *cathodic areas separated by the finite distance between which the current flows.*

The *anodic reactions are always oxidation reaction* (*i.e.* liberation of electrons) and, therefore, tend to destroy the anode metal by causing it either to dissolve as an ion or to revert to a combined state such as an oxide.

The cathodic reactions are *always reduction reactions* and usually do not affect the cathode part of the metal since most metals cannot be further reduced. The electrons that are produced by the anodic reaction flow throughout the metal, are used up in the cathodic reaction (Fig. 21.52).

The *deposition of the reaction products is decisive in controlling the rate of corrosion.* If the corrosion product goes into solution or is evolved as a gas, the corrosion reaction may go uninterrupted. On the other hand, if the corrosion product is an insoluble compound, that may cover the metal surface, thereby reducing the rate of further corrosion. However, if insoluble compound is formed at a distance from the corroding area, it will have little protective value.

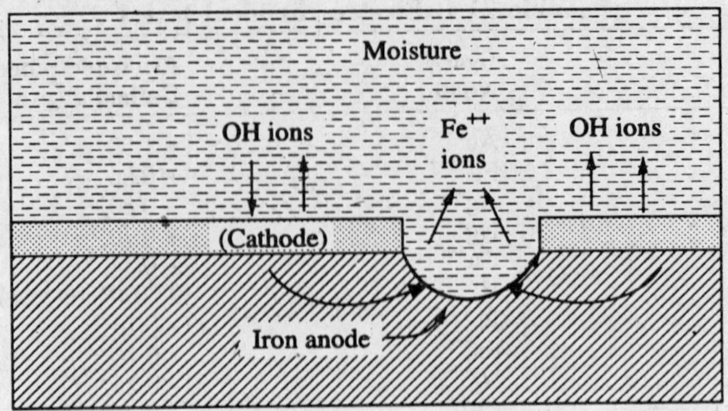


Fig. 21.52

### 21.6.3 Various Types of Corrosion

Depending upon the nature or the various ways in which it may occur, corrosion may be classified as follows:

1. Galvanic corrosion.
2. Corrosion due to differential aeration.
3. Pitting corrosion.
4. Uniform corrosion.
5. Intergranular corrosion.
6. Stress corrosion.
7. Erosion corrosion.
8. Fretting corrosion.

### 21.6.4 Control and Prevention of Corrosion

The various methods by which corrosion can be *controlled* and *prevented* are given below:

1. Protection by design and fabrication procedure.
2. Modifying the corrosive environment.
3. Application of inhibitors.
4. Cathodic protection.
5. Alloying.
6. Application of protective coatings.

**1. Protection by design and fabrication procedure:**

Corrosion rate can be reduced by proper design and fabrication. The design, in general, should be such that uniform conditions are maintained throughout. For better corrosion resistance the following *points are worth noting*:

(*i*) The use of two dissimilar metals contact should be avoided because in the presence of an electroyte such a contact may result in galvanic corrosion. If their use is a must they should be as close as possible to each other in galvanic series.

(*ii*) Recesses and sharp corners should be avioded.

(*iii*) Excessive stress concentration should be avoided (by proper fabrication and design).

(*iv*) Design should be such as to provide uniform temperature gradient, because due to non-uniform temperature distribution uneven heating results which may increase the corrosion at these points.

(*v*) Wall thickness in pipes, tanks etc. should be designed in such a way as to make allowances for reduction in thickness to uniform corrosion.

**2. Modifying the corrosive environment:**

Corrosive resistance is not an inherent property of materials but largely depends on the nature of environment. Corrosive environment can be modified in the following three ways:

(*i*) By alteration of envirnoment variables such as temperature, pressure, concentration, velocity etc.

(*ii*) By eliminating corrosive ingredients such as humidity, oxygen, oxidizing agent, and solid contaminants.

(*iii*) Introduction of inhibitors

**3. Application of inhibitors:**

*An inhibitor may be defined as any substance which when added in small quantities to the environment decreases the corrosion rate.* Inhibitors are used in liquid media, gaseous media as well as solid and semi-solid materials such as paint films, packing materials and protective greases.

The inhibitors can be classified as follows:

1. Filming Inhibitors (*i*) Anodic (*ii*) Cathodic (*iii*) Mixed (Anodic and Cathodic). 2. Absorption inhibitor 3. Vapour phase inhibitor.

**4. Cathodic protection:**

In electro-chemical corrosion there is flow of current from anode patch to cathode patch of a surface. The anode is corroded and at its expense the cathode is protected. This suggests that corrosion can be controlled by making the whole surface cathodic to an extraneous surface which acts as anode. This is called *cathodic protection*. It is achieved in two ways:

1. Sacrifical anode method.
2. External voltage method.

**Sacrificial anode** *is so called as the anodic metal is eaten away and has to be replaced from to time.* To get uniform protection of large surface it is necessary to employ a number of anodes distributed at different points. *This method is commonly used for protection of non-ferrous structure and also for ferrous structures such as steam boilers and ship's hull.*

The Fig 21.53 (*a, b, c*) shows three examples of sacrificial anode method, magnesium and zinc are being used as sacrificial anodes.

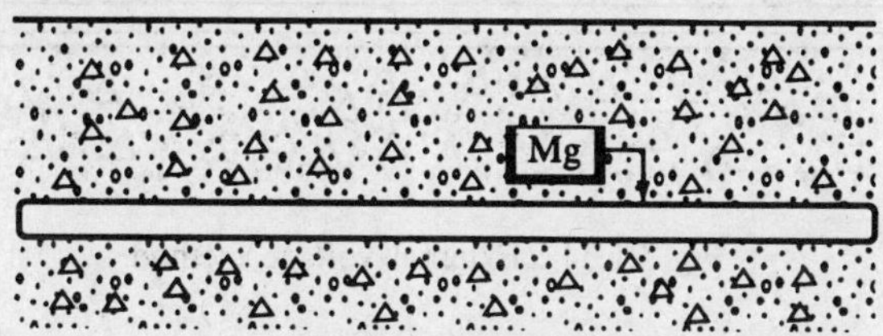


(*a*) Undeground pipe

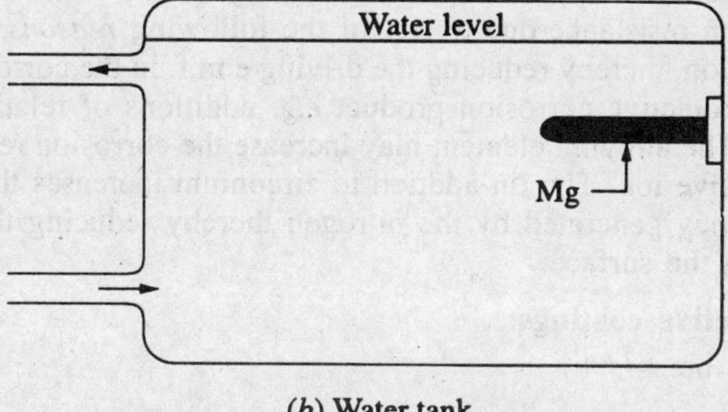


(*b*) Water tank

Fig. 21.53 : Sacrificial anode method (*a*) and (*b*).

An advantage of such procedures is that the spent anode can be replaced quite easily. For example the magnesium plates (Fig. 21.53) can be replaced at a fraction of cost of replacing the underground pipe.

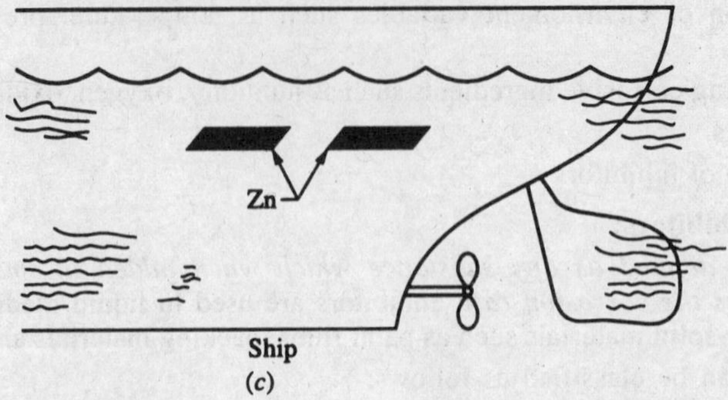


Fig. 21.53 : Sacrificial anode method (*c*).

**External voltage method** like sacrificial anode method involves the same protection principle; that is, *extra electrons* are supplied so that the *metal becomes the cathode* and the corrosion reactions do not proceed. Fig. 21.54 shows an underground pipe being protected against corrosion by applying d.c. voltage which will provide adequate electrons to make the equipment cathode. This method is also used for protection of *ferrous structures such as steam boilers, condensers, oil and water pipes, and tanks.*

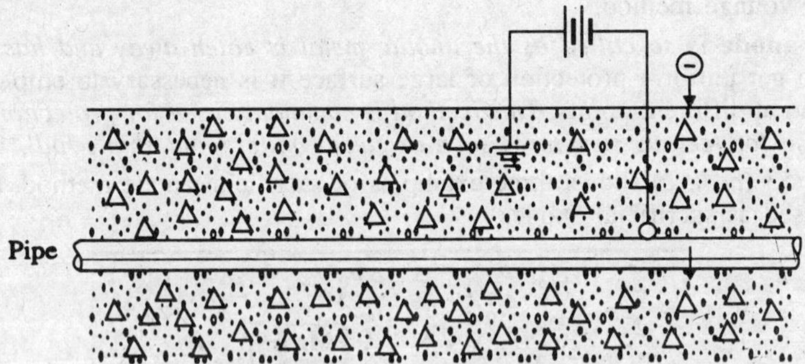


Fig. 21.54 : External voltage method.

**5. Alloying:**

Corrosion resistance of most commercial metals can be increased by alloying them with suitable alloying elements. For maximum corrosion resistance alloy should form solid solution. Alloying may increase the corrosion resistance due to any of the following *reasons:* (*i*) The alloying may raise mobility of the solution, thereby reducing the driving e.m.f. in the corrosion cell, (*ii*) Alloying may give rise to more protective corrosion product *e.g.* additions of relatively small amount of Cu, Cr, Ni to steels, (*iii*) The alloying element may increase the corrosion resistance by decreasing the mobility of the corrosive ion. The tin addtion to zirconium increases the corrosion resistance by neutralizing the tendency generated by the nitrogen thereby reducing the mobility of oxygen ions in the oxide film on the surface.

**6. Application of protective coatings:**

The applied coating may be:

(*i*) Metallic or

(*ii*) Nonmetallic

The protective coatings (both metallic and nonmetallic) should possess the following *characteristics:*

(*a*) Good corrosion resistance against medium with which it comes in contact.

(*b*) Perfect adherence to the underlying metal.

(*c*) Continuous, so as to cover the surface completely.

(*i*) **Metallic coating:** *Metallic coatings* are usually applied in one of the following ways:

(*a*) Electroplating

(*b*) Dipping

(*c*) Spraying

(*d*) Cladding

(*e*) Cementation.

(*ii*) **Non-metallic coating:** Some of the important *non-metallic coatings* are:

(*a*) Vitreous or porcelain enamels

(*b*) Anodised coatings

(*c*) Surface conversion or chemical-dip coatings.

## 21.7 MANUFACTURING PROCESSES

The materials which are covered under the scope of material science are available either from nature or industry. However these materials cannot be used in raw form (whatever the source may be) for useful purposes. They have to be shaped and formed into articles through different manufacturing processes. Besides there are some processes which improve material properties. In some processes materials are changed into their primary forms for some selected parts. In some cases the materials are suitably finished for commercial uses. In other cases, neither surface finish nor the dimensions are satisfactory for the final product, and further work is necessary. However *the selection of the best process for a given product requires a knowledge of all possible production methods.*

Some important manufacturing processes are listed below:

1. **Cold Working:**

(*i*) Drawing (*ii*) Squeezing
(*iii*) Bending (*iv*) Shearing
(*v*) Hobbing (*vi*) Shot peening.

2. **Hot Working:**

(*i*) Rolling (*ii*) Forging
(*iii*) Pipe welding (*iv*) Hot piercing
(*v*) Hot drawing (*vi*) Hot spinning
(*vii*) Hot extruding.

3. **Forging:**

(*i*) Hand forging (*ii*) Machine forging.

4. **Casting:**

(*i*) Sand casting (*ii*) Shell moulding
(*iii*) Permanent mould casting (*iv*) Die casting
(*v*) Centrifugal casting (*vi*) Investment casting
(*vii*) Plaster casting (*viii*) Flush casting.

## OBJECTIVE TYPE QUESTIONS

**Choose the Correct Answer/Fill in the Blanks**

1. An element is a substance which

(*a*) can be decomposed into other substances
(*b*) cannot be decomposed into other substances
(*c*) is composed of many compounds
(*d*) is composed of two molecules only.

**2.** The value of permittivity of space is equal to
(*a*) $8.9 \times 10^{-12}$ farad/m (*b*) $7 \times 10^{-12}$ farad/m
(*c*) $8.854 \times 10^{-12}$ farad/m (*d*) $9.2 \times 10^{-12}$ farad/.m

**3.** The value of Planck's constant is equal to
(*a*) $6.6 \times 10^{-32}$ J–s (*b*) $6.5 \times 10^{-34}$ J–s
(*c*) $6.625 \times 10^{-34}$ J-s (*d*) $7.2 \times 10^{-34}$ J-s.

**4.** The value of 1Å is equal to
(*a*) $10^{-6}$ cm (*b*) $10^{-8}$ cm
(*c*) $10^{-9}$ cm (*d*) $10^{-4}$ cm.

**5.** A proton is a
(*a*) positively charged particle (*b*) negatively charged particle
(*c*) neutral particle (*d*) any of the above.

**6.** The mass of an electron is about
(*a*) 1/1700 that of the neutron (*b*) 1/1836 that of proton
(*c*) 1/1836 that of neutron (*d*) 1/1936 that of neutron.

**7.** The nucleus comprises of
(*a*) electrons only (*b*) protons only
(*c*) electrons and protons (*d*) protons and neutrons.

**8.** Electrons are...................charged particles.
(*a*) positively (*b*) negatively
(*c*) positively-negatively. (*d*) neutral.

**9.** The diameter of the nucleus is ................... of the atom as a whole.
(*a*) 1/100th (*b*) 1/1000th
(*c*) 1/10000th (*d*) 1/100000th.

**10.** The number of electrons which surround the nucleus of a neutral atom is ...................
(*a*) equal to (*b*) less than
(*c*) more than (*d*) any of the above.

**11.** The formula that determines the number of electrons that can be accommodated in any level is given by:
(*a*) $2\ n^3$ (*b*) $2\ n^2$
(*c*) $2\ n^4$ (*d*) $2\sqrt{n}$.

**12.** The tiny block formed by the arrangement of small group of atoms is called
(*a*) unit cell (*b*) space lattice
(*c*) lattice point (*d*) none of *a, b, c*.

**13.** The idea of space lattice was introduced by Bravais in
(*a*) 1800 (*b*) 1820
(*c*) 1840 (*d*) 1880.

**14.** According to Bravais there are ........... possible types of space lattices in the seven basic crystal systems.
(*a*) 8 (*b*) 14
(*c*) 20 (*d*) 24.

**15.** There are ................... basic crystal systems.
(*a*) two (*b*) four
(*c*) six (*d*) seven.

**16.** A crystal is said to have a simple form if its all faces are
(*a*) similar (*b*) dissimilar

(c) partly similar (d) partly dissimilar.

**17.** The co-ordination number of simple cubic structure is

(a) two (b) four

(c) six (d) eight.

**18.** The co-ordination number of face centred cubic structure is

(a) four (b) six

(c) eight (d) twelve.

**19.** Atomic radius is defined as .................... the distance between nearest neighbours in a crystal of a pure element.

(a) one-third (b) one-fourth

(c) half (d) none of a, b, c.

**20.** Atomic packing factor of a simple cube is approximately equal to

(a) 0.4 (b) 0.42

(c) 0.52 (d) 0.8.

**21.** Total number of atoms in a body-centred cube is

(a) one (b) two

(c) three (d) four.

**22.** Total number of atoms in a face centred cube is

(a) two (b) three

(c) four (d) six.

**23.** Atomic packing factor of a face centred cube is equal to

(a) 0.6 (b) 0.7

(c) 0.74 (d) 0.84.

**24.** The distance '$d$' between adjacent planes of a set of parallel planes of the indices $(h, k, l)$ is given by

(a) $d = \dfrac{a}{\sqrt{h+k+1}}$ (b) $\dfrac{a^2}{\sqrt{h^2+k^2+l^2}}$

(c) $\dfrac{a}{\sqrt{h^2+k^2+l^2}}$ (d) $\dfrac{a^3}{\sqrt{h^2+k^2+l^2}}$.

where '$a$' is the edge of the cube.

**25.** For a B.C.C. lattice structure

(a) $\dfrac{1}{d_1} : \dfrac{1}{d_2} : \dfrac{1}{d_3} :: 1 : \sqrt{2} : \sqrt{3}$ (b) $\dfrac{1}{d_1} : \dfrac{1}{d_2} : \dfrac{1}{d_3} :: 1 : \sqrt{2} : \dfrac{1}{\sqrt{3}}$

(c) $\dfrac{1}{d_1} : \dfrac{1}{d_2} : \dfrac{1}{d_3} :: 1 : \dfrac{1}{\sqrt{2}} : \sqrt{3}$ (d) none of the above.

**26.** In metals the atoms and molecules are held together by force of

(a) attraction (b) repulsion

(c) attraction and repulsion (d) any of the above.

**27.** Primary bonds are the

(a) strongest bonds (b) weakest bonds

(c) both a, b (d) none of the above.

**28.** Secondary bonds are also called

(a) ionic bonds (b) covalent bonds

(c) molecular bonds (d) metallic bonds.

**29.** Ionic bonds are mainly formed in

(a) inorganic compounds (b) organic compounds

(c) metals (d) alloy steels.

**30.** Ionic compounds are

(a) soft (b) rigid

(*c*) neither soft nor rigid (*d*) any of the above.

**31.** The covalent bond is formed by ........................ between atoms.
(*a*) sharing of electrons (*b*) transfer of electrons
(*c*) sharing and transfer of electrons (*d*) any of the above.

**32.** Covalent compounds are generally..................electrically
(*a*) conductors (*b*) semiconductors
(*c*) insulators (*d*) all above.

**33.** A metallic bond results from
(*a*) sharing of electrons
(*b*) transfer of electrons
(*c*) sharing of variable number of electrons by a variable number of atoms
(*d*) none of above.

**34.** Metallic compounds are
(*a*) very bad conductors of electricity (*b*) very good conductors of electricity
(*c*) both (*a*) and (*b*) (*d*) none of the above.

**35.** Metallic compounds are .............. in nature.
(*a*) crystalline (*b*) non-crystalline
(*c*) liquid (*d*) none of the above.

**36.** Metallic compounds have
(*a*) high reflectivity and lustre (*b*) low reflectivity and no lustre
(*c*) high reflectivity and no lustre (*d*) none of the above.

**37.** Dipole bonds are ............. than ionic bonds.
(*a*) much weaker (*b*) much stronger
(*c*) either of the above (*d*) none of the above.

**38.** Ionic solids have
(*a*) low melting point (*b*) high melting point
(*c*) very low melting point (*d*) none of the above.

**39.** Metallic solids are generally
(*a*) hard and brittle (*b*) tough and ductile
(*c*) soft and plastically deformable (*d*) none of the above.

**40.** Molecular solids have
(*a*) very high boiling points (*b*) very low melting points
(*c*) fairly low melting points (*d*) none of the above.

**41.** In metals the plastic deformation generally takes place by the process of ................

**42.** ................. is defined as a shear deformation that moves atoms by many interatomic distances in one crystal plane over the atoms of another crystal plane.

**43.** The shear stress required to produce slip on a crystal plane is called the ................ shear stress.

**44.** The critical resolved shear stress, $\tau_{cr} = \frac{P}{A} \cos \alpha \cos \beta$. This equation is known as .................. law.

**45.** ....................... is that process by which a portion of the crystal takes up an orientation which makes that protion a mirror image of the parent crystal.

**46.** ....................... is produced suddenly and is accompanied with sound.

**47.** The well known cry of tin is the result of .......................

**48.** ....................... occurs over wide planes.

**49.** In ....................... every atomic plane is involved.

**50.** ....................... takes place in several milliseconds.

**51.** In ....................... the orientation of the crystal above and below the slip plane is same after deformation as before.

**52.** ....................... takes place in a few micro-seconds.

**53.** ....................... defects are imperfect point like regions in a crystal.

54. A ....................... is the simplest point defect and involves a missing atom within a metal.
55. ....................... imperfections involve vacancies of pair of ions of opposite charges.
56. Both vacancies and schottky defects facilitate ....................... diffusion.
57. The vacancy and ....................... are inverse phenomena.
58. An ion dislodged from the lattice into an interstitial site is called ....................... defect.
59. A linear disturbance of atomic arrangement, which can move very easily on the slip plane through the crystal is known as .......................
60. ....................... is an operation or combination of operations, involving heating and cooling of a metal or alloy in its solid state with the object of changing the characteristics of the material.
61. ....................... does not harden when cooled rapidly.
62. ....................... is a definite carbide of iron which is extremely hard, being harder than ordinary hardened steel or glass.
63. ....................... comprises of alternate layers of ferrite and cementite in steel.
64. ....................... is a eutectoid of steel.
65. ....................... steels are mixtures of pearlite and cementite.
66. ....................... is a hard brittle mass of fibrous or needle like structure and is the chief constituent of hardened steels.
67. ....................... is not as tough as austenite.
68. ....................... is a solid solution of iron-carbon which is stable only within a particular range of composition and temperature.
69. The main object of annealing is to ....................... the metals.
70. ....................... is a type of annealing which causes practically carbides in the steel to agglomerate in the form of small globules or spheroids.
71. ....................... is the method of rapid cooling of a metal in a bath of liquid during heat treatment.
72. The magnitude of the critical cooling rate depends on the stability of the .......................
73. When the risk of distortion is great, quenching must be carried out in .......................
74. ....................... test is one of the most useful and widely accepted method for measuring the hardening response of steel to heat treatment.
75. ....................... process decomposes the martensite into a ferrite-cementite mixture.
76. In ....................... process nitrogen is introduced to steel by passing ammonia gas through a muffle furnace containing the steel to be nitrided.
77. The articles which are ....................... heated have no scale effect.
78. ....................... baths transmit heat quickly and uniformly and afford a protection to the steel during the heating cycle.
79. Young's modulus is indicative of the property called .......................
80. The property of spring back is a function of modulus of .......................
81. ....................... is the property that enables the formation of a permanent deformation in a material.
82. The ability of a metal to withstand elongation or bending is known as

(*a*) Ductility (*b*) Stiffness
(*c*) Brittleness (*d*) Melleability.

83. The property by virtue of which a material may be hammered or rolled into thin sheets without rupture is known as

(*a*) Elasticity (*b*) Plasticity
(*c*) Ductility (*d*) Malleability.

84. ....................... is the strength with which the material opposes rupture.
85. Lack of ductility is .......................
86. ....................... is usually defined as resistance of a material to penetration.
87. In ....................... testing a component does not break and even after being tested, so it can be uses for the purpose for which it was made.
88. ....................... rays are emitted during the disintegration of radioactive material.
89. Gamma rays are ....................... penetrating than X-rays.
90. Gamma rays are used for detecting defects in castings ....................... than those inspected by

*X*-rays.

**91.** *X*-ray method is ...................... rapid than gamma-ray method.

**92.** The sound waves whose frequency is above the upper pitch limit of the human ear are called ......................

**93.** ...................... testing and inspection is one of the most useful non-destructive methods in metal testing.

**94.** Ultrasonic waves are usually generated by the ...................... effect.

**95.** The microscopic properties of materials under applied forces or loads are broadly classed as:

(*a*) Electrical properties (*b*) Magnetic properties

(*c*) Mechanical properties (*d*) Dielectric properties.

**96.** The maximum stress at which stress remains directly proportional to strain is called

(*a*) Elastic limit (*b*) Yield point stress

(*c*) Proportional limit (*d*) Rupture stress.

**97.** The elastic limit is the maximum stress which the material can withstand without causing ...................... deformation which remains after removal of stress.

**98.** The stress at which a material exhibits a specified limiting permanet set is called ...................... strength.

**99.** The ...................... point is the stress at which there first occurs a marked increase in strain without an increase in stress.

**100.** The yield point most commonly is observed in ...................... steels, although it has been detected in a few other alloys.

**101.** ...................... strength is calculated by dividing the maximum load carried by the specimen during a tension test by the original cross-sectional area of the specimen.

**102.** ...................... is determined by dividing the load at the time of fracture by the original cross-sectional area.

**103.** Resistance to elastic deformation is more commonly called ......................

**104.** ...................... is an instrument or device by which the changes in length of specimen under test can be precisely measured.

**105.** ...................... stress is the load at any elongation divided by the cross-sectional area at that elongation.

**106.** ...................... strain is the change in length with reference to the instanteneous gauge length rather than the original length.

**107.** ...................... test signifies toughness of material that is ability of material to absorb energy during plastic deformation.

**108.** ......................takes into account both the strength and ductility of the material.

**109.** ...................... test uses a cantilever test piece of 10 mm × 10 mm section specimen having standard 45° notch 2 mm deep.

**110.** ...................... test uses simply supported test piece of 10 mm × 10 mm section.

**111.** For material like mild steel the impact values amply improve if angle exceeds ................. degrees.

**112.** In an impact test, as the sharpness of the notch increases the energy required to rupture the specimen ....................

**113.** The Brinell hardness number (B.H.N.) is found out by the following equation

(*a*) $\dfrac{2P}{\pi D^2 (D - \sqrt{D^2 - d^2})}$ (*b*) $\dfrac{3P}{\pi D^2 (D - \sqrt{D^2 - d^2})}$

(*c*) $\dfrac{2P}{\pi D (D - \sqrt{D^3 - d^2})}$ (*d*) $\dfrac{2P}{\pi D (D - \sqrt{D^2 - d^2})}$.

where *D* and *d* are the diameters of ball and indentation respectively and *P* is the load.

**114.** Pyramid hardness number (D.P.N.) is given by

(*a*) $\dfrac{P \sin \theta/2}{d^2}$ (*b*) $\dfrac{3\,P \sin \theta/2}{d^2}$

(c) $\frac{2P \sin \theta/2}{3d^2}$ (d) $\frac{2P \sin \theta/2}{d^2}$

where $d$ = mean length of two diagonals in mm

$\theta$ = angle between the opposite faces, and

$P$ = load in kg.

**115.** .................... is a gradual chemical or electro-chemical attack on a metal by its surroundings such that the metal is converted into an oxide, salt or some other compound.

**116.** The deposition of reaction products is decisive in controlling the rate of ..................

**117.** .................... is usually caused by the breakdown or cracking of the protective film on a metal at specific points.

**118.** Pitting exercises considerable effect on fatigue properties but no effect on ....................properties.

**119.** The term ...................... corrosion describes the combined effect of mechanical stresses and corrosive environment on a metal.

**120.** In .................... corrosion the stresses should be tensile, static and of sufficient magnitudes.

**121.** The damage due to .................... corrosion is more in soft materials than in hard materials.

**122.** An ...................... is any substance which when added in small quantities to the environment decreases the corrosion rate.

**123.** In ...................... protection the anode is corroded.

**124.** For maximum corrosion resitance alloy should form ......................

**125.** .................... is the simplest process of applying a metallic coating.

**126.** .................... is the method of obtaining a coherent coating of zinc or iron.

**127.** .................... coating are applied by either the wire gun or powder metal method.

**128.** .................... consists in placing of a comparatively thick lining upon the metal surface.

**129.** In ...................... process base metal is heated in the powder of the metal which forms the coating.

**130.** .................... consists in giving a thin coating of aluminium on steel.

## ANSWERS

**1.** (b) **2.** (c) **3.** (c) **4.** (b) **5.** (a)
**6.** (c) **7.** (d) **8.** (b) **9.** (c) **10.** (a)
**11.** (b) **12.** (a) **13.** (d) **14.** (b) **15.** (d)
**16.** (a) **17.** (c) **18.** (d) **19.** (c) **20.** (c)
**21.** (b) **22.** (c) **23.** (c) **24.** (c) **25.** (c)
**26.** (a) **27.** (a) **28.** (c) **29.** (a) **30.** (b)
**31.** (a) **32.** (c) **33.** (c) **34.** (b) **35.** (a)
**36.** (a) **37.** (a) **38.** (b) **39.** (b) **40.** (c)
**41.** Slipping, **42.** Slip **43.** Critical resolved,
**44.** Schimd's law **45.** Twinning **46.** Twinning
**47.** Twinning **48.** Slip **49.** Twinning
**50.** Slip **51.** Slip **52.** Twinning
**53.** Point **54.** Vacancy **55.** Schottky
**56.** Atomic **57.** Interstitialcy **58.** Frenkel,
**59.** Dislocation, **60.** Heat treatment **61.** Ferrite
**62.** Cementite; **63.** Pearlite; **64.** Pearlite
**65.** Hard; **66.** Martensite; **67.** Martensite; **68.** Austenite; **69.** soften
**70.** Spheroidization; **71.** Quenching **72.** Austenite; **73.** Air blast
**74.** Jominy; **75.** Tempering; **76.** Nitriding; **77.** Induction;
**78.** Molten salt **79.** Stiffness **80.** elasticity **81.** Plasticity
**82.** (a) **83.** (d) **84.** Tenacity or toughness.
**85.** brittleness **86.** Hardness **87.** Non-destructive **88.** Gamma

| | | | | | | | |
|---|---|---|---|---|---|---|---|
| **89.** | more | **90.** | thicker | **91.** | more | **92.** | ultrasonics |
| **93.** | Ultrasonics | **94.** | Piezoelectric | **95.** | (*c*) | **96.** | (*c*) |
| **97.** | permanent | **98.** | yield | **99.** | yield | **100.** | mild |
| **101.** | Ultimate | **102.** | Rupture strength | | | **103.** | stiffness |
| **104.** | Extensometer | **105.** | True | | | **106.** | True |
| **107.** | Impact | **108.** | Toughness | | | **109.** | Izod |
| **110.** | Charpy | **111.** | 60 | | | **112.** | decreases |
| **113.** | (*d*) | **114.** | (*d*) | | | **115.** | Corrosion |
| **116.** | corrosion | **117.** | Pitting | | | **118.** | Mechanical |
| **119.** | Stress | **120.** | Stress | | | **121.** | Fretting |
| **122.** | Inhibitor | **123.** | Cathodic | | | **124.** | Solid solution |
| **125.** | Dipping | **126.** | Galvanising | | | **127.** | Sprayed |
| **128.** | Cladding | **129.** | Cementation | | | **130.** | Calorising |

## THEORETICAL QUESTIONS

1. Define the following terms:
   Atom, Electron, Proton, Nucleus, Atomic number and Isobars.
2. What are the constituents of an atom?
3. Explain briefly the following:
   (*i*) Electrons
   (*ii*) Protons
   (*iii*) Neutrons.
4. State the physical properties of electrons.
5. Explain atomic weight and atomic number. What is the significance between these two?
6. What is the meaning of isotope?
7. What is the maximum number of electrons allowable in any single major energy level?
8. Give a concise account of atomic structure, pointing out the characteristics which govern the physical and chemical properties of atoms.
9. What are molecules? Explain saturated and unsaturated molecules.
10. Discuss briefly the general usefulness of the periodic table in reference to atomic structure.
11. How are materials broadly classified?
12. What is the difference between crystalline and non-crystalline solids?
13. What is 'unit cell'?
14. What do you mean by 'space lattice'?
15. What is meant by 'lattice parameter'?
16. What are crystallographic planes?
17. What are lattice parameters of a unit cell?
18. Describe briefly the basic seven crystal systems.
19. Enumerate and sketch the unit cells of Bravais lattices.
20. Explain briefly the following:
    (*i*) Body centred cubic structure (B.C.C).
    (*ii*) Face centred cubic structure (F.C.C.)
    (*iii*) Hexagonal close packed structure (H.C.P).
21. Give the comparison between ionic, covalent and metallic bonds.
22. Name the four types of atomic bonds observed in materials. Describe any two of them giving examples.
23. In what respects secondary or molecular bonds differ from primary bonds?
24. Explain briefly three types of molecular bonds.
25. (*a*) Distinguish between plastic deformation and fracture.

(*b*) Explain with neat sketches the mechanism for 'dislocation' and 'twinning' as related to plastic deformation.

**Ans.** *Plastic deformations*, in simple words, are those in which strain occurs simultaneously with the application of stress but does not vanish if the stress is removed. *Fracture* is that state in which physical separation takes place.

**26.** What are metallurgical advantages of hot working process over cold working process?

**27.** Distinguish between the term 'recovery' and 'recrystallisation' involved in the process of heating cold worked metal.

**28.** (*a*) Explain with simple sketches the following: Edge dislocation and screw dislocation.

(*b*) Differentiate between 'Slip and Twinning'.

**29.** (*a*) What is cold working? How important properties like 'hardness' and 'electrical resistance' change typically for a metal when it is cold worked?

(*b*) Explain the effect of recovery, recrystallisation and grain growth on the above properties of a cold worked metal on its subsequent annealing.

**30.** Write short notes on Edge and Screw dislocations.

**31.** Define the term "heat-treatment" and state its objectives.

**32.** Enumerate the constituents of iron and steel and explain any three of them in detail.

**33.** What is "austenite" and how is it formed?

**34.** What is "martensite" and how is it formed?

**35.** Draw the 'iron-carbon equilibrium diagram' and explain it.

**36.** Enumerate various heat-treatment processes and explain any two of them briefly.

**37.** State the objects of "annealing".

**38.** Explain briefly the following annealing processes:

(*i*) Full annealing;

(*ii*) Isothermal annealing;

(*iii*) Sub-critical annealing.

**39.** What is accomplished by normalising?

**40.** Describe the "critical rate" of quenching.

**41.** Name and explain the various quenching media used for heat treatment of steel.

**42.** Enumerate the tests you will carry out for identification of metals and alloys.

**43.** Explain the procedure for preparing the specimen for micro-examination.

**44.** Describe with a neat diagram the construction and working of a metallurgical microscope. State its limitations also.

**45.** Explain an 'electron microscope' in respect of the following:

Principle of operation, preparation of specimen, construction and operation, important features and limitations.

**46.** List the purposes for which materials are tested.

**47.** How do you classify tests on materials?

**48.** Enumerate various methods used for non-destructive testing and explain in detail any two of them.

**49.** Describe briefly any two of the following non-destructive tests:

(*i*) X-ray radiography

(*ii*) Magnetic particle inspection

(*iii*) Ultrasonic testing

(*iv*) Damping testing.

**50.** Discuss the significance of the following terms in a tensile test:

(*i*) Limit of proportionality

(*ii*) Elastic limit

(*iii*) Yield point

(*iv*) Maximum stress

(*v*) Breaking stress.

**51.** Name the important mechanical tests which gives valuable information about metals and alloys.

**52.** What does impact test signify? Explain with necessary formulation, the procedures to be adopted in the impact test, conduoted on a pendulum type impact testing machine.

**53.** What are different hardness tests of interest to the testing engineer under various situations?

**54.** Explain the Brinell hardness testing of mild steel specimen.

**55.** What is a '*rebound test*'? How is it carried out?

**56.** What is a fatigue failure? How is a fatigue test carried out?

**57.** Explain briefly the procedure of conducting a torsion test.

**58.** What is creep? Draw a typical creep curve and explain the different stages of creep.

**59.** (*a*) What do you understand by the term 'Corrosion'? How corrosion can be prevented?
(*b*) Explain briefly, how iron tanks corrode faster if they are filled with water.

**60.** Under what conditions iron will or will not trust. Explain why?

**61.** What are the different types of coatings applied to prevent corrosion?
Compare their advantages and disadvantages.

**62.** (*a*) Define the term corrosion of metals. Explain corrosion by 'oxidation' or 'galvanic' corrosion.

22

# Ceramic Materials

22.1. Introduction. 22.2. Classification of ceramics. 22.3. Advantages of ceramic materials. 22.4. Applications of ceramics. 22.5. Properties of ceramic materials. 22.6. Structure of crystalline ceramics. 22.7. Silicate structures. 22.8. Polymorphism. 22.9. Glass. 22.10. Advanced ceramics. 22.11. Rocks and building stones. 22.12. Refractories. 22.13. Cements and concrete. Highlights-Objective Type Questions-Theoretical Questions.

## 22.1. INTRODUCTION

***"Ceramic materials"** are defined as those containing phases that are compounds of metallic and nonmetallic elements.*

The *science of ceramics,* nearly as old as mankind, is the processing *of earthly materials by* heat. The crude cooking utensils of early man were the first appication of the materials now used in jet engines and atomic reactors. All the early ceramic products were made from *clay* because the ware could be easily formed. It was then dried and fired to develop the permanent structure. Because the other ceramic materials lacking plasticity also have desirable properties, other methods of forming and processing have been developed. Other forming methods used for ceramic materials are injection moulding, sintering and hot pressing. In other cases the formed materials are allowed to harden on the job by the addition of water, as in case of cements.

## 22.2. CLASSIFICATION OF CERAMICS

**A. Classification of Ceramic Materials:**

Ceramic materials are classified as follows :

**1. Functional Classification :**

(*i*) Abrasives : Alumina, carborundum

(*ii*) Pure oxide ceramics : MgO, $Al_2O_3$, $SiO_2$

(*iii*) Fire-clay products : Bricks, tiles, procelain etc.

(*iv*) Inorganic glasses : Window glass, lead glass etc.

(*v*) Cementing materials : Portland cement, lime etc.

(*vi*) Rocks : Granites, sandstone etc.

(*vii*) Minerals : Quartz, calcite, etc.

(*viii*) Refractories : Silica bricks, magnesite, etc.

2. **Structural Classification:**

   (*i*) *Crystalline ceramics:* Single-phase like MgO or multi-phase form the MgO to $Al_2O_3$ binary system.

   (*ii*) *Non-crystalline ceramics:* Natural and synthetic inorganic glasses.

   (*iii*) *"Glass-bonded" ceramics:* Fire clay products-crystalline phases are held in glassy matrix.

   (*iv*) *Cements:* Crystalline or crystalline and non-crystalline phases.

**B. Classification of Ceramic Products:**

A general classification of 'ceramic products' is difficult to make because of the great versatility of these materials, but the following list includes the major groups:

1. Whitewares.
2. Bricks and tiles.
3. Chemical stonewares.
4. Cements and concretes.
5. Abrasives.
6. Glass.
7. Insulators.
8. Porcelain enamel.
9. Refractories.
10. Electrical porcelain.
11. Mineral ores.
12. Slags and fluxes.

## 22.3. ADVANTAGES OF CERAMIC MATERIALS

The ceramic materials entail the following **advantages:**

1. The ceramics are hard, strong and dense.
2. They have high resistance to the action of chemicals and to the weathering.
3. Possess a high compression strength compared with tension.
4. They have high fusion points.
5. They offer excellent dielectric properties.
6. They are good thermal insulators.
7. They are resistant high temperature creep.
8. Availability is good.
9. Good sanitation.
10. Better economy.

## 22.4. APPLICATIONS OF CERAMICS

The applications of ceramics are listed below :

1. **The whitewares (older ceramics)** are largely used as/in :
   - Tiles;
   - Sanitary wares;
   - Low and high voltage insulators;
   - High frequency applications;
   - *Chemical industry*–as crucibles, jars and components of chemical reactors;
   - *Heat resistant applications*–as pyrometers, burners, burner tips, and radiant heater supports.

2. **Newer ceramics** (*e.g.*, borides, carbides, nitrides, single oxides, mixed oxides, silicates, metalliod and intermetallic compounds) which have the *high hardness values* and *heat and oxidation values* are largely used as/in:
   - *Refractories* for industrial furances.
   - *Electrical and electronic industries-as* insulators, semiconductors, dielectrics, ferro-electric crystals, piezo-electric cystals, glass, porcelain alumina, quartz and mica etc.
   - *Nuclear applications*–as fuel elements, fuel containers, moderators, control rods and structural parts. Ceramics such as $UO_2$, UC, $UC_2$ are employed for all these purposes.
   - *Ceramic metal cutting tools–made* from glass free $Al_2O_3$.
   - *Optical applications-Ytralox,* a comparative newcomer in the ceramic material field, is useful since it is as transparent as window glass and can resist very high temperature.
3. **Advanced ceramics** (*e.g.*, SiC, $Si_3N_4$, $ZrO_2$, $B_4C$, SiC, $TiB_2$ etc.)

   The advanced ceramics are untilized as/in:
   - Internal combustion engines and turbines, as armor plate;
   - Electronic packaging;
   - Cutting tools;
   - Energy conversion, storage and generation.

## 22.5. PROPERTIES OF CERAMIC MATERIALS

### 1. Mechanical Properties:

The ceramic materials possess the following *mechanical properties* :

(*i*) The compressive strength is several times more than the tensile strength.

(*ii*) *Non-ductile/brittle.* Stress concentration has little or no effect on compressive strength.

(*iii*) The ceramic materials possess ionic and convalent bonds which impart high modulus of elasticity. The modulus decreases with increase in temperature (due to increase in interatomic distance at elevated temperature).

(*iv*) As compared to pure metals, more force is required to cause slip in diatomic ceramic materials, because diatomic material consists of a mixture of positively and negatively charged ions which have strong forces of attraction between them.

(*v*) Below recrystallisation temperature, non-crystalline ceramics are fully brittle. The cleavage failure occurs along crystallographic planes and propagation of the crack takes place at high speed.

(*vi*) At high temperature rigidity is high.

(*vii*) In case of alloy consisting of two or metals, each phase may have appreciable difference of coefficient of thermal expansion which generates stress. This stress may then cause the metal to fail.

### 2. Electrical Properties:

The electrical properties of ceramic products vary from the low loss, high frequency dielectrics to semiconductors. Electrical insulators fall into two general classifications, the classical electrical porcelain for both high and low tension service and the special bodies such as steatite, rutile, cordierite, high alumina, and clinoestatite for high frequency insulation

**Dielectric constant:**

- ***"Dielectric constant"*** is the *ratio of the capacitance of a dielectric compared to the capacitance of air under the same conditions.*

- A low dielectric constant contributes to low power loss and low loss factor; a high dielectric constant permits small physical size.
- The dielectric constant for electrical porcelain varies between 4.1 and 11.0. Some special bodies have reported values of several thousands.
- Porcelain has large positive temperature coefficient.
- Rutile bodies have large negative coefficients.
- By combining capacitor dielectrics having different temperature coefficients it is possible to reduce effect of the temperature change.

**Dielectric strength:**

- The ***"dielectric strength"*** of a material is defined as the *ability of a material to withstand electrical breakdown.*
- The specific values of dielectric strength vary from 100 V per mil for low-tension electrical porcelain to 500 V per mil for some special bodies.
- Rutile bodies show higher breakdown strength at higher frequencies.

**Volume and surface resistivity :**

- A *volume resistivity* of $10^6$ ohms/cm$^3$ is considered the lower limit for an insulating material. At room temperature practically all ceramic materials exceed this lower limit. As the temperature of ceramic materials is raised, the volume resistivity decreases; the volume resistivity of sida-lime glasses decreases rapidly with temperature, whereas some special bodies are good insulators (above $10^6$ ohm/cm$^3$) at 700°C. Crystallised alumina has a volume resistivity of 500 ohms/cm$^3$ at 1600°C.
- *Surface resistivity* for dry, clean surface is $10^{12}$ ohms/cm$^2$. At 98% humidity, the surface resistivity may be $10^{11}$ ohms/cm$^2$ for a glazed piece or $10^9$ ohms/cm$^2$ for an unglazed piece. The presence of dissolved gases and other deposits also tends to decrease the surface resistivity of ceramic materials.

## 3. Thermal Properties:

Since the ceramic materials contain relatively few electrons, and ceramic phases are transparent to radiant type energy, their thermal properties differ amply from that of metals. The following are the most important thermal properties of ceramic materials (which vary from material to material and from condition to condition):

(*i*) Thermal capacity
(*ii*) Thermal conductivity
(*iii*) Thermal shock resistance.

### (*i*) Thermal capacity:

- The specific heats of fine clay bricks are 0.25 and 0.297 at 1000°C and 1400°C respectively.
- Carbon bricks possess specific heats of about 0.812 at 200°C and 0.412 at 1000°C.

### (*ii*) Thermal conductivity:

- The ceramic materials possess a vary low thermal conductivity since they do not have enough electrons (for bringing about thermal conductivity). The conduction of heat takes place by phonon conductivity and the interaction of lattice vibration, while at elevated temperatures conduction takes place by the transfer of radiant energy.
- The impurity content, porosity and temperature decrease the thermal conductivity.
- In order to have maximum thermal conductivity, it is imperative to have maximum density which most of the ceramic materials do not possess.

**(iii) Thermal Shock:**

***"Thermal shock resistance"*** is the *ability of a material to resist cracking or disintegration of the material under abrupt or sudden changes in temperature.*

Thermal shock is developed primarily because of thermal expansion or contraction, which is largely a function of internal structure particularly the inter-atomic bonding. Loosely packed structures can provide *internal expansion.* Thus the coefficient of expansion is low.

- *Lithium compounds* are used in many ceramic compounds to *reduce thermal expansion* and *to provide excellent thermal shock resistance.*
- Common ceramic materials graded in order of decreasing thermal shock resistance are given below:

(*i*) Silicon nitride
(*ii*) Fused silica
(*iii*) Cordierite
(*iv*) Zircon
(*v*) Silicon carbide
(*vi*) Beryllia
(*vii*) Alumina
(*viii*) Porcelain
(*ix*) Steatite.

**4. Chemical, Optical and Nuclear Properties :**

**Chemical Properties**

- Several ceramic products are highly resistant to all chemicals except hydrofluoric acid and to some extent, hot caustic solutions. They are not affected by the organic solvents.
- Oxidic ceramics are completely resistant to oxidation, even at very high temperatures.
- Zirconia, magnesia, alumina, graphite etc., are resistant to certain molten metals and are thus employed for making crucibles and furance linings.
- Where resistance to attack from acids, bases and salt solutions is required, ceramics like glass are employed.

**Optical Properties:**

- Several types of glasses have been employed for the production of windows, subjected to high temperatures and optical lenses.
- Special glasses, in large number, have also been used for selective transmission or absorption of particular wavelengths such as infrared and ultraviolet.

**Nuclear Properties:**

As ceramics are refractory, chemically resistant and because different compositions offer a wide range of neutron capture and scatter characteristics, they are finding nuclear applications and are being used as : *Fuel elements, moderators, controls* and *shielding.*

## 22.6. STRUCTURE OF CRYSTALLINE CERAMICS

Most ceramic phases, like metals, have *crystalline structure.* Ceramic crystals are formed by either a pure ionic bond, a pure covalent bond or by bonds that possess the ionic as well as covalent characteristics.

- **Ionic bonds** *give ceramic materials of relatively high stability.* As a class, they have a much higher melting point, on the average, than do metals or organic materials. Generally speaking, they are also *harder and more resistant to chemical reaction.*
- **Covalent crystals** *usually also possess high hardness, high melting point and low electrical conductivity at room temperature.*

The ceramics crystals structures are, however, invariably more complex as compared to those of metals, since atoms of different sizes and electronic configurations are assembled together.

Common crystal structures found in crystalline ceramics particularly those of *oxide type* include the following :

(*i*) Rock salt structure

(*ii*) Cerium chloride structure

(*iii*) Zinc blende structure

(*iv*) Wurzite structure

(*v*) Spinel structure

(*vi*) Fluorite structure

(*vii*) Ilmenite structure.

## 22.7. SILICATE STRUCTURES

The silicates are *co-ordinate structures based upon large anions arranged about small cations.* The dimensions of the lattice in general are controlled by the anions rather than cations because of the larger sizes of the former. Most important are the $Si^{+4}$ and $O^{-2}$ ions. In all silicates the basic unit is the *$SiO_4$ tetrahedron.* This appears to remain essentially unaltered regardless of the other materials present.

**Silicates** are important constituents of most of the ceramic materials since they are *plentiful, cheap and have certain distinct properties, necessary for certain engineering applications.*

- *Portland cement is the most widely known silicate.* It has the very advantage of forming a hydraulic bond.
- Certain other construction materials made of silicates are brick, tile, glass, vitreous enamels etc.
- Silicates are also used as *reinforcing glass fibres, chemical wares* and *electrical insulators.*

## 22.8. TYPES OF SILICATE STRUCTURES

The various silicate structures are :

1. Silicon-oxygen tetrahedron $(SiO_4)^{4-}$ structure.
2. Double and poly-tetrahedral structures.
3. Chain structures.
4. Sheet structures,.
5. Framework structures.
6. Vitreous structures.

**1. Silicon-oxygen tetrahedron $(SiO_4)^{4-}$ structure :**

Refer to Fig. 22.1. In this structure (primary structural unit of silicates) one silicon atoms fits interstitially among four oxygen atoms.

*Example:* Forsterite ($Mg_2SiO_4$), a mineral which is a high temperature refractory.

**2. Double and poly-tetrahedral structures:**

Refer to Fig. 22.2.

- This type of structure results when three or more tetrahedral units link together (a ring type structure is produced). One of the oxygens is a member of two units.
- The composition of a three poly-hedral unit is $Si_3O_9$ which produces $(Si_3O_9)^{6-}$ ions.

*Example* : Polysilicates (double tetrahedral structure).

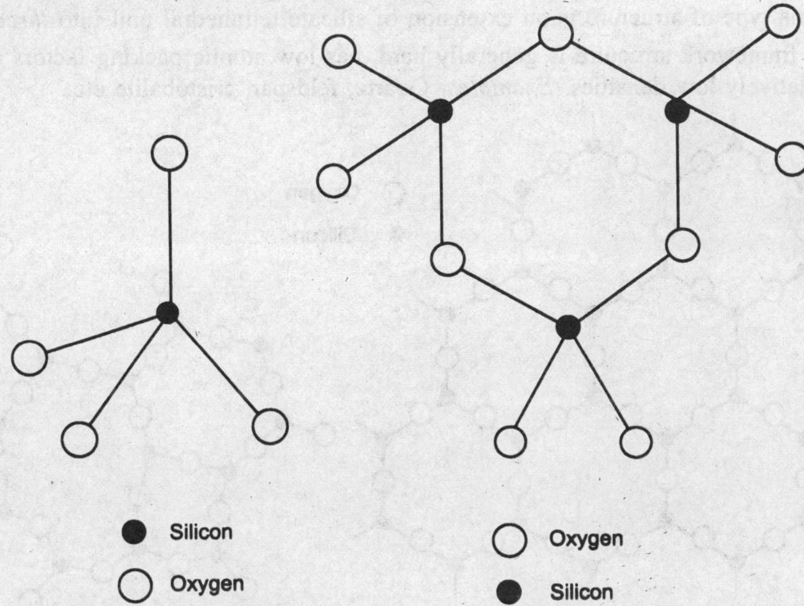


Fig. 22.1 Silicon-oxygen tetrahedron $(SiO_4)^{4-}$ Fig. 22.2 Poly-tetrahedral silicate $(Si_3O_9)^{-6}$

3. **Chain structure:**

- A chain structure is formed when two corners of each tetrahedra are linked.
- A *single chain structure* (Fig. 22.3) can be noticed in ***proxenes***.
- A *double chain structure* (Fig. 22.4) results when two parallel identical chains are polymerized by sharing oxygen to every alternate tetrahedranol. *Example* : Amphiboles.
- Theoretically, the length of these chain structures can be almost infinite.

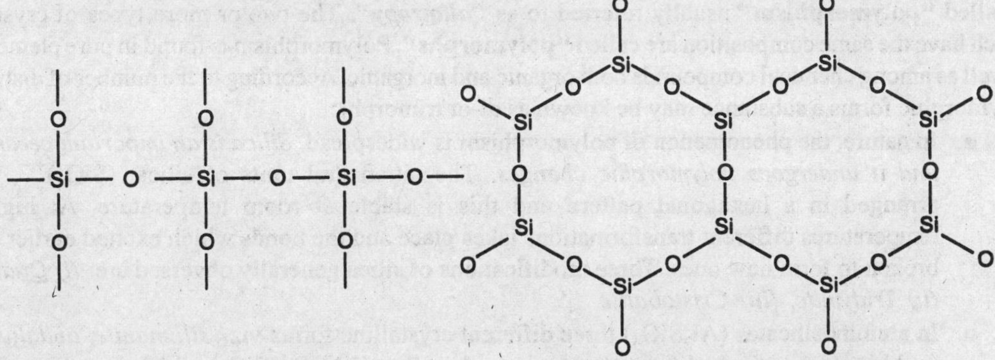


Fig. 22.3. Single chain structure. Fig. 22.4. Double chain structure (Si-silicon, O-Oxygen).

4. **Sheet structure:**

Refer to Fig. 22.5.

- When the double chain structure extends infinitely in a two dimensional plane, a sheet structure results.
- This structural arrangement provides certain important properties; *e.g.* the lubricating characteristics of talc and plasticity of clay, the cleavage of mica etc.
- The sheet structure is found in ceramic materials which as clays, micas and talc.

5. **Framework structure:**

- This type of structure is an extension of silicate tetrahedral unit into *three dimensions.*
- A framework structure is generally hard, has low atomic packing factors and possesses relatively low densities. *Examples:* Quartz, feldspar, cristobalite etc.

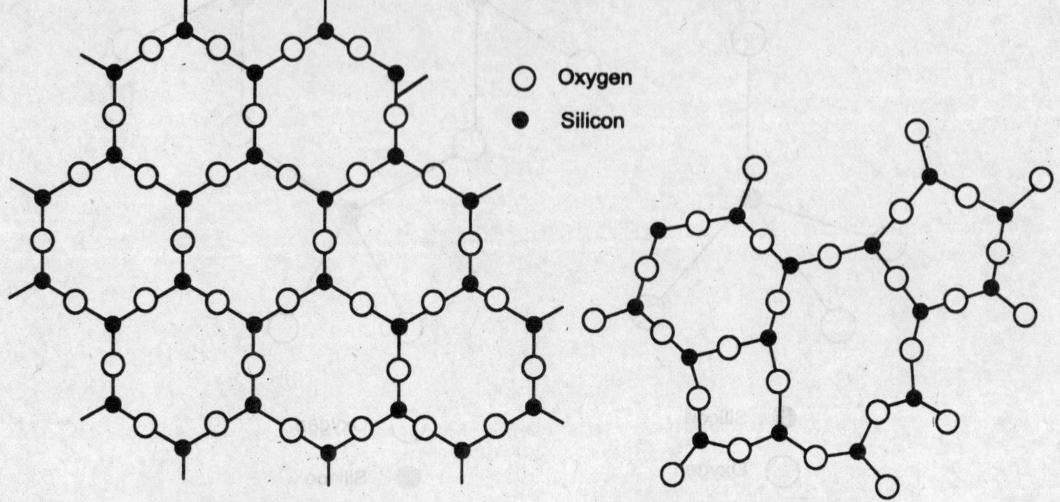


Fig. 22.5. Sheet structure. Fig. 22.6. Vitreous structure.

6. **Vitreous structure:**

- Fig. 22.6 shows a vitreous form of silica (glass) in two dimensions.
- Glass is a vitreous silicate, having a vitreous structure. Glass has a three-dimensional frame work structure containing covalent bonds.

## 22.9. POLYMORPHISM

*Depending on the conditions one and the same substance may crystallize in different forms,* this is called **"polymorphism"** usually referred to as ***"allotropy".*** The two or more types of crystals which have the same composition are called **"polymorphs".** Polymorphism is found in pure elements as well as among chemical compounds both organic and inorganic. According to the number of distinct polymorphic forms a substance may be known as di-or trimorphic.

- In nature, the phenomenon of polymorphism is widespread. *Silica is an important ceramic and it undergoes polymorphic changes.* The tetrahedral units of silica, $(SiO_4)^{4-}$, are arranged in a hexagonal pattern and this is stable at room temperature. At higher temperatures different transformations takes place and the bonds which excited earlier are broken to form new ones. Three modifications of silica generally obversed are: *(i) Quartz, (ii) Tridymite, (iii) Cristobalite.*
- In aluminosilicates $(Al_2SiO_5)$ three different crystalline forms viz., *sillimanite, andalusite and kyanite* are formed from the compound mullite $(3Al_2O_32SiO_2)$ which is stable at the liquidus.
- Similarly, alpha and gamma forms of alumina are obtained from fused alumina.
- Among metals, the most important example of polymorphism is *iron.*
- Among *non-metallic elements, polymorphism is found in phosphorus and carbon.*

*Polymorphic transformations are usually brought about by temperature changes but several have been induced by high pressure or drastic cold work treatment also.*

## 22.10. GLASS – (Refer to Chapter 14.)

## 22.11. ADVANCED CERAMICS

**1. Glass Ceramics:**

- These are special glass compositions that are thermally treated prior to forming operations to *devitrify or precipitate a crystalline phase from the material;* this phase gives that material special properties such as *zero thermal expansion* for applications involving *high thermal-shock application.*
- The compositions (typical of glasses) in which nucleation and crystallization have been commercially produced are : $MgO$-$Al_2O_3$- $SiO_2$; $LiO_2$-$Al_2O_3$-$SiO_2$:; $LiO$-$MgO$-$SiO_2$.

**Characteristics:**

(*i*) Very low coefficient of thermal expansion.

(*ii*) Relatively high mechanical strengths.

(*iii*) High thermal conductivities.

(*iv*) Can be easily fabricated (conventional glass-forming techniques may be employed conveniently in the mass production of nearly pore-free ware).

**Uses:**

(*i*) Owing to their excellent resistance to thermal shock and their high conductivity, glass ceramics are used as *ovenware* and *tableware.*

(*ii*) As insulators.

(*iii*) As substrates for printed circuit boards.

**2. Dielectric Ceramics:**

- The use of ceramic materials is made both as *electrical insulators* and as *functional parts* of an electrical circuit. Since the electrical insulators can breakdown under high electrical voltages, the insulators are *designed with lengthened surface paths to decrease the possibility of surface shorting.* Since internal pores and cracks provide opportunity for additional *surface* failure, the insulators are *glazed to* make them non-absorbent.
- *Non-linear dielectric ceramics* are suitable in the miniaturization of electronic parts which have led to the development of increasingly sophisticated electrical circuitry. ,
- These ceramics are also used in capacitors.
- Some typical non-linear dielectric ceramics are : Lead zirconate-titanate, lead niobates, barium titanate, etc.

**3. Electronic Ceramics:**

Ferrites ferro-electric ceramics etc. are the ceramic materials with unusual properties that are of specific use in electronic circuits.

- *Ferrites are* mixed-metal-oxide ceramics (almost completely crystalline). They assimilate high electric resistivity and strong magnetic properties. *Soft ferrites* can be used for specific uses such as *memory cores for computers* and cores for radio and television loop antennas. *Barium and lead ferrites* are widely used in permanent-magnet motors in automobiles, portable electrical tools and small appliances.
- *Ferro-electric ceramics* can convert electrical signal into mechanical energy (such as sound); and can also change sound, pressure or motion into electrical signals. Thus they function as *transducers.*

***Examples :*** *Barium titanate* (most common), *tantalates, zirconates, niobates,* etc.

**4. Cermets:**

- Cermets are ceramic-metal composites.
- Cermets contain alumina ($Al_2O_3$) and chromium in varying proportions.

- These are used in *brake shoe linings, oxidation-resistant parts and inject engines.*
- The most common cermet is *cemented carbide* and such like composites are extensively used as cutting tools for hardened steels.

## 22.12. REFRACTORIES

### 22.12.1. General Aspects

*The term "**refractories**" embraces all materials used in the arts for the construction of heat resisting containers, using the word in its broadest sense, whether it be to afford space for the evolution of gases in combustion processes or the holding of molten charges or of solids undergoing heat-treatment.* The two principal functions involved in the use of refractory materials are to those of *thermal insulation and conduction.* In the outside walls of a furnace, refractories serve the purpose of *confining the heat and preventing excessive loss to the atmosphere;* in a muffle or retort, they serve to *conduct the heat through the walls to the charge.*

Although, in general, the heat resisting quality of refractories is of paramount importance, this is by no means the only requirement and may at times be only a secondary consideration. Refractories may be expected to be relatively unaffectd by high temperatures under stress whether negligible or heavy; to resist mechanical abrasion at various temperatures; to resist the intrusion of molten metals, slags or metallic vapours, as well the action of superheated steam and hydrocarbons, sulphurous oxide, chlorine or other gases; and to withstand sudden temperature changes. Under one set of conditions, high thermal conductivity may be required and under another high insulation value, while in still other cases high electrical resistance at moderately high temperatures may be demanded. It is obvious that no single refractory will completely fulfil all these functions, and hence the proper selection of a suitable material is often a complicated task.

### 22.12.2. Properties of Refractories

A good refractory material should have the following *properties*:

1. It should be able to withstand high temperatures generated in the furance.
2. It should be able to withstand sudden alternating heating and cooling, *i.e.*, thermal shocks.
3. It should be able to withstand abrasion and rough usage.
4. Its contraction and expansion due to the inevitable temperature variation should be minimum possible.
5. It should be able to withstand fluxing action of the slags and the corrosive action of gases.
6. It should have good heat insulating properties.
7. It should be chemically inactive at elevated temperatures.
8. It should be impermeable to gases and liquids as far as possible.
9. If used in electric furnaces, it must have low electrical conductivity.

### 22.12.3. Types of Refractories

According to their chemical behaviour, the refractory materials which are mainly oxides or carbides of metals or non-metals or their mixtures, are classified as follows:

1. Acid refractories.
2. Basic refractories.
3. Neutral refractories.

### 22.12.4. Acid Refractories

***"Acid refractories"** are those which really combine with bases and are therefore, termed "Acid". They are materials consisting of silica or containing silica as their chief constituent greatly* in excess of the bases present. The important acid refractories are *quartz, sand, ganister, dinas* rock etc. Most fire-clays contain $SiO_2$ in excess of the amount indicated by the formula $Al_2O_3.\ 2SiO_2.\ 2H_2O$ and are therefore, classed as acid.

**Silica** : When pure silica fuses at a very high temperature, viz., 1713°C, but when heated in contact with basic substances it forms silicates, some of which fuse easily. Therefore, the presence of bases in silica to be used as refractory materials is to be guarded against. Silica occurs as quartzite, Dinas rock and ganister. *Ganister* is a naturally occurring high siliceous rock (98% $SiO_2$). In the prepared state it is used in the form of bricks. Good deposits of silica in India occur in the Gaya and Monghyr districts of Bihar.

*Silica bricks* are made from hard, dense fine grained quartzite (known as ganister). After crushing, the rock is ground with water and mixed with about 2 percent lime. From this mixture, bricks may be moulded by hand or by power press; after drying in hot air, they are baked for proper time and at the proper temperature. The lime being base reacts chemically with acidic $SiO_2$ and fuses at a number of places in the brick and gives its strength. The bricks expand 3.5 percent after the baking is finished.

*Silica bricks* possess the following properties:

1. Low porosity and free from air pockets.
2. Good thermal expansion and conductivity.
3. Remarkable load bearing capacity especially at high temperatures.
4. Not susceptible to thermal spalling at temperatures above 400°C.
5. Ability to withstand thermal shock poor.

Silica bricks are *extremely suitable* for those parts of a furnace which are subjected to *uniformly high temperature, e.g.,* fire bridges, roofs of Acid and Basic Open and Hearth, Electric furnaces, copper smelting and refining furnaces etc.

**Fire Clay** : The main constituent of *fire-clay is kaolinite,* $Al_2O_3 . 2SiO_2 . 2H_2O$. The fire clays, however, contain varying proportions of alumina and silica. Impurities in fire clays such as alkalies, sand, oxides of iron and silicates of calcium and magnesium adversely affect their properties. At high temperature the fire clays lose their water of hydration and mainly consist of alumina and silica.

The fireclay *bricks are manufactured in the same manner as silica bricks.* The clay in a finely crushed state is mixed with a definite amount of water in a pugmill. The mixture is then pressed into moulds, dried and finally burnt. To control the shrinkage, and accelerate the rate of production about 20 to 80% of burnt or calcined clay called **Grog** is mixed in, the clay, while using flint clay, plastic clay is employed as binder. By altering the properties of flint clay and plastic clay in the brick mixture the properties of the bricks made can be controlled.

Important properties *of fire-clay bricks* are:

1. Coefficient of thermal expansion low.
2. Sufficient *strength* at high temperature.
3. Resist spalling.
4. Not so refractory as silica bricks but much cheaper.

Fire clay bricks are *used* for the following:

1. Linings of blast furnaces for the melting of iron, copper, lead ores etc.
2. The linings of flues and shocks.
3. Heat treatment furnaces.
4. Glass furnaces.
5. Checker-work of regenerative furnaces.
6. Reverberatory furnaces.
7. Pottery kilns.

### 22.12.5. Basic Refractories

***"Basic refractoris" consists mainly of basic oxides without free silica and resist the action of bases.*** The most common basic refractories are *Magnesite, Dolomite, lime* etc. But lime, because of its dehydrating tendency is never used as a *refractory.*

**Magnesite**: This is most important basic refractory material. It is made by "dead burning" the mineral magnesite which is essentially magnesium carbonate. At any temperature above about 700°C, magnesite is dissociated to magnesia and carbon dioxide, but the properties of the resultant product differ grealy according to the temperature of burning. Below 900°C the product is "caustic magnesia", which is readily hydrated to magnesium hydroxide, and is used in cement. Microscopic examination shows no crystals, so that it is often called amorphous magnesia. Prolonged firing at 1800°C gives an insoluble, unreactive "dead burnt" magnesite, easily recognized as the crystalline mineral.

In the manufacture, magnesite is first "dead burnt" to remove shrinkage as much as possible, then crushed, graded for minimum porosity, bonded, and moulded under hydraulic pressure. Bonding is done by adding an organic binder, or by use of a little caustic magnesia, which sets to a hydrated cement with water, or the crushed, deal burnt magnesite itself is allowed to stand in water for some days until the surface of the particles has undergone hydration.

*Magnesite possesses the following properties* :

1. Thermal conductivity greater than that for fire-clays and silica.
2. Highly resistant to the action of basic slags and iron oxide.
3. Starts losing strength at temperatures above 1500°C.

Magnesite bricks are not capable of resisting sudden changes of temperature and show a tendency to spall under such conditions. They are constlier than silica and fireclay bricks and used in the construction of those parts of a furnace which are required to *withstand the corrosive action of basic slags.*

They are *used* for:

1. Hearths of basic open-hearth and copper reverberatory furnaces.
2. Electric arc and induction furnaces.
3. Lining of L.D. crucible (in the manufacture of steel).

**Dolomite:** It is a *double carbonate of calcium and magnesium* and occurs abundantly in India. It is burnt generally in shaft furnaces. Dolomitic bricks have low conductivity and refractoriness as compared to magnesite bricks. Calcium oxide present in the bricks has a tendency to combine with water and carbon dioxide present in the atmosphere, and causes disintegration of the bricks. They are cheaper than magnesite bricks but are poor in performance. It is *extensively used for repairing the banks and the bottoms of the Basic Open-hearth furnaces.*

Burned bricks made from dolomite have not been successful in service. It is possible to produce brick that will not hydrate under reasonable treatment, but the stabilizing agents added cause high shrinkage when the brick is heated to high temperature. Such bricks are also sensitive to spalling influences.

### 22.11.6. Neutral Refractories

***"Neutral refractories"*** *are substances which do not combine with either acidic or basic oxides* and for this reason constitute the *most satisfactory furnace lining, e.g.,* chromite and carbon in graphite form.

**Chromite** : Chromite refractories are made from core ores which mainly contain chromite, $FeO.Al_2O_3$. The ore is finely powdered and mixed with a small amount of binding material, such as fire-clay, bauxite and magnesia. The mixture is moulded into bricks, dried and burnt in a kiln at a temperature of about 1500°C to 1700°C.

Chrome bricks possess the following *properties* :

1. Low porosity.
2. High resistant to both acid and basic slags

3. High thermal conductivity.
4. Low resistance to spalling.

Chrome bricks are extensively used in steel industry for lining open hearths and also in the bottom of soaking pits.

**Graphite** : It is *a form of pure carbon and is natural product.* It occurs mixed with calcareous and siliceous rocks. It is employed in making graphite crucibles using clay as the bond.

Graphite bricks have the *following properties :*

1. Reduced oxidising action (on the metal) at high temperatures.
2. Resist acid and basic slags.
3. Free from scaffolds formation due to the adherence of metals to the lining.

**Kaynite** : Its composition is $Al_2O3$. $SiO_2$ (Disthene) and crystals are triclinic. It decomposes at a temperature of 1350°-1330°C. Its, hardness is 5 to 7 mhos.

**Sillimanite** : It occurs naturally in Assam and Rewa in India. Its composition is $Al_2O_3$. $Si_2$ (fibrolite) and occurs as long needle-shaped cyrstals. It has high softening point (above 1800°C). It has very low electrical conductivity. It is employed for making blocks, bricks, crucibles, refractory fittings for electrical goods and tubes for surface combustion.

In the Table 22.1 are shown the common refractory materials with their chemical composition, melting point etc.

**Table 22.1. Common Refractory Materials**

| *Name* | *Formula* | *Composition* | *Melting point* | *Class* |
|---|---|---|---|---|
| Silica | $SiO_2$ | 100% $SiO_2$ | 1713°C | Acid |
| Kaolin (Fire-clay) | $Al_2O_3 . 2SiO_2$ $2H_2O$ | 4.50% $Al_2O_3$ 54.1%$SiO_2$ | About 1775°C | Acid (when pure) |
| Sillimanite | $Al_2O_3 . SiO_2$ | 63% $Al_2O_3$ 37% $SiO_2$ | About 1800°C | Acid |
| Mulite | $3Al_2O_3 . 2SiO_2$ | 72% $Al_2O_3$ 28% $SiO_2$ | 1900°C | Acid |
| Alumina (Corrundum) | $Al_2O_3$ | 100% $Al_2O_3$ | 2050°C | Acid |
| Lime | CaO | 100% CaO | 2570°C | Basic |
| Magnesia | MgO | 100% Mgo | 2800°C | Basic |
| Chromite | $FeO . Cr_2O_3$ | 32% FeO 68% $Cr_2O_3$ | 2180°C | Neutral |
| Zirconia | $ZrO_2$ | 100% $ZrO_2$ | 2700°C | Neutral |
| Graphite | C | 100% C | 3000°C | Neutral |

### 22.11.7. Refractory Fibres

- These are made from *high purity alumina and silica grains melted in electric furnace and blasted by high velocity gases into light fibres.*
- Ceramic fibres are available in the form of felts, blocks and blankets.
- The heat losses through ceramic fibres are low compared to other refractory materials.

- They have *excellent temperature resistance, outstanding thermal stability and good resistance to vibration.*
- Refractory fibres, in general, are considered in four broad categories namely :
  (*i*) Alumina silica fibres,
  (*ii*) Pura silica fibres,
  (*iii*) Pure metal oxide fibres,
  (*iv*) High silica leached fired glass fibres.

**USES:**

1. Because of *low thermal conductivity* the refractory ceramic fibres are fast replacing and are used for *lining of furnaces, kilns and other high temperature equipments.*
2. *Silica fibres* are very expresive, hence, limited industrial use. They find applications in *aerospace industry.*
3. *Alumina silica fibres* are generally used from 1000 to 1500°C temperature because of their low cost and good thermal properties (they have average diameter ranging from 1.5 to 3.0 microns).
4. *Metallic oxide, alumina fibres* are used for 1500 to 1600°C.

## 22.13. CEMENTS AND CONCRETE (Refers to Chapter 7 & 8)

### HIGHLIGHTS

1. *Ceramic Materials* are defined as those containing phases that are compounds of metallic and non-metallic elements.
2. *Ionic bonds* give ceramic materials relatively high stability. 'Covalent crystals' usually also possess high hardness, high melting point and low electrical conductivity at room temperature.
3. The silicates are coordinate structures based upon large anions arranged about small cations. Portland cement is the most widely known silicate.
4. Depending on the conditions one and the same substance may crystallize in different forms, this is called "polymorphism" usually referred to as allotrophy.
5. The germ 'glass', as ordinarily used refers to material which is made by fusion of mixture of silica, basic oxides and a few other compounds that react either with silica or with the basic oxides.
6. The term *'Refractories'* embraces all materials used in the arts for the construction of heat resisting container, using the word in its broadest sense, whether it be to afford space for the evolution of gases in combustion processes or the holding of molten charges or of solids undergoing heat-treatment.
7. *Refractories* are classified as follows:
   (*i*) Acid refractories (*ii*) Basic refractories
   (*in*) Neutral refractories.
8. Abrasives are substances or objects that are employed to wear off or rub down objects, in order to give them desired size, shape or finish.
9. Cements may be natural or artificial. There are two processes employed for the manufacture of cement: (*i*) Wet process, and (*ii*) Dry process.
10. Concrete is a mixture of cement (or lime), sand, brick or stone ballast and water, which when placed in forms and allowed to cure, becomes hard like stone.

## OBJECTIVE TYPE QUESTIONS

**Fill in the Blanks/Say 'Yes' or 'No' :**

1. ________ materials are defined as those containing phases that are compounds of metallic and non-metallic elements.
2. The science of ceramics is the processing of earthly materials by heat.
3. 'Ionic bonds' give ceramic materials relatively low stability.
4. The ____ are coordinate structures based upon large anions arranged about small cations,
5. ________ cement is the most widely used known silicate.
6. The two or more types of crystals which have the same composition are called ________.
7. ________ is the ratio of the capacitance of a dielectric compared to the capacitance of air under the same condition.
8. Dielectric strength is the ability of the material to withstand electrical breakdown.
9. Glass is weak in tension and it always fails in tension no matter how stress is applied.
10. ________ glass is used in situation where light is required without transparency.
11. The ________ rocks are formed by cooling of molten lava on or inside earth' *s* surface during the volcanic eruption.
12. Stratified rocks are those which do not show any sign of strata and cannot be easily split into slabs
13. In ________ rocks clay or alumina in the main constituent.
14. ________ rocks have calcium carbonate or lime as their leading constituent.
15. Marble is used for carving and decoration work.
16. The main constituent of fire clay is kaolinite.
17. ________ bricks are made from hard, dense, fine grained quartzite.
18. ________ refractories consist mainly of basic oxides without free silica and resist the action of bases.
19. Chrome bricks are extensively used in steel industry for lining open hearths and also in the bottom of soaking pits.
20. ________ are substances or objects that are employed to wear off or rub down objects, in order to give them desired size, shape or finish.
21. Corrundum is least used for the manufacture of wheels used in the metal industry.
22. Emery is a natural aluminium oxide.
23. Quartz is mostly used for making sand paper, scouring powder and soap.
24. ________ cement exhibits increase in volume on hardening.
25. ________ resisting Portland cement is used at places where sulphate action is severe.

## ANSWERS

| | | | | |
|---|---|---|---|---|
| **1.** Ceramic | **2.** Yes | **3.** No | **4.** Silicates | **5.** Portland |
| **6.** Polymorphs | **7.** Dielectric constant | **8.** Yes | **9.** Yes | **10.** Ground |
| **11.** Igneous | **12.** No | **13.** Argillaceous | **14.** Calcarious | **15.** Yes |
| **16.** Yes | **17.** Silica | **18.** Basic | **19.** Yes | **20.** Abrasives |
| **21.** No | **22.** Yes | **23.** Yes | **24.** Expanding | **25.** Sulphate |

## THEORETICAL QUESTIONS

1. What are ceramic materials ? Name some of important ceramic materials.
2. State the advantages of ceramic materials.
3. Explain briefly the electrical properties of ceramic materials.
4. How is glass manufactured ?
5. State briefly the different varieties of glass.
6. How are rocks classified ?
7. State the common rock forming minerals.
8. What are the characteristics of good building stones ?
9. Name the rocks to which the following stones belong:
   (*i*) Granite (*ii*) Slate
   (*iii*) Chalk (*iv*) Kankar.
10. Define the term 'refractories'. State their properties.
11. How are the refractories classified ?
12. State the important properties of the fire clay bricks. Differentiate between ordinary fire clay bricks and highly refractory fire clay bricks.
13. Explain briefly the following common refractory materials: Kaolin, Alumina, Magnia and Graphite.
14. What are 'Abrasives' ?
15. How are abrasives classified ?
16. Explain briefly the following abrasives : Corrundum, silicon carbide and aluminiuim oxide.
17. What is 'cement' ?
18. How are cements classified ?
19. Explain briefly the 'wet process' for the manufacture of Portland cement.
20. Enumerate various types of Portland cement.
21. Write a short note on 'storage of cement'.
22. What is concrete ?
23. What are the constituents of concrete ?
24. What are the functions of concrete constituents ?
25. What is 'Reinforced cement concrete' ?
26. What are the advantages and properties of 'Reinforced cement concrete' ?
27. Explain briefly the principle of hardening of cement concrete.

❑❑❑

## *SECTION :*
## SHORT ANSWER QUESTIONS

**Q. 1.** *What is the use of studying 'engineering materials'?*

**Ans.** 'Engineering materials' subject deals with the study of materials in respect of the

– Sources, composition and properties;

– Manufacturing methods and testing;

– Utility in the various fields of engineering and technology;

– Modern techniques being developed for handling and using materials to materialise economical and safer design of structures and machines etc.

**Q. 2.** *How are engineering materials classified?*

**Ans.** The engineering materials may broadly *classified* as follows:

1. *"Civil engineering materials"*:

*Examples*: Building stones, bricks and clay products, lime cement, concrete, mortar timber, etc.

2. "Electrical engineering materials":

*Examples*: "Conductors" (aluminium, iron, steel, etc.); "Semiconductors" (silicon, germanium, etc.), "Insulator" (Asbertas, baketite, mica, varnishes, air, etc.); "Magnetic materials" (Iron, cobalt, etc.)

3. *"Mechanical engineering materials"*:

*Examples*: Cast iron, steel, lubricating materials, etc. These may also be classified as:

(*i*) Metals; (*ii*) Non-metals; (*iii*) Metals and alloys; Ceramics; Organic polymer.

**Q. 3.** *Enumerate the various properties of 'building materials'.*

**Ans.** The properties of building materials may be classified as follows:

(*i*) Mechanical properties; (*ii*) Chemical properties;
(*iii*) Electrical properties; (*iv*) Magnetic properties;
(*v*) Optical properties; (*vi*) Thermal properties;
(*vii*) Physical properties.

**Q. 4.** *From where do we get stones?*

**Ans.** The ***stones*** are *derived from rocks* which form the earth's crust and have no definite shape or chemical composition but are *mixtures of two or more minerals.*

- The *"mineral"* is a substance which is formed by the natural inorganic process and possesses a definite chemical composition and molecular structure

**Q. 5.** *How are 'igneous, sedimentary and metamorphic rocks' formed?*

**Ans.** ***Igneous rocks*** are formed by *cooling the molten lava on or inside the earth's surface during the volcanic eruption.*

***Sedimentary rocks*** are formed by gradual deposition of disintegrated rocks (due to the atmospheric action such as rain, wind and temperature), *vegetable matter and clay at the bottom of rivers, lakes or sea.* These are also called *stratified* rocks because these are formed in layers.

***Metamorphic rocks*** are the rocks that are formed when *sedimentary or even igneous rocks are subjected to great heat and pressure inside the earth.*

**Q. 6.** *Name the common rock forming minerals.*

**Ans.** The common rock forming minerals are:

Silica; Felspar; Mica; Hornblende; Calcite; Dolomite.

**Q. 7.** *Enumerate the various agencies which bring about the deterioration of stones.*

**Ans.** The various agencies causing deterioration of stones are:

Rain; Temperature; Wind; Frost; Atmospheric impurities; Vegetable growth; Living organism; Movements of chemicals.

**Q. 8.** *What test are used to determine the suitability of stone for its use in engineering works?*

**Ans.** The various test used are:

(*i*) Hardness test
(*ii*) Crushing test
(*iii*) Impact test
(*iv*) Five resistance test
(*v*) Attrition test
(*vi*) Acid test
(*vii*) Electrical resistance/water absorption
(*viii*) Smiths test
(*ix*) Crystallization test
(*x*) Microscopic test
(*xi*) Freezing and thawing test.

**Q. 9.** *What is 'natural bed of stone'?*

**Ans.** ***Natural bed of stone*** is the *original position occupied by the stratified rock from which the stone is obtained.*

**Q. 10.** *What is the difference between 'quarrying' and 'quarry'?*

**Ans.** ***Quarrying*** is an *art of extracting* from the rock beds, the stones of different varieties as used for general building work and broken stones for roads and concrete work, etc.

***Quarry*** is the place from which the stone is obtained (by digging or blasting).

**Q. 11.** *Enumerate the different methods of quarrying.*

**Ans.** The various methods of quarrying are

(*i*) By digging;
(*ii*) By heating;
(*iii*) By wedging
(*i*) By blasting.

**Q. 12.** *What do you mean by 'dressing of stones'?*

**Ans.** The art of cutting the stones to shape required for use in structure is known as ***dressing.***

**Q. 13.** *What is 'kaoline'?*

**Ans.** 'Brick earth' is derived by the *disintegration of igneous rocks.* Potash felspars, orthoclase or microlime are mainly responsible for fielding clay minerals in the earth. The mineral decomposes to yield 'kaolinite' a silicate of alumina which on hydration gives a clay deposit $Al_2 O_3 . 2H_2O$ known as ***kaoline.***

**Q. 14.** *What is the composition of a good brick earth?*

**Ans.** A good brick earth contains the following:

(*i*) Alumina or clay = 20 – 30 percent by weight.

(*ii*) Silica or sand = 35 – 50 percent by weight.

(*iii*) Silt = 20 – 25 percent by weight.

The total content of clay and silt may preferably be not less than 50 percent by weight.

(*iv*) Remaining ingredients : Iron oxide, Magnesia (Mgu), Lime (CaO), Sodium potash, etc.

**Q. 15.** *What are the harmful ingredients in brick earth?*

**Ans.** The harmful ingredients in brick earth are:

Limestone and kanbasn modules; Alkation, Pebbles of stone and gravel; Iron pyrites; Kalbar or reh; Vegetation and organic matter.

**Q. 16.** *Name the tests which are generally performed to determine the suitability of the brick earth.*

**Ans.** The various tests performed on the brick earth are:

(*i*) Consistency test;

(*ii*) Test for moulding properties;

(*iii*) Test for deformation and shrinkage on burning.

**Q. 17.** *Enumerate the steps involved in the manufacture of bricks.*

**Ans.** The manufacture of bricks involves the following steps:

(*i*) Preparation of clay;

(*ii*) Moulding of bricks;

(*iii*) Drying of bricks;

(*iv*) Burning of bricks.

**Q. 18.** *Name the various types of kiln in which the bricks may be burnt.*

**Ans.** The bricks may be burnt in any one of the following kilns:

(*i*) Clamp or open kilns;

(*ii*) Intemitten kilns;

(*iii*) Continuous kilns:

– Bull's trench kiln

– Hoffman's kiln

– Tunnel kiln

**Q. 19.** *What are the tests to which a brick is generally subjected to find its suitability for the construction work?*

**Ans.** The various tests are:

(*i*) Water absorption test

(*ii*) Compressive strength test

(*iii*) Efflorescence test

(*iv*) Structure test

(*v*) Shape and size test

(*vi*) Soundness test

(*vii*) Hardness test

**Q. 20.** *What are the various varieties of refractory bricks?*

**Ans.** The various types of refractory bricks are:

(*i*) *Acid bricks:* (*a*) Fire bricks; (*b*) Silica bricks

(*ii*) *Basic bricks:* (*a*) Magnesite bricks; (*b*) Dolomite bricks; (*c*) Bauxite bricks

(*iii*) *Neutral bricks:* (*a*) Chrome bricks; (*b*) Chrome-magnetsite bricks; (*c*) Spinel; (*d*) Forsterite bricks

**Q. 21.** *What are building tiles?*

**Ans.** ***Tiles*** are thin slabs of brick earth, burnt in kiln.

**Q. 22.** *What is 'tera-cotta'?*

**Ans.** ***Tera-cotta*** is a kind of *high quality earthenware* which is used as a substitute for stone for *ornamentation of buildings*.

**Q. 23.** *Define the term 'glazing'.*

**Ans.** ***Glazing*** is the process of *forming some transparent film* over the surfaces of bricks, tiles, earthenware or stoneware to improve upon their appearance and also to protect them from the action of weather, sewage, chemicals or other destroying agents.

Glazing may be transparent like glass or opaque like enamels.

**Q. 24.** *How is 'lime' produced?*

**Ans.** ***Lime*** is produced by *heating limestone* which is move or less pure 'calcium carbonate'. During heating carbon-dioxide is driven off and non-volatile impurities such as oxides of silicon, magnesium, aluminium, iron and manganese are left behind in the line.

**Q. 25.** *State the properties of good quality lime.*

**Ans.** A good quality lime entails the following properties:

(i) Easily workable

(ii) Provide strength to the masonry

(iii) Possesses good plasticity

(iv) Offers good resistance to moisture

(v) Stiffness early

(vi) Line masonry proves durable due to low shrinkage in drying.

**Q. 26.** *What is 'quick lime'?*

**Ans.** The lime obtained by *calcining* (the process of heating limestone to redness in contact with air) *pure limestone is called* ***Quick lime*** (CaO).

**Q. 27.** *What is 'slaked lime'?*

**Ans.** ***Slaked lime*** hydrate of lime is formed by the *absorption of water by quick lime* and is the hydrated oxide of calcium, $Ca(OH)_2$.

A plastic mass of lime which results from the slaked lime in the presence of a sufficient quantity of water is called ***putty.***

**Q. 28.** *Define the term 'slaking'.*

**Ans.** ***Slaking*** is the process of *chemical combination* of quick lime with required quantity for *billing the heat and getting into powder form for use.*

$$CaO + H_2O = Ca\,(OH)_2$$

(Quick lime) (Hydrate of lime)

**Q. 29.** *What do you mean by the term 'hydraulicity'?*

**Ans.** ***Hydraulicity*** is the property due to which lime (or current) will *set in damp places* or under water or in thick masonry walls where there is no free access of air.

**Q. 30.** *What is 'pozzolona'?*

**Ans.** ***Pozzolona*** is the volcanic dust containing 80 percent clay with lime, magnesia and iron oxide in varying proportion.

**Q. 31.** *Enumerate the constituents of limestones.*

**Ans.** Following are the constituents of limestones/main source of getting lime:

(i) Clay

(ii) Soluble silica

(iii) Magnesium carbonate

(iv) Alkalies and metallic oxide

(v) Sulphates

(vi) Iron compounds

(vii) Carbonaceous matter.

**Q. 32.** *How are limes classified?*

**Ans.** Limes are classified as follows:

(i) Fat lime

(ii) Hydraulic lime

(iii) Poor lime

**Q. 33.** *List the operations involved in the manufacture of lime.*

**Ans.** Following operations are involved in the manufacture of lime:

(i) Collection of raw materials

(ii) Burning of limestone

(iii) Slaking of burnt lime

– Air slaking

– Basket slaking

– Platform slaking

– Tank slaking

**Q. 34.** *Name the tests which are generally performed on limestones.*

**Ans.** The various tests performed on limestones:

(*i*) Physical test

(*ii*) Chemical tests:

– Heat test

– Hydraulic test-slaking

– Acid test.

**Q. 35.** *What are 'pozzolinic materials'?*

**Ans.** ***Pozzolonic materials*** are the siliceous and aluminous mineral substances which do not possess cementitious properties themselves but *form cementitious compounds* by reacting with lime in the presence of water.

– 'Pozzolonic' is a *volconic ash* 80 per cent clay.

**Q. 36.** *Name the various types of pozzolonic materials.*

**Ans.** Various types of pozzolonic materials are:

(*i*) Natural occuring pozzolonic:

– Clay and shales, (viz, kaolinite etc.)

– Opolime materials, (viz, opalime shalers, etc.)

– Volcanic tuffs and pumicites (viz., Rhyolite, bide site and Basaltic types.)

(*ii*) Atificial pozzolannas:

– Flyash

– Burnt clay or ground bricks

– Slags.

**Q. 37.** *What is 'cement'?*

**Ans.** ***Cement*** may be prescribed as a ***material with adhesive and cohesive properties which make it capable of bonding mineral fragments into a compact whole.*** The definition embraces a large variety of cementing materials.

For 'constriction purposes' the meaning of the term cement is restricted to the *bonding materials* used with stones, bricks, building blocks etc. The *principle constituents* of this type of cement are *compounds of lime,* so that in building and civil engineering we are concerned with calcareous cement. The cement of interest in the making of *concrete* have property of setting and of interest in *under water* by *virtue of a chemical reaction with it* and therefore called *"hydraulic cement"*.

– Cements may be classified as *"Natural cement"* and *"Artificial cement"*.

**Q. 38.** *What is 'Portland cement'?*

**Ans.** ***Portland cement*** is so named because a paste of cement and water, after it *sets hard, resembles in colour and hardness a 'Portland stone', a limestone quarried in Dorset.*

**Q. 39.** *What are the properties of a good cement?*

**Ans.** The properties of a good cement are:

(*i*) Provides strength to masonry

(*ii*) Stiffens and hardens early

(*iii*) Possesses good plasticity

(*iv*) An excellent building material

(*v*) Easily workable

(*vi*) Good moisture-resistant.

**Q. 40.** *Enumerate the chemical constituents of cement.*

**Ans.** Percentage of various ingredients for the manufacture of Portland, cement should be as follow:

Lime (CaO), 63%; Silica ($SiO_2$) = 22%; Alumina ($Al_2O_3$) = 6%; Iron oxide ($Fe_2O_3$) = 3%;

Magnesium oxide (MgO) = 2.5%; Sulphur trioxide ($SO_3$) = 1.75%; Loss on ignition = 1.5%; Insoluble residue = 0.25%.

- Besides the above ingredients, calcium sulphate ($CaSO_4$), commonly known as *gypsum is also added by 3 to 4%* during grinding process for *controlling the initial setting time of cement* (larger to proportion of gypsum, more will be the initial setting time and vice versa).

**Q. 41.** *What 'steps' are involved in the manufacture of 'Portland cement'?*

**Ans.** The following two processes are employed for the manufacture of the Portland cement:

1. Wet process; 2. Dry process

Both the processes involve the following ***steps*** in the manufacture of Portland cement:

(*i*) Collection of raw materials
(*ii*) Crushing, grinding and mixing of raw materials
(*iii*) Burning
(*iv*) Grinding of clinker.

**Q. 42.** *What do you mean by 'cement hydration'?*

**Ans.** The chemical reaction between cement and water is called *hydration of cement* or simply ***cement hydration.***

**Q. 43.** *What is the difference between 'setting of cement' and 'hardening of cement'?*

**Ans.** When cement is mixed with water (varying from 25 to 35 percent by weight), a *stiff* and *sticky paste* is formed which remains plastic for a short period. With the passage of time, the plasticity gradually disappears and the cement paste becomes stiff *due to initial hydration of cement. This phenomenon by virtue of which the plastic cement changes into a solid of mass* is known as **setting of cement.**

On setting, the cement binds the aggregates into a solid mass which gains strength as the time clapses, till the hydration of cement is complete. Thus, the *phenomenon by virtue of which the cement paste,* which is finally set, *develops strength* is known as ***hardening of cement.***

**Q. 44.** *Name the tests (laboratory) which are carried out for normal setting of Portland cement.*

**Ans.** The various tests conducted are:

(*i*) Chemical composition
(*ii*) Fineness test
(*iii*) Consistency test
(*iv*) Setting time test
(*v*) Soundness test
(*vi*) Tensile strength test
(*vii*) Compressive strength test.

**Q. 45.** *What are 'admixtures'?*

**Ans.** ***Admixtures*** are the *materials which are added in cement mortar or concrete to improve upon their quality.* Various admixtures under different trade names are available in the market.

**Q. 46.** *How cement should be stored?*

**Ans.** Cement should be stored in *special water-tight shed* with a dry damp-proof floor, water-proof wall and leak-proof roof on the top. Cement bags should be stached close together to *avoid free circulation of air.*

**Q. 47.** *What is 'mortar'?*

**Ans.** ***Mortar*** is a *mechanical mixture in varying proportion of a binding material* like cement or lime and an *inert material* or *fine aggregate* like sand, surktri, etc.

Mortars are normally named according to the binding material used in their preparation *e.g.,* cement-mortar, lime-mortar, etc.

**Q. 48.** *What are the functions of mortar?*

**Ans.** The functions of mortar are:

(*i*) It binds together stones or bricks properly.

(*ii*) In any concrete, it holds coarse aggregates together.

(*iii*) In stone masonry and brick masonry, it fills up empty joints; a thin liquid mortar used for such purposes is termed as 'grout'.

(*iv*) It does pointing or plastering to the structure.

**Q. 49.** *What are the ingredients of mortars?*

**Ans.** The ingredients of various mortars used for different engineering purposes are:

(*i*) Binding or cementing materials such as cement and lime;

(*ii*) Fine aggregates such as sand, surktri, ashes, cinder, etc.

(*iii*) Water.

**Q. 50.** *What are the various types of mortars?*

**Ans.** In general mortars are classified as follows:

(*i*) Cement mortars

(*ii*) Lime mortars

(*iii*) Lime-cement mortars (or composite or gauge mortar);

(*iv*) Special mortars: Mud mortars, cement-clay mortar, hight and heavy mortars, decorative mortars, gypsum motars, etc.

**Q. 51.** *How are lime mortars classified?*

**Ans.** Lime mortars are classified as:

(*i*) Non-hydraulic lime mortars

(*ii*) Hydraulic lime mortars

(*iii*) Block mortars

**Q. 52.** *What is 'Plastering'?*

**Ans.** ***Plastering*** is an art of covering rough walls and uneven surfaces in the construction of houses and other structures with a plastic composition, called *"plaster"*.

Plastering, in general, is not required for good brickwork, made of well burnt bricks or neat stone masonry constructed of durable stones and good mortar.

**Q. 53.** *Name the defects in plaster work.*

**Ans.** Some of the defects which usually occur in a plaster are:

(*i*) Cracks

(*ii*) Falling of plaster

(*iii*) Blistering or blowing of plaster

(*iv*) Efflorescence.

**Q. 54.** *What do you mean by the term 'Pointing'?*

**Ans.** ***Pointing*** is the art of finishing the mortar joints of the walls or similar structure with either cement mortar or lime mortar in order to protect the joints from atmospheric agencies and also to improve the appearance of the structure.

There are various types of pointing such as flush point pointing 'recrossed joint pointing', keyed joint tuck pointing, etc.

**Q. 55.** *What is 'concrete'?*

**Ans.** ***Concrete is mixture of cement (or lime), sand brick or stone ballast and water, which when placed in forms and allowed of cure, becomes hard like stone.***

- Cement when used by itself is known as *"mass concrete"*.
- Concrete is *much stronger in compression then tension.*

**Q. 56.** *What are the ingredients of plain cement concrete?*

**Ans.** The various ingredients of plain cement concrete are:

(*i*) Cement

(*ii*) Sand (fine aggregate)

(*iii*) Coarse aggregates (broken stone, gravel, broken brick, etc.)

(*iv*) Water

**Q. 57.** *What do you understand by 'Bulking of sand'?*

**Ans.** *The increase in volume of sand due to presence of surface moisture upto some extent is called* ***bulking of sand.***

- The extent of bulking depends on the percentage of moisture present in the sand and on its fineness.

**Q. 58.** *What is meant by 'Water-cement ratio'?*

**Ans.** ***Water-cement ratio*** is the *ratio of water to cement* and is expressed as the ratio of the weight or volume of *water* to the weight or volume of *cement* in concrete mixture.

**Q. 59.** *What is 'proportioning of concrete'?*

**Ans.** The process of *determining the proportion of cement or fine aggregate and coarse aggregate for concrete mix is called* ***proportioning of concrete.***

The various methods of proportioning concrete are:

(*i*) Arbitrary method

(*ii*) Trial mixtures methods

(*iii*) Minimum voids method

(*iv*) Fineness modulus method.

**Q. 60.** *Enumerate the methods which affect workability of fresh concrete.*

**Ans.** The following factors affect the workability of fresh concrete:

(*i*) Mix proportions

(*ii*) Aggregate properties

(*iii*) Environmental conditions

(*iv*) Time.

**Q. 61.** *What is 'curing of concrete'?*

**Ans.** ***Curing*** is the process of *keeping the set concrete damp for some days* in order to enable the concrete *gain more strength.*

– For 'Portland cement', the curing period is about 7 to 14 days.

– In cold weather, the concrete should be cured for at least 14 days.

**Q. 62.** *Enumerate the various methods of curing.*

**Ans.** Depending upon the type of construction work, the concrete is cured by any of the following methods:

(*i*) By ponding

(*ii*) By using wet sacks and springing water frequently

(*iii*) By steam

(*iv*) By alternating current

(*v*) By sealing curing compounds

(*vi*) By chemicals.

**Q. 63.** *What is 'Reinforced cement concrete (RCC)?*

**Ans.** The cement concrete in which the *reinforcement is provided for taking tensile or excessive compressive and shear stresses is called reinforced cement concrete.*

**Q. 64.** *What is 'Prestressed concrete'?*

**Ans.** ***Prestressed concrete*** is the concrete in which *high compressive stresses are artificially induced before its use;* the prestress is commonly induced by tensioning the steel reinforcement.

**Q. 65.** *What is 'Ferrocement'?*

**Ans.** ***Ferrocement*** may be considered as *a type of thin reinforced concrete construction where cement mortar matrix is reinforced with many layers of continuous and relatively small diameter wive-meshes.* While the mortar provides the mass, the wire-mesh imparts tensile strength and ductility to the material.

**Q. 66.** *What is an 'admixture'?*

**Ans.** An ***admixture*** is a material other than water, aggregates and cement and is added to batch immediately before or during its mixing.

Admixtures are used to improve or give special properties to concrete.

**Q. 67.** *How are admixtures classified?*

**Ans.** Admixtures are classified as follows:

(*i*) Air-entraining agents (*ii*) Retarders
(*iii*) Accelerators (*iv*) Water-proofers
(*v*) Pozzolanas (*vi*) Pigments
(*vii*) Workability agents.

**Q. 68.** *Define the term 'formwork'.*

**Ans.** ***Formwork*** (also called *false work, centring,* and *shuttering*) may be defined as the *timber or other materials which are used to form moulds into which concrete is poured at site.*

**Q. 69.** *What is 'timber'?*

**Ans.** ***Timber*** is the wood suitable for building or engineering purposes and it is applied to trees measuring not less than 0.6 m in girth.

**Q. 70.** *What is the difference between 'heart wood' and 'sap wood'?*

**Ans.** ***Heart wood*** consists of the *inner angular rings round the pith.* It is *compact, strong, durable,* and *darker in colour.* This part provides the best timber for engineering works or structures.

The outer annual rings between heart-wood and cambium layer is known as the sapwood. It is less durable.

**Q. 71.** *What do you mean by 'seasoning of timber'?*

**Ans.** ***Seasoning of timber*** is the process of *drying timber or removing moisture sap, present in a freshly felled timber, under more or less controlled conditions.*

**Q. 72.** *What are the methods of seasoning of timber?*

**Ans.** 1. "Natural seasoning":

(*i*) Air drying/seasoning; (*ii*) Water seasoning

2. "Artificial seasoning":

(*i*) Kiln seasoning; (*ii*) Chemical seasoning; (*iii*) Electrical seasoning.

**Q. 73.** *Enumerate the different methods of preservation of timber.*

**Ans.** The different methods are:

Tarrin; Charming; Painting, Creosoting, Ascue treatment; Fire proofing; Abel's process.

**Q. 74.** *What are 'veneers'?*

**Ans.** ***Veneers*** are thin sheets or slices of wood of superior quality having thickness varying from 0.4 mm to 6 mm or more.

**Q. 75.** *How is 'plywood' made?*

**Ans.** ***Plywood*** is made by cementing together several layers of wood which may be thin veneers or thicker boards.

**Q. 76.** *What are 'Biomaterials' ?*

**Ans.** **Biomaterials** are *employed in components implanted into the human body for replacement of diseased or damaged body parts.*

- These materials *must not produce toxic substances* and must be compatible with body tissues (*i.e.,* must not cause adverse biological reactions).
- All of the above materials–metals, ceramics, polymers, composites and semi-conductors may be used as biomaterials.

**Q. 77.** *What are 'Advanced materials' ?*

**Ans.** *Materials that are utilised in high-technology (or high-tech) applications are sometimes* ***Advanced materials****.* By high technology we mean a device or product that operates or functions using relatively intricate and sophisticated principles: *Examples* include *electronic equipment* (VCRs, CD players etc.) *computers, fiber-optic systems, spacecraft, aircraft, and militory rocketry.*

- These advanced materials are typically *either traditional materials whose properties have been enhanced or newly developed, high-performance materials*. Furthermore, they may be of all materials types (*e.g.* metals, ceramics, polymers) and are normally *relatively expensive.*

**Q. 78.** *What are 'Smart materials' ?*

**Ans.** ***Smart (or intelligent) materials*** are a group of new and state-of-the-art materials now being developed that will have a significant influence on many of our technologies. The adjective ***smart*** *implies that these materials are able to sense changes in their environments and then respond to these changes in predetermined manners-traits that are also found in living organisms.* In addition, this "*smart*" concept is *being extended to rather sophisticated systems that consist of both smart and traditional materials.*

Components of a smart material (or system) include some types of *sensor* (that detects an input signal), and an *actuator* (that performs a responsive and adaptive function). Actuators may be called upon to change shape, position, frequency, or mechanical characteristics in response to changes in temperature, electric fields, and/or magnetic fields.

**Q. 79.** *Name the four types of materials (made from the components of a smart material) commonly used for 'actuators'.*

**Ans.** Following four types of materials are commonly used for ***actuators:***

(*i*) *Shape memory alloys*. These are metals that, after having been deformed, revert back to their original shapes when temperature is changed.

(*ii*) *Piezoelectric ceramics*. These expand and contract in response to an applied electric field (or voltage); conversely they also generate an electric field when their dimensions are altered.

(*iii*) *Magnetostrictive materials*. The behaviour of these materials is analogous to that of piezoelectrics, except that they are responsive to magnetic fields.

(*iv*) *Electroheological/magnetorheological fluids*. These are liquids that experience dramatic changes in viscosity upon the application of electric and magnetic fields, respectively.

*Materials/devices* employed as **sensors** include the following:

(*i*) Optical fibers

(*ii*) Piezoelectric materials (including some polymers)

(*iii*) Micro-electromechanical devices.

*Example*: One type of smart system is used in helicopters to reduce aerodynamic cockpit noise that is created by the rotating rotor blades. *Piezoelectric sensors* inserted into the blades, monitor blade stresses and deformations; feedback signals from these sensors are fed into a computer-controlled adaptive device which generates noise-cancelling antinoise.

**Q. 80.** *What is 'Nanotechnology'?*

**Ans.**
- The general procedure utilised by scientists to understand the chemistry and physics of materials, until recent times, has been to begin by studying large and complex structures, and then to investigate the fundamental building blocks of these structure that are smaller and simpler. This approach is sometimes termed "*top-down*" science.
- However, with the advent of scanning probe microscopes, which permit observation of individual atoms and molecules, it has become possible *to manipulate and move atoms and molecules to form new structurres, and thus design new materials that are built from atomic level constituents,* (*i.e., "materials by design"*). This ability to carefully arrange atoms provides opportunities to develop mechanical, electrical, magnetic and other properties that are not otherwise possible. This is termed as "*bottom-up*" *approach and the study of the properties of these materials is termed* ***nanotechnology***; the "*nano*" prefix denotes that the dimensions of these structure entities are on the order of a nanometre ($10^{-9}$ m), as a rule, less than 100 nanometres (equivalent to approximately 500 atom diameters).

**Q. 81.** *Name as asked for*

(*i*) *The source energy of atom-bomb is a nuclear-process called nuclear fission. The source of energy in Sun is also a nuclear process; name this process.*

(*ii*) *Einstein in 1905 gave a theory of relativity called 'special theory of relativity' which talks about mass increase, length contraction and time dilatior, and $E = mc^2$ etc. He in 1913 gave another theory of relativity which talks about that gravity is there because of curvature of space; what (name) this theory of relativity is called?*

**Ans.** (*i*) Nuclear fusion.

(*ii*) General theory of relativity.

**Q. 82.** *Write materials for*

(*i*) *Engine cylinder blocks;* (*ii*) *Fire-proof clothing*

**Ans.** (*i*) Grey cast iron, aluminium alloy, nickel or chromium alloyed cast iron.

(*ii*) Chemically bonded ceramic, Bromine based fabric.

**Q. 83.** *What are the points on which 'material engineering' focuses?*

**Ans.** ***Material engineering*** focuses on the synthesis of materials in useful quantities and on processing of component materials into engineering products.

**Q. 84.** *Mention some advances due to new materials.*

**Ans.** (*i*) Hundred-fold increase in the strength of permanent magnets.

(*ii*) Breakthrough in the superconductivity transition temperature to a range suitable for engineering applications.

(*iii*) Hundred-fold increase in the strength to density ratio of composite materials.

(*iv*) Reduction in the size of integrated circuit devices to less than one micron, and an increase in the size of flat screen displays more than one metre.

(*v*) Hundred fold increase in the speed of cutting tools (due to enormous increase in the hardness and chemical resistance of thin film coatings).

**Q.85.** *What is an alloy?*

**Ans.** A metallic substance is composed of two or more elements.

**Q.86.** *Name the advanced materials used in the manufacture of the following:*

(*i*) *Electricity generation* (*ii*) *Hyper-high speed computers*
(*iii*) *Aircraft components* (*iv*) *Military tanks*
(*v*) *Magnetic bubbles (memory devices)* (*vi*) *Sensor elements for guided missiles*
(*vii*) *Computer memory cores* (*viii*) *Lightweight structure and collapsible bridges*
(*ix*) *Cryogenic industries and cryogenic engines*
(*x*) *Missile wings.*

**Ans.**

| Components | Materials |
|---|---|
| (*i*) Electricity generation | Selenium (in solar battery) |
| (*ii*) Hyper-high speed computers | New generation 'Al Ga As' chips |
| (*iii*) Aircraft components | Hybrid composites |
| (*iv*) Military tanks | Rolled armour steel |
| (*v*) Magnetic babbles (memory devices) | Iron gasket material grown on $Sm_3Ga_5O_{12}$ |
| (*vi*) Sensor elements for guided missiles | Hg-Cd-Te crystals |
| (*vii*) Computer memory cores | Ferrites of Cu-Mn, Mg-Mn and Zn-Mn |
| (*viii*) Light weight structures and collapsible bridges | Cellular concrete |
| (*ix*) Cryogenic industries and cryogenic engines | Nickel base cryogenic steel |
| (*x*) Missile wings | Magnesium alloy plates. |

**Q. 87.** *Name the three basic components of an intelligent material.*

**Ans.** 1. *Sensors* such as piezoelectric polymers (polyvinyldene), optical, fibres.

2. *Processors* such as conductive electroactive polymers, microchips.

3 *Actuators* such as shape memory alloys (Ni-Ti *i.e.,* nitinol) chemically responding polymers (polypyrole).

**Q. 88.** *What are the characteristics of metals ?*

**Ans.**
- '*Metals*' are characterised by their *high thermal and electrical conductivities.*
- They are *opaque*, and usually they can be polished to high luster.
- Commonly, but not always, they are relatively heavy and deformable.

**Q. 89.** *What are ceramics ?*

**Ans.** ***Ceremics*** are compounds between metallic and non-metallic elements; they are most frequently oxides, nitrides and carbides.

**Q. 90.** *What are polymers ?*

**Ans.** ***Polymers*** include the *plastic and rubber materials*. Many of them are organic compounds that are chemically based on carbon, hydrogen, and other non-metallic elements, furthermore, they have *very large molecular structures*.

- These materials typically have low densities and may be extremely flexible.

**Q. 91.** *What is a element ?*

**Ans.** A substance which cannot be decomposed into other substances is known as ***element***.

**Q. 92.** *What is an atom?*

**Ans.** The smaller particle of an element which takes part in chemical reaction is known as an ***atom***.

- An '*atom*' consists of a very small 'nucleus' composed of *protons* and *neutrons*, which is encircled by moving *electrons*.

**Q. 93.** *What is an 'electron'?*

**Ans.** An *electron* is a *negatively charged particle present* in an atom. Its mass is about 1/1836 that of the neutron but its size is roughly comparable to that of proton or neutron.

**Q. 94.** *Define the term 'Atomic number'.*

**Ans.** The number of protons contained in the nucleus of a normal atom is known as the ***atomic number***.

**Q. 95.** *What are 'Isotopes'?*

**Ans.** All atoms having different atomic weights but belonging to the same element are called ***Isotopes***.

**Q. 96.** *What are 'Isobars'?*

**Ans.** Atoms with same mass but belonging to different chemical elements are called ***Isobars***. The first pair of isobars is argon and calcium.

**Q. 97.** *What are 'Isotones'?*

**Ans.** Atoms of those elements in which number of neutrons in nucleus are equal, are known as ***Isotones***.

**Q. 98.** *What is Avogadro's number'?*

**Ans.** This is the number of atoms in a gamma-atom of any substance. It is a universal constant. It is designated as N and its value is $6.02 \times 10^{23}$ $mol^{-1}$.

**Q. 99.** *What is 'Atomic weight'?*

**Ans.** The sum of protons and neutrons in the nucleus of an atom is known as ***atomic weight***.

**Q. 100.** *What does Pauli's exclusion principle state?*

**Ans.** Paul's exclusion principle states that *no two electrons in an atom can have the same quantum state*.

**Q. 101.** *What are 'Primary bonds'?*

**Ans.** Primary bonds, also known as '*attractive bonds*', are the strongest bonds between atoms by virture of their *interatomic nature*. The attractive forces are directly associated with the valence electrons in the respective orbitals.

**Q. 102.** *Why are secondary bonds also called intermolecular bonds?*

**Ans.** Since they result from intermolecular or dipole attractions.

**Q. 103.** *What are the examples of secondary bonds?*

**Ans.** Ven der Waals bonds and hydrogen bonds.

**Q. 104.** *Name three common ionic crystal structures.*

**Ans.** (*i*) Sodium chloride, (*ii*) Caesium chloride, (*iii*) Zinc blende structures.

**Q. 105.** *What is an excellent example of covalent bond?*

**Ans.** Chlorine molecule.

**Q. 106.** *Why covalent solids do not form close packed structures?*

**Ans.** Because the covalent bonds are very strong and rigid, and their directional nature must be maintained.

**Q. 107.** *What are the characteristics of secondary compounds?*

**Ans.** (*i*) Generally transparent to light and good insulators with the exception of water.

(*ii*) Low melting points.

**Q. 108.** *Where does mixed ionic covalent bonding occur?*

**Ans.** In the structure of quartz, glass or silica.

**Q. 109.** *Which bonds are responsible for the unusual physical properties of ice and water?*

**Ans.** Hydrogen bonds.

**Q. 110.** *Why do metals possess ductility?*

**Ans.** Due to the shifting of electrons.

**Q. 111.** *Why metallic bonded materials are opaque and lustrous ?*

**Ans.** Because of the fact that impinging radiant energy is easily absorbed by the valence electrons.

**Q. 112.** *What is an 'amorphous solid'?*

**Ans.** An ***amorphous solid*** is a substance whose constituents do not possess an orderly arrangement. The size of ordered region in these substance is limited to a few molecular distances. Amorphous solids are called non-crystalline solids.

**Q. 113.** *What is a 'crystalline solid'?*

**Ans.** A ***crystalline solid*** is a substance whose constituents possess an orderly arrangement in a definite geometric pattern.

**Q. 114.** *What do you understand by 'Isotropic' 'Anisotropic' and 'Orthotropic materials'?*

**Ans.**
- ***Isotropic materials*** obey Hooke's law. All metals are isotropic.

  ***Anisotropic materials*** such as composites obey generalised Hooke's law which is different from Hooke's law.

  ***Orthotropic material*** is a special kind of anisotropic material. An *unidirectional fibre composite* is the example. It has three orthogonal planes of symmetry.
- The nature of identical properties in all the directions is called *isotropy* while dissimilar properties in all the directions is known as *anisotropy*. When properties are identical is some direction and dissimilar of others, the nature is called *orthotropy*.

**Q. 115.** *What is a 'Bi-modulus material'?*

**Ans.** A material, whose stress-strain diagram has different slopes in tension and compression, is known as ***bi-modulus material***. Such a material possesses unequal values of Young's moduli in tension and compression (Elastomer and wood may keep a bi-moduli character).

**Q. 116.** *What do you mean by the term 'Fatigue'? What are its main effects on the properties of materials ?*

**Ans.**
- The behaviour of materials under fluctuating and reversing loads (or stresses) is known as ***fatigue***.
- The main effect of fatigue on the properties of materials are: (*i*) Loss of ductility; (*ii*) Loss of strength; (*iii*) Enhanced uncertainty in strength and the service life of materials.

**Q. 117.** *State the steps which lead to fatigue fracture.*

**Ans.** (*i*) Nucleation of crack; (*ii*) Crack growth; (*iii*) Crack growing to a tiny size; (*iv*) Fracture.

**Q. 118.** *What is 'Creep'?*

**Ans.** The permanent deformation (strain) of a material under steady load as a function of time is called ***creep***.

**Q. 119.** *How does creep occur ?*

**Ans.** Occurence of creep in materials is supposed to be the effect of following phenomena; (*i*) Vacancy diffusion; (*ii*) Edge dislocation climb-up or climb-down; (*iii*) Grain boundary sliding; (*iv*) Screw dislocations cross-slip; (*v*) Elastic aftereffect.

**Q. 120.** *What are 'Non-destructive tests'?*

**Ans.** ***Non-destructive tests*** may be defined as those which in a specific context would *not* damage the material being examined to such an extent such that it is rendered useless for future for which it was originally meant.

**Q. 121.** *What is the principle on which a 'crack detector' operates ?*

**Ans.** A ***crack detector*** operates on the principle that if a crack occurs anywhere within the pieces it interferes with the flow of electric current through the metal, therefore, increasing its overall resistance. This holds true regardless of the shape of the piece.

**Q. 122.** *What is an 'Extensometer' ?*

**Ans.** An ***extensometer*** is an instrument or device by which the changes in length of specimen under test can by precisely measured.

**Q. 123.** *Define the terms: 'True stress' and 'True strain'.*

**Ans.** ***True stress*** is the load at any elongation divided by the cross-sectional area at that elongation and ***true strain*** is the change in length with reference to the instantaneous gauge length rather than the orginal length.

**Q. 124.** *What is 'Strain hardening coefficient' ?*

**Ans.** The gradient of the straight part of the true stress stain curve beyond maximum elastic stress is called the ***strain hardening coefficient***.

**Q. 125.** *Which is the material which exhibits a marked yield point ?*

**Ans.** Structural steel.

**Q. 126.** *What does impact test signify ?*

**Ans.** An impact test signifies *toughness* of material, that is ability of material to absorb energy during plastic deformation. Thoughness takes into account both the strength and ductility of the material.

**Q. 127.** *What is 'Corrosion fatigue' ?*

**Ans.** ***Corrosion fatigue*** is the simultaneous action of fatigue (cyclic stress) and corrosion (chemical attack).

**Q. 128.** *List the characteristics on which the tests for preliminary identification of materials are based.*

**Ans.** (*i*) Appearance and colour; (*ii*) Sound; (*iii*) Weight; (*iv*) Magnetic properties; (*v*) Bending; (*vi*) Hardenability; (*vii*) Spark; (*viii*) Filing; (*ix*) Microstructure.

**Q. 129.** *What purpose does a bending test serve ?*

**Ans.** 'Bending test' serves to determine the capability of a metal (or alloy) for plastic deformation and to reveal flaws.

**Q. 130.** *What are the steps for preparing the speciment for micro-examination ?*

**Ans.** (*i*) Selection of specimen; (*ii*) Cutting of specimen; (*iii*) Mounting the specimen; (*iv*) Obtaining the flat specimen surface; (*v*) Grinding; (*vi*) Rough polishing; (*vii*) Fine polishing (*viii*) Etching.

**Q. 131.** *What is the use of a 'metallograph'?*

**Ans.** A 'metallograph' is used to produce a photographic record of the microstructure.

**Q. 132.** *Name the various methods used to measure grain size.*

**Ans.** (*i*) Comparison method; (*ii*) Intercept method; (*iii*) Planimetric method.

**Q. 133.** *How perfect crystals of proper external shape can be obtained ?*

**Ans.** Perfect crystals of proper external shape can be obtained only if crystallisation develops under conditions when the *degree of supercooling is very slight and the metal has a very high purity.*

**Q. 134.** *What are dendrities ?*

**Ans.** Branched or tree-like crystals obtained during solidification are called ***dendrities***.

**Q. 135.** *What is an 'allotropy'?*

**Ans.** Existence of a given metal in two or more stable but different crystal structures is known as ***allotropy***.

**Q. 136.** *What is a 'grain'?*

**Ans.** A ***grain*** is a crystal with almost any external shape, but with an internal atomic structure based upon the space lattice with which it was born. The grain size is controlled by several factors out of which temperature and time of heating are impcrtant.

**Q. 137.** *How can the solidification of metal alloys be clearly understood ?*

**Ans.** By meants of *phase diagrams*, also known as *equilibrium diagrams* or *constitution diagrams*.

**Q. 138.** *What are 'Phase diagrams' ?*

**Ans.** ***Phase diagrams*** are maps that give the relationships between phases in *equilibrium* in a system as a function of temperature, pressure and composition.

**Q. 139.** *What is a 'solid solution'?*

**Ans.** A ***solid solution*** may be defined as a solution in the solid state which consists of two kinds of atoms combined in one type of space lattice.

**Q. 140.** *State Hume-Rothery's rule which govern the formation of substitutional solid solution.*

**Ans.** Two metals can be *completely soluble in each other only if they have:*

(*i*) Same lattice patterns; (*ii*) Nearly equal atom diameters; (*iii*) Equal number of valency electrons.

**Q. 141.** *What is a 'system'?*

**Ans.** A ***system*** is a substance (or group of substances) so isolated from surroundings that it is unaffected by these and is subjected to changes in overall composition, temperature, pressure or total volume only to the extent followed by the investigation.

**Q. 142.** *What is an 'alloy system' ?*

**Ans.** An ***alloy system*** is a combination of two or more elements forming the alloys which are considered within a specified range of temperature, pressure and concentration.

**Q. 143.** *What is a 'component'?*

**Ans.** A ***component*** is a unit of the composition variable of the systems.

**Q. 144.** *What is a 'phase' ?*

**Ans.** A ***phase*** is a physically and chemically homogeneous portion of a system separated from the other portions by a surface, the interface.

**Q. 145.** *What is an 'eutectic' ?*

**Ans.** An ***eutectic*** is a structural constituent of the alloy.

**Q. 146.** *What is 'Gibb's phase rule'?*

**Ans.** ***Gibb's phase rule*** establishes the relationship between the number of degrees of freedom ($F$), the number of components ($C$), and the number of phases ($P$). It is expressed mathematically as:

$$F = C + n - P$$

where, $n$ is the number of external factors (e.g., temperature and pressure)

**Q. 147.** *How does tie-line rule differ from lever rule ?*

**Ans.** The ***tie-line rule*** gives the composition of the co-existing phases, whereas the ***lever rule*** gives the fractions (or amounts) of the phases.

**Q. 148.** *What are ferrous metals?*

**Ans.** The ***ferrous metals*** are iron base metals which include all varieties of irons and steels.

**Q. 149.** *What is the product of a 'blast furnace'?*

**Ans.** Crude and impure iron (pig iron).

**Q. 150.** *Why is blast furnace so named ?*

**Ans.** Blast furnace is so named as a very high temperature is developed inside it by means of forming a blast of heated air.

**Q. 151.** *What operations are involved in the manufacturer of wrought iron ?*

**Ans.** (*i*) Refining; (*ii*) Puddling; (*iii*) Shingling; (*iv*) Rolling.

**Q. 152.** *What is the quality of steel produced by Bessemer process ?*

**Ans.** Bessemer process produces an inferior quality steel associated with blow holes and inclusions.

**Q. 153.** *Can cast iron be forged or welded ?*

**Ans.** No, cast iron can neither be forged nor welded.

**Q. 154.** *What are the uses of 'tungsten steel' ?*

**Ans.** Tungsten steel is used for making permanent magnets and high speed cutting tools.

**Q. 155.** *Define the term 'Heat treatment'.*

**Ans.** ***Heat treatment*** is defined as an operation or combination of operations, involving heating and cooling of a metal or alloy in its solid state with the object of changing the characteristics of the material.

**Q. 156.** *List the commonly occurring microscopic constituents of iron and steel ?*

**Ans.** Ferrite, Cermentite, Pearlite, Martensite, Austenite, Troositite, Sorbite.

**Q. 157.** *What is 'Ferrite' ?*

**Ans.** Iron which contains little or no carbon.

**Q. 158.** *What is 'Pearlite' ?*

**Ans.** ***Pearlite*** is the name given to a mixture of about 87.5 percent ferrite and 12.5 percent cementite. It is eutectoid of steel.

**Q. 159.** *What is 'Martensite' ?*

**Ans.** It is a hard brittle mass of fibrous or needle like structures and is the chief constituent of hardened steel.

**Q. 160.** *What is 'Austenite' ?*

**Ans.** It is a solid solution of iron-carbon which is stable only within a particular range of composition and temperature and is non-magnetic.

**Q. 161.** *What is 'Troositite' ?*

**Ans.** It is a structure in steel ($\alpha$-iron) produce either by tempering a martensitic steel between 250°C and 450°C or by quenching steel at a speed insufficient to suppress the thermal change point fully.

**Q. 162.** *What is 'Sorbite' ?*

**Ans.** It is a structure which consists of evenly distributed carbide of iron particles in a mass of ferrite, formed when a fully hardened steel is tempered between 550°C and 650°C.

**Q. 163.** *What is 'Cementite' ?*

**Ans.** This is a definite carbide of iron ($Fe_3C$) which is extremely hard, being harder than ordinary hardened steel or glass.

**Q. 164.** *What are TTT diagrams ?*

**Ans.** Isothermal transformation of austenite at sub-critical temperatures (less than 723°C for 0.8 W/°C) is known as "Time-temperature-transformation" (***TTT***) diagrams.

**Q. 165.** *What is the importance of TTT diagram ?*

**Ans.** TTT diagram is the principal source of information on the actual process of austenite decomposition under non-equilibrium conditions.

**Q. 166.** *What is 'Bainite'?*

**Ans.** ***Bainite*** is the name given to the structures that form an isothermal transformation at

**Q. 167.** *What is the characteristic feature of martensitic transformation ?*

**Ans.** A characteristic feature of martensitic transformation is that it is *practically never completed.*

**Q. 168.** *How retained austenite can be eliminated ?*

**Ans.** (*i*) Tempering; (*ii*) Cold treatment (which simply requires cooling the steel below $M_f$).

**Q. 169.** *What is 'Hardening' process ?*

**Ans.** ***Hardening*** is a process in which steel is heated to a temperature above the critical point, held at this temperature and quenched (rapidly cooled) in water, oil or molten salt baths.

**Q. 170.** *What is 'Mass effect'?*

**Ans.** ***Mass effect*** is the variation in hardness across a section of the components having higher thickness, through heat treatment.

**Q. 171.** *What is the difference between 'Hardening power' and 'Hardenability' ?*

**Ans.** Whereas '*Hardening power*' refers to the maximum hardness obtainable with steel of given composition, '*hardenability*' refers to the depth of hardening in a given thickness of steel.

**Q. 172.** *What is 'Case hardening' ?*

**Ans.** ***Case hardening*** consists in heating a steel in the presence of solid, liquid, or gas, rich in carbon in order to enable the surface to be hardened, while retaining a tough ductile core.

**Q. 173.** *What is 'cyaniding' ?*

**Ans.** ***Cyaniding*** is a process of superficial case hardening in which the steel is heated in molten cyanide salt at about 850°C followed by quenching. Both carbon and nitrogen are absorbed in this process.

**Q. 174.** *What is 'Critical cooling rate' ?*

**Ans.** The minimum cooling rate at which all of the austenite is supercooled to point $M_s$ and is transformed into martensite is called the ***critical cooling rate***.

**Q. 175.** *What is 'Bauxite' ?*

**Ans.** '*Bauxite*' is the commercial ore of *aluminium*. It is hydrated aluminium oxide generally believed to consist of mixture of monohydrate $Al_2O_3.H_2O$ and trihydrate $Al_2O_3.3H_2O$.

**Q. 176.** *Name the process by which metallic magnesium is prepared commercially.*

**Ans.** The *chloride* and *oxide* process.

**Q. 177.** *What are the main constituents of 'brasses' and 'bronzes' ?*

**Ans.** The *brasses* are primarily alloys of copper and zinc and *bronzes* of copper and tin.

**Q. 178.** *What is 'coinage bronze' ?*

**Ans.** The bronze containing 4 to 8 percent tin is called 'coinage bronze'.

**Q. 179.** *What is 'bell metal' ?*

**Ans.** The copper-tin series of alloys containing 15 to 25 percent of tin is known as *bell metal*.

**Q. 180.** *What is the composition of 'duralumin' ?*

**Ans.** Al = 94%, Cu = 4%; Mg, Mn, Si, Fe 0.5% each.

**Q. 181.** *What is the composition of Y-alloy ? What are its applications ?*

**Ans.**
- Al = 92.5%, Cu = 4%, N = 2%, Mg = 1.5%.
- Piston, cylinder heads and crankcases of I.C. engines, pump rods etc.

**Q. 182.** *Name important nickel alloys.*

**Ans.** Iconol, monel metal, Nichrome Bright rays alloys, Manganese nickel.

**Q. 183.** *What is the 'Monel metal' ?*

**Ans.** *Monel metal* can be said to be two thirds nickel, one third copper with small percentage of elements iron, silicon, manganese, and carbon.

**Q. 184.** *What is 'Babbit metal'?*

**Ans.** It is a white metal bearing alloy. Its approximate composition is : Tin = 89.2%, copper = 3.68%, antimony = 7.0%.

**Q. 185.** *What are 'Magnetic materials'?*

**Ans.** These are those materials in which *state of magnetisation can be induced.*

**Q. 186.** *What is 'Curie temperature'?*

**Ans.** The temperature above which a ferromagnetic material becomes paramagnetic.

**Q. 187.** *How does ferromagnetism arise ?*

**Ans.** Ferromagnetism arises out of *electronic structure with the atoms.*

**Q. 188.** *What are the various names of a 'Magnetisation curve'?*

**Ans.** B-H curve; magnetic saturation curve; virgin curve; or simply the saturation curve.

**Q. 189.** *Do the ferromagnetic materials exhibit spontaneous magnetisation below their temperatures.*

**Ans.** Yes.

**Q. 190.** *What is 'Magnetostriction'?*

**Ans.** When a ferromagnetic material is magnetised small changes in dimensions occur, the effect being known as ***magnetostriction***.

**Q. 191.** *What is 'Diamagnetism'?*

**Ans.** ***Diamagnetism*** is the property of material due to which, when it is placed in a magnetic field becomes weakly magnetised in a direction opposite to the magnetisation of the external field.

**Q. 192.** *What are the characteristics of 'soft-magnetic materials'?*

**Ans.** (*i*) High permeability; (*ii*) Magnetic energy stored in not high; (*iii*) Negligible coercive force; (*iv*) Low remanance.

**Q. 193.** *What is the percentage composition of 'Mumetal'?*

**Ans.** Ni = 75.4%; Cu = 4%; Cr = 1.5% and the remainder iron.

**Q. 194.** *What are the distinguishing features of 'ferrites'?*

**Ans.** Fairly constant permeability, and a narrow hysteresis loop; for this reason they are used in making cores for coils operating at elevated and high frequencies (in telephony and radio).

**Q. 195.** *What are the characteristics of 'Hard magnetic materials' ?*

**Ans.** (*i*) High value of BH product; (*ii*) High retentivity; (*iii*) High coercivity; (*iv*) Strong magnetic reluctance (*v*) Hysteresis loop is more rectangular in shape.

**Q. 196.** *What is 'refractoriness'?*

**Ans.** ***Refractoriness*** is the ability of a material to withstand the action of heat without appreciable deformation or softening.

**Q. 197.** *How are refractories classified according to chemical behaviour?*

**Ans.** (*i*) Acidic refractories; (*ii*) Basic refractories; (*iii*) Neutral refractories.

**Q. 198.** *What is 'Dielectric constant'?*

**Ans.** ***Dielectric constant*** is the ratio of the capacitance of a dielectric compared to the capacitance of air under the same conditions.

**Q. 199.** *What is 'thermal shock' resistance'?*

**Ans.** ***Thermal shock resistance*** is the ability of a material to resist cracking or disintegration of the material under abrupt or sudden changes in temperature.

**Q. 200.** *State the nuclear applications of ceramics.*

**Ans.** Fuel elements, moderators, controls and shielding.

**Q. 201.** *What are 'silicates'?*

**Ans.** The *silicates are co-ordinate* structures based upon large anions arranged about small cations. In all silicates, the basic unit is the $SiO_4$ tetrahedron.

**Q. 202.** *What is glass?*

**Ans.** *Glass* in any substance or mixture of substances that has solidified from the liquid state without crystallisation.

**Q. 203.** *What are the constituents of glass?*

**Ans.** (*i*) Silica; (*ii*) sodium or potassium carbonate; (*iii*) Lime; (*iv*) Manganese dioxide; (*v*) Cullet; (*vi*) Colouring substance.

**Q. 204.** *What are polymers?*

**Ans.** ***Polymers*** are organic materials having carbon as the common element in their make up.

**Q. 205.** *What are monomers?*

**Ans.** The polymers are composed of a large number of *repeating units* (small molecules) called '*monomers*'. A polymer is made up of thousands of monomers joined together to form a large molecule of colloidal dimension, called *macromolecule*.

**Q. 206.** *What is a plastic?*

**Ans.** A ***plastic*** in broadest sense, is defined as any non metallic material that can be moulded to shape.

**Q. 207.** *What are 'thermoplastic materials'?*

**Ans.** ***Thermoplastic materials*** are those plastics which soften on the application of heat with or without pressure, but they require cooling to set them to shape.

**Q. 208.** *Name some commercial thermoplastics?*

**Ans.** Polyethene, Polyvinyl chloride (PVC); Polystrene; Polytetrafluoroethylene (PTFE) etc.

**Q. 209.** *What are 'thermosetting materials'?*

**Ans.** These are those plastics which require heat and pressure to mould them into shape.

**Q. 210.** *Name some commercial thermosetting plastics?*

**Ans.** Phenolics, polyesters, epoxies, silicones etc.

**Q. 211.** *What is the general name of polyamides?*

**Ans.** Nylon

**Q. 212.** *What is 'calendering'?*

**Ans.** It is the production of sheet of materials by rolling the plastics between multiple rollers.

**Q. 213.** *Which moulding process is employed mainly for thermosetting resins?*

**Ans.** Compression moulding.

**Q. 214.** *What are plasticisers?*

**Ans.** Low molecular weight (approx 300) materials blended with polymers are called ***plasticisers***

**Q. 215.** *What is anelasticity?*

**Ans.** The time dependent elastic behaviour is known as ***anelasticity***.

**Q. 216.** *What is a composite material?*

**Ans.** A ***composite material*** is a combination of two or more materials having compositional variations and depicting properties distinctly different from those of the individual materials of the composite.

**Q. 217.** *What do you mean by 'Fracture'?*

**Ans.** *Fracture* is the breaking of a metal to yield an irregular surface.

**Q. 218.** *Name the three distinction modes of fracture in a polycrystalline material.*

**Ans.** (*i*) Inter-granular fracture by separation of crystals at grain boundaries; (*ii*) Fracture due to shear between the crystallographic planes; (*iii*) Cleavage fracture through pulling apart of crystallographic planes.

**Q. 219.** *How are fractures classified according to the fracture mechanisms?*

**Ans.** (*i*) Ductile fractures; (*ii*) Brittle fractures.

**Q. 220.** *How is 'ductile fracture' characterised?*

**Ans.** In case of *ductile fracture* there occurs an appreciable plastic deformation prior to failure and the fractured surfaces give cup and cone appearance. Metal with a high yield strength gives a smaller cup. The fracture faces are dull, irregular and fibrous in appearance.

**Q. 221.** *How does failure of metal occur in case of 'brittle fracture'?*

**Ans.** In case of *brittle fracture* failure of metal occurs when the fracture crack propagates through the cross section without an appreciable plastic deformation. Such a fracture is more likely to occur in metal with poor plasticity and low temperatures.

**Q. 222.** *What is 'transition temperature'?*

**Ans.** The ***transition temperature*** is a narrow temperature interval in which there is a drastic decrease in the percentage reduction of area and elongation and the energy absorbed during impact loading of metals with BCC structure are more susceptible to ductile-brittle transition than those of FCC.

**Q. 223.** *What is 'Griffith's theory of fracture'?*

**Ans.** According to ***Griffith's theory*** there are many fine elliptical cracks and there are high concentrations of stress at the tips of such cracks. With such stress concentration the theoretical cohesive strength can be obtained in this localised region when the body of a material is under a fairly low applied tensile stress.

**Q. 224.** *What is 'corrosion'?*

**Ans.** ***Corrosion*** is a gradual chemical or electro-chemical attack on a metal by its surroundings such that the metal is converted into an oxide, salt or some other compound.

**Q. 225.** *Name the factors which influence corrosion?*

**Ans.** (*i*) Chemical nature of metal; (*ii*) Environment; (*iii*) Internal structure (*iv*) Nature of engineering applications for which it is used.

**Q. 226.** *Name the two general types of corrosion?*

**Ans.** (*i*) Chemical or dry corrosion; (*ii*) Electro-chemical corrosion.

**Q. 227.** *What is 'Hydrogen embrittlement'?*

**Ans.** ***Hydrogen embrittlement*** is the action of hydrogen on metal at ordinary temperature. It is due to the formation of atomic hydrogen as result of chemical or electrolytic action occurring on metal surfaces under specific environments.

**Q. 228.** *What is 'Electrode potential'?*

**Ans.** When a piece of metal is partly dipped in a salt solution a potential ofference is spontaneously setup between the metal surfaces and solution. This is called *electrode potential* and its magnitude depends upon the nature of the metal, the concentration of solution and the temperature.

**Q. 229.** *State the conditions under which wet or electro-chemical corrosion occurs?*

**Ans.** (*i*) Where the liquid with metal is conducting; (*ii*) When their exists a difference of potentials either between two dissimilar metals or between different areas on the surface of the metal or alloy.

# Index

**C**